Methods in Enzymology

Volume 222
PROTEOLYTIC ENZYMES IN COAGULATION,
FIBRINOLYSIS, AND COMPLEMENT ACTIVATION
Part A
Mammalian Blood Coagulation Factors and Inhibitors

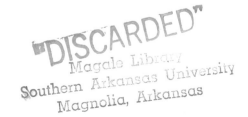

METHODS IN ENZYMOLOGY

EDITORS-IN-CHIEF

John N. Abelson   Melvin I. Simon

DIVISION OF BIOLOGY
CALIFORNIA INSTITUTE OF TECHNOLOGY
PASADENA, CALIFORNIA

FOUNDING EDITORS

Sidney P. Colowick and Nathan O. Kaplan

Methods in Enzymology

Volume 222

Proteolytic Enzymes in Coagulation, Fibrinolysis, and Complement Activation

Part A

Mammalian Blood Coagulation Factors and Inhibitors

EDITED BY

Laszlo Lorand

DEPARTMENT OF BIOCHEMISTRY, MOLECULAR AND CELL BIOLOGY
NORTHWESTERN UNIVERSITY
EVANSTON, ILLINOIS

Kenneth G. Mann

DEPARTMENT OF BIOCHEMISTRY
THE UNIVERSITY OF VERMONT
COLLEGE OF MEDICINE
BURLINGTON, VERMONT

ACADEMIC PRESS, INC.

A Division of Harcourt Brace & Company

San Diego   New York   Boston   London   Sydney   Tokyo   Toronto

Academic Press, Inc.
1250 Sixth Avenue, San Diego, California 92101-4311

United Kingdom Edition published by
Academic Press Limited
24–28 Oval Road, London NW1 7DX

International Standard Serial Number: 0076-6879

International Standard Book Number: 0-12-182123-4

PRINTED IN THE UNITED STATES OF AMERICA
93 94 95 96 97 98   MM   9 8 7 6 5 4 3 2 1

Table of Contents

v

Contributors to Volume 222

Article numbers are in parentheses following the names of contributors.
Affiliations listed are current.

DEBRA H. ALLEN (16), *Department of Biochemistry, University of Vermont College of Medicine, Burlington, Vermont 05405*

JAN ASTERMARK (24), *Department of Clinical Chemistry, University of Lund, Malmö General Hospital, S-214 01 Malmö, Sweden*

FRANK A. BAGLIA (5), *The Sol Sherry Thrombosis Research Center, Temple University School of Medicine, Philadelphia, Pennsylvania 19140*

S. PAUL BAJAJ (6), *Departments of Medicine, Pathology and Biochemistry, St. Louis University Medical Center, St. Louis, Missouri 63110*

KATHLEEN L. BERKNER (26), *Department of Cell Biology, Cleveland Clinic Foundation, Research Institute, Cleveland, Ohio 44195*

JENS J. BIRKTOFT (6), *Department of Biochemistry and Molecular Biophysics, Washington University School of Medicine, St. Louis, Missouri 63110*

INGEMAR BJÖRK (30), *Department of Veterinary Medical Chemistry, Swedish University of Agricultural Sciences, S-751 23 Uppsala, Sweden*

PAUL E. BOCK (27), *Department of Pathology, Vanderbilt University School of Medicine, Nashville, Tennessee 37232*

GEORGE J. BROZE, JR. (11), *Division of Hematology/Oncology, The Jewish Hospital at Washington University Medical Center, St. Louis, Missouri 63110*

SAULIUS BUTENAS (10), *Department of Biochemistry, University of Vermont College of Medicine, Burlington, Vermont 05405*

SHI-HAN CHEN (8), *Department of Pediatrics, University of Washington, Seattle, Washington 98104*

WILLIAM R. CHURCH (23), *Department of Biochemistry, University of Vermont College of Medicine, Burlington, Vermont 05405*

ROBERT W. COLMAN (4), *The Sol Sherry Thrombosis Research Center, Temple University School of Medicine, Philadelphia, Pennsylvania 19140*

MARGARET F. DOYLE (17), *Haematologic Technologies, Inc., Essex Junction, Vermont 05452*

THOMAS S. EDGINGTON (12), *The Scripps Research Institute, La Jolla, California 92037*

CHARLES T. ESMON (21), *Howard Hughes Medical Institute, Oklahoma Medical Research Foundation, Oklahoma City, Oklahoma 73104*

NAOMI L. ESMON (21), *Department of Pathology, University of Oklahoma Health Sciences Center, Oklahoma City, Oklahoma 73104*

DAVID N. FASS (7), *Hematology Research Section, Mayo Clinic/Foundation, Rochester, Minnesota 55904*

PHILIP J. FAY (7), *Hematology Unit, Department of Medicine, University of Rochester School of Medicine and Dentistry, Rochester, New York 14642*

BARBARA C. FURIE (25), *Center for Hemostasis and Thrombosis Research, Division of Hematology-Oncology, New England Medical Center, Boston, Massachusetts 02111*

BRUCE FURIE (25), *Center for Hemostasis and Thrombosis Research, Division of Hematology-Oncology, New England Medical Center, Boston, Massachusetts 02111*

THOMAS J. GIRARD (11), *The Monsanto Company, Chesterfield, Missouri 63017*

ix

PAUL E. HALEY (17), *Haematologic Technologies, Inc., Essex Junction, Vermont 05452*

RUTH ANN HENRIKSEN (18), *Department of Medicine, East Carolina University School of Medicine, Greenville, North Carolina 27858*

AKIKO HIJIKATA-OKUNOMIYA (19), *School of Allied Medical Sciences, Kobe University, Kobe 654-01, Japan*

LEON W. HOYER (9), *Holland Laboratory, American Red Cross, Rockville, Maryland 20855*

AKITADA ICHINOSE (3), *Department of Molecular Pathological Biochemistry, Yamagata University School of Medicine, Yamagata 990-23, Japan*

BRADFORD A. JAMESON (5), *Thomas Jefferson University, Philadelphia, Pennsylvania 19107*

RICHARD JENNY (23, 29), *Haematologic Technologies, Inc., Essex Junction, Vermont 05452*

JONG-MOON JEONG (2), *Department of Biochemistry, Molecular Biology and Cell Biology, Northwestern University, Evanston, Illinois 60208*

ARTHUR E. JOHNSON (21), *Department of Chemistry and Biochemistry, University of Oklahoma, Norman, Oklahoma 73019*

HIROSHI KAETSU (3), *The Chemo-Sera-Therapeutic Research Institute, Kumamoto 860, Japan*

MICHAEL KALAFATIS (13), *Department of Biochemistry, University of Vermont College of Medicine, Burlington, Vermont 05405*

RANDAL J. KAUFMAN (14), *Genetics Institute, Inc., Cambridge, Massachusetts 02140*

KAREN J. KOTKOW (25), *Center for Hemostasis and Thrombosis Research, Division of Hematology/Oncology, New England Medical Center, Boston, Massachusetts 02111*

SRIRAM KRISHNASWAMY (10, 13, 15), *Department of Medicine, Division of Hematology/Oncology, Emory University, Atlanta, Georgia 30322*

JEFFREY H. LAWSON (10), *Department of Surgery, Duke University Medical Center, Durham, North Carolina 27710*

BERNARD F. LE BONNIEC (21), *Oklahoma Medical Research Foundation, Oklahoma City, Oklahoma 73104*

SIDNEY D. LEWIS (20), *Biological Chemistry Department, Merck Research Laboratories, West Point, Pennsylvania 19486*

PETE LOLLAR (7), *Department of Medicine, Division of Hematology/Oncology, Emory University, Atlanta, Georgia 30322*

LASZLO LORAND (Introduction, 2), *Department of Biochemistry, Molecular and Cell Biology, Northwestern University, Evanston, Illinois 60208*

KENNETH G. MANN (Introduction, 10, 13, 15, 28, 29), *Department of Biochemistry, University of Vermont College of Medicine, Burlington, Vermont 05405*

TERRI L. MESSIER (23), *Department of Biochemistry, University of Vermont College of Medicine, Burlington, Vermont 05405*

DAVID J. MILES (12), *The Scripps Research Institute, La Jolla, California 92037*

MICHEAL E. NESHEIM (15), *Department of Biochemistry, Queen's University, Kingston, Ontario, Canada K7L 3N6*

ASSUNTA S. NG (20), *Biological Chemistry Department, Merck Research Laboratories, West Point, Pennsylvania 19486*

ANN-KRISTIN ÖHLIN (24), *Department of Clinical Chemistry, University of Lund, University Hospital, S-22185 Lund, Sweden*

SHOSUKE OKAMOTO (19), *Kobe Research Projects on Thrombosis and Haemostasis, Kobe 655, Japan*

STEVEN T. OLSON (30), *Division of Biochemical Research, Henry Ford Hospital, Detroit, Michigan 48202*

LAURIE A. OUELLETTE (23), *Department of Biochemistry, University of Vermont College of Medicine, Burlington, Vermont 05405*

LÁSZLÓ PATTHY (1), *Institute of Enzymology, Biological Research Center, Hungarian Academy of Sciences, H-1113 Budapest, Hungary*

EGON PERSSON (24), *Department of Clinical Chemistry, University of Lund, Malmö General Hospital, S-214 01 Malmö, Sweden*

DEBRA D. PITTMAN (14), *Genetics Institute, Inc., Cambridge, Massachusetts 02140*

ROBIN A. PIXLEY (4), *The Sol Sherry Thrombosis Research Center, Temple University School of Medicine, Philadelphia, Pennsylvania 19140*

THOMAS J. PORTER (25), *Center for Hemostasis and Thrombosis Research, Division of Hematology/Oncology, New England Medical Center, Boston, Massachusetts 02111*

EDWARD L. G. PRYZDIAL (15), *Research Department, Canadian Red Cross Society, Ottawa, Ontario, Canada K1G 4J5*

JAMES T. RADEK (2), *Department of Biochemistry, Molecular Biology and Cell Biology, Northwestern University, Evanston, Illinois 60208*

MATTHEW D. RAND (13), *Department of Biochemistry, University of Vermont College of Medicine, Burlington, Vermont 05405*

ALNAWAZ REHEMTULLA (12), *Genetics Institute, Inc., Cambridge, Massachusetts 02140*

ROYCE A. ROBINSON (16), *Department of Medicine, Washington University School of Medicine, St. Louis, Missouri 63110*

DAVID A. ROTH (25), *Center for Hemostasis and Thrombosis Research, Division of Hematology/Oncology, New England Medical Center, Boston, Massachusetts 02111*

WOLFRAM RUF (12), *The Scripps Research Institute, La Jolla, California 92037*

JULES A. SHAFER (20), *Biological Chemistry Department, Merck Research Laboratories, West Point, Pennsylvania 19486*

JOSEPH D. SHORE (30), *Division of Biomedical Research, Henry Ford Hospital, Detroit, Michigan 48202*

JOHAN STENFLO (24), *Department of Clinical Chemistry, University of Lund, Malmö General Hospital, S-214 01 Malmö, Sweden*

KOJI SUZUKI (22), *Department of Molecular Biology on Genetic Disease, Mie University School of Medicine, Mie 514, Japan*

ARTHUR R. THOMPSON (8), *Department of Medicine, University of Washington, Seattle, Washington 98104; and Puget Sound Blood Center, Seattle, Washington 98104*

PAULA B. TRACY (16), *Department of Biochemistry, University of Vermont College of Medicine, Burlington, Vermont 05405*

RUSSELL P. TRACY (29), *Departments of Pathology and Biochemistry, University of Vermont College of Medicine, Burlington, Vermont 05405*

CARMEN VALCARCE (24), *Department of Clinical Chemistry, University of Lund, Malmö General Hospital, S-214 01 Malmö, Sweden*

PETER N. WALSH (5), *Departments of Medicine and Biochemistry, The Sol Sherry Thrombosis Research Center, Temple University School of Medicine, Philadelphia, Pennsylvania 19140*

E. BRADY WILLIAMS (28, 29), *Department of Chemistry, College of St. Catherine, St. Paul, Minnesota 55105*

JAMES WILSON (2), *Department of Biochemistry, Molecular Biology and Cell Biology, Northwestern University, Evanston, Illinois 60208*

LAURA A. WORFOLK (16), *Department of Cell and Molecular Biology, University of Vermont College of Medicine, Burlington, Vermont 05405*

Preface

The field of proteolytic enzymes and inhibitors has grown to such an extent that is is necessary to restrict one's focus to areas which are important and large enough by themselves to warrant a specialized volume in the *Methods in Enzymology* series. This volume of Proteolytic Enzymes in Coagulation, Fibrinolysis, and Complement Activation (Volume 222, Part A) deals with mammalian blood coagulation factors and inhibitors, subjects of interest to biochemists, molecular biologists, pharmacologists, and hematologists. Its companion Volume 223 (Part B) includes the related topics of complement activation, fibrinolysis, and the nonmammalian blood coagulation factors and inhibitors.

LASZLO LORAND
KENNETH G. MANN

METHODS IN ENZYMOLOGY

VOLUME 73. Immunochemical Techniques (Part B)
Edited by JOHN J. LANGONE AND HELEN VAN VUNAKIS

VOLUME 74. Immunochemical Techniques (Part C)
Edited by JOHN J. LANGONE AND HELEN VAN VUNAKIS

VOLUME 75. Cumulative Subject Index Volumes XXXI, XXXII, and XXXIV–LX
Edited by EDWARD A. DENNIS AND MARTHA G. DENNIS

VOLUME 76. Hemoglobins
Edited by ERALDO ANTONINI, LUIGI ROSSI-BERNARDI, AND EMILIA CHIANCONE

VOLUME 77. Detoxication and Drug Metabolism
Edited by WILLIAM B. JAKOBY

VOLUME 78. Interferons (Part A)
Edited by SIDNEY PESTKA

VOLUME 79. Interferons (Part B)
Edited by SIDNEY PESTKA

VOLUME 80. Proteolytic Enzymes (Part C)
Edited by LASZLO LORAND

VOLUME 81. Biomembranes (Part H: Visual Pigments and Purple Membranes, I)
Edited by LESTER PACKER

VOLUME 82. Structural and Contractile Proteins (Part A: Extracellular Matrix)
Edited by LEON W. CUNNINGHAM AND DIXIE W. FREDERIKSEN

VOLUME 83. Complex Carbohydrates (Part D)
Edited by VICTOR GINSBURG

VOLUME 84. Immunochemical Techniques (Part D: Selected Immunoassays)
Edited by JOHN J. LANGONE AND HELEN VAN VUNAKIS

VOLUME 85. Structural and Contractile Proteins (Part B: The Contractile Apparatus and the Cytoskeleton)
Edited by DIXIE W. FREDERIKSEN AND LEON W. CUNNINGHAM

VOLUME 86. Prostaglandins and Arachidonate Metabolites
Edited by WILLIAM E. M. LANDS AND WILLIAM L. SMITH

VOLUME 87. Enzyme Kinetics and Mechanism (Part C: Intermediates, Stereochemistry, and Rate Studies)
Edited by DANIEL L. PURICH

VOLUME 88. Biomembranes (Part I: Visual Pigments and Purple Membranes, II)
Edited by LESTER PACKER

VOLUME 89. Carbohydrate Metabolism (Part D)
Edited by WILLIS A. WOOD

VOLUME 90. Carbohydrate Metabolism (Part E)
Edited by WILLIS A. WOOD

VOLUME 91. Enzyme Structure (Part I)
Edited by C. H. W. HIRS AND SERGE N. TIMASHEFF

VOLUME 92. Immunochemical Techniques (Part E: Monoclonal Antibodies and General Immunoassay Methods)
Edited by JOHN J. LANGONE AND HELEN VAN VUNAKIS

VOLUME 93. Immunochemical Techniques (Part F: Conventional Antibodies, Fc Receptors, and Cytotoxicity)
Edited by JOHN J. LANGONE AND HELEN VAN VUNAKIS

VOLUME 94. Polyamines
Edited by HERBERT TABOR AND CELIA WHITE TABOR

VOLUME 95. Cumulative Subject Index Volumes 61–74, 76–80
Edited by EDWARD A. DENNIS AND MARTHA G. DENNIS

VOLUME 96. Biomembranes [Part J: Membrane Biogenesis: Assembly and Targeting (General Methods; Eukaryotes)]
Edited by SIDNEY FLEISCHER AND BECCA FLEISCHER

VOLUME 97. Biomembranes [Part K: Membrane Biogenesis: Assembly and Targeting (Prokaryotes, Mitochondria, and Chloroplasts)]
Edited by SIDNEY FLEISCHER AND BECCA FLEISCHER

VOLUME 98. Biomembranes [Part L: Membrane Biogenesis: (Processing and Recycling)]
Edited by SIDNEY FLEISCHER AND BECCA FLEISCHER

VOLUME 99. Hormone Action (Part F: Protein Kinases)
Edited by JACKIE D. CORBIN AND JOEL G. HARDMAN

VOLUME 100. Recombinant DNA (Part B)
Edited by RAY WU, LAWRENCE GROSSMAN, AND KIVIE MOLDAVE

VOLUME 101. Recombinant DNA (Part C)
Edited by RAY WU, LAWRENCE GROSSMAN, AND KIVIE MOLDAVE

VOLUME 102. Hormone Action (Part G: Calmodulin and Calcium-Binding Proteins)
Edited by ANTHONY R. MEANS AND BERT W. O'MALLEY

VOLUME 103. Hormone Action (Part H: Neuroendocrine Peptides)
Edited by P. MICHAEL CONN

VOLUME 104. Enzyme Purification and Related Techniques (Part C)
Edited by WILLIAM B. JAKOBY

VOLUME 105. Oxygen Radicals in Biological Systems
Edited by LESTER PACKER

VOLUME 106. Posttranslational Modifications (Part A)
Edited by FINN WOLD AND KIVIE MOLDAVE

VOLUME 124. Hormone Action (Part J: Neuroendocrine Peptides)
Edited by P. MICHAEL CONN

VOLUME 125. Biomembranes (Part M: Transport in Bacteria, Mitochondria, and Chloroplasts: General Approaches and Transport Systems)
Edited by SIDNEY FLEISCHER AND BECCA FLEISCHER

VOLUME 126. Biomembranes (Part N: Transport in Bacteria, Mitochondria, and Chloroplasts: Protonmotive Force)
Edited by SIDNEY FLEISCHER AND BECCA FLEISCHER

VOLUME 127. Biomembranes (Part O: Protons and Water: Structure and Translocation)
Edited by LESTER PACKER

VOLUME 128. Plasma Lipoproteins (Part A: Preparation, Structure, and Molecular Biology)
Edited by JERE P. SEGREST AND JOHN J. ALBERS

VOLUME 129. Plasma Lipoproteins (Part B: Characterization, Cell Biology, and Metabolism)
Edited by JOHN J. ALBERS AND JERE P. SEGREST

VOLUME 130. Enzyme Structure (Part K)
Edited by C. H. W. HIRS AND SERGE N. TIMASHEFF

VOLUME 131. Enzyme Structure (Part L)
Edited by C. H. W. HIRS AND SERGE N. TIMASHEFF

VOLUME 132. Immunochemical Techniques (Part J: Phagocytosis and Cell-Mediated Cytotoxicity)
Edited by GIOVANNI DI SABATO AND JOHANNES EVERSE

VOLUME 133. Bioluminescence and Chemiluminescence (Part B)
Edited by MARLENE DELUCA AND WILLIAM D. McELROY

VOLUME 134. Structural and Contractile Proteins (Part C: The Contractile Apparatus and the Cytoskeleton)
Edited by RICHARD B. VALLEE

VOLUME 135. Immobilized Enzymes and Cells (Part B)
Edited by KLAUS MOSBACH

VOLUME 136. Immobilized Enzymes and Cells (Part C)
Edited by KLAUS MOSBACH

VOLUME 137. Immobilized Enzymes and Cells (Part D)
Edited by KLAUS MOSBACH

VOLUME 138. Complex Carbohydrates (Part E)
Edited by VICTOR GINSBURG

VOLUME 139. Cellular Regulators (Part A: Calcium- and Calmodulin-Binding Proteins
Edited by ANTHONY R. MEANS AND P. MICHAEL CONN

VOLUME 158. Metalloproteins (Part A)
Edited by JAMES F. RIORDAN AND BERT L. VALLEE

VOLUME 159. Initiation and Termination of Cyclic Nucleotide Action
Edited by JACKIE D. CORBIN AND ROGER A. JOHNSON

VOLUME 160. Biomass (Part A: Cellulose and Hemicellulose)
Edited by WILLIS A. WOOD AND SCOTT T. KELLOGG

VOLUME 161. Biomass (Part B: Lignin, Pectin, and Chitin)
Edited by WILLIS A. WOOD AND SCOTT T. KELLOGG

VOLUME 162. Immunochemical Techniques (Part L: Chemotaxis and Inflammation)
Edited by GIOVANNI DI SABATO

VOLUME 163. Immunochemical Techniques (Part M: Chemotaxis and Inflammation)
Edited by GIOVANNI DI SABATO

VOLUME 164. Ribosomes
Edited by HARRY F. NOLLER, JR., AND KIVIE MOLDAVE

VOLUME 165. Microbial Toxins: Tools for Enzymology
Edited by SIDNEY HARSHMAN

VOLUME 166. Branched-Chain Amino Acids
Edited by ROBERT HARRIS AND JOHN R. SOKATCH

VOLUME 167. Cyanobacteria
Edited by LESTER PACKER AND ALEXANDER N. GLAZER

VOLUME 168. Hormone Action (Part K: Neuroendocrine Peptides)
Edited by P. MICHAEL CONN

VOLUME 169. Platelets: Receptors, Adhesion, Secretion (Part A)
Edited by JACEK HAWIGER

VOLUME 170. Nucleosomes
Edited by PAUL M. WASSARMAN AND ROGER D. KORNBERG

VOLUME 171. Biomembranes (Part R: Transport Theory: Cells and Model Membranes)
Edited by SIDNEY FLEISCHER AND BECCA FLEISCHER

VOLUME 172. Biomembranes (Part S: Transport: Membrane Isolation and Characterization)
Edited by SIDNEY FLEISCHER AND BECCA FLEISCHER

VOLUME 173. Biomembranes [Part T: Cellular and Subcellular Transport: Eukaryotic (Nonepithelial) Cells]
Edited by SIDNEY FLEISCHER AND BECCA FLEISCHER

VOLUME 174. Biomembranes [Part U: Cellular and Subcellular Transport: Eukaryotic (Nonepithelial) Cells]
Edited by SIDNEY FLEISCHER AND BECCA FLEISCHER

VOLUME 175. Cumulative Subject Index Volumes 135–139, 141–167

VOLUME 176. Nuclear Magnetic Resonance (Part A: Spectral Techniques and Dynamics)
Edited by NORMAN J. OPPENHEIMER AND THOMAS L. JAMES

VOLUME 177. Nuclear Magnetic Resonance (Part B: Structure and Mechanism)
Edited by NORMAN N. OPPENHEIMER AND THOMAS L. JAMES

VOLUME 178. Antibodies, Antigens, and Molecular Mimicry
Edited by JOHN J. LANGONE

VOLUME 179. Complex Carbohydrates (Part F)
Edited by VICTOR GINSBURG

VOLUME 180. RNA Processing (Part A: General Methods)
Edited by JAMES E. DAHLBERG AND JOHN N. ABELSON

VOLUME 181. RNA Processing (Part B: Specific Methods)
Edited by JAMES E. DAHLBERG AND JOHN N. ABELSON

VOLUME 182. Guide to Protein Purification
Edited by MURRAY P. DEUTSCHER

VOLUME 183. Molecular Evolution: Computer Analysis of Protein and Nucleic Acid Sequences
Edited by RUSSELL F. DOOLITTLE

VOLUME 184. Avidin-Biotin Technology
Edited by MEIR WILCHEK AND EDWARD A. BAYER

VOLUME 185. Gene Expression Technology
Edited by DAVID V. GOEDDEL

VOLUME 186. Oxygen Radicals in Biological Systems (Part B: Oxygen Radicals and Antioxidents)
Edited by LESTER PACKER AND ALEXANDER N. GLAZER

VOLUME 187. Arachidonate Related Lipid Mediators
Edited by ROBERT C. MURPHY AND FRANK A. FITZPATRICK

VOLUME 188. Hydrocarbons and Methylotrophy
Edited by MARY E. LIDSTROM

VOLUME 189. Retinoids (Part A: Molecular and Metabolic Aspects)
Edited by LESTER PACKER

VOLUME 190. Retinoids (Part B: Cell Differentiation and Clinical Applications)
Edited by LESTER PACKER

VOLUME 191. Biomembranes (Part V: Cellular and Subcellular Transport: Epithelial Cells)
Edited by SIDNEY FLEISCHER AND BECCA FLEISCHER

VOLUME 192. Biomembranes (Part W: Cellular and Subcellular Transport: Epithelial Cells)
Edited by SIDNEY FLEISCHER AND BECCA FLEISCHER

Introduction: Blood Coagulation

By Kenneth G. Mann and Laszlo Lorand

Thrombin generation, initiated by cellular and vascular damage, is the central biochemical reaction in the phenomena of normal hemostasis and thrombosis. Production of thrombin is rapid (≤ 5 min) and autocatalytic in nature because of its action as a feedback activator of other coagulation components (factors V, VIII, and VII). Thrombin is the key agent for triggering the important events of platelet aggregation, the transformation of soluble fibrinogen into the clot-forming substrate, fibrin, and the activation of factor XIII for stabilizing the blood clot.

Fibrinogen is a large, nearly symmetrical protein that is a doublet of three disulfide-bonded constituent chains in an $A\alpha_2 B\beta_2 \gamma_2$ structure. Reaction with thrombin removes the NH_2-terminal fibrinopeptide moieties, amounting to about 3% of the total weight of the parent protein.[1,2] The resulting fibrin monomers [$\alpha_2\beta_2\gamma_2$], representing ~97% of the parent molecule, have a tendency to self-assemble in a half-staggered and side-to-side array into a fibrin clot. This limited proteolytic pattern of converting protein A to generate protein B with specific properties has become the hallmark for activating the precursors in the coagulation, fibrinolytic, and complement pathways.

Within the context of the actual clotting process in blood, thrombin performs the dual function of catalyzing not only the conversion of fibrinogen to fibrin but also regulating the rate of production of factor XIIIa, a transamidating enzyme that serves the purpose of strengthening the clot by cross-linking fibrin by N^ε-(γ-glutamyl))lysine bridges within the required physiological time frame. Cross-linking by factor XIIIa augments the mechanical rigidity of the clot structure and also greatly increases its resistance to lysis[3]; the latter may be related to the finding that a certain proportion of the α_2 plasmin inhibitor becomes covalently incorporated into the fibrin network by XIIIa. In any case, the hereditary absence of factor XIII or the sudden appearance of an acquired, mostly autoimmune, inhibitor (either against activation of the factor or the functioning of the activated cross-linking enzyme) often gives rise to life-threatening hemorrhage. Non-

---

[1] L. Lorand, *Nature (London)* **167,** 192 (1951).

[2] M. W. Mosesson and R. F. Doolittle (eds.), *in* "Molecular Biology of Fibrinogen and Fibrins," p. 408. N.Y. Academy of Sciences, 1983.

[3] L. Lorand, M. S. Losowsky, and K. J. M. Miloszewski, *Prog. Hemostasis Thromb.* **5,** 245 (1980).

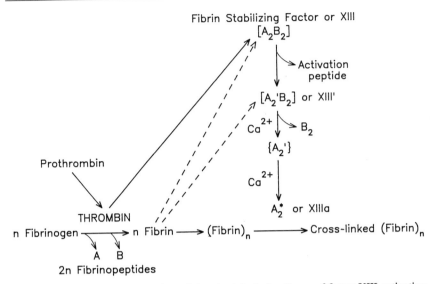

FIG. 1. A schematic presentation of the physiological pathway of factor XIII activation and fibrin stabilization in blood clotting. The dashed lines emphasize the "feed-forward" regulatory aspects played by fibrin in promoting the reaction of factor XIII with thrombin and in reducing the Ca^{2+} requirement for the dissociation of the hydrolytically modified A_2B_2 allozymogen.

cross-linked clots undergo fibrinolysis with greater ease, not allowing sufficient time for tissue repair.

Whereas all other enzymes, including thrombin, generated in blood coagulation belong to the trypsin family of serine proteinases, which can be inhibited, for example, by diisopropylphosphofluoridate, factor XIIIa carries a cysteine thiol active center, which can be blocked, for example, by iodoacetamide, with complete abolition of its enzymatic activity. The factor XIII zymogen also differs from all the other monomeric precursors in coagulation by the fact that it is a heterooligomeric entity with an A_2B_2 structure, where A represents the subunits associated with catalytic potential and the B subunits serve the role of plasma carrier.

Conversion of the factor XIII zymogen occurs in two consecutive and distinct stages; the first requires thrombin and the second depends on Ca^{2+} ions. Both steps are promoted enormously by the presence of fibrin.[4,5] In the reconstructed scheme of the pathway of factor XIII conversion and fibrin stabilization (Fig. 1), the dashed lines emphasize the "feed-forward" regulatory aspects played by fibrin in promoting the reaction of factor XIII

[4] M. G. Naski, L. Lorand, and J. Shafer, *Biochemistry* **30**, 934 (1991).
[5] L. Lorand and J. T. Radek, *in* "Thrombin: Structure and Function" (L. J. Berliner, ed.), p. 257. Plenum Press, New York, 1992.

with thrombin and in reducing the Ca^{2+} requirement for the dissociation of the hydrolytically modified A_2B_2 allozymogen.

The promotion of the thrombin-catalyzed activation of factor XIII by fibrin is of regulatory significance from the point of view of physiological controls in blood clotting. Promotion of the initial hydrolytic step in the activation of factor XIII by polymeric fibrin ensures that significant amounts of factor XIII are not activated until its physiological substrate, polymeric fibrin, is present. This method of control would also minimize the wasteful and premature generation of factor XIIIa and thus the possibly dangerous cross-linking of other plasma proteins by the transamidase. Moreover, the promoting activity of polymeric fibrin is rapidly lost after the catalytically competent factor XIIIa is produced. This finding is consistent with the view that the factor XIIIa-mediated cross-linking of fibrin inactivates fibrin as a promoter for the release of activation peptide from the factor XIII zymogen by thrombin. Clearly, such a feedback shutoff regulation would serve the purpose of a safeguard against the continued generation of factor XIIIa after its fibrin substrates have been cross-linked, thereby avoiding the overproduction of the catalytically active factor XIIIa species.

The formation of a clot is followed by fibrinolysis and tissue repair. The damage repair processes probably cycle continuously at subthreshold levels without noticeable physiologic sequelae related to either significant blood loss or vascular occlusion.

The waterfall/cascade[5a,5b] description of blood clotting provides the logic required to define deficiencies associated with fibrin formation in plasma *in vitro*, but does not give a fully satisfactory description of blood clotting *in vivo*. In addition, although useful for diagnostic analysis, clotting assays still lack the quantitative information necessary to understand the nuances of the blood clotting mechanism *in vivo*. Current literature supports the conclusion that all the reactions involved in thrombin generation are propagated by the formation of complex enzymes comprising a serine protease and some cofactor protein(s) that act in concert on a membrane or other surface to express catalytic activity, as depicted in Fig. 2.[6,7,8] The product of each enzymatic complex provides the serine protease component required for the assembly and activity of the successive enzyme complex, ultimately leading to the formation of thrombin (factor IIa).

Schemes of blood clotting *in vitro*, summarized in Fig. 2, have provided

[5a] R. G. Macfarlane, *Nature (London)* **202**, 498 (1964).

[5b] E. W. Davie and O. D. Ratnoff, *Science* **145**, 1310 (1964).

[6] C. M. Jackson and Y. Nemerson, *Annu. Rev. Biochem.* **49**, 765 (1980).

[7] K. G. Mann, R. J. Jenny, and S. Krishnaswamy, *Annu. Rev. Biochem.* **57**, 915 (1988).

[8] B. Furie, *Cell (Cambridge, Mass.)* **53**, 505 (1988).

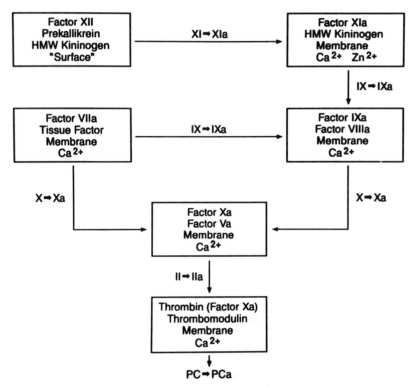

FIG. 2. The multiprotein complexes of the coagulation cascade are each presented as complex enzymes that involve a serine protease, one or more cofactor proteins, divalent cations, and a surface. Roman numeral notations for proteins are used by convention. Each of the coagulation proteins is designated in its active form by an "a"; PC, protein C; PCa, activated protein C. For expression of all the complexes, with the exception of the contact pathway (factor XII complex), an active enzyme is required. In the case of the contact pathway, activated factor XII (factor XIIa) is generated as a consequence of binding to a surface. Reciprocal activations involving kallikrein and factor XIIa lead to the generation of substantial amounts of factor XIIa, which activates factor XI to factor XIa. All of the other complexes involve a phospholipid membrane surface and a cofactor protein acting in concert with a serine protease to convert the appropriate zymogen substrate to its respective product. (K. G. Mann, Normal Hemostasis, in "Textbook of Internal Medicine," 2nd Ed., Vol. 1, pp. 1240–1245. J. B. Lippincott, Co., Philadelphia, PA, 1992.)

useful tools for the elucidation of blood plasma clotting in a glass tube. However, these descriptions have not consistently predicted the pathology associated with hemostasis and thrombosis. The "circuit diagram" of Fig. 2 would imply that factor XII, prekallikrein, and high-molecular-weight kininogen, all of which are essential for clotting of plasma *in vitro,* are associated with underlying hemostatic defects. However, patients with these laboratory abnormalities do not require replacement therapy follow-

ᵘrgery. Thus, although these
ⁿ *vitro,* the *in vivo* signifi-
ᵉ questioned. The hemo-
ₐcy does not usually require
. However, patients with this
ₑent therapy following hemostatic
ₛs suggests an accessory rather than a
ₒrmal hemostasis. In contrast, individ-
ᵣ IX have profound hemostatic defects, as
factor V. Factor VII deficiency is not ade-
ₐ basis because of its rarity and other complexi-
ₙboplastin species and assay sensitivity. Patients
ₑncy, identified using human thromboplastins, have
ᵥe extensive bleeding. Protein C deficiency is associated
risk.

ₑexes of the vitamin K-dependent enzymes clearly associated
ₛtatic or thrombotic risk are depicted in Fig. 3.[9] Each of these
ₑs involves a serine protease and a cofactor protein assembled on a
ᵣane surface in the presence of calcium ions. Each of these is thought
ᵦe essential for normal blood clotting homeostasis. However, the de-
ₛcriptions of these, as shown in Fig. 3, are still inadequate to describe the
hemostatic defects associated with hemophilia A and hemophilia B. Figure
3 shows two mechanisms by which factor X can be converted to factor Xa.
The reaction can be catalyzed by the tissue factor–factor VIIa complex
and by the factor VIIIa–factor IXa complex. Further, although the tissue
factor–factor VIIa complex is capable of activating both factor X and
factor IX to their respective enzyme products in purified reaction systems,
factor X has been reported to be the preferred substrate for the tissue
factor–factor VIIa complex. The tissue factor–factor VIIa complex is
capable of activating factor X to factor Xa, thus appearing to bypass the
requirement for the factor VIIIa–factor IXa complex. The latter would
appear superfluous to thrombin generation. The qualitative description
provided by Fig. 3 thus does not provide a rationale to explain why patients
with hemophilia A (i.e., deficiency of factor VIII) or hemophilia B (i.e.,
deficiency of factor IX) show a bleeding diathesis following hemostatic
challenge. Recent quantitative data suggest that significant enzymatic rate
differentials and cooperation between factor VIIa–tissue factor and factor
Xa in factor IX activation can account for the preferred formation of factor
Xa by the factor VIII–factor IXa complex.[10]

[9] K. G. Mann, M. E. Nesheim, W. R. Church, P. Haley, and S. Krishnaswamy, *Blood* **76,** 1 (1990).
[10] K. G. Mann, S. Krishnaswamy, and J. H. Lawson, *Semin. Hematol.* **29,** 213 (1992).

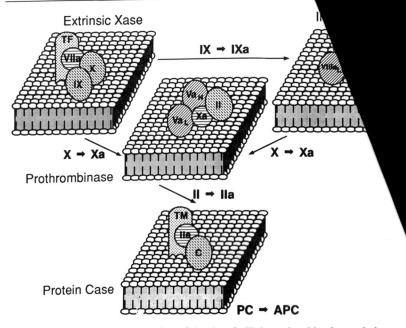

FIG. 3. A schematic representation of the vitamin K-dependent blood coagulation enzyme complexes. Each complex is represented assembled on a phospholipid surface. The cofactors, tissue factor (TF), thrombomodulin (TM) factor VIIIa, and factor Va, are represented associated with their respective complementary serine proteases, factor VIIa, factor Xa, factor IXa, and thrombin (IIa). The substrates of the complexes, factor IX, factor X, prothrombin (II), and protein C (C), are also represented; APC corresponds to the activated form of protein C. The reactants and products associated with each reaction complex are also represented.

The functional enzymes depicted in Fig. 3 are assembled from the cellular and plasma-derived cofactors and the vitamin K-dependent proteases of the coagulant (thrombin) and anticoagulant (protein C) pathways. Although each of these complexes exhibits discrete substrate and proteolytic specificity, they share several common features:

1. The complexes are functionally analogous, with structurally homologous constituents.
2. The enzyme complexes exhibit similar requirements for assembly and activity.
3. In each case, the assembly of the enzyme complex leads to a significant enhancement in the localized catalytic rate of activation of the substrate.

Three key regulatory events are associated with the formation of the vitamin K-dependent enzymatic complexes:

1. The conversion of a vitamin K-dependent zymogen to a serine protease.
2. The proteolytic activation of plasma-derived procofactor to an active cofactor in the case of factors V and VIII or the membrane expression of an integral membrane cofactor (tissue factor or thrombomodulin).
3. The presentation of the appropriate membrane surface to accommodate the protein-binding interactions.

Most of the reactions illustrated in Fig. 3 have been studied using synthetic phospholipids as mimics for a cell membrane, and all reactions display a preference for anionic phospholipid mixtures. Although these lipid vesicle systems serve as useful models, they are fundamentally unregulated in the expression of a coagulant-active membrane. *In vivo,* the regulated expression of the cell-derived membrane sites for complex assembly is an essential step in providing for enzyme complex formation, and the nature of the cellular sites and their regulation is presently unexplained.

The protein–membrane binding processes associated with the formation of the catalytic complexes leads to localization of proteolytic activity at the point at which damaged vascular cells or activated adherent peripheral blood cells provide the membrane site required for enzyme complex assembly. The processes of localization on a membrane surface and the formation of the enzyme complex also provide significant increases in reaction rates for the complexed serine proteases when compared with the equivalent serine proteases acting on physiologic substrates in bulk solution. The reaction rates measured for the proteases factor VIIa, factor IXa, and factor Xa converting their respective protein substrates to products are so small as to be almost insignificant. When these same serine proteases are bound to a membrane surface in the presence of the appropriate cofactor, enormous changes (10^5- to 10^6-fold) in reaction efficiency are observed. Thus the uncomplexed serine proteases display virtually no activity when compared with the catalytic efficiencies of the complex enzymes. Thus, complex formation is itself an "on switch" that leads to the rapid generation of proteolytic activity. Dissociation of any constituent protein from the membrane-bound complex similarly acts as an "off switch" that stops the reaction. The enzyme complex requirement for catalytic activity attenuates the propagation of coagulation reaction downstream from the reaction site.

The complex interplay of serine protease and cofactor in the expression of procoagulant activity on a membrane surface also provides significant opportunities for intermodulation of the reaction series. The procofactors (factor V and factor VIII) are converted to the active cofactors (factors Va and VIIIa) both by thrombin and by factor Xa. Factor VII is activated to factor VIIa by factors XIIa, Xa, and IXa as well as by thrombin. The condensation of multiple-reaction complexes in series on an individual membrane surface may also provide increased efficiency for local generation of thrombin activity.

The principal targets for thrombin activity are the activation of platelets through proteolytic cleavage of a surface glycoprotein[11] and the conversion of soluble circulating fibrinogen to the insoluble fibrin matrix. The latter reaction product is stabilized by inter- and intramolecular cross-links introduced by thrombin-activated factor XIIIa. Thrombin also participates in feedback activation of a number of components of the coagulation pathway, including factors V, VIII, XI, and VII. These thrombin-mediated reactions are vital to the explosive generation of the ultimate thrombin activity produced. Finally, thrombin initiates its own down-regulation by the proteolytic activation of protein C.

The termination reactions of the blood clotting process involve both constitutive inhibition processes and clotting-initiated, or enzyme-regulated, termination reactions. A variety of plasma inhibitor proteins, including fibrin, can interfere with the function of most of the proteases in the coagulation process. The principal constitutive protein inhibitor of coagulation is antithrombin III.[12] This protein has the capacity to interfere with the coagulant activity of all the serine proteases in the coagulation system. The interaction of antithrombin III with coagulation proteases is greatly enhanced by the binding of antithrombin III to cell surface heparan sulfate glycosaminoglycans. The binding of antithrombin III to heparin and heparan sulfate polymers leads to a conformational change in the antithrombin III protein that causes it to react effectively with factor Xa. In reactions with thrombin, the inhibition process is accelerated by the simultaneous binding of antithrombin III and thrombin to the heparan sulfate molecule.

Before explicit knowledge of these molecular processes was available it was recognized that heparin polysaccharides derived from animal tissues were potent anticoagulants. All of these heparin materials express their anticoagulant functions by interactions with antithrombin III and pro-

[11] H. V. Thein-Khai, D. T. Hung, V. I. Wheaton, and S. R. Coughlin, *Cell (Cambridge, Mass.)* **64**, 1057 (1991).

[12] R. D. Rosenberg, and J. S. Rosenberg, *J. Clin. Invest.* **74**, 1 (1984).

teases. Because the interaction of antithrombin III with different proteases involves interactions of antithrombin III with heparin, as well as simultaneous interactions of antithrombin III and the subject proteases with heparin, it is possible to differentiate antithrombin III inhibitory functions somewhat by the use of high- and low-molecular-weight heparinoid materials. In normal hemostasis, however, the constitutive endothelial cell-bound heparan sulfate appears to participate in the localized vascular reaction environment by binding and activating antithrombin III. In the immediate coagulation reaction site, heparinoids are (probably) neutralized by released platelet factor 4, the antiheparin protein secreted by the activated platelet.

Unlike the other proteases in the blood coagulation process, factor VIIa is poorly inhibited by antithrombin III–heparin complexes. The inhibition of factor VIIa by this inhibitor is significantly enhanced in the presence of tissue factor and heparin. Factor VIIa levels in blood appear to be regulated by the tissue factor pathway inhibitor (TFPI), also referred to as extrinsic pathway inhibitor (EPI) and the lipid-associated coagulation inhibitor (LACI).[13] TFPI inhibits factor VIIa by forming a calcium ion-stabilized noncovalent complex between factor VIIa–TFPI–tissue factor and factor Xa. The complex thus neutralizes one molecule of factor Xa and one molecule of factor VIIa. Because the complex is stabilized by calcium, chelation of calcium ions leads to dissociation of the complex and the release of factor VIIa and factor Xa.

The activation of protein C is an important reaction associated with the proteolytic termination of the coagulation response.[14] This reaction is directly linked to the formation of α-thrombin and hence the level of activation of protein C is proportionate to the extent of the coagulation response. The discovery of γ-carboxyglutamate as the hallmark of vitamin K action led to the identification of the plasma proteins C and S. Subsequently, these proteins were determined to have anticoagulant functions. Individuals with hereditary reduced levels of proteins C and S display a predisposition toward familial thrombosis. The zymogen protein C is activated by thrombin complexed to the endothelial cell-bound thrombomodulin. Presumably, thrombin escaping from the coagulation environment binds to thrombomodulin present on surrounding vascular endothelial cells and activates plasma protein C (Fig. 3). The response of protein C activation is proportional under normal circumstances to the amount of thrombin generated; thus, protein C activation corresponds to a proportionally regulated termination reaction. Activated protein C provides its

[13] G. J. Broze, T. J. Girard, and W. F. Novotny, *Biochemistry* **29,** 7539 (1990).
[14] C. T. Esmon, *J. Biol. Chem.* **263,** 4743 (1989).

anticoagulant function by cleavage of selected peptide bonds in factors Va and VIIIa, a process that inactivates these two vital plasma cofactor proteins. Once cleaved by activated protein C, neither factor Va nor factor VIIIa can perform their binding functions for their respective enzymes (factor Xa or IXa) or their substrates (factor X or prothrombin). When factor Xa is bound to factor Va, activated protein C is ineffective in binding to and inactivating factor Va. The protective effect of factor Xa can be suppressed by protein S. Thus, protein S serves as an accessory factor in the inactivation process.

[1] Modular Design of Proteases of Coagulation, Fibrinolysis, and Complement Activation: Implications for Protein Engineering and Structure–Function Studies

By László Patthy

Introduction

Dramatic progress has been made in the biochemistry and molecular biology of the proteases involved in blood coagulation, fibrinolysis, and complement activation. At the time of the last review of the subject in this series, only the primary structures of prothrombin, plasminogen, factor IX, and factor X were known.[1,2] By the end of the 1980s the amino acid sequences of most proteases of the plasma effector systems had been determined and/or deduced from the nucleotide sequences of cDNA clones. With a few exceptions, the exon–intron organization of their genes is also known (Table I).

The characterization of the amino acid sequences and genes of the proteases of blood permitted the analysis of their molecular evolution. These studies revealed some surprising facts about the domain organization and modular design of the proteases, as well as the evolutionary origin and mechanism of formation of their genes. This chapter will briefly review the evidence supporting modular evolution of the proteases of blood clotting, fibrinolysis, and complement activation, will summarize the principles and rules that emerged from these studies, and will illustrate how these rules can facilitate structure–function studies and protein engineering of such modular proteins.

---

[1] R. R. Porter, this series, Vol. 80, p. 80.
[2] E. W. Davie, this series, Vol. 80, p. 153.

Evolution of Proteases of Blood Coagulation, Fibrinolysis, and
Complement Activation by Assembly from Modules

Evidence for Modular Assembly

Analysis of the amino acid sequences of the blood coagulation, fibrinolytic, and complement proteases revealed that their catalytically active
regions are homologous with those of trypsinlike serine proteases. In contrast with the zymogens of simple proteases of the trypsin family (pancreatic proteases, glandular kallikreins, mast cell proteases, etc., in which only
a signal peptide is attached to the amino-terminal end of the enzyme
region) in the complex regulatory proteases, very large segments are inserted between the signal peptide and the zymogen activation domain of
the trypsin–homolog region. From structure–function studies it is known
that, in general, the nonprotease parts are involved in interactions with
macroscopic structures, cofactors, substrates, and inhibitors that regulate
the activity and activation of these enzymes and their zymogens. The
nonprotease parts display remarkable variation in size and structure, in
harmony with their involvement in specific interactions regulating the
individual proteases. The structural variation of the nonprotease parts has
raised puzzling questions about the origin of this diversity.

The amino-terminal extensions of some proteases (prothrombin, plasminogen, coagulation factors IX and X) have long been known to be
organized into structural–functional domains (kringle domains, vitamin
K-dependent calcium-binding domains, epidermal growth factor-related
structures). Studies on isolated kringles have shown that they correspond to
independent structural and functional units that are also autonomous with
respect to folding.[3-8] These observations raised the possibility that the
independence of these "miniproteins" is a relic of their evolutionary autonomy: they may have evolved outside the family of serine proteases and
may have been fused to the catalytic regions individually (Fig. 1).

The first proof for the hypothesis that the nonprotease parts of regulatory proteases were constructed from modules was the finding that a region
of tissue-type plasminogen activator is homologous with the finger domains (type I domains) of fibronectin, an extracellular matrix protein that

[3] A. Varadi and L. Patthy, *Biochem. Biophys. Res. Commun.* **103,** 97 (1981).
[4] Z. Vali and L. Patthy, *J. Biol. Chem.* **257,** 2104 (1982).
[5] M. Trexler, Zs. Vali, and L. Patthy, *J. Biol. Chem.* **257,** 7401 (1982).
[6] Z. Vali and L. Patthy, *J. Biol. Chem.* **259,** 13690 (1984).
[7] M. Trexler and L. Patthy, *Proc. Natl. Acad. Sci. U.S.A.* **80,** 2457 (1983).
[8] M. Trexler and L. Patthy, *Biochim. Biophys. Acta* **787,** 275 (1984).

TABLE I
MODULAR ORGANIZATION OF COAGULATION, FIBRINOLYTIC, AND COMPLEMENT
ACTIVATION PROTEASES AND RELATED PROTEINS

Protein[a]	M[a,b]	Gene structure
Plasminogen	K	c
Apolipoprotein (a)	K	d
Hepatocyte growth factor	K	e
Urokinase	G K	f
Tissue plasminogen activator	F G K	g
Factor XII	F G K	h
Prothrombin	C K	i
Factor VII	C G	j
Factor IX	C G	k
Factor X	C G	l
Protein C	C G	m
Protein Z	C G	—
Factor XI	CF	n
Plasma prekallikrein	CF	o
Complement component C1s	B C1r G	p
Complement component C1r	B C1r G	p
Haptoglobin	B	q
Complement factor I	C7 LDL SC	—
Complement factor B	B vW	r
Complement factor C2	B vW	r
Limulus factor C	B G LN	s

[a] The list of references for the amino acid sequences of proteases, their homologies, and the definition of modules may be found in earlier reviews [L. Patthy, *Cell (Cambridge, Mass.)* **41**, 657 (1985); *Semin. Thromb. Hemostasis* **16**, 245 (1990); *Blood Coagulation Fibrinolysis* **1**, 153 (1990); *Curr. Opin. Struct. Biol.* **1**, 351 (1991)].

[b] Abbreviations of class 1–1 modules: B, complement B type; C, vitamin K-dependent calcium binding; C1r, complement C1r type; C7, complement C7 type; CF, contact factor; F, finger; G, growth factor; K, kringle (type II module); LDL, LDL receptor; LN, C-type lectin; SC, scavenger receptor; vW, von Willebrand factor.

[c] T. E. Petersen, M. R. Martzen, A. Ichinose, and E. W. Davie, *J. Biol. Chem.* **265**, 6104 (1990).

[d] C. Lackner, E. Boerwinkle, C. C. Leffert, T. Rahmig, and H. H. Hobbs, *J. Clin. Invest.* **87**, 2153 (1991).

[e] K. Miyazawa, A. Kitamura, and N. Kitamura, *Biochemistry* **30**, 9170 (1991); S. Han, L. A. Stuart, and S. J. Friezner Degen, *Biochemistry* **30**, 9768 (1991).

[f] A. Riccio, G. Grimaldi, P. Verde, G. Sebastio, S. Boast, and F. Blasi, *Nucleic Acids Res.* **13**, 2759 (1985); S. Friezner Degen, J. L. Heckel, E. Reich, and J. L. Degen, *Biochemistry* **26**, 8270 (1987).

[g] T. Ny, F. Elgh, and B. Lund, *Proc. Natl. Acad. Sci. U.S.A.* **81**, 5355 (1984); S. Friezner Degen, B. Rajput, and E. Reich, *J. Biol. Chem.* **261**, 6972 (1986).

[h] D. E. Cool and R. T. A. MacGillivray, *J. Biol. Chem.* **262**, 13662 (1987).

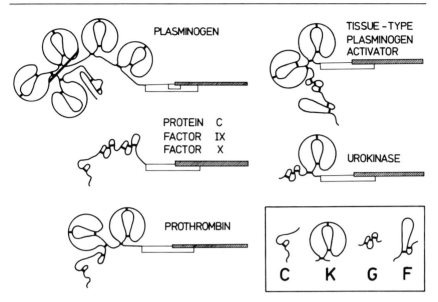

FIG. 1. Modular design of proteases of blood coagulation and fibrinolysis. The bars represent the catalytic domains. The inset shows the schemes of vitamin K-dependent calcium-binding module (C), kringle module (K), growth factor module (G), and finger module (F). Reproduced with permission from Patthy.[12]

[i] S. Friezner Degen and E. W. Davie, *Biochemistry* **26**, 6165 (1987); D. M. Irwin, K. A. Robertson, and R. T. A. MacGillivray, *J. Mol. Biol.* **200**, 31 (1988).

[j] P. J. O'Hara, F. J. Grant, B. A. Haldeman, C. L. Gray, M. Y. Insley, F. S. Hagen, and M. J. Murray, *Proc. Natl. Acad. Sci. U.S.A.* **84**, 5158 (1987).

[k] D. S. Anson, K. H. Choo, D. J. G. Rees, F. Giannelli, K. Gould, J. A. Huddleston, and G. G. Brownlee, *EMBO J.* **3**, 1053 (1984).

[l] S. P. Leytus, D. C. Foster, K. Kurachi, and E. W. Davie, *Biochemistry* **25**, 5098 (1986).

[m] D. C. Foster, S. Yoshitake, and E. W. Davie, *Proc. Natl. Acad. Sci. U.S.A.* **82**, 4677 (1985).

[n] R. Asakai, E. W. Davie, and D. W. Chung, *Biochemistry* **26**, 7221 (1987).

[o] G. Beaubien, I. Rosinski-Chupin, M. G. Mattei, M. Mbikay, M. Chrétien, and N. G. Seidah, *Biochemistry* **30**, 1628 (1991).

[p] M. Tosi, C. Duponchel, T. Meo, and E. Couture-Tosi, *J. Mol. Biol.* **208**, 709 (1989).

[q] N. Maeda, *J. Biol. Chem.* **260**, 6698 (1985).

[r] N. Ishikawa, M. Nonaka, R. A. Wetsel, and H. R. Colten, *J. Biol. Chem.* **265**, 19040 (1990).

[s] T. Muta, T. Miyata, Y. Misumi, F. Tokunaga, T. Nakamura, Y. Toh, Y. Ikehara, and S. Iwanaga, *J. Biol. Chem.* **266**, 6554 (1991).

is otherwise unrelated to serine proteases.[9] The kringle domains of proteases were also shown to be related to the type II domains of fibronectin.[10] The finding that epidermal growth factor is derived from a precursor that is not related to proteases also suggested that the growth factor-related structures of proteases could be borrowed from some other proteins.[11]

On the basis of these initial observations we have proposed that the nonprotease parts of regulatory proteases of blood coagulation and fibrinolysis were assembled from modules.[12] This hypothesis assumes that the common ancestor of regulatory proteases was similar to the simple trypsinlike proteases and that all regulatory modules were inserted individually between the signal peptide and zymogen activation domains. Comparison of the evolutionary history of the individual modules with the genealogy of the protease domains has confirmed this assumption and permitted the reconstruction of the assembly process.[12]

In agreement with the predictions of our hypothesis, the proteases of the complement, blood coagulation, and fibrinolytic cascades sequenced more recently were also found to be mosaics assembled from modules. The generality of this evolutionary strategy is further supported by the observation that most nonprotease constituents of the plasma effector systems, components of the extracellular matrix, and receptors were also shown to have been assembled from modules.[13-15] Of the module types found in the regulatory proteases of blood, the contact factor module (present in plasma prekallikrein and factor XI) is the only one that has not yet been found in other mosaic proteins. (Table I shows the module organization of the proteases of blood coagulation, fibrinolysis, and complement activation.)

The fact that the majority of the regulatory proteases of the plasma effector systems evolved by assembly from modules underlines the importance of this evolutionary strategy. The value of this mechanism is that acquisition of new modules can endow the recipient protein with novel binding specificities and can lead to dramatic changes in its regulation and targeting. Furthermore, in modular proteins numerous distinct binding specificities may coexist, making such proteins ideal members of regulatory networks where multiple interactions are critical.[13-15]

The same principle may be exploited in protein engineering when novel

[9] L. Banyai, A. Varadi, and L. Patthy, *FEBS Lett.* **163**, 37 (1983).
[10] L. Patthy, M. Trexler, Zs. Vali, L. Banyai, and A. Varadi, *FEBS Lett.* **171**, 131 (1984).
[11] R. F. Doolittle, D. F. Feng, and M. S. Johnson, *Nature (London)* **307**, 558 (1984).
[12] L. Patthy, *Cell (Cambridge, Mass.)* **41**, 657 (1985).
[13] L. Patthy, *Semin. Thromb. Hemostasis* **16**, 245 (1990).
[14] L. Patthy, *Blood Coagulation Fibrinolysis* **1**, 153 (1990).
[15] L. Patthy, *Curr. Opin. Struct. Biol.* **1**, 351 (1991).

interactions changing regulation, specificity, and targeting of proteases are desirable.

Structure–Function Studies and Protein Engineering of Plasma Proteases by Module Shuffling

The modular assembly hypothesis has obvious implications for structure–function studies and protein engineering investigations of regulatory proteases: the assembly process can be mimicked with the tools of molecular biology.

Research on fibrinolytic proteases (plasminogen and urokinase-type and tissue-type plasminogen activators) has extensively employed such domain-shuffling experiments to create chimeric proteins, variants with module deletions, duplications, and rearrangements, to clarify structure–function aspects and to engineer novel fibrinolytic enzymes.[16-18] The general success of this approach illustrates the power of the modular assembly principle and provides convincing evidence regarding the hypothesis that the modules correspond to autonomous units.

Module Shuffling by Exon Shuffling

The modular assembly hypothesis implies that the gene pieces encoding the modules display a remarkable mobility during evolution: they undergo repeated insertions, tandem duplications, and exchanges. To explain the unusual mobility of these domains it had to be assumed that some special features of their genes contributed to the frequent genetic rearrangements. Exon shuffling was one of the obvious theoretical mechanisms that could be held responsible for module shuffling.[19] The fact that, in the gene of tissue-type plasminogen activator, introns were found exactly where homology predicted module boundaries supported our suggestion that "exon-shuffling was the mechanism whereby these otherwise unrelated proteins acquired" homologous finger domains.[9,20] The correlation between the mosaic structure of tissue-type plasminogen activator and the exon–intron organization of its gene thus provided the first unquestionable evidence for modular exchange by exon shuffling.

Subsequent analysis of the exon–intron organization of the genes of regulatory proteases (and mosaic proteins containing related modules)

[16] T. J. R. Harris, *Protein Eng.* **1,** 449 (1987).
[17] H. Pannekoek, C. de Vries, and A. J. van Zonneveld, *Fibrinolysis* **2,** 123 (1988).
[18] E. Haber, T. Quertermous, G. R. Matsueda, and M. S. Runge, *Science* **243,** 51 (1989).
[19] W. Gilbert, *Nature (London)* **271,** 501 (1978).
[20] T. Ny, F. Elgh, and B. Lund, *Proc. Natl. Acad. Sci. U.S.A.* **81,** 5355 (1984).

revealed that, in general, introns are found at the module boundaries, a finding consistent with our hypothesis that intronic recombination/exon shuffling played a major role in the assembly of these genes.[9,12,21] Because plasma proteases were the first proteins where evidence for exon shuffling was overwhelming, the analysis of their evolutionary history provided unique opportunities to study the mechanisms and rules of exon shuffling. These studies have revealed some unusual features: the introns found at both boundaries of the modules used in the construction of regulatory proteases were always phase 1, i.e., they split the reading frame between the first and second nucleotide of a codon.[21] In principle, if we classify domains with respect to the phase class of introns (phases 0, 1, and 2) at their 5' and 3' boundaries, we arrive at domain classes that differ markedly in their suitability to be shuffled by intronic recombination. We have shown that only symmetrical domains (i.e., classes 1–1, 2–2, and 0–0, which have introns of the same phase at both their ends) can be inserted into introns (of the same phase class), can undergo tandem duplication into adjacent introns, or can be deleted by intronic recombination without disrupting the reading frame.[21] Nonsymmetrical domains (classes 1–2, 0–1, etc.), if inserted, duplicated, or deleted by intronic recombination, would shift the reading frame, thereby destroying downstream structures of the protein.[21] It seems clear that the symmetry of the class 1–1 modules (growth factor, kringle, and finger modules, etc.) is critical for the frequent shuffling and duplication of these modules during the evolution of mosaic proteins.[13,15,21]

A survey of the gene structures of all trypsinlike serine proteases (proteases of the regulatory cascades as well as the simple proteases of the pancreas, proteases of the glandular kallikrein family, and mast cell proteases, etc.) has shown that a phase 1 intron is present at the boundary of the protease region (zymogen activation and catalytic domains) and the nonprotease part (modular region and/or signal peptide). It is obvious that the common ancestor of these proteases also possessed a phase 1 intron in this position and, according to our hypothesis, this intron of the ancestral protease was the recipient of modules during the assembly process. The splice junctions of this intron were phase compatible with class 1–1 modules, permitting their insertion during the construction of plasma proteases.[13–15,21]

The implicit predictions of this hypothesis have been borne out by studies on proteases of the blood coagulation, fibrinolytic, and complement cascades inasmuch as the assembly of their nonprotease parts followed the same rules.[13–15] The novel module types used in the construction of these proteases were all found to belong to class 1–1 as dictated by the

[21] L. Patthy, *FEBS Lett.* **214**, 1 (1987).

[1]   MODULAR DESIGN OF PROTEASES   17

phase 1 intron at the boundary of the signal peptide and catalytic domains of the ancestral serine protease. Studies on genes of plasma proteins, extracellular matrix proteins, and membrane-associated proteins that have been assembled from class 1–1 modules confirm the general validity of these simple rules of modular assembly of mosaic proteins.[13-15]

Correlation between Modular Organization of Mosaic Proteins and Structure of Their Genes

During modular assembly of the regulatory proteases, insertions and duplications of class 1–1 modules divide, duplicate, and thus proliferate phase 1 introns, eventually leading to gene structures in which all intermodule introns belong to the phase 1 class. The dominance of a single intron-phase class in the noncatalytic chains of plasma proteases and related proteins is thus a necessary consequence of their evolution by exon shuffling: nonrandom intron-phase usage is a diagnostic sign of gene assembly by exon recruitment.[21] In other words, gene assembly by exon shuffling is reflected not only in a correlation between the domain organization of the protein and exon–intron organization of its gene, but also in the nonrandom phase usage of intermodule introns.[21]

The validity of this rule is illustrated by the fact that the mosaic nature of several proteins could be recognized on the basis of the nonrandom intron-phase usage of their genes.[13-15] It should be pointed out, however, that there is now ample evidence that during evolution of gene families introns may be lost from or inserted into genes.[15] Such changes in the exon–intron organization of mosaic proteins may sometimes obscure or eliminate the original correlation between the modular structure of protein and its gene.

Folding Autonomy of Modules

As pointed out above, the modules used in the construction of the regulatory parts of the blood coagulation, fibrinolytic, and complement activation proteases display remarkable functional and structural independence and our studies on isolated kringles have shown that this module type also corresponds to an autonomous folding unit.[7,8] Folding of kringle 4 of plasminogen was found to be an unusually well-directed process: in its folding pathway native intermediates are dominant.[7] It is tempting to assume that the remarkably simple, well-directed folding pathway of kringles reflects the utmost importance of folding autonomy of modules in multidomain proteins. The folding of the module must be unambiguously determined in order to minimize the influence of neighboring domains and to ensure that folding of the module is not deranged when inserted into a novel protein environment. It seems likely that the high degree of conservation of certain residues reflects their importance for the autonomous folding and stability of the kringle structure.[7]

Structure – Function Correlations of Modules

The different kringles of prothrombin, plasminogen, urokinase, tissue-type plasminogen activator, and factor XII retained the same gross architecture, but apparently diverged to bind different proteins or low-molecular-weight compounds. It may thus be predicted that if the amino acid sequences of kringles possessing different binding functions are compared, the residues involved directly in the diverse binding functions may show great variability, whereas the residues found in all or most of the kringles are essential for the autonomous folding and structure of the kringles.[7] According to this view, the kringle architecture conserved in all kringles serves as a scaffold and the variable binding sites are determined predominantly by variable peptide regions not reserved for the residues that determine folding.[7]

Binding sites of individual kringles may thus be predicted for regions noted for their variability. Localization of residues directly involved in ligand binding has indeed confirmed this plausible expectation.[5-7,22,23]

The conclusion that residues highly conserved in kringles are essential for folding and stability is supported by NMR spectroscopy and X-ray crystallography of various kringles.[22-36] These studies have shown that the

[22] M. Trexler, L. Banyai, L. Patthy, N. D. Pluck, and R. J. P. Williams, *FEBS Lett.* **154**, 311 (1983).
[23] M. Trexler, L. Banyai, L. Patthy, N. D. Pluck, and R. J. P. Williams, *Eur. J. Biochem.* **152**, 439 (1985).
[24] A. DeMarco, N. D. Pluck, L. Banyai, M. Trexler, R. A. Laursen, L. Patthy, M. Llinas, and R. J. P. Williams, *Biochemistry* **24**, 748 (1985).
[25] V. Ramesh, M. Gyenes, L. Patthy, and M. Llinas, *Eur. J. Biochem.* **159**, 581 (1986).
[26] V. Ramesh, A. M. Petros, M. Llinas, A. Tulinsky, and C. H. Park, *J. Mol. Biol.* **198**, 481 (1987).
[27] A. Motta, R. A. Laursen, M. Llinas, and A. Tulinsky, *Biochemistry* **26**, 3827 (1987).
[28] A. M. Petros, M. Gyenes, L. Patthy, and M. Llinas, *Arch. Biochem. Biophys.* **264**, 192 (1988).
[29] A. M. Petros, M. Gyenes, L. Patthy, and M. Llinas, *Eur. J. Biochem.* **170**, 549 (1988).
[30] A. Tulinsky, C. H. Park, B. Mao, and M. Llinas, *Proteins* **3**, 85 (1988).
[31] A. Tulinsky, C. H. Park, and E. Skrzypczak-Jankun, *J. Mol. Biol.* **202**, 885 (1988).
[32] T. Thewes, K. Constantine, I. L. Byeon, and M. Llinas, *J. Biol. Chem.* **265**, 3906 (1990).
[33] T. P. Seshadri, A. Tulinsky, E. Skrzypczak-Jankun, and C. H. Park, *J. Mol. Biol.* **220**, 481 (1991).
[34] A. M. Mulichak, A. Tulinsky, and K. G. Ravichandran, *Biochemistry* **30**, 10576 (1991).
[35] I. L. Byeon, R. F. Kelley, and M. Llinas, *Eur. J. Biochem.* **197**, 155 (1991).
[36] R. F. Kelley, A. M. De Vos, and S. Cleary, *Proteins* **11**, 35 (1991).

conserved residues are usually involved in homologous interactions that stabilize the kringle fold.[22-36] Mutagenesis in such structurally conserved positions is likely to affect the folding or stability of kringles.[36]

The lesson from structure–function studies on kringles is that modules may show great variation in function while retaining the capacity to acquire the same three-dimensional architecture. Comparison of the sequences of modules with diverse binding specificities can help identify the structurally important regions of the proteins and localize the regions that are likely to determine binding specificity. This information may be used to predict the three-dimensional structure of modules and to guide site-directed mutagenesis and protein engineering of modules.

Homologies of Plasma Proteases: Implications for Structure and Function

Implications of Homologies for Protein Structure

The identification of homologies has greatly facilitated studies on the structure and function of regulatory proteases and other mosaic proteins. Recognition of homologies proved to be especially useful for the prediction of the structure of mosaic proteins. The lack of X-ray crystallographic information on multidomain regulatory proteases underlines the importance of this approach.

Because modules used in the construction of regulatory proteases are autonomous structural and folding units, the mosaic structure of the protease is obviously reflected in its domain organization. Furthermore, because the gross architecture of related modules of different proteins is conserved, the structure of modules may be predicted on the basis of homology.[26,30] The domain organization and key structural features of regulatory proteases may be thus predicted if the architecture of homologous domains and modules is known. Combined with information on the spatial arrangement and interactions of modules (obtained by neutron scattering, X-ray scattering studies, and cross-linking experiments), molecular modeling on the basis of homology provides a valuable substitute for crystallographic data.[37] This molecular modeling approach requires that the structure of at least one prototype of each module family should be known. Kringles have been studied in great detail by NMR spectroscopy and X-ray crystallography; similar studies on members of the growth factor

---

[37] M. Baron, D. G. Norman, and I. D. Campbell, *Trends Biochem. Sci.* **16**, 13 (1991).

module family, finger modules, complement B-type modules, and type II modules are well under way.[37-41]

Even in the absence of experimental information on the architecture of modules, structural predictions may be greatly facilitated by the availability of numerous homologous sequences. The principle of this approach is that when comparing the sequences of distantly related modules (which have vastly divergent functions but which still have the same protein fold), conservation of residues reflects their importance in the protein fold. In multiple alignments of homologous sequences, regions essential for the structural integrity of the protein (e.g., regions that form regular secondary structures) are likely to be conserved, whereas regions that correspond to external loops connecting structural motifs are likely to be variable in sequence and tolerant to deletions and insertions.[10,42] In view of the expected similarity of three-dimensional structures, predictions of secondary structures of homologous proteins should reflect this similarity, a criterion that greatly improves the reliability of predictions.[10] The power of predictions based on multiple alignment of homologous sequences is illustrated by the fact that it correctly predicted the major structural features (β-sheets, β-turns, external loops) of kringles.[10]

Implications of Homologies for Protein Function

Modules used for the construction of regulatory proteases (growth factor, finger, kringle, and complement B-type modules, etc.) have been shown to be present in numerous mosaic proteins, where they may fulfill vastly divergent functions.[13-15] In view of such functional versatility of modules it is clear that homology does not necessarily imply similarity of function. Nevertheless, homologies may provide useful hints at possible functions that can be tested experimentally.[9,43]

Detecting Distant Homologies of Modules

Recognition of homologies of mosaic proteins provides valuable information on their domain organization and facilitates prediction of their three-dimensional structure and structure–function aspects. Homology of mosaic proteins, however, is not readily recognized by conventional com-

[38] D. G. Norman, P. N. Barlow, M. Baron, A. J. Day, R. B. Sim, and I. D. Campbell, *J. Mol. Biol.* **219,** 717 (1991).
[39] P. N. Barlow, M. Baron, D. G. Norman, A. J. Day, A. C. Willis, R. B. Sim, and I. D. Campbell, *Biochemistry* **30,** 997 (1991).
[40] M. Baron, D. Norman, A. Willis, and I. D. Campbell, *Nature (London)* **345,** 642 (1990).
[41] K. L. Constantine, V. Ramesh, L. Banyai, M. Trexler, L. Patthy, and M. Llinas, *Biochemistry* **30,** 1663 (1991).
[42] L. Patthy, *Acta Biochim. Biophys. Hung* **24,** 3 (1989).
[43] L. Banyai and L. Patthy, *FEBS Lett.* **282,** 23 (1991).

puter programs because the modules are usually distantly related. Even though they retain homologous three-dimensional structures their sequence similarity may be restricted to a few key residues; in the alignment of their sequences, gaps are common. To detect such distant homologies special procedures are required to decide whether the low degree of sequence similarity is due to conservation of key features of the protein fold (and thus reflects homology) or is due to chance similarity of unrelated proteins.

As noted above, when the sequences of a protein family (module family) are compared, the pattern of accepted mutations (the pattern of conserved residues, variable segments, and regions that tolerate gap events) reflects the three-dimensional structure of the protein fold and is thus characteristic of the given family. We have shown that consensus sequences incorporating key features of a protein fold may provide powerful tools when searching for distantly related members of a protein family.[44]

In the first step of this consensus sequence procedure the sequences of established members of a protein family are aligned and the pattern of conserved residues and variable and gap regions characteristics of the family are determined and formulated as consensus sequences.[44] In the second step, the data base of protein sequences is searched with the consensus sequences to identify proteins or protein families that have a related pattern of key residues, variable segments, and gap regions, i.e., which qualify as new members of the protein family.[44] The power of this procedure derives from the fact that it weighs key residues, permits the position-dependent scoring of similarities and gaps, and ignores similarities in variable regions, thereby eliminating background noise that may be due to chance similarity.

The rules of evolution by exon shuffling can provide additional tools for the detection of distant homologies of mosaic proteins: they can help in the recognition of mosaic proteins/mosaic genes and in identification of their modules, and can delineate the group of sequences in which the presence of homologous sequences may be expected.[45] With the application of these rules it is possible to concentrate homology search to a limited group of sequences, thereby simplifying the search.

The utility of the procedure exploiting the rules of evolution of mosaic proteins and their modules is illustrated by the fact that it can identify several new module types and can detect homologies of mosaic proteins that are missed by conventional methods of sequence comparison.[9,10,44-48]

[44] L. Patthy, *J. Mol. Biol.* **198**, 567 (1987).
[45] L. Patthy, *J. Mol. Biol.* **202**, 689 (1988).
[46] L. Patthy, *Cell (Cambridge, Mass.)* **61**, 13 (1990).
[47] L. Patthy, *FEBS Lett.* **289**, 99 (1991).
[48] N. Behrendt, M. Ploug, L. Patthy, G. Houen, F. Blasi, and K. Dano, *J. Biol. Chem.* **266**, 7842 (1991).

[2] Human Plasma Factor XIII: Subunit Interactions and Activation of Zymogen*

By Laszlo Lorand, Jong-Moon Jeong, James T. Radek, and James Wilson

Introduction

Blood coagulation factor XIII, or fibrin-stabilizing factor,[1] is the plasma zymogen of the enzyme responsible for strengthening the clot network[2-5] and for endowing it with a high resistance to lysis.[6-8] Under physiological conditions, activation of the zymogen is brought about by thrombin and Ca^{2+}.[9] Generation of the enzyme (denoted as factor XIII_a or XIIIa; related to EC 2.3.2.13, protein-glutamine:amine γ-glutamyltransferase), a trans-amidase, occurs in two consecutive steps readily resolvable in the test tube into a Ca^{2+}-independent hydrolysis by thrombin followed by a Ca^{2+}-specific, but thrombin-independent, change of the proteolytically cleaved factor XIII' intermediate:

$$XIII \xrightarrow{\text{thrombin}} XIII' \xrightarrow{Ca^{2+}} XIIIa$$

Factor XIII is made up of two different subunits in an AB protomeric structure, thought to associate into an A_2B_2 tetramer.[10-13] Only the A subunits are modified by thrombin, releasing an N-terminal activation

* Portions of this article are reprinted with permission from *Biochemistry* **32**, 3527–3534 1993. Copyright 1993 American Chemical Society.

[1] L. Lorand, M. S. Losowsky, and K. J. M. Miloszewski, *in* "Progress in Hemostasis and Thrombosis" (T. H. Spaet, ed.), p. 245. Grune & Stratton, New York, 1980.

[2] W. W. Roberts, L. Lorand, and L. F. Mockros, *Biorheology* **10**, 29 (1973).

[3] L. F. Mockros, W. W. Roberts, and L. Lorand, *Biophys. Chem.* **2**, 164 (1974).

[4] L. L. Shen, R. P. McDonagh, J. McDonagh, and J. Hermans, Jr., *Biochem. Biophys. Res. Commun.* **56**, 793 (1974).

[5] L. Shen and L. Lorand, *J. Clin. Invest.* **71**, 1336 (1983).

[6] L. Lorand and A. Jacobsen, *Nature (London)* **195**, 911 (1962).

[7] J. Bruner-Lorand, T. R. E. Pilkington, and L. Lorand, *Nature (London)* **210**, 1273 (1966).

[8] L. Lorand and J. L. G. Nilsson, *Drug Des.* **3**, 415 (1972).

[9] L. Lorand and K. Konishi, *Arch. Biochem. Biophys.* **105**, 58 (1964).

[10] M. L. Schwartz, S. V. Pizzo, R. L. Hill, and P. A. McKee, *J. Biol. Chem.* **246**, 5851 (1971).

[11] M. L. Schwartz, S. V. Pizzo, R. L. Hill, and P. A. McKee, *J. Biol. Chem.* **248**, 1395 (1973).

[12] H. Bohn, *Ann. N. Y. Acad. Sci.* **202**, 256 (1972).

[13] N. A. Carrel, H. P. Erickson, and J. McDonagh, *J. Biol. Chem.* **264**, 551 (1989).

[14] Y. Mikuni, S. Iwanaga, and K. Konishi, *Biochem. Biophys. Res. Commun.* **54**, 1393 (1973).

[15] S. Nakamura, S. Iwanaga, T. Suzuki, Y. Mikuni, and K. Konishi, *Biochem. Biophys. Res. Commun.* **58**, 250 (1974).

[16] T. Takagi and R. F. Doolittle, *Biochemistry* **13**, 750 (1974).

peptide (AP). The primary cleavage site in the A subunit[14-16] is at Arg-37; secondary cleavage sites for thrombin were reported[17] at Lys-513 and Ser-514. Both α- and γ-thrombin can activate factor XIII, but the specificity constant for the release of AP with γ-thrombin is about five times lower than with α-thrombin.[18]

The hydrolytically cleaved A subunit, denoted as A', carries the active site cysteine necessary for the expression of the transamidating activity. However, in the heterologous $A_2'B_2$ combination of factor XIII', the critical cysteine residues are still buried, and unmasking requires Ca^{2+} ions.[19-21] The effect of Ca^{2+} is twofold; first, it brings about the heterologous dissociation of $A_2'B_2$, then it promotes a conformational change in the separated A_2' species that induces the enzymatically active configuration (designated as A_2^*):

$$A_2'B_2 \xrightarrow[\;B_2\;]{Ca^{2+}} A_2' \xrightarrow{Ca^{2+}} A_2^*$$

A minimum representation for the physiological conversion of factor XIII to XIIIa is as follows:

$$A_2B_2 \xrightarrow[\;2AP\;]{thrombin} A_2'B_2 \xrightarrow[\;B_2\;]{Ca^{2+}} A_2' \xrightarrow{Ca^{2+}} A_2^*$$

Accordingly, during the last phase of normal blood coagulation, thrombin exerts a dual role in catalyzing not only the conversion of fibrinogen into fibrin through the release of fibrinopeptides A and B (FPA and FPB), but also in controlling the rate of production of the transamidating enzyme (XIIIa), which cross-links (XL) the clot network by a few critical N^{ε}-(γ-glutamyl)lysine side-chain bridges.

$$n \text{ fibrinogen} \xrightarrow{2n(FPA + FPB)} n \text{ fibrin} \rightleftharpoons (\text{fibrin})_n \text{ clot}$$

$$\text{factor XIII} \xrightarrow[\substack{2AP}]{thrombin} \underset{A_2'B_2}{XIII'} \xrightarrow[\;B_2\;]{Ca^{2+}} \underset{A_2^*}{XIIIa} \;\Big|\; Ca^{2+}$$

$$\downarrow$$

$$XL (\text{fibrin})_n \text{ clot}$$

[17] N. Takahashi, Y. Takahashi, and F. W. Putnam, *Proc. Natl. Acad. Sci. U.S.A.* **83**, 8019 (1986).

[18] S. D. Lewis, L. Lorand, J. W. Fenton, II, and J. A. Shafer, *Biochemistry* **26**, 7597 (1987).

[19] C. G. Curtis, P. Stenberg, C.-H. J. Chou, A. Gray, K. L. Brown, and L. Lorand, *Biochem. Biophys. Res. Commun.* **52**, 51 (1973).

[20] C. G. Curtis, K. L. Brown, R. B. Credo, R. A. Domanik, A. Gray, P. Stenberg, and L. Lorand, *Biochemistry* **13**, 3774 (1974).

[21] L. Lorand, A. J. Gray, K. Brown, R. B. Credo, C. G. Curtis, R. A. Domanik, and P. Stenberg, *Biochem. Biophys. Res. Commun.* **56**, 914 (1974).

A fraction of the α_2-plasmin inhibitor (α_2PI) and of fibronectin circulating in plasma becomes covalently cross-linked to fibrin in the XIIIa-catalyzed reaction, which might further contribute to the stabilization of the clot.[22,23] The inherited absence of factor XIII or the appearance of various inhibitors (often an autoimmune antibody) directed against the pathway of clot stabilization can give rise to very severe, frequently fatal, hemorrhage.[1]

Fibrin was shown to be an important modulator by greatly accelerating the thrombin-catalyzed activation of factor XIII[24,25] (see also [Introduction] in this volume). In the physiological setting, such a "feed-forward" mechanism would ensure that factor XIII was not activated to a significant degree until its substrate, polymerized fibrin, was produced. Interestingly, the promoting effect of fibrin is rapidly lost after formation of factor XIIIa, suggesting also a "shutoff" mechanism, which would safeguard against the continued generation of XIIIa after its fibrin substrate has been cross-linked, thereby avoiding the overproduction of the cross-linking enzyme.

The Ca^{2+} requirement for the heterologous dissociation of the $A_2'B_2$ structure of the thrombin-activated factor XIII (i.e., XIII') is markedly reduced in the presence of fibrin[ogen],[26,27] which is yet another important physiological controlling feature in the sequence of events in the scheme above. The relevant domain on fibrin[ogen] responsible for interacting with factor XIII in this manner was located in the α chain of the protein[27] (residues Aα 242–424).

Human platelets and placenta contain a protein considered to be identical to the A_2 subunits of plasma factor XIII.[10-12] This protein, too, can be activated by thrombin and Ca^{2+} to cross-link fibrin,[28] but its physiological function remains unclear. Also, its activation kinetics are quite different from that of plasma factor XIII. However, on combining it with the B_2 subunits purified from plasma, the activation properties become indistinguishable from those of the A_2B_2 ensemble of plasma factor XIII.[21]

Both the amino acid and cDNA sequences for human placental A_2 protein are known[17,29,30] and recombinant products expressed in yeast have been obtained by several investigators.[31-33] Such material was kindly fur-

[22] A. Ichinose and N. Aoki, Biochim. Biophys. Acta **706**, 158 (1982).
[23] D. F. Mosher, J. Biol. Chem. **250**, 6614 (1975).
[24] S. D. Lewis, T. J. Janus, L. Lorand, and J. A. Shafer, Biochemistry **24**, 6772 (1985).
[25] M. G. Naski, L. Lorand, and J. Shafer, Biochemistry **30**, 934 (1991).
[26] R. B. Credo, C. G. Curtis, and L. Lorand, Proc. Natl. Acad. Sci. U.S.A. **75**, 4234 (1978).
[27] R. B. Credo, C. G. Curtis, and L. Lorand, Biochemistry **20**, 3770 (1981).
[28] K. Buluk, T. Januszko, and J. Olbromski, Nature (London) **191**, 1093 (1961).
[29] A. Ichinose, L. E. Hendrickson, K. Fujikawa, and E. W. Davie, Biochemistry **25**, 6900 (1986).
[30] U. Grundmann, E. Amann, G. Zettlmeissl, and H. A. Küpper, Proc. Natl. Acad. Sci. U.S.A. **83**, 8024 (1986).

nished to us by Dr. Paul Bishop of ZymoGenetics, Seattle, WA. We developed various procedures for measuring the interaction of the recombinant protein (rA_2) with the native B_2 carrier protein isolated from human plasma.[34] The results indicate a very tight binding between rA_2 and B_2 in forming the rA_2B_2 structure, and also show that cleavage of rA_2 by thrombin greatly weakens the heterologous association of subunits. In this chapter, we also provide a description for separating the A_2 and B_2 subunits of the purified human plasma factor XIII by using the Ca^{2+}-specific, thrombin-independent pathway[26,34,35] for directly dissociating the zymogen.

Association of rA_2 with B_2 and Activation of rA_2B_2

Human factor XIII (FXIII) is purified from outdated blood bank plasma.[34,36,37] The recombinant protein rA_2[31] is stored at 4° in a buffer of 75 mM Tris-HCl, 1 mM glycine, 0.5 mM EDTA, and 0.2% sucrose, pH 7.5. The carrier B_2 subunits are isolated from human plasma according to Lorand et al.[34] Human α-thrombin (a gift of Dr. J. W. Fenton II, of the New York State Department of Health, Albany, NY) is stored in 75 mM Tris-HCl, pH 7.5 (500 NIH units/ml). Protein concentrations[38] are expressed in terms of a Pierce Chemical Co. (Rockford, IL) bovine serum albumin standard.

Nondenaturing Electrophoresis Showing Complex Formation between rA_2 and B_2

Nondenaturing electrophoresis was employed earlier for studying the heterologous dissociation of the FXIII ensemble during activation of the zymogen,[20] and the same method has been used for demonstrating the phenomenon of association between the recombinant rA_2 and the native B_2 subunits. Experiments are performed at 15° employing the PhastSystem (Pharmacia, Uppsala, Sweden: Separation Technique File No. 120), using gels with a continuous gradient from 8 to 25% and 2% cross-linking. The

[31] P. D. Bishop, D. C. Teller, R. A. Smith, G. W. Lasser, T. Gilbert, and R. L. Seale, Biochemistry 29, 1861 (1990).
[32] U. Rinas, B. Risse, R. Jaenicke, M. Bröker, H. E. Karges, H. A. Küpper, and G. Zettlmeissl, Bio Technology 8, 543 (1990).
[33] P. Jagadeeswaran and P. Haas, Gene 86, 279 (1990).
[34] L. Lorand, R. B. Credo, and T. J. Janus, this series, Vol. 80, p. 333.
[35] T. J. Janus, R. B. Credo, C. G. Curtis, L. Haggroth, and L. Lorand, Fed Proc., Fed. Am. Soc. Exp. Biol. 40, 1585, Abstr. 258 (1981).
[36] L. Lorand and T. Gotoh, this series, Vol. 19, p. 770.
[37] C. G. Curtis and L. Lorand, this series, Vol. 45, p. 177.
[38] O. H. Lowry, N. J. Rosebrough, A. L. Farr, and R. Randall, J. Biol. Chem. 193, 265 (1951).

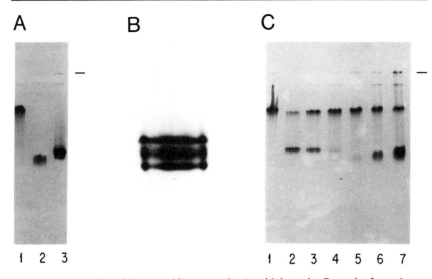

FIG. 1. Association of the recombinant protein rA_2 with its native B_2 carrier from plasma, studied by nondenaturing electrophoresis. (A) Lane 1, reference plasma factor XIII (2.4 pmol); lane 2, reference B_2 protein purified from human plasma (1.5 pmol); lane 3, rA_2 (ZymoGenetics, Seattle, WA; 1.5 pmol). (B) Enlargement of the band seen with specimens of rA_2 shows evidence of microheterogeneity for the recombinant product. (C) Lane 1, reference human plasma factor XIII (2.4 pmol); lanes 2–6, analysis of mixtures of 1.5 pmol of rA_2 with varying amounts of B_2 of 0.4, 0.8, 1.0, 1.5, and 6.0 pmol, respectively. Formation of the rA_2B_2 structural equivalent of factor XIII seems to be optimal in lanes 4 and 5. In addition, higher ordered complexes (rA_2B_n, where $n > 2$) form in the presence of excess B_2 (lanes 6 and 7). Reprinted with permission from *Biochemistry* **32**, 3527 (1993); Copyright [1993] American Chemical Society.

gel buffer is 0.112 M Tris at pH 6.4; the upper and lower buffer strips are 2% agarose in 0.88 M L-alanine and 0.25 M Tris, pH 8.8. Samples are prepared in 2.5 mM glycine, 1.3 mM EDTA, and 0.5% sucrose at pH 7.6, with 0.01% bromphenol blue added. Separations are obtained with approximately 6 mA current (after 325 V-hr). Mixtures of rA_2 and B_2 subunits are incubated for 15 min at room temperature prior to application to the gels.

Under the above conditions, the reference specimen of FXIII (i.e., A_2B_2) moves much slower (Fig. 1A, lane 1) than either B_2 (lane 2) or rA_2 (lane 3); B_2 is slightly ahead of rA_2. Higher magnification of the stained band in lane 3 reveals a certain degree of microheterogeneity within rA_2 (Fig. 1B). Whatever consequences this circumstance might entail for the potential enzyme activity that might be generated from the rA_2 material (i.e., are all subspecies of rA_2 equally active?), as demonstrated by the results presented in Fig. 1C, all components in the preparation are able to associate with native B_2 subunits. When rA_2 and B_2 are mixed prior to electrophoresis in various mole ratios (i.e., 1.5 pmol of rA_2 to 0.4, 0.8, 1.0,

1.5, 3.0, and 6.0 pmol of B_2, shown in Fig. 1C, lanes 2–7), a slow-moving species indistinguishable from the reference FXIII (Fig. 1C, lane 1) appears. In lanes 2 and 3 (Fig. 1C) an excess of rA_2 is evident, whereas in lanes 6 and 7 (Fig. 1C) free B_2 remains, suggesting that conditions for the one-to-one complexing of rA_2 and B_2 to yield rA_2B_2 are best met with the mixtures shown in lanes 4 and 5 (Fig. 1C). However, with excess carrier B_2, the electrophoretic gels (lanes 6 and 7, Fig 1C) also show the presence of higher orders of oligomers (rA_2B_n, where $n > 2$). The possibility of forming such unusual complexes must be carefully analyzed prior to considering use of the rA_2 product for infusion into patients as a substitute for factor XIII.

HPLC Gel Filtration Demonstrating Association of rA_2 with B_2

Complex formation between rA_2 and B_2 can also be readily demonstrated by HPLC, using two 0.75×30-cm TSK-3000 SW columns in series, equipped with a 0.75×10-cm GSWP guard column (Beckman Industries, Berkeley, CA). Samples ($5–10~\mu l$) are injected through a Beckman 210 injector, and elution is performed at 0.8 ml/min with a buffer of 20 mM sodium acetate, 0.15 M sodium chloride at pH 6.5 using a LDC minipump (Rainin Instrument Co., Woburn, MA). The effluent is monitored at 220 nm with a Waters Associates 450 variable-wavelength detector (Milford, MA) connected to a Hewlett-Packard 3390A Reporting Integrator (Palo Alto, CA), set to an attenuation of 6. The premixed samples are incubated for 35 min at room temperature prior to analysis.

Peaks which elute from the HPLC column are collected and analyzed by SDS-PAGE. Samples are first dialyzed against three changes of water for 15 hr at 4° and then brought to dryness with a Speed Vac Concentrator (Savant Instruments, Inc., Hicksville, NY). Electrophoresis is performed in the PhastSystem (Pharmacia Separation Technique File No. 110) using gels with a continuous gradient from 10 to 15% and 2% cross-linking. The gel buffer is 0.112 M sodium acetate and 0.112 M Tris, pH 6.4, while the upper and lower buffer strips are 2% agarose containing 0.2 M Tricine, 0.2 M Tris (pH 7.5), and 0.55% SDS. Samples are prepared in 10 mM Tris-HCl, 1 mM EDTA, pH 8.0, with 2.5% SDS, and are heated to 100° for 10 min prior to application to the gel. Electrophoretic separations are carried out at about 10 mA (70 V-hr). The gel is silver stained for protein using the photochemical procedure of Oakley *et al.*[39] and the Bio-Rad Silver Stain Kit (Bio-Rad, Richmond, CA).

The gel filtration data are presented in Fig. 2, and these, too, prove the high affinity of rA_2 for B_2. When 31.2 pmol of rA_2 is mixed prior to injection with increasing amounts of B_2 (6.2, 15.6 and 31.2 pmol, shown in

[39] B. R. Oakley, D. R. Kirsch, and N. R. Morris, *Anal. Biochem.* 105, 361 (1980).

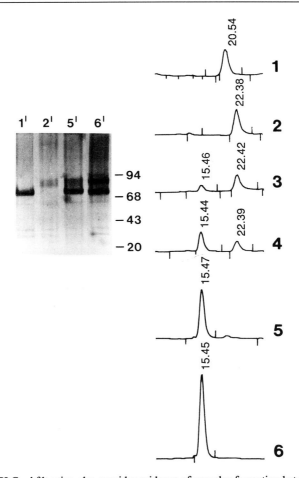

FIG. 2. HPLC gel filtration also provides evidence of complex formation between rA_2 and native B_2. Elution profiles 1 and 2 pertain to 31.2 pmol of the B_2 and rA_2 proteins, whereas profiles 3–5 represent mixtures of 31.2 pmol of rA_2 with increasing amounts of B_2, i.e., 6.2, 15.6, and 31.2 pmol, respectively. A reference profile for human plasma factor XIII is shown in graph 6. The SDS–PAGE patterns on the left (indicated with molecular weight markers $\times 10^{-3}$) reflect the protein composition of some of the peaks collected from the HPLC column: lane 1′, the 20.54-min peak from profile 1; lane 2′, the 22.38-min peak from profile 2; lane 5′, the 15.47-min peak from profile 5; and lane 6′, the 15.45-min peak from profile 6. The protein composition of the complex of rA_2 and B_2, isolated as in profile 5, seems to be qualitatively similar to that of human plasma factor XIII. Reprinted with permission from *Biochemistry* **32**, 3527 (1993); Copyright [1993] American Chemical Society.

graphs 3–5, Fig. 2), there is a progressive generation of a faster emerging peak (~15.5 min), which, in fact, elutes at the position of reference FXIII (graph 6, Fig. 2). Moreover, as the mole ratio of B_2 to rA_2 increases, the free rA_2 (which, according to graph 2, Fig. 2, would appear at ~22.3 min) diminishes in graphs 3 and 4 (Fig. 2), and essentially disappears with the approximately equimolar mixing of rA_2 and B_2 in graph 5 (Fig. 2). However, a very small peak corresponding to the position of free B_2 (~ 20.5 min, see graph 1, Fig. 2) can still be seen in graph 5 (Fig. 2).

SDS-PAGE analysis (Fig. 2, inset) confirms that the new ensemble created by the near equimolar mixing of rA_2 and B_2, and isolated by gel filtration chromatography, really contains both types of subunits.

Immunoblotting and ELISA Procedures for Measuring Binding of rA_2 to B_2

Immunoblotting

Various amounts of rA_2 (0–1700 ng) are spotted onto a 0.6 × 9.5-cm nitrocellulose sheet (BA83, 0.2-μm pore; Schleicher and Schuell, Keene, NH). Prior to overlaying with B_2 subunits, the sheet is immersed for two periods of 20 min in 4 ml of 0.5% bovine serum albumin (BSA) in phosphate-buffered saline (PBS; 10 mM sodium phosphate buffer plus 140 mM NaCl, pH 7.4). It is then incubated for 1 hr with a 3-ml solution of 0.5% BSA in PBS containing 30 μg of B_2. This is followed by washing in 4 ml of 0.5% BSA in PBS (3×, 10 min each time). Bound B_2 is assayed with a polyclonal rabbit antibody (IgG fraction) raised against B_2. Dilutions of this antibody (in 4 ml of PBS with 0.5% BSA) are applied overnight and the sheet is then washed with 4 ml of 0.5% BSA in PBS (3×, 10 min each time). An antirabbit IgG–alkaline phosphatase-conjugated goat IgG (Promega, Madison, WI) is used as secondary antibody (1:5000 in 4 ml of PBS with 0.5% BSA, incubated for 1 hr). The paper is washed (3×, 10 min each time) with BSA–PBS and rinsed with alkaline phosphatase buffer (100 mM Tris-HCl, 100 mM NaCl, 5 mM MgCl$_2$, pH 9.5). Color development[40] is initiated by immersion in 0.37 mM 5-bromo-4-chloro-3-indolyl phosphate and 0.39 mM nitro blue tetrazolium in alkaline phosphatase buffer and is stopped by placing the sheet in water.

With a 1:2,000,000 dilution of the rabbit antibody to B_2, binding of B_2 to rA_2 (≥ 50 ng) can be detected. The control sheet without B_2 shows no reaction.

ELISA

Wells of a microtiter plate are coated with rA_2 (1 μg in 100 μl of PBS) for 2 hr at room temperature on an orbit shaker (Lab Line, Melrose Park, IL). Simultaneously, B_2 (300 ng in 50 μl of PBS) is mixed with 50 μl of 0–500 ng of rA_2 and incubated for 30 min at room temperature on the orbit shaker. Each mixture B_2 with rA_2 is diluted with 900 μl of 0.5% BSA in PBS and 100-μl aliquots are applied to the rA_2-coated wells. After 1 hr,

[40] M. S. Blake, K. H. Johnston, G. J. Russell-Jones, and E. C. Gotschlich, *Anal. Biochem.* **136**, 175 (1984).

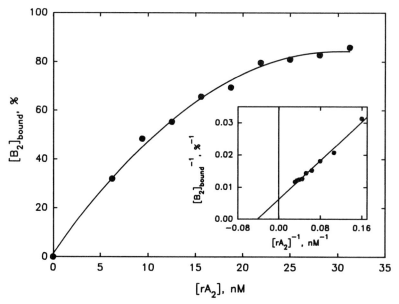

FIG. 3. An ELISA procedure for measuring the association of rA_2 with B_2. Various amounts of rA_2 (0–500 ng) were incubated with a fixed amount of B_2 (300 ng) for 30 min at room temperature. The concentrations of rA_2 in mixtures with B_2 are given on the abscissa of graph. The residual amount of free B_2 remaining in the various mixtures with rA_2 was determined by the immunoassay described in the text and, from this, the percentage of B_2 complexed to rA_2 was calculated. This is presented on the ordinate as $[B_2]_{bound}$ after normalizing measurements for development of color without rA_2. Wells with 0.5% BSA in PBS instead of rA_2B_2 mixture were used as blanks. Inset: Double-reciprocal presentation of the data from which an apparent binding constant of $4 \times 10^7 M^{-1}$ could be derived for the association of rA_2 with B_2. Reprinted with permission from *Biochemistry* **32**, 3527 (1993); Copyright [1993] American Chemical Society.

the wells are washed (3×, 5 min each time) with 200 μl of 0.5% BSA in PBS. Rabbit antibody to B_2 subunit is diluted 1:2,000,000 with 0.5% BSA in PBS and 200 μl of the diluted antibody is added to each well. The plate is incubated for 1 hr and washed (3×, 5 min each time) with 0.5% BSA in PBS. This is followed by a 1-hr incubation (200 μl per well) with a goat antirabbit IgG–alkaline phosphatase conjugate diluted 1:5,000 with 0.5% BSA in PBS. The plate is washed (2×, 5 min each time) with 0.5% BSA in PBS and rinsed with alkaline phosphatase buffer (100 mM Tris-HCl/ 100 mM NaCl/5 mM MgCl$_2$, pH 9.5; 200 μl per well).[40] One hundred microliters of the p-nitrophenyl phosphate substrate (Sigma, St. Louis, MO; dissolved to 1 mg/ml in the alkaline phosphatase buffer) is added to each well. After 40 min at room temperature, absorbancy is measured at 410 nm on a Dynatech plate reader (Chantilly, VA) (Fig. 3).

Fluorescence Depolarization Studies

Earlier experiments, using fluorescein isothiocyanate (FITC)-labeled plasma factor XIII (i.e., $A_2^F B_2^F$), showed the applicability of depolarization techniques for studying the thrombin plus Ca^{2+}-dependent dissociation of the labeled zymogen ensemble[41]: $A_2^F B_2^F$. We have now explored this methodology to examine the association of the recombinant rA_2 protein, as well as its thrombin-modified form, rA_2', with the FITC-labeled native plasma carrier,[42,43] B_2^F. The approach of labeling only one type of subunit allows for a less ambiguous interpretation of the dissociation of $rA_2 B_2^F$ with thrombin plus Ca^{2+} and that of $rA_2' B_2^F$ with Ca^{2+}.

The B_2 subunits (200–300 μg) are dialyzed against 100 mM Tris-HCl and 1 mM EDTA at pH 9.0 overnight at 4°. A sixfold molar excess of FITC (Sigma Chemical Co., St. Louis, MO) is allowed to react[41] with the protein for 90 min at 23° in the dark, and the unbound reagent is removed by filtration through a Sephadex G-50 column (1 × 17 cm; medium-grade beads; Sigma), equilibrated in 100 mM Tris-HCl, 1 mM EDTA, pH 7.5. The labeled protein is then concentrated on an Amicon PM30 filter (Amicon Corp., Lexington, MA) and is dialyzed overnight against 75 mM Tris-HCl, 1 mM EDTA, pH 7.5, at 4° with 2% activated charcoal (Norit A, MCB, Norwood, OH).

Labeling efficiency (~3 mol of fluorescein per mole of B_2 of 2 × 80,000) is estimated by measuring A_{490} using an extinction coefficient of 3.4×10^4 liter mol^{-1} cm^{-1}.[44]

Fluorescence polarization measurements are carried out on an SLM 8000C double-emission spectrofluorometer (SLM Aminco, Urbana, IL) equipped with Glan–Thompson calcite prism polarizers.

The labeled proteins (10–20 μg/ml in a 2-ml volume) are placed in a 1 × 1-cm quartz cuvette at 37° and stirred at ~1500 rpm. Additions are made at specified times through an injection port with a glass syringe (Hamilton, Reno, NV). Total volume is not allowed to increase from the initial 2 ml by more than 50 μl on additions. Excitation is fixed at 490 nm and emissions are measured at 530 nm. Polarizers in the two emission channels are set perpendicular to each other; only the polarizer on the excitation side is rotated.

Polarization (P) is calculated using the equation $P = [(R_V/R_H) - 1]/[(R_V/R_H) + 1]$, where R_V is the ratio of intensities in the emission channels when the excitation polarizer is in the vertical position and R_H is the ratio

---

[41] J.-M. Freyssinet, B. A. Lewis, J. J. Holbrook, and J. D. Shore, *Biochem. J.* **169**, 403 (1978).
[42] J. T. Radek, J. Wilson, and L. Lorand, *FASEB J.* **4(7)**, Abstr. No. 3376 (1990).
[43] J. T. Radek and L. Lorand, *Biophys. J.* **61**, Abstr. No. 1891 (1992).
[44] J. E. Churchich, *Biochim. Biophys. Acta* **147**, 511 (1967).

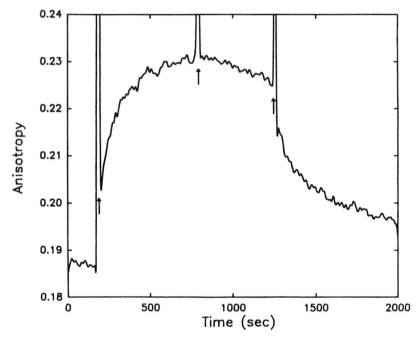

FIG. 4. A fluorescence depolarization study of the association of rA_2 with fluorescein-labeled B_2 (i.e., B_2^F) to form the $rA_2B_2^F$ equivalent of plasma factor XIII; heterologous dissociation of the zymogen with Ca^{2+} following treatment by thrombin. A 75 mM Tris-HCl buffer with 0.15 M NaCl and 2 mM EDTA at 37° was used. B_2^F (47 nM) was mixed with rA_2 (48 nM; arrow at 180 sec). Human thrombin (5 NIH units) was injected first (at 800 sec), followed by 30 mM Ca^{2+} (at 1250 sec).

when the excitation polarizer is set in the horizontal position. Anisotropy is defined as $2P/(3 - P)$. Data are subjected to a smoothing routine as recommended by the SLM Operator's Manual by using 10 passes of a fixed bandwidth (16 nm), sharp cutoff, and three-point low-pass linear digital filter. An IBM Personal System 2-Model 50Z computer, with software provided by SLM Aminco, is employed for data collection and storage.

Adding increasing amounts of rA_2 to a fixed amount of B_2^F gives rise to stepwise increases in fluorescence anisotropy, reaching a point of saturation with a mixture of approximately 1 mol of rA_2 to B_2^F, clearly indicating the formation of $rA_2B_2^F$. The illustration in Fig. 4 pertains to the situation of mixing rA_2 and B_2^F in equimolar amounts.

The addition of thrombin does not cause a significant change in anisotropy until the subsequent injection of Ca^{2+}, which makes the anisotropy value of the system gradually drop to near that of the starting B_2^F alone. This finding confirms the idea that the hydrolytic conversion of $rA_2B_2^F$ by thrombin to $rA_2'B_2^F$ does not by itself lead to the heterologous dissociation of subunits, which, however, occurs readily by subsequent exposure to Ca^{2+}. (It should be noted that the same concentration of Ca^{2+} does not produce such an effect on the $rA_2B_2^F$ ensemble.)

Quantitative measurements for the equilibrium $rA_2 + B_2^F \rightleftharpoons rA_2B_2^F$ yield an association constant of $5.3 \times 10^7 \ M^{-1}$, whereas the much lower value of $3.4 \times 10^6 \ M^{-1}$ is obtained for the binding of the thrombin-modified rA_2' to B_2^F: $rA_2' + B_2^F \rightleftharpoons rA_2'B_2^F$. Thus, the removal of N-terminal activation peptides from the A subunits by thrombin would weaken the stability of subunits of the factor XIII zymogen by about 2 kcal/mol. This loss of stability is also reflected in a markedly increased sensitivity of the thrombin-modified $rA_2'B_2^F$ structure to dissociate in the presence of Ca^{2+}.

Interestingly, removal of the sialic acid residues from the B subunits by neuraminidase[1,34] does not affect the ability of the desialylated B_2^F to bind rA_2.

Thrombin-Independent Mode of Activating Factor XIII[44a]

Breaking the heterologous association of subunits seems to be the key for activating factor XIII. In the physiological conversion ($\mu = 0.15$), the limited proteolytic attack by thrombin produces an intermediate (XIII' = $A_2'B_2$) in which this heterologous attraction is considerably weaker, so that, in the presence of fibrin[ogen], dissociation and unmasking of the active center cysteine ensues at the 1.5 mM concentration of free Ca^{2+} in plasma.[27] However, purified human plasma factor XIII can become activated to express full transamidating enzyme activity even without the prior removal of the N-terminal AP segments from the A subunits by thrombin. Exposure of the zymogen to high enough concentrations of Ca^{2+} (0.1 M) alone is sufficient to bring about this conversion.[26,34,35] Specificity for Ca^{2+} was demonstrated by the markedly greater efficacy of Ca^{2+} over Ba^{2+} and Sr^{2+} ($\sim$7-fold) and Mg^{2+} ($>$ 10-fold). The Ca^{2+} requirement can be substantially reduced and the thrombin-independent conversion of the zymogen can be highly accelerated by the addition of solutes often referred to as "chaotropic," with the relative efficacies of p-toluene sulfonate $\geq$ thiocyanate $>$ iodide $>$ bromide. At 37°, pH 7.5, and with 0.05 M Ca^{2+} and 0.2 M p-toluene sulfonate ($\mu = 0.4$) or KSCN ($\mu = 1.2$), for example, complete conversion is achieved in about 10 min. Because, under these conditions, the AP moieties would still remain attached to the A subunits, the process should be distinguished from the thrombin-dependent physiological pathway of producing the A_2^* enzyme. Therefore, the active species, generated in the thrombin-independent direct conversion of factor XIII, is

---

[44a] This section is prepared from the work of C. G. Curtis, R. B. Credo, T. J. Janus, and L. Lorand.

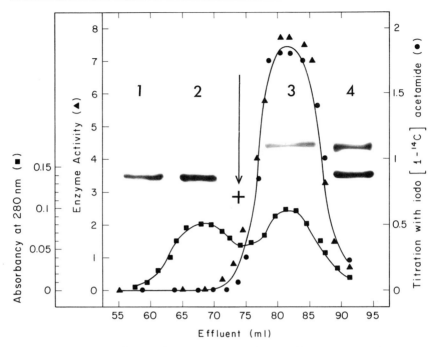

FIG. 5. Separation of the native A_2 and B_2 constituents of human plasma factor XIII, following dissociation of the A_2B_2 zymogen by Ca^{2+} at high ionic strength. Chromatography of factor XIII (7.6×10^{-9} mol) was performed following a brief exposure (5 min at 37°) of the protein to 0.5 M Ca^{2+}. Two tandem columns (0.9×95 cm) of Sepharose 6B were equilibrated with a solution of 0.01 M $CaCl_2$, 0.5 M KCl, and 0.05 M Tris-HCl, pH 7.5. Elution was carried out with the same at 4°. Each fraction was assayed for absorbance at 280 nm, for susceptibility to alkylation with iodoacetamide, and for transamidase activity. Transamidase activity was measured by the incorporation of radioactive putrescine into N,N-dimethylcasein. Prior activation with thrombin (5×10^{-8} M) was carried out for 30 min at 25° in mixtures of 0.075 ml, which contained 0.025 ml of the column fractions, 0.033 M $CaCl_2$, 0.17 M KCl, and 0.05 M Tris-HCl, pH 7.5. Other aliquots of the collected fractions, which were not subjected to activation with thrombin, were treated in an identical manner but with the omission of thrombin. Transamidase activity was then assayed at 37° in solutions of 0.1 ml containing 0.025 ml of the fraction, 3.75×10^{-8} M thrombin, 0.125 M KCl, 1.6×10^{-4} M N,N-dimethylcasein, 3.3×10^{-5} M [1-^{14}C]putrescine, 2.5×10^{-3} M $CaCl_2$, and 0.05 M Tris-HCl, pH 7.5. Aliquots of 10 μl were removed after 30 min to quantitate covalently bound isotope.[45] Prior to alkylation with iodo[1-^{14}C]acetamide, activation with thrombin was performed for 30 min at 25° in solutions of 0.105 ml containing 0.1 ml of the column fraction to be assayed, 3.75×10^{-8} M thrombin, 0.476 M KCl, 9.5×10^{-3} M $CaCl_2$, and 0.05 M Tris-HCl, pH 7.5. Alkylation was then carried out at 37° in solutions of 0.11 ml containing 0.1 ml of the fraction, 3.58×10^{-8} M thrombin, 0.455 M KCl, 9.1×10^{-3} M $CaCl_2$, and 0.05 M Tris-HCl, pH 7.5. After 30 min of reaction time, aliquots of 10 μl were removed to quantitate covalently bound isotope.[19,20] Absorbance at 280 nm (■); transamidase activity given as moles of [1,4-^{14}C]putrescine incorporated in 30 min per mole of N,N-dimethylcasein (▲); susceptibility to alkylation with iodo[1-^{14}C]acetamide expressed as

designated as A_2°. Gel-filtration chromatography showed that heterologous dissociation of subunits took place; however, it could be reversed on removal of Ca^{2+}:

$$A_2B_2 \underset{-Ca^{2+}}{\overset{+Ca^{2+}}{\rightleftharpoons}} A_2^\circ + B_2$$

This mode of dissociating the zymogen can be employed for separating its constituent A_2 and B_2 subunits in their native forms (Fig. 5).

With regard to the titratability of the catalytic cysteine sulfhydryl groups with iodo[^{14}C]acetamide, essentially full unmasking of the active center ($\sim$ 1 mol of thiol per A$^\circ$ subunit) was found. The catalytic nature of A$^\circ$ was proved by the incorporation of isotopic putrescine into N,N-dimethylcasein and also by rigorous measurements with the synthetic substrate pair of β-phenylpropionylthiocholine and methanol.[45] Steady-state analysis of the latter system allowed a meaningful comparison between the catalytic properties of the A_2° species and the A_2^* produced in the thrombin-dependent pathway. Both enzymes displayed purely Michaelian kinetics, without any indication of A$^\circ$ to A$^\circ$ or A* to A* cooperativities. Significantly, apparent Michaelis constants for the acyl group-containing substrate (β-phenylpropionylthiocholine) as well as for the nucleophile (methanol) were identical for the A_2° and A_2^* enzymes, and the molar turnover numbers (k_{cat}) were also indistinguishable. These findings clearly indicate that the presence of the N-terminal segment containing 37 amino acids in the A subunit would not per se prevent expression of catalytic activity. Dissociation of A_2B_2 and the necessary conformational change to generate the active A_2° enzyme can apparently be accomplished by moving this N-terminal flap out of its original position.

Acknowledgments

This work was aided by a USPHS Research Career Award (HL-03512) and by grants from the National Institutes of Health (HL-02212 and HL-16346).

[45] K. N. Parameswaran and L. Lorand, *Biochemistry* 20, 3703 (1981).

---

moles of isotope incorporation in 30 min ($\bullet$). The SDS–PAGE tracks represent purified B subunit as control (track 1), the protein fraction collected at 68 ml of effluent (track 2), the fraction at 82 ml of effluent (track 3), and the parent factor XIII zymogen as control (track 4). Migration is indicated by the vertical arrow, its top representing the origin. It is clear that the first peak (emerging at 68 ml) represents the B_2 carrier protein, devoid of transamidating enzyme activity, whereas the second peak (at 82 ml) corresponds to A_2 with the catalytic potential.

[3] Molecular Approach to Structure – Function Relationship of Human Coagulation Factor XIII

By Akitada Ichinose and Hiroshi Kaetsu

Introduction

Human coagulation factor XIII (fibrin-stabilizing factor, a zymogen of transglutaminase; EC 2.3.2.13, protein-glutamine : amine γ-glutamyltransferase) consists of two a and two b subunits, each of which has unique functions.[1-3] The a subunit contains an active site, releases an activation peptide when activated by thrombin in the presence of calcium ions, binds to fibrin, and forms the a_2 homodimer and a_2b_2 heterotetramer. The b subunit forms the b_2 homodimer and a_2b_2 heterotetramer and stabilizes the a subunit. In order to define the structure – function units or modules in factor XIII, the primary structure of both the a and b subunits was first determined by cDNA cloning and amino acid sequence analysis. Then, the structures of the genes coding for the a and b subunits were determined by isolation and characterization of genomic clones employing cDNAs as probes for screening. Chromosomal localization of the genes for the a and b subunits was determined by both *in situ* hybridization using the cDNAs and somatic cell hybrid analysis with the polymerase chain reaction (PCR). Genomic DNAs from patients with a or b subunit deficiencies have been analyzed by Southern blotting, and PCR and nucleotide sequencing, employing amplification primers which were designed from the genomic nucleotide sequence of each subunit. These experiments were performed in order to find possible defects in the sites and amino acids essential for the functions of factor XIII as well as to understand genetic regulatory mechanisms for its biosynthesis. The cDNAs were also utilized for expression and characterization of recombinant a and b subunits in yeast and mammalian cell systems. A number of deletion mutants for the a subunit lacking each exon have been made by a site-directed mutagenesis method in order to search for the functional regions in the molecule.

[1] J. E. Folk and J. S. Finlayson, *Adv. Protein Chem.* **31**, 1 (1977).

[2] L. Lorand, M. S. Losowsky, and K. J. M. Miloszewski, *Prog. Hemostasis Thromb.* **5**, 245 (1980).

[3] D. Chung and A. Ichinose, *in* "The Metabolic Basis of Inherited Disease" (C. R. Scriver, A. L. Beaudet, W. S. Sly, and D. Velle, eds.), p. 2135. McGraw-Hill, New York, 1989.

Primary Structure of Factor XIII

Preparation of Factor XIII

Factor XIII is purified from human plasma[4] using DEAE-Sephacel or DEAE-Sepharose CL-6B (Pharmacia, Piscataway, NJ) because the elution profiles are sharper than that obtained using DEAE-cellulose. The purified factor XIII is activated by bovine thrombin, and the thrombin is subsequently inactivated by hirudin to prevent further degradation of the activated *a* subunit of factor XIII. The activated *a* subunit and *b* subunit are separated in the presence of calcium ions by gel filtration using the BioGel A-5m column (Bio-Rad, Richmond, CA).[5] The purified subunits are used to raise polyclonal antibodies in rabbits, to make two affinity columns in order to purify the specific antibodies against the *a* and *b* subunits, and to obtain cyanogen bromide fragments for amino acid sequence analysis.

Amino Acid Sequence of b Subunit

Screening and Subcloning

The *b* subunit is selected as the first target because the *b* subunit free from the *a* subunit can be easily obtained by DEAE-Sephacel chromatography.[6] A human liver cDNA library is employed for screening because, at the time of this experiment, both the *a* and *b* subunits were believed to be synthesized in liver. Nine positive clones are isolated from 2 million plaque-forming unit (pfu) phage using the ^{125}I-labeled, affinity-purified antibody against the *b* subunit.[7] cDNA inserts are released by restriction digestion with *Eco*RI from a λgt11 phage vector. Two clones with the largest inserts (2.2 and 1.6 kb; Fig. 1) are selected for further analysis. Each of two *Eco*RI fragments is ligated into a pUC plasmid and M13 sequencing vector with *Eco*RI cloning sites.

Identification of cDNA Clone

A DNA sequencing reaction is performed by the dideoxy method[8] employing the Klenow fragment of DNA *Pol*I (BRL, Gaithersburg, MD). When one of the *Eco*RI ends of the 2.2-kb cDNA is sequenced, a long open reading frame is found to include a stretch of hydrophobic amino acids and a sequence which matches that of the amino terminus of the mature *b*

[4] C. G. Curtis and L. Lorand, this series, Vol. 45, p. 177.
[5] S. I. Chung, M. S. Lewis, and J. E. Folk, *J. Biol. Chem.* **249**, 940 (1974).
[6] A. Ichinose, B. A. McMullen, K. Fujikawa, and E. W. Davie, *Biochemistry* **25**, 4633 (1986).
[7] R. A. Young and R. W. Davis, *Proc. Natl. Acad. Sci. U.S.A.* **80**, 1194 (1983).
[8] F. Sanger, S. Nicklen, and A. R. Coulson, *Proc. Natl. Acad. Sci. U.S.A.* **74**, 5463 (1977).

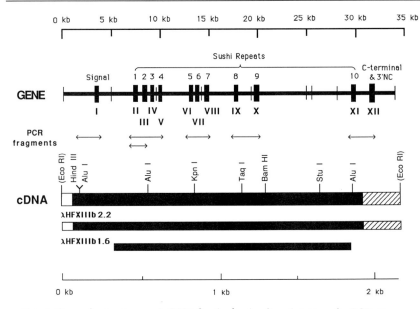

FIG. 1. Genomic structure and cDNA for the *b* subunit and strategy for PCR. Exons are indicated by wide vertical bars and Roman numerals. Vertical lines are *Eco*RI sites in the gene. The *Eco*RI sites in parentheses at the ends of the cDNA were created by a linker for cloning. Horizontal arrows represent DNA fragments amplified by PCR for analysis of genomic DNAs from *b* subunit deficiency. The 5'-noncoding, coding, and 3'-noncoding regions of the cDNA are shown by open, closed, and hatched boxes, respectively. Two cDNA clones contain the 2.2- and 1.6-kb inserts, respectively.

subunit. Nucleotide sequence analyses reveal that the other *Eco*RI end of the 2.2-kb cDNA contains an AATAAA sequence and poly(A) tail. Thus, the 2.2-kb clone turns out to be a nearly full-length cDNA coding for the *b* subunit of factor XIII. On the other hand, both *Eco*RI ends of the 1.6-kb cDNA have open reading frames, and these nucleotide sequences have been found to match the internal sequence of the 2.2-kb clone. Accordingly, the 2.2-kb cDNA has been chosen for further analysis and subcloned into a pUC vector.

Nucleotide Sequence Analysis

Various cDNA fragments are produced by available restriction enzymes, such as *Hin*dIII, *Alu*I, *Kpn*I, *Taq*I, *Bam*HI, and *Stu*I (Fig. 1), and have been sequenced after being ligated into M13 vectors with corresponding subcloning sites. When a convenient restriction site is not available between two sites, a blunt end is created by controlled digestion with *Bal*31 (BRL) in order to produce overlapping sequences. Both strands are sequenced completely by this strategy. DNA sequences are analyzed by the DNA Inspector (Textco, West Lebanon, NH) or Genepro program (Riverside Scientific Enterprises, Seattle, WA) employing an Apple Macintosh or IBM PC/AT-compatible computer. All amino acid sequences of 10 cyanogen bromide fragments of the *b* subunit completely match that of the longest open reading frame of the 2.2-kb cDNA.

The *b* subunit turns out to be composed of 10 tandem repetitive structures called the Sushi domains.[9] These domains are also found in more than 25 additional proteins. Thus, the *b* subunit is a member of one of the largest protein superfamilies.

Amino Acid Sequence of a Subunit

Amino Acid Sequencing

Once activated by thrombin, the *a* subunit becomes unstable, and its yield is relatively low. Therefore, an *a* subunit-enriched fraction is obtained from the purified factor XIII (a_2b_2 tetramer)[10] and is subjected to amino acid sequence analysis. Because the complete sequence of the *b* subunit has already been determined, several cyanogen bromide fragments that originate from the *b* subunit are easily recognized and discarded. As a result, 11 cyanogen bromide fragments are found to be unique. Characterization of full-length cDNAs for the *a* subunit later revealed that these fragments are part of the sequence of the *a* subunit.[10]

cDNA Cloning

A polyclonal antibody against the *a* subunit is made as described above. Because screening of two liver libraries for the *a* subunit failed, a placenta cDNA library was used in the third trial. More than 100 clones in 2 million pfu phage gave a positive signal with a ^{125}I-labeled antibody. Therefore, a total of 36 clones are selected by intensities of signals (12 each of strong, medium, and weak), and are dot-blotted onto a nitrocellulose membrane.

Detection of 5' End Clones

In order to find cDNA clones which extend to the 5' end or near the amino terminus, the membrane is hybridized with two oligonucleotide probes that were designed from the amino acid sequence in the activation peptide of the *a* subunit.[11] Six clones are positive with one (5'-

[9] A. Ichinose, R. E. Bottenus, and E. W. Davie, *J. Biol. Chem.* **265**, 13411 (1990).
[10] A. Ichinose, L. E. Hendrickson, K. Fujikawa, and E. W. Davie, *Biochemistry* **25**, 6900 (1986).
[11] T. Takagi and R. F. Doolittle, *Biochemistry* **13**, 750 (1974).

CTCCACGGTGGGCAGGTCGTCCTCG-3′) of two ^{32}P-labeled oligonucleotide probes, and turn out to share a nucleotide sequence with an open
reading frame including a Met residue and the amino acid sequence of the
mature amino terminus of the *a* subunit. Four of these six clones span
about 3.8 kb and are composed of four *Eco*RI fragments. Therefore, each
of the four *Eco*RI fragments is subcloned into pUC vectors and various
lengths of cDNA fragments are produced by controlled digestion with the
*Bal*31 enzyme in order to obtain overlapping sequences. Three *Eco*RI
junctions of the cDNA are confirmed by producing and sequencing three
unique fragments containing *Eco*RI sites inside.

Search for Possible Signal Peptide

Because these cDNAs for the *a* subunit lack a signal peptide for secretion from cells, 12 additional clones are isolated from the remaining
positive clones employing the ^{32}P-labeled *Eco*RI–*Sma*I fragment (0.2 kb)
obtained from the 5′ end of the cDNA. However, nucleotide sequences of
all of the 12 clones start from similar positions in the 5′-noncoding region,
which is followed by the coding sequence for an initiator Met and the
amino terminus of the *a* subunit. Although nonactivated and activated
monocyte cDNA libraries (2 million pfu each) have also been screened
with a probe of the 1.2-kb *Eco*RI fragment, no positive clones extend to the
coding sequence for the activation peptide. Thus, no cDNA clones extend
farther upstream of the 5′ ends of the existing clones, and no signal peptide
for secretion has been found.

The amino acid sequence of the *a* subunit has also been reported by
two other groups.[12,13] The unique sequence of the *a* subunit of factor XIII
is homologous to tissue transglutaminase, red blood cell band 4.2 protein,
and epidermal transglutaminase.

Genomic Structure of Factor XIII

Gene for a Subunit

First Screening with All cDNA Fragments

In order to search for a possible exon coding for a signal peptide
upstream of the 5′ end of the cDNA for the *a* subunit and to define its
exons and structural units, the gene for the *a* subunit has been character-

---

[12] U. Grundmann, E. Amann, G. Zettlemeissl, and H. A. Kupper, *Proc. Natl. Acad. Sci.
U.S.A.* **83,** 8024 (1986).
[13] N. Takahashi, Y. Takahashi, and F. W. Putnam, *Proc. Natl. Acad. Sci. U.S.A.* **83,** 8019
(1986).

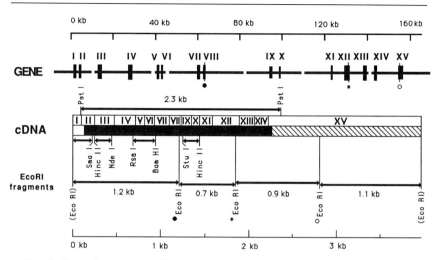

FIG. 2. Genomic structure, cDNA, and cDNA fragments of the *a* subunit. Exons are indicated by wide vertical bars and Roman numerals. Vertical lines with closed circles, asterisks, and open circles correspond to the *Eco*RI sites in the cDNA and the gene for the *a* subunit. Horizontal arrows represent cDNA fragments employed for screening of genomic libraries, for Southern blotting, and for construction of an expression vector. The 5'-noncoding, coding, and 3'-noncoding regions of the cDNA are shown by an open, closed, and hatched boxes, respectively.

ized.[14] A commercial human leukocyte genomic library (2.8 million pfu) is screened first by employing four *Eco*RI fragments (Fig. 2) that are ^{32}P-labeled in a mixture. All 12 isolated clones hybridize with the 1.1- and 0.9-kb *Eco*RI fragments, but not with the 0.2-kb *Eco*RI–*Sma*I fragment derived from the 5' end of the cDNA as shown by dot-blotting analysis. The 12 clones are digested with several restriction enzymes, including *Eco*RI and *Sal*I, and are electrophoresed and transferred onto a nylon membrane. Exon-containing fragments are detected by Southern blotting analysis using each of four ^{32}P-labeled *Eco*RI fragments; however, only the 1.1- and 0.9-kb *Eco*RI fragments give positive bands. The restriction digestion pattern and Southern blotting analysis indicate that there are three unique clones among the 12. The exon-containing fragments are ligated into pUC and M13 sequencing vectors with the *Eco*RI and/or *Sal*I cloning sites. Nucleotide sequence analysis reveals that the three unique clones contain only exons XIV and/or XV, confirming that the third and fourth *Eco*RI fragments of 0.9 and 1.1 kb hybridize with the positive phage plaques most efficiently.

[14] A. Ichinose and E. W. Davie, *Proc. Natl. Acad. Sci. U.S.A.* **85,** 5829 (1988).

Second Screening Employing Selected Probes

The first and second *Eco*RI fragments of 1.2 and 0.7 kb, respectively, are ^{32}P-labeled separately and are used as probes separately for the next screening of the same library. By employing a combination of the 1.2-kb *Eco*RI probe and its 0.2-kb *Eco*RI–*Sma*I fragment (Fig. 2), eight unique clones containing exons I, II, IV, VII, and VIII are isolated, while only one unique clone containing exon XI is obtained by screening with the 0.7-kb *Eco*RI probe. Thus, exons III, V, VI, IX, X, XII, and XIII are missing.

Change of Libraries and Use of Exon-Specific Probes

Exons XII and XIII in four unique clones are obtained by rescreening of the leukocyte library (one clone) and a human fibroblast genomic library (three clones), with the 0.7-kb *Eco*RI probe. In order to search selectively for exons III, V and VI, and IX and X, the *Hin*cII–*Nde*I fragment (0.16 kb) and *Rsa*I–*Bam*HI (0.29 kb) are produced from the 1.2-kb *Eco*RI fragment, and the *Stu*I–*Hin*cII fragment (0.16 kb) is made from the 0.7-kb *Eco*RI fragment. These three probes are separately ^{32}P labeled and used for screening of the fibroblast genomic library and a fetal liver genomic library. Exon III is found in one unique clone obtained from the fibroblast library, and exons IX and X are found in two unique clones from the fetal liver library. Although five additional unique clones have also been isolated from the fibroblast library, these clones code for exons IV (three clones) and VII and/or VIII (two clones).

Size Estimate of Missing Part and Use of Size-Fractionated Library

At this point, exons V and VI are still missing. In order to estimate the sizes of *Eco*RI fragments containing these two exons, genomic Southern blotting analysis is performed after genomic DNAs are digested with *Eco*RI and are blotted onto a nylon membrane. A single band of about 4.5 kb hybridizes with the radiolabeled *Rsa*I–*Bam*HI fragment. Therefore, a size-fractionated library (3.9–5.0 kb) is screened using the *Rsa*I–*Bam*HI fragment as a probe. Finally, one unique clone is isolated and confirmed to contain exons V and VI by nucleotide sequence analysis. All 15 exons and their boundaries are sequenced completely on both strands. A total 93 clones are isolated and 25 are unique. These clones cover about 160 kb of the gene for the *a* subunit; however, there are gaps in several introns (Fig. 2).

Gene Coding for b Subunit

First Screening

The gene coding for the *b* subunit has been characterized in order to define structural and genetic repetitive units and to search for a possible genomic sequence(s) that would account for the duplication of the 10 tandem Sushi domains. A commercial leukocyte genomic library is screened by using a ^{32}P-labeled 2.2-kb cDNA coding for the b subunit. Two unique clones are isolated first and contain 5'- and 3'-end exons of the gene for the b subunit,[15] respectively.

Second Screening

Because exons for the middle portion of the gene are missing, two other genomic libraries, a human fibroblast library and human fetal liver library, are employed for the second screening. Four and six additional clones are isolated from the fetal liver and fibroblast libraries, respectively. Four unique clones among 12 are identified by the restriction digest patterns, and are further characterized by Southern blotting analysis employing ^{32}P-labeled cDNA fragments and oligonucleotides and by nucleotide sequence analysis of all *Eco*RI fragments. The four clones cover the entire gene for the b subunit, including all 12 exons and 11 introns. Continuity of the *Eco*RI fragments is confirmed by sequencing analysis of the DNA fragments obtained either by cleavage at other restriction sites outside the *Eco*RI sites or by PCR[16] employing pairs of primers designed from the genomic sequence. A total 33,207 nucleotides have been determined, and about 54% of the sequence was confirmed on both strands.

This study revealed that each of 10 Sushi domains in the b subunit is coded by a single exon, which is also the case with nearly all genes known to contain Sushi domains thus far. Although many repetitive sequences have been identified in the gene for the b subunit, none of them explain how duplication of the Sushi domains occurred.

Chromosomal Localization of Genes for a and b Subunits

In Situ Hybridization

In order to investigate possible relationships between the gene loci for the a and b subunits, chromosomal localization of both genes is performed. The cDNAs for the a subunit (1.2-kb *Eco*RI fragment) and the b subunit (2.2-kb *Eco*RI fragment) are labeled with tritiated dATP, dCTP, and dTTP (Amersham, Piscataway, NJ) by nick translation, and are hybridized separately *in situ* to chromosome preparations banded with 5-bromodeoxyuridine.[17,18] Probed slides are autoradiographed for 6–7 days. The genes for

[15] R. E. Bottenus, A. Ichinose, and E. W. Davie, *Biochemistry* **29**, 11195 (1990).
[16] R. K. Saiki, D. H. Belfand, S. Stoffel, S. J. Scharf, R. Higuchi, G. T. Horn, K. B. Mullis, and H. A. Erlich, *Science* **239**, 487 (1988).
[17] P. G. Board, G. C. Webb, J. McKee, and A. Ichinose, *Cytogenet. Cell Genet.* **48**, 25 (1988).
[18] G. C. Webb, M. Coggan, A. Ichinose, and P. G. Board, *Hum. Genet.* **81**, 157 (1989).

the *a* and *b* subunits are localized to chromosome bands 6p24-25 and 1q31-32.1, respectively.

In Vitro Amplification

Because cDNA probes may cross-hybridize to other genes coding for homologous proteins, such as other transglutaminases and Sushi domain-containing proteins, the chromosomal localization of the genes for the *a* and *b* subunits has been reexamined by PCR employing gene-specific primers and human–hamster hybrid cells. The gene-specific primers are designed from the genomic DNA sequences[14,15]: for exon XV of the *a* subunit, 5′ side, GATCAAGCTTCCGAACCTCTCCTCTCTTTTC, and 3′ side, CAATAAGCTTGAGTGTTGCACCTGCTTTCTT; for exons XI and XII of the *b* subunit, 5′ side, CAAGAATTCCTGTTCT-TACCAAGATGTAACAAG, and 3′ side, TATGAATTCTAGTCAATG-GGCATTAGGAAATGA; for exons II and III of the *b* subunit, 5′ side, ATTGGATCCTGATTACAAATTTATGTTTTTAGATTTG,   and   3′ side, TCAGGATCCTGCATTGTAGACATAATGAAAAATAA. In principle, the amplification primers are approximately 35-mers and their corresponding melting temperatures are set at 60–68°. Convenient restriction sites (underlined) are added for subcloning at the 5′ ends of the primers downstream of a protector of three to four nucleotides against possible exonuclease activity. These oligonucleotides are prepared with a synthesizer purchased from Applied Biosystems (Foster City, CA). Genomic DNA samples (0.1 μg) from human, hamster, and the human–hamster hybrid cell lines (Panel I, BIOS Corp., New Haven, CT) are amplified in 50-μl reaction mixtures employing 1.5–2.5 units of *Thermus aquaticus (Taq)* DNA polymerase (Perkin-Elmer Cetus, Norwalk, CT or Stratagene, La Jolla, CA).

Gene for *a* Subunit

After 30 cycles of amplification at 67° (annealing) for exon XV of the *a* subunit (Fig. 2), 10 μl of each reaction mixture is applied to an 0.8% (w/v) agarose gel containing ethidium bromide and is electrophoresed at 80 V for 2 hr. Buffer and water are added as negative controls without genomic DNA. A single band of 0.9 kb is obtained from the human DNA and from DNAs of the 860 and 909 somatic cell hybrids, which contain human chromosome 6, while no bands are observed with the hamster DNA and DNAs of other cell lines (Fig. 3, top). Another set of cell hybrids (Panel II, BIOS Corp.) was also examined, and the same band was obtained only from the human DNA and DNAs of the 756, 860, 904, and 909 cell lines,

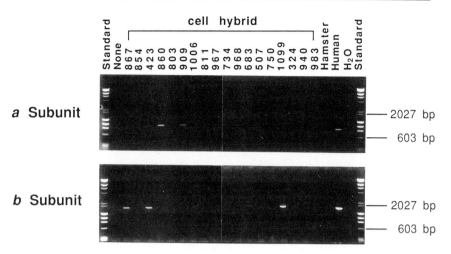

Fig. 3. Chromosomal localization of the genes for the *a* and *b* subunits. Amplified products of genomic DNA samples from human, hamster, and human–hamster hybrid cell lines [Panel I (see text); shown by numbers] were electrophoresed in an 0.8% agarose gel. "None" and "H₂O" are negative controls without DNA.

which contain human chromosome 6. Thus, the gene for the *a* subunit has been localized to chromosome 6 with 0% discordance (0 out of 25 cell lines), which is consistent with the result obtained by *in situ* hybridization.

Gene for *b* Subunit

The DNA samples of Panel I are amplified 30 times at 62° (annealing) by employing the primers for exons XI and XII (Fig. 1). A single band of 1.7 kb is obtained from the human DNA and DNAs of the 867 and 1099 somatic cell hybrids, which contain human chromosome 1 (Fig. 3, bottom). However, this band is also observed with DNAs of the 423 and 683 cell lines, which contain human chromosome 3 and several chromosomes, respectively. Another set of the cell hybrids (Panel II) was also examined and the same band was obtained from DNAs of the 867 and 937 cell lines, which contain human chromosome 1, and from DNA of the 683 cell line.

The experiment is repeated by 30 cycles of amplification at 61° (annealing) employing both Panels I and II and the primers for exons II and III of the *b* subunit. A single band of 1.5 kb is obtained from the human DNA and DNAs of the 867, 937, 1099, 423, and 683 cell lines. This result is exactly the same as that of the previous experiment.

The 423 cell line also gives a single positive band when amplified by gene-specific primers for human protein Z, factor VII, and factor X, which map to chromosome 13,[19] although the 423 cell line is supposed to contain only chromosome 3. Because the 423 and 683 cell lines are consistently positive for the *b* subunit, other lots of genomic DNAs from these cell lines (two each) have been examined. Because both of the two different lots of DNAs from the 423 cell line turn out to be negative, it is likely that its previous DNA preparations were contaminated with preparations of other cell lines that contain human chromosome 1. On the contrary, the two new lots of DNAs from the 683 cell line remain positive. It may contain part or all of chromosome 1. Thus, the discordancy is 4% (1 out of 25 cell lines) for chromosome 1, while the discordancy for other chromosomes is 20–76%.

The DNA sequences of the 0.9- and 1.7-kb bands are determined by employing the Sequenase kit (United States Biochemical, Cleveland, OH) in order to confirm the genomic nucleotide sequences for the *a* and *b* subunits, respectively. This approach indicates that the previous assignment of the loci of the genes for the *a* and *b* subunits is correct.

Characterization of Abnormal Genes for Factor XIII

a Subunit Deficiency

Genomic Southern Blotting

To analyze genes of patients with factor XIII deficiencies, genomic DNAs are purified from leukocytes after informed consent has been obtained. In order to see a possible large deletion(s), insertion(s), and rearrangement(s), genomic Southern blotting analysis is performed first employing the *Eco*RI cDNA fragments. The genomic DNAs obtained from four patients with *a* subunit deficiency showed exactly the same bands as those of normal individuals. Because this method is incapable of finding small changes in the patients' DNA, a negative finding does not mean that a patient's gene is intact.

In Vitro Amplification and Nucleotide Sequencing

To search for a possible small deletion(s), insertion(s), and mutation(s) by PCR, 16 pairs of amplification primers (two for exon XV) have been designed utilizing the genomic sequence coding for the *a* subunit. Oligonucleotides are prepared according to the principle described above, with a synthesizer (Applied Biosystems). Because the sizes of introns of the gene

---

[19] A. Ichinose, unpublished data (1991).

for the *a* subunit are large (Fig. 2), a single exon, including its boundaries, is amplified one by one. Approximately 1 µg of the genomic DNA is amplified in 100-µl reaction mixtures employing 2.5–5.0 units of *T. aquaticus* DNA polymerase. After 30 cycles of amplification at 65° (annealing), 10 µl of each reaction mixture is applied to an 0.8% agarose gel containing ethidium bromide. The amplified DNA fragments are digested with proper restriction enzymes to generate corresponding ends for ligation into M13 sequencing vectors.

Thus far, the DNA samples from two patients have been analyzed completely by nucleotide sequencing, and several overt mutations have been found.[20] Arg(CGT)-260 was replaced by Cys(TGT), and Ala(GCT)-394 by Val(GTT). However, the significance of these amino acid substitutions is not clear, because the physiological mechanism(s) of biosynthesis of the *a* subunit of factor XIII has not yet been established. Future plans include expression of mutant proteins in the mammalian cell systems described below. This *in vitro* amplification method of the gene for the *a* subunit can be applied for detection of polymorphisms among normal individuals. In fact, two of the polymorphisms in the *a* subunit[14] have been confirmed by the replacement of Leu(CTG)-564 with Pro(CCG), and Val(GTT)-650 with Ile(ATT) in the genomic DNAs of the patients.

b Subunit Deficiency

Patients with *b* subunit deficiency have also been found in Japan.[21] The lack of the *b* subunit may result in the loss of the *a* subunit by an unknown mechanism(s), and the patient with *b* subunit deficiency manifest a bleeding tendency like those with *a* subunit deficiencies. Utilizing the nucleotide sequence of the gene coding for the *b* subunit, five pairs of amplification primers have been designed. Because the introns of this gene are relatively small, DNA fragments containing more than two adjacent exons and their boundaries can be amplified simultaneously (Fig. 1; shown by arrows). Amplification of the gene for the *b* subunit is performed as described above for the characterization of the genomic DNAs for the *a* subunit.

The patient with the *b* subunit deficiency is a phenotypic homozygote; however, the sequence analysis reveals that she is a genotypic compound heterozygote based on the two alleles in her genomic DNA.[22] The results indicate that an adenosine at the intron A/exon II splice junction is missing and a point mutation that replaces a Cys residue with Phe has occurred

[20] A. Ichinose, J. Takamatsu, and H. Saito, unpublished data (1989).
[21] M. Saito, H. Asakura, T. Yoshida, K. Ito, K. Okafuji, T. Yoshida, and T. Matsuda, *Br. J. Haematol.* **74,** 290 (1990).
[22] T. Hashiguchi, M. Saito, E. Morishita, T. Matsuda, and A. Ichinose, *Blood,* in press.

in exon VIII. Loss of the Cys residue leads to destruction of a disulfide bridge in the seventh Sushi domain, and may result in instability of the molecule and/or prevent its secretion from cells.

Expression of a Subunit for Factor XIII and Its Mutants in Yeast

Construction of Expression Vector

Because the *a* subunit of factor XIII is a transglutaminase, it is important to express the *a* subunit in large quantity and apply it for clinical purposes without risk of viral contamination. The following features are consistent with the *a* subunit being a typical cytoplasmic protein[9]: the *a* subunit lacks a signal peptide for secretion, its amino-terminal residue is acetylated,[11] and the molecule has no carbohydrate attached in spite of the presence of the consensus sequences. These features make it easy to express the *a* subunit in yeast cells on a large scale. In order to construct an expression vector with the cDNA coding for the *a* subunit, a *Pst*I fragment of 2.3 kb (Fig. 2) is excised from a phage clone and is ligated into a yeast vector called RPOT.[23] The 3'-noncoding region of the cDNA is removed later in order to increase its expression level. The *a* subunit produced in yeast shares the same properties with the native molecule purified from plasma, including molecular weight, amino acid composition, dimer structure, and ability to be activated by thrombin.

Preparation of Mutants

Various mutants can be produced by *in vitro* mutagenesis[24] in order to examine possible effects of mutations on the functions of the *a* subunit. Oligonucleotides for mutagenesis are 30 to 40-mers so that the melting temperatures for both the 5' and 3' ends of the target sites can be kept at approximately 40°. DNA sequences of the mutant clones are confirmed by sequencing for the intended mutation and the absence of unexpected nucleotide substitutions. This method is extremely efficient, because about 75% of the tested clones have the correct mutations. Each of the exons or several adjacent exons are removed (Fig. 2), and several amino acids are replaced by others. A total of 25 mutants have been made and characterized.[25] Preliminary data suggest that the carboxy-terminal sequence may be important for the stability of the *a* subunit, that a Gly residue in the center of a putative calcium binding site is not essential for the enzymatic activity

[23] P. D. Bishop, D. C. Teller, R. A. Smith, G. W. Lasser, T. Gilbert, and R. L. Seale, *Biochemistry* **29**, 1861 (1990).
[24] K. L. Nakamaye and F. Eckstein, *Nucleic Acids Res.* **14**, 9679 (1986).
[25] A. Ichinose, unpublished data (1989).

of the *a* subunit, and that a replacement of the Pro-Arg sequence at the activation site with Phe-Lys makes the *a* subunit activatable by plasmin instead of thrombin.

Expression of Factor XIII in Mammalian Cells

Expression of Functional a Subunit

Construction of Expression Vector

In order to study a mechanism(s) for release of the *a* subunit of factor XIII from cells, an expression vector for the *a* subunit is constructed by employing the cDNA coding for the *a* subunit and a mammalian expression vector called ZMB4. This and another vector, ZMB3, were kind gifts from Ms. C. Sprecher and Dr. D. Foster (ZymoGenetics, Seattle, WA). ZMB4 permits insertion of DNA fragments into an *Eco*RI site downstream of the adenovirus promoter and also carries an expression unit for the dihydrofolate reductase (DHFR) gene driven by the SV40 early promoter. The expression vector for the *a* subunit, named ZMB4/XIIIA, has been confirmed to contain the entire coding region for the *a* subunit in the proper orientation with respect to the adenovirus promoter by restriction enzyme mapping and DNA sequence analysis. DNA sequence analysis confirms that the inserted cDNA does not contain any unexpected mutations.

Transfection and Selection

Baby hamster kidney (BHK) cells are grown in Dulbecco's modified Eagle's medium containing 10% (v/v) fetal bovine serum, L-glutamine, and an antibiotic mixture of penicillin, streptomycin, and neomycin (PSN, GIBCO, Grand Island, NY). Approximately 1 million BHK cells are transfected with 10 μg of the expression plasmids by the calcium phosphate method.[26] For selection of stable clones, the cells are divided 48 hr after transfection in the culture medium containing a selective agent, 10 μM methotrexate (MTX). Positive clones are detected for production of the recombinant *a* subunit (rXIIIA) after about 7 days by an immunofilter assay[27] and are grown individually for protein analysis.

Characterization of Recombinant a Subunit

The rXIIIA is characterized by a transglutaminase assay after thrombin treatment, Western blotting analysis after SDS-PAGE, and gel filtration employing a Sepharose CL-6B (Pharmacia) column in the absence and

---

[26] F. L. Graham and A. J. van der Eb, *J. Virol.* **52**, 456 (1973).
[27] A. A. McCracken and J. L. Brown, *BioTechniques* **2**, 82 (1984).

presence of the *b* subunit. The rXIIIA is indistinguishable from the native *a* subunit purified from plasma, in terms of its enzymatic activity, molecular weight, and capability to form the a_2 homodimer and a_2b_2 heterotetramer.[28,29,29a] Only small amounts of the rXIIIA are found in the culture medium and most of the rXIIIA remains in the cytoplasma. The release of the *a* subunit cannot be blocked by the addition of a potent inhibitor of the conventional secretory pathway, Brefeldin A, suggesting the presence of a unique mechanism(s) for the release of the *a* subunit from the cells.

This system is being used for characterization of the mutants of the *a* subunit, which have been detected as described above in the genomic DNAs obtained from those patients with *a* subunit deficiencies.

Expression of b Subunit

Secretion of Recombinant b Subunit

The cDNA coding for the *b* subunit of factor XIII is inserted into a mammalian expression vector, named ZMB3, which differs from ZMB4 in carrying an expression unit for the neomycin resistance gene as a selectable marker. The expression vector for the *b* subunit is named ZMB3/XIIIB, and the orientation and nucleotide sequence of the insert have been confirmed. After transfection of ZMB3/XIIIB into BHK cells in the culture medium containing a selective agent of 5 mg/ml G418 (GIBCO), positive clones are detected for the recombinant *b* subunit (rXIIIB) by an immunofilter assay and are grown individually. An ELISA assay reveals that most of the rXIIIB is secreted into the culture medium, and this secretion is completely blocked by Brefeldin A, indicating that the *b* subunit is released from the cells through the conventional secretory pathway. The rXIIIB has been confirmed to have the same molecular weight as the native *b* subunit purified from plasma by Western blotting after SDS-PAGE, and to retain its capability to form the b_2 homodimer and a_2b_2 heterotetramer by gel filtration.[28]

Coexpression of Both Subunits

In order to investigate whether the *b* subunit promotes extracellular release of the *a* subunit, rXIIIA and rXIIIB are coexpressed in the same BHK cells. For this coexpression study, the stable clones expressing the rXIIIB are transfected a second time with ZMB4/XIIIA, and transfectants are selected in culture medium containing both 1 μM MTX and 5 mg/ml G418. Resistant colonies are screened for production of the rXIIIA and are

[28] H. Kaetsu and A. Ichinose, unpublished data (1991).
[29] J. T. Radek, J.-M. Jeong, J. Wilson, and L. Lorand, *Biochemistry* **32,** 3527 (1993).
[29a] This Volume [2].

cultured individually for analysis. Preliminary results suggest that there is no correlation between the secretion (or release) of the *a* and *b* subunits.

This expression system for the *b* subunit is also being used for the characterization of the mutants of the *b* subunit, which has been identified as described above by the study of the genomic DNA obtained from the patient with *b* subunit deficiency.[22]

Acknowledgments

The authors thank Professor E. W. Davie for support throughout this study, E. Espling for assistance in experiments, J. Harris for synthesis of the oligonucleotides, Drs. D. Foster and T. Hashiguchi for useful discussion, and Ms. L. Boda for help in preparation of the manuscript.

[4] Factor XII: Hageman Factor

By ROBIN A. PIXLEY and ROBERT W. COLMAN

Introduction

Factor XII is the name given to a protein that was found lacking in a railroad brakeman from Cleveland in 1955, John Hageman.[1] The coagulation defect is localized to the intrinsic pathway, which becomes activated when blood or plasma is exposed to "foreign" surfaces. Factor XII circulates in blood as a zymogen of a serine protease. It is present in plasma at concentrations ranging from 0.3 to 0.5 μM (29–40 μg/ml).[2] Because factor XII is able to autoactivate when exposed to an anionic surface, it is probably the first factor activated during contact activation, and is implicated in the pathways leading to inflammation, coagulation, and fibrinolysis.[3]

Structure

Factor XII is a single-chain glycoprotein of β-globulin mobility with a pI of 6.3, composed of 596 amino acids, containing 16.8% carbohydrate with a molecular weight of 76,000–80,000 (Fig. 1). The entire primary

[1] O. D. Ratnoff, *in* "Blood, Pure and Eloquent" (M. M. Wintrobe, ed.), p. 601–607. McGraw-Hill, New York, 1980.

[2] S. D. Revak, C. G. Cochrane, A. P. Johnson, and T. E. Hugli, *J. Clin. Invest.* **54,** 619 (1974).

[3] R. W. Colman, *J. Clin. Invest.* **72,** 1249 (1984).

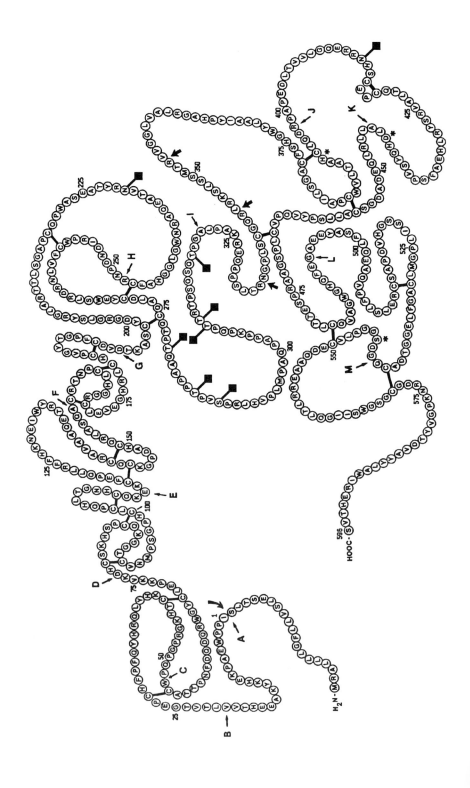

amino acid sequence of factor XII has been reported[4,5] and confirmed by nucleotide sequence analysis of cDNA clones.[6-9] Analysis of the primary sequence has revealed that the light chain and the heavy chain have structural features that, by homology, imply further physiological functions of the molecule. The light chain of factor XIIa, with a molecular weight of 28,000 (Fig. 1, Val^{354}-Ser^{596}), is homologous to plasmin, an enzyme that participates in fibrinolysis, and to tissue plasminogen activator (tPA), a protease that converts plasminogen to plasmin in the fibrinolytic pathway. The sequence homology is highest at the activation regions and near the active sites of these proteases. The light chain contains one N-linked carbohydrate chain on Asn-414. The heavy chain (Ile^1-Arg^{353}) has extensive sequence homology with both urokinase and tissue plasminogen activator. The heavy chain contains an amino acid sequence that shares sequence homology with type II regions of fibronectin and is composed of approximately 60 residues, including four half-cysteine residues (Fig. 1, Lys^{13}-Cys^{69}). The type II homologies probably comprise the collagen-binding site in fibronectin and, by analogy, may be responsible for reported collagen-binding properties of factor XII. Two regions of factor XII are homologous to a sequence resembling epidermal growth factor (Fig. 1, Cys^{79}-Cys^{111} and Cys^{159}-Cys^{190}) that could represent regions of factor XII that interact with cells such as neutrophils or monocytes. Many proteins contain a similar domain, including transforming growth factor type 1, tPA, factor IX, protein C, and several coagulation proteins. Another homologous region in the heavy chain is a single kringle domain

[4] B. A. McMullen and K. Fujikawa, *J. Biol. Chem.* **260**, 5278 (1985).

[5] K. Fujikawa and B. A. McMullen, *J. Biol. Chem.* **258**, 10925 (1983).

[6] D. E. Cool, C.-J. S. Endgell, G. V. Louie, M. J. Zoller, G. D. Brayer, and R. T. A. MacGillivray, *J. Biol. Chem.* **260**, 1366 (1985).

[7] D. E. Cool and R. T. A. MacGillivray, *J. Biol. Chem.* **262**, 13662 (1987).

[8] B. G. Que and E. W. Davie, *Biochemistry* **25**, 1525 (1986).

[9] M. Tripodi, F. Ciarella, S. Guida, P. Galeffi, A. Fantoni, and R. Cortese, *Nucleic Acids Res.* **14**, 3146 (1986).

FIG. 1. Schematic diagram of human factor XII protein-coding domains and the positions of introns. The solid black bars indicate the disulfide bonding between cysteine residues based on protein homologies of other proteins. The numbers are the amino acid positions. The arrows designated A–M indicate the positions of the 13 intron–exon junctions in the cDNA coding sequence. The curved arrow is the site of signal peptidase cleavage. The thick straight arrows are the cleavage sites of factor XII, which occur by autoactivation and kallikrein activation. The solid squares indicate carbohydrate linkage on the amino acid. The asterisks represent the catalytic triad amino acids that participate in catalysis. (Reproduced from a drawing generously supplied by Ross T. A. MacGillivray, Dept. of Biochemistry, The University of British Columbia, Vancouver, Canada.)

(Fig. 1, Cys^{198}-Cys^{276}). Multiple kringles have been found in prothrombin, plasminogen, urokinase, and tPA. The function of the kringle regions is unclear, but in plasminogen and tPA, they participate in the binding to fibrinogen and fibrin. Asn-230 in this region contains an N-linked carbohydrate. The heavy chain also contains a proline-rich region (33% of the sequence) and includes six of the eight carbohydrate chain linkages of factor XII, all of which are O-linked (Fig. 1, Cys^{276}-Arg^{334}). The significance of this region is presently unclear.

The structural gene encoding human factor XII is located on chromosome 5 at location 5q33-qter.[10] The gene is approximately 12 kilobase pairs (kbp) in length and consists of 13 introns and 14 exons. The intron–exon junctions have been located and are indicated by letters A–M in the correlated amino acid sequence shown in Fig. 1.[7]

Characterization

In vitro activation of zymogen factor XII, either by autoactivation or by plasma kallikrein, initially results in cleavage at the Arg^{353}-Val^{354} peptide bond, resulting in a disulfide-linked two-chain molecule of identical molecular weight, designated factor XIIa (also known as α-factor XIIa and HFa). The enzyme is composed of a carboxyl-terminal light chain, molecular weight 28,000, containing the catalytic triad, Ser-544, His-393, and Asp-442 (Fig. 1), and is linked by a disulfide bond to the amino-terminal heavy chain, molecular weight 52,000, containing the surface binding region. *In vitro,* kallikrein can further transform human factor XIIa by cleavage at Arg^{334}-Asn^{335} and Arg^{343}-Leu^{344}, to two forms of a smaller enzyme, molecular weight approximately 30,000, designated factor XIIf (also known as β-factor XIIa and HFf), composed of the 28K catalytic region disulfide-linked to either a 10- or 19-amino acid carboxyl-terminal peptide derived from the 52K heavy chain.[11] Factor XIIf retains its ability to cleave rapidly and to activate some contact system zymogens such as prekallikrein, but not others, such as factor XI.

The activity of factor XII in plasma is not fully expressed unless a negatively charged activating "surface," high-molecular-weight kininogen, and kallikrein are all contiguous. *In vitro* activation of factor XII occurs on negatively charged surfaces by autoactivation, by proteolytic cleavage, by conformational change, or by some combination of these mechanisms.

[10] N. J. Royle, M. Nigli, D. Cool, R. T. MacGillivray, and J. L. Hamerton, *Somatic Cell Mol. Genet.* **14,** 217 (1988).
[11] M. Silverberg and A. P. Kaplan, this series, Vol. 163, p. 68.

These changes occur on substances with negatively charged activating surfaces, including glass, kaolin, Celite, dextran sulfate, and ellagic acid. Biological substances that activate factor XII include sulfatides, chondroitin sulfate, and some mast cell proteoglycans.[12] *In vivo*, the activator may be present in the subendothelial vascular basement membrane,[13] or on the stimulated endothelial cell surface; however, the responsible component(s) is not clear.

Once bound factor XII is activated on a surface, it can proteolyze factor XI, which is associated with high-molecular-weight kininogen (HK) in a stoichiometric noncovalent complex in plasma. A second substrate of surface-bound factor XIIa is prekallikrein, a molecular weight of 88,000 proenzyme that is noncovalently complexed with HK. Both factor XIIa and XIIf can cleave and activate prekallikrein to kallikrein. Zinc is reported to accelerate both the autoactivation and kallikrein activation of factor XII.[14] Kallikrein can cleave HK, releasing the vasoactive nonapeptide bradykinin, as well as factor XIIa at multiple sites, forming factor XIIf.[3,15] Other actions of activated factor XII that have been reported are the activation of C1 of the complement system,[16] activation of factor VII,[17,18] and the down-regulation of monocyte FcγRI receptors for IgG.[19]

The prime modulator of the activity of activated factor XII in plasma is the plasma protease inhibitor, C$\overline{1}$ inhibitor, which contributes ~92% of the inhibitory activity of normal plasma.[20] C$\overline{1}$ inhibitor inactivates purified factor XIIa with a second-order rate of 222 mM^{-1} min^{-1}.[20] Activating surfaces such as kaolin, dextran sulfate, and sulfatides are able to protect against inhibition of factor XIIa by C$\overline{1}$ inhibitor thereby creating a protected environment for the protease to activate and to cleave its substrates.[21]

[12] Y. Hojima, C. G. Cochrane, R. C. Wiggins, K. F. Austen, and R. L. Stevens, *Blood* **63**, 1453 (1984).

[13] O. D. Ratnoff, *Prog. Hematol.* **5**, 204 (1966).

[14] J. D. Shore, D. E. Day, P. I. Bock, and S. T. Olson, *Biochemistry* **26**, 2250 (1987).

[15] G. Tans and J. Rosing, *Semin. Thromb. Hemostasis* **13**, 1 (1987).

[16] A. P. Kaplan, M. Silverberg, and B. Ghebrehiwet, *Adv. Exp. Med. Biol.* **198**, 11 (1986).

[17] V. Seligsohn, B. Osterud, J. F. Brown, J. H. Griffin, and S. I. Rapaport, *J. Clin Invest.* **64**, 1056 (1979).

[18] E. M. Gordon, J. Douglas, and O. D. Ratnoff, *J. Clin. Invest.* **72**, 1833 (1983).

[19] P. Chien, R. A. Pixley, L. G. Stumpo, R. W. Colman, and A. D. Schreiber, *J. Clin. Invest.* **82**, 1554 (1988).

[20] R. A. Pixley, M. Schapira, and R. W. Colman, *J. Biol. Chem.* **260**, 1723 (1985).

[21] R. A. Pixley, A. Schmaier, and R. W. Colman, *Arch. Biochem. Biophys.* **256**, 490 (1987).

Assay Procedures

Coagulant Assay

The functional activity of factor XII or factor XIIa can be determined via the correction of the abnormal clotting time of human factor XII-deficient plasma by comparing the times obtained through the addition of pooled normal plasma (PNP).

Reagents. Factor XII-deficient plasma and pooled normal plasma can be purchased from George King Biomedicals (Overland Park, KS) or Baxter Healthcare Corporation (Miami, FL). Rabbit brain cephalin (Sigma Chemical Company, St. Louis, MO) is first reconstituted with saline according to the manufacturer's instructions. Aliquots (1 ml) of the stock may be frozen for longer storage. When assaying, the stock aliquot is then diluted 1:20 to 1:100 with buffer A (0.05 M Tris, 0.15 M NaCl, pH 7.4, 0.02% NaN$_3$), depending on the obtained clotting times. CaCl$_2$ is dissolved at 30 mM in H$_2$O. Acid washed Kaolin (Fischer, Pittsburg, PA) is suspended in saline at 5 mg/ml.

Reagents are added to polystyrene tubes (10 × 75 mm) at room temperature, in the following order; 100 μl of kaolin (5 mg/ml in saline), 100 μl rabbit brain cephalin (in buffer A), 100 μl of buffer A, and 100 μl factor XII-deficient plasma. The mixture is vortexed for 1 sec, then 1–20 μl of standard or sample containing factor XII [undiluted or diluted in buffer A containing 2% (w/v) bovine serum albumin (BSA)] is added. The solution is vortexed for 1 sec and each tube is placed in a 37° water bath for exactly 8 min. At the precise time, 100 μl of CaCl$_2$ solution is added to each tube, vortexed for 1 sec, replaced in the water bath, and the timing begun again. The end point, a solid clot observed by eye, is determined by continuously tilting the tube every few seconds. The concentration of factor XII in 1 ml of pooled normal plasma is defined as 1 unit/ml and is equivalent to 0.37 μM. PNP is added directly to generate a standard curve, using 1, 2.5, 5, 7.5, 10, 12.5, and 15 μl of plasma to represent 0.1–1.5 unit/ml of factor XII, and gives typical clotting times of 300–60 sec (the clotting time of 10 μl is defined as 1 unit/ml of factor XII). The PNP may be diluted 1:10 with 2% BSA in buffer A to produce a curve in the 0.01–0.1 U/ml range. A standard curve is constructed by plotting the log of the clotting time against the log of the factor XII (U/ml) and fitting the data linearly, by computer using an iterative procedure or by hand. Typically, a 10-μl aliquot, in duplicate or triplicate, of an unknown containing factor XII is examined. The value of the unknown is interpolated from the standard curve and, if appropriate, corrected for dilution.

Microtiter Chromogenic Assay for Factor XII in Plasma

There is a commercially available microtiter plate assay (Channel Diagnostics, Walmer, Kent, UK) that uses the substrate S-2222 and a selective inhibitor of kallikrein.[22] A similar assay has been described.[23]

Quantitative Immunoblotting Assay of Factor XII

A sensitive immunodetection assay of factor XII in plasma using an electroblotting technique from SDS-PAGE coupled with an immunoassay using a polyclonal antibody and radiolabeled factor XII has been described and is sensitive to the presence of 5 ng of factor XII.[24]

Chromogenic Assay of Factor XIIa or Factor XIIf

Factor XIIa or XIIf activity of purified samples is assayed by measuring the initial velocity of the cleavage of the chromogenic substrate, S-2302.

Reagent. D-Prolyl-L-phenylalanyl-L-arginine-*p*-nitroaniline dihydrochloride (S-2302) is available from Kabi Pharmacia (Franklin, OH).

Assay buffer (200 μl; 0.05 M Tris, 0.14 M NaCl, 1 mM EDTA, pH 7.8) is added to a glass microcuvette along with 30 μl of S-2302 stock (4 mM in H$_2$O) to give a final substrate concentration of 0.5 mM. The cuvette is placed in a thermostatted cuvette chamber of a Cary 210 spectrophotometer (Varian Instruments, Sugarland, TX) for 2 min to reach the equilibrium temperature of 37°. Then 10 μl of sample to be assayed is added to the cuvette and quickly mixed with a plastic stirrer. The absorbance change at 405 nm is recorded for 2–5 min to determine the initial velocity of the reaction. The concentration of factor XIIa can be computed: 1 U/ml of factor XIIa (not factor XIIf), as measured by coagulant assay, gives a change of absorption per minute of 0.057 under these conditions. Soybean trypsin inhibitor, type II-S (SBTI; Sigma, St. Louis, MO) at 50 μg/ml may be added to the chromogenic buffer if samples contain kallikrein. Addition of corn trypsin inhibitor (CTI; Enzyme Research Laboratories, Inc., South Bend, IN) to the S-2302 buffer (CTI is a specific inhibitor of the amidolytic activity of factor XIIa/XIIf[25]) at a concentration of 50 μg/ml may be used to confirm the absence of kallikrein in the sample.

[22] K. J. Walsh, I. J. Mackie, M. Gallimore, and S. J. Machin, *Thromb. Res.* **47**, 365 (1987).
[23] J. Sturzebecher, L. Svendesen, R. Eichenberger, and F. Markwardt, *Thromb. Res.* **55**, 709 (1989).
[24] B. Lämmele, M. Berrettini, H. P. Schwarz, M. J. Heeb, and J. Griffin, *Thromb. Res.* **41**, 747 (1986).
[25] E. P. Kirby and P. J. McDivitt, *Blood* **61**, 652 (1983).

Purification Procedures

Factor XII, factor XIIa, and other coagulation proteins may be purchased from Enzyme Research Laboratories Inc. (South Bend, IN).

Purification of Factor XII by Zinc Chelate Chromatography

Plastic containers and columns are used throughout the purification procedure. All dialysis tubing and plastic containers are prerinsed with a 2-mg/ml solution of Polybrene (hexadimethrine bromide, Sigma, St. Louis, MO) then rinsed with H_2O. All steps are carried out at room temperature except where indicated. Concentration of factor XII is performed by negative-pressure dialysis at 4°. Zinc chelate agarose may be generated according to an established procedure,[26] or purchased from Pharmacia/LKB (Piscataway, NJ).

Fresh-frozen plasma (4 liters) containing 4% (w/v) sodium citrate as anticoagulant is thawed at 37° in a polypropylene container containing SBTI and Polybrene to achieve a final concentration of 100 μg/ml and 360 μg/ml, respectively.

Ammonium Sulfate Precipitation. Crystalline ammonium sulfate (144 g/liter plasma, 25%) is slowly dissolved in the plasma and stirred for 30 min. The solution is centrifuged at 13,680 *g* for 30 min and the precipitate is discarded. Ammonium sulfate (158 g/liter, 50%) is slowly dissolved in the decanted supernatant at room temperature and stirred for 60 min. The mixture is centrifuged at 13,680 *g* for 60 min. The precipitate is dissolved in minimal amounts (1 liter) of 0.025 *M* Na_2HPO_4, 0.8 *M* NaCl, 0.2 mg/ml SBTI, 0.36 mg/ml Polybrene, 0.02% NaN_3, pH 6.5, and is dialyzed overnight against two changes of 20 liters of the same buffer without SBTI at 4°. Recovery of factor XII activity from the 25–50% ammonium sulfate fractionation is excellent at 88% (Table I) and ranged from 87 to 96% in other preparations. A major contaminant removed by this first step is primarily fibrinogen, a protein that strongly binds to zinc chelate-Sepharose.

Zinc Chelate Chromatography: Column 1. The solution is then centrifuged for 10 min at 4000 *g* to remove precipitate formed during dialysis. Equilibrated zinc chelate-Sepharose (300 ml) is placed in a plastic Büchner funnel under low vacuum. The solution is slowly allowed to flow through the resin and is collected. The resin is washed with equilibrating buffer and collected in 500-ml fractions until the absorbance reading at 280 nm is below 0.1 (approximately 11 liters). The resin is then washed with 2 to 3 liters cacodylate buffer (0.02 *M* sodium cacodylate, 0.15 *M* NaCl, 0.1 mg/ml SBTI, 0.03 mg/ml Polybrene, 0.02% NaN_3, pH 5.5) until the

[26] R. A. Pixley and R. W. Colman, *Thromb Res.* **41**, 89 (1986).

TABLE I
PURIFICATION OF FACTOR XII BY ZINC CHELATE METHOD[a]

Purification step	Total activity (units)	Total protein (mg)	Specific activity (units/mg)	Purification-fold	Recovery (%)
Plasma	4000	85,710	0.047	1	100
Ammonium sulfate	3500	10,220	0.342	7	88
Zinc chelate: column 1	2280	190	12.0	255	57
Zinc chelate: column 2	1360	48	28.3	602	34
Gel filtration	1200	20	60.0	1277	30

[a] Factor XII activity was determined by coagulant assay. The original plasma factor XII activity is before addition of inhibitors. Protein was determined by Bradford dye-binding assay with BSA as standard. Reprinted with permission from Pixley and Colman,[26] Copyright 1986, Pergamon Press plc.

absorbance readings of the fractions are below 0.1. Factor XII fractions are eluted with 10 liters of acetate buffer (0.1 M sodium acetate, 0.8 M NaCl, 0.1 mg/ml SBTI, 0.03 mg/ml Polybrene, 0.02% NaN_3, pH 4.5). The fractions are assayed for factor XII coagulant activity. The factor XII fractions are pooled and dialyzed overnight against two changes of 20 liters of phosphate–acetate buffer (0.025 M Na_2HPO_4, 0.005 M sodium acetate, 0.8 M NaCl, 0.001 mg/ml Polybrene, 0.02% NaN_3, pH 6.5) at 4°.

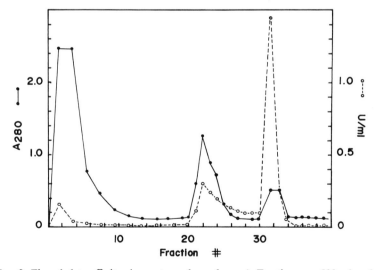

FIG. 2. Zinc chelate affinity chromatography: column 1. Fractions are 500 ml each and collected under low vacuum using a plastic Büchner funnel at room temperature. Protein is estimated by absorbance at 280 nm. Factor XII activity is determined by coagulant assay. Reprinted with permission from Pixley and Colman,[26] Copyright 1986, Pergamon Press plc.

The elution profile of zinc chelate affinity chromatography (column 1) using stepwise elution is presented in Fig. 2. The pH of the cacodylic acid buffer in this step must be precisely 5.5 to separate most of the α_2-macroglobulin from factor XII. A lower pH will result in the majority of the factor XII eluting in the cacodylic acid buffer. The majority of the factor XII activity was recovered in the acetate buffer, giving an overall recovery of 57% (Table I) with a range of 43–65% in other preparations.

Zinc Chelate Chromatography: Column 2. The dialyzed factor XII fractions are applied to a column (2.5 × 25 cm) containing 120 ml of phosphate–acetate buffer equilibrated zinc chelate-agarose and washed overnight with 2 liters of the same buffer at a flow rate of 100 ml/hr. A pH gradient of 250 ml each of the phosphate–acetate buffer at pH 6.5 and 4.0, followed by a 100-ml wash of pH 4.0 buffer, is applied to the column and 5-ml fractions are collected. Aliquots of 500 μl are then placed in polypropylene Eppendorf tubes for analysis and the fractions and aliquots are frozen at −70°.

The gradient aliquots are analyzed for protein, factor XII coagulant activity, and S-2302 amidolytic activity and are subjected to SDS-PAGE. The fractions determined to contain factor XII are pooled and concentrated in portions by negative-pressure dialysis against phosphate–acetate

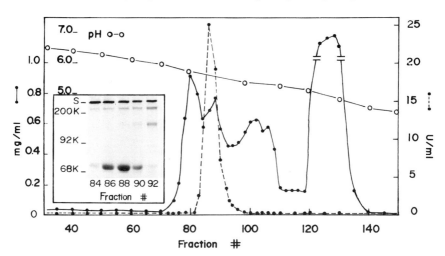

FIG. 3. Zinc chelate affinity chromatography: column 2. After applying the pooled and dialyzed factor XII fractions from the previous column, the factor XII is eluted with a 500-ml pH gradient of phosphate–acetate buffer, pH 6.5 to pH 4.0. The fractions were assayed for protein concentration, factor XII coagulant activity, and pH. Chromatography is carried out at room temperature at a flow rate of 100 ml/hr. Fractions are 5 ml each. Insert: Nonreduced SDS-PAGE of fractions. "S" indicates the stacking gel interface, which contains α_2-macroglobulin. Reprinted with permission from Pixley and Colman,[26] Copyright 1986, Pergamon Press plc.

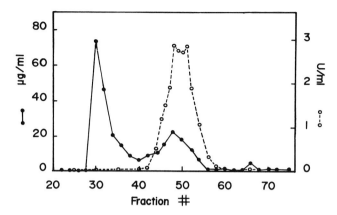

FIG. 4. Sieving chromatography on BioGel A-0.5m. Factor XII fractions 84–90 from Fig. 3 were pooled and vacuum concentrated against phosphate–acetate buffer, pH 6.5. A 1-ml aliquot containing 50 U of factor XII is applied to a 1.5 × 88-cm column equilibrated with the same buffer. Chromatography is at room temperature with a flow rate of 0.5–0.8 ml/min, collected in 2-ml fractions. Fractions are assayed for protein content and factor XII coagulant activity. Reprinted with permission from Pixley and Colman,[26] Copyright 1986, Pergamon Press plc.

buffer, pH 6.5. The elution profiles of zinc chelate affinity chromatography (column 2) using a pH gradient with an accompanying SDS-PAGE of some fractions are seen in Fig. 3. The elution results in four protein peaks eluting at pH 5.8, 5.5, 5.2, and 5.0. Factor XII is recovered in the second peak at pH 5.5 with an overall recovery of 34% (Table I) and a range of 16–34% with other preparations. None of the fractions in the gradient elution exhibits any amidolytic activity against S-2302, indicating that factor XII is still in its zymogen form at this purification step. SDS-PAGE also indicates factor XII is still a zymogen at 80,000 Da under reducing and nonreducing conditions. The major contaminant of factor XII at this step is α_2-macroglobulin.

Gel Filtration. The factor XII fractions containing contaminating α_2-macroglobulin (tubes 84–90) are pooled and subjected to negative-pressure dialysis to concentrate the proteins for gel filtration. Although time consuming, negative-pressure dialysis, using high salt and Polybrene, results in negligible loss of factor XII activity, as compared to other methods of concentrating, such as using Amicon concentrators. Aliquots containing 50–100 U of factor XII coagulant activity are applied to a 1.5 × 88-cm column of BioGel A-0.5m (Bio-Rad) equilibrated with phosphate–acetate buffer at a flow rate of 1 ml/min. Then 2-ml fractions are collected and analyzed for protein, coagulant activity, and S-2302 amidolytic activity and are subjected to SDS-PAGE. The gel-filtration profile, which separates factor XII from α_2-macroglobulin, is indicated in Fig. 4. α_2-Macroglobulin

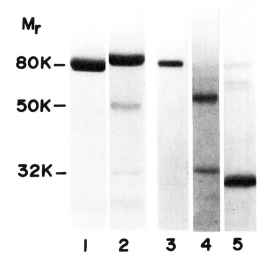

Fig. 5. SDS-PAGE of purified factor XII from zinc chelate purification method and activated factor XII products. Lane 1, factor XII nonreduced; lane 2, factor XII reduced; lane 3, factor XIIa nonreduced; lane 4, factor XIIa reduced; lane 5, factor XIIf nonreduced. Reprinted with permission from Pixley and Colman,[26] Copyright 1986, Pergamon Press plc.

is eluted as a protein peak in the void volume and factor XII is eluted as an included protein peak. The recovery of factor XII in this step ranges from 88 to 98% with an overall recovery of 30% with a range of 13 to 30%.

The fractions determined to contain purified factor XII are concentrated by negative-pressure dialysis against phosphate–acetate buffer, pH 6.5, and are frozen in appropriate aliquots (1–3 ml) until used.

Just before use, the aliquots are thawed and placed into the buffer of choice by using a 5-ml desalting column equilibrated with buffer (Pharmacia PD-10 column) and reassayed for functional activity.

Activation of factor XII with kallikrein results in factor XIIa and factor XIIf with full amidolytic activity against S-2302. On examination by SDS-PAGE (see Fig. 5), several species of factor XIIa are observed, including 80,000 nonreduced (lane 3, Fig. 5), and 50,000 and 32,000 reduced (lane 4, Fig. 5). Factor XIIf under nonreducing conditions exhibited a molecular weight of 28,000 (lane 5, Fig. 5).

Purification of Factor XII by Immunoaffinity Chromatography Using Monoclonal Antibody B7C9

Monoclonal antibody (Mab) B7C9 is an $IgG_1\kappa$ antibody directed against the heavy chain of factor XII with a K_d of 9.8 nM.[27] The antibody is

[27] R. A. Pixley, L. G. Stumpo, K. Birkmeyer, L. Silver, and R. W. Colman, J. Biol. Chem. 262, 10141 (1987).

purified on protein A agarose and coupled to an active ester agarose (Affi-Gel 10, Bio-Rad) as per instructions of the manufacturer, using a concentration of antibody of 5–7 mg/ml per milliliter of agarose. The antibody–agarose is then packed into a polycarbonate column (to avoid the use of glass) fitted with adaptors to attach to the Pharmacia/LKB fast protein liquid chromatography (FPLC) system. A 1- to 2-cm layer of Sepharose 4B-CL is placed on top of the resin to protect against contaminant accumulation. The column is equilibrated with filtered and degassed equilibration buffer composed of 0.05 M Tris, 0.5 M NaCl, and 1 nM EDTA, pH 7.8. The column may be regenerated from time to time by removing and replacing some of the Sepharose 4B-CL and by using 5–10 ml of 2 M guanidine hydrochloride to wash the resin.

Ammonium Sulfate Precipitation. Ammonium sulfate precipitation of factor XII from fresh-frozen plasma is performed in a manner similar to that described in the zinc chelate purification method, except the 50% precipitate is redissolved in 0.05 M Tris/HCl, 0.5 M NaCl, and 1 mM EDTA, pH 7.8, containing 100 μg/ml SBTI, and is dialyzed overnight against the same buffer without SBTI. The dialyzed solution is then frozen in 50-ml aliquots.

The solution (100 ml) is quickly thawed in a 37° waterbath and delipidated using 1,1,2-trichlorotrifluoroethane (Aldrich Chem. Co., Milwaukee, WI) by mixing two volumes of the dissolved precipitate with one volume of the delipidating reagent in a 50-ml polypropylene tube. The mixture is gently rocked for 5–10 min at room temperature, centrifuged for 5 min at 5000 rpm, and the top aqueous solution containing factor XII is removed into a separate polypropylene container. The solution is then degassed by gently bubbling through helium delivered using a cut polypropylene transfer pipette.

The prepared solution is then loaded onto the equilibrated affinity column, containing 5 ml of immunoaffinity matrix, using a separate peristaltic pump with polyvinyl chloride and polyethylene tubing at the rate of 12 ml/hr and monitored at 280 nm. After application, the column is washed extensively with a buffer (0.01 M Tris, 0.5 M NaCl, pH 7.5) at 0.5 ml/min, until the absorbance reaches baseline values. The factor XII is eluted at a flow rate of 0.5 ml/min using 0.1 M sodium citrate, pH 2.2, and is collected into 5-ml fractions containing 1.5 ml of 1 M Tris base and mixed as soon as possible to adjust the pH of the solution. The column is reequilibrated with starting buffer for the next cycle. The factor XII, detected by coagulant assay, is recovered in the first 10–25 ml of the eluant with typical concentrations of 10–20 U/ml (290–580 μg/ml).

Some preparations contain contaminating human IgG, which seems to associate with the mouse IgG but elutes off at a higher pH than factor XII. This contamination is avoided by having a slight initial pH gradient in which this associated IgG is removed in the first 5–10 ml. This procedure slightly dilutes the purified product.

The purification results in return of 70–80% of the applied factor XII with a 63–72% recovery of the factor XII in plasma. A typical run using 5 ml of affinity matrix yields 6–10 mg of zymogen factor XII with a specific coagulant activity of 80 U/mg.

Samples can be prepared for use on a desalting column as described earlier.

Other Purification Methodologies

Conventional chromatography procedures have been described for factor XII.[11,28–31] An affinity purification method using popcorn inhibitor has been described by Ratnoff *et al.*[32]

Activation of Factor XII

Purified factor XII is autoactivated to factor XIIa (α-factor XIIa) by incubation of zymogen factor XII in a low-ionic-strength buffer using 500-kDa dextran sulfate (Sigma, St. Louis, MO) as activating agent. Zymogen factor XII at 5–50 U/ml (0.15–1.5 mg/ml) is pipetted into a polypropylene test tube in a buffer of 0.02 M Tris (or HEPES) and 0.02 M NaCl, pH 8. From a stock solution of 10 mg/ml dextran sulfate in H_2O, a volume is added to obtain a concentration of 20–40 μg/ml. The mixture is incubated at 37° for 30–40 min, testing 5-μl aliquots of solution by the initial-velocity chromogenic assay for factor XII (described above) every 10 min to monitor the reaction. When the initial velocity is constant, the reaction mixture is added to 3–10 ml of QAE-agarose packed in a plastic chromatography column equilibrated with the incubation buffer. Factor XII, factor XIIa, factor XIIf (β-factor XIIa), and dextran sulfate bind to the ion-exchange resin. The column is washed with a small amount of 0.02 M Tris and 0.1 M NaCl, pH 8.0 (10 ml). Primarily factor XIIa and some factor XIIf are consecutively eluted in a NaCl gradient of 24 ml of buffer containing 0.1–0.6 M NaCl. Dextran sulfate remains bound on the QAE column, which is discarded. The fractions are then assayed for protein, coagulant activity, and chromogenic activity and are characterized by SDS-PAGE.

[28] F. Van Der Graaf, G. Tans, B. N. Bouma, and J. H. Griffin, *J. Biol. Chem.* **257**, 300 (1982).
[29] F. Espana and O. D. Ratnoff, *J. Lab. Clin. Med.* **102**, 31 (1983).
[30] Y. C. Chan and H. Z. Movat, *Thromb. Res.* **8**, 337 (1976).
[31] K. Fujikawa and E. W. Davies, this series, Vol. 80, p. 198.
[32] O. D. Ratnoff, B. Everson, V. H. Donaldson, and B. H. Mitchell, *Blood* **67**, 1550 (1986).

Purification of Factor XIIf (β-factor XIIa)

Factor XIIf may be isolated as a by-product of factor XIIa preparations, by extraction of the heavy-chain-containing portions with kaolin in a low-ionic-strength buffer (0.01 *M* Tris, pH 8.0), and concentrating down factor XIIf in an Amicon concentrator with a 10,000 molecular weight cutoff. Factor XIIf can be isolated directly from dextran sulfate-activated plasma using the methodology described in de Agostini *et al.*[33] A more detailed procedure for factor XIIf is described by Silverberg and Kaplan.[11]

[33] A. de Agostini, H. R. Lijnen, R. A. Pixley, R. W. Colman, and M. Schapira, *J. Clin. Invest.* **73**, 1542 (1984).

[5] Factor XI: Structure–Function Relationships Utilizing Monoclonal Antibodies, Protein Modification, Computational Chemistry, and Rational Synthetic Peptide Design

By PETER N. WALSH, FRANK A. BAGLIA, and BRADFORD A. JAMESON

I. Introduction

A. Factor XI Biochemistry

Coagulation factor XI is a plasma glycoprotein involved in the initiation of the intrinsic coagulation pathway. It is present in plasma at a concentration of approximately 30 n*M* and requires proteolytic activation to develop serine protease activity.[1–6] Factor XI is unique among coagulation proteins because it is a disulfide-linked homodimer with each identical monomer containing 607 amino acids.[7,8] The zymogen is cleaved by factor XIIa at an internal Arg[369]-Ile[370] bond to yield a heavy chain that contains four tandem repeat sequences of 90–91 amino acids designated Apple domains, each of which contains three internal disulfide bonds.[7,8] These

[1] K. D. Wuepper, *Fed. Proc., Fed. Am. Soc. Exp. Biol.* **31**, 624 (1972).
[2] H. Z. Movat and A. H. Ozge-Anwar, *J. Lab. Clin. Med.* **84**, 861 (1974).
[3] L. W. Heck and A. P. Kaplan, *J. Exp. Med.* **140**, 1615 (1974).
[4] K. Kurachi and E. W. Davie, *Biochemistry* **16**, 5831 (1977).
[5] B. N. Bouma and J. H. Griffin, *J. Biol. Chem.* **252**, 6432 (1977).
[6] H. Saito and G. Goldsmith, *Blood* **50**, 377 (1977).
[7] K. Fujikawa, D. W. Chung, L. E. Hendrickson, and E. W. Davie, *Biochemistry* **25**, 2417 (1986).
[8] B. A. McMullen, K. Fujikawa, and E. W. Davie, *Biochemistry* **30**, 2056 (1991).

four Apple domains contain amino acid sequences that are 58% identical to the corresponding region of plasma prekallikrein and have 23–34% identity with one another.[7,8]

B. Genetics

The structure and sequence of the mature factor XI protein have been deduced from the isolation and sequencing of a cDNA obtained from human liver,[7,8] and the organization of the gene has been subsequently determined.[9] The factor XI gene is 23 kilobases (kb) in length and consists of 15 exons and 14 introns, of which the first two code for the 5' untranslated region and the signal peptide. The four tandem repeat Apple domains comprising the heavy chain are encoded by exons III–X, with two exons coding for each Apple domain and with the introns separating each of these exon pairs, being located at the same position in each of the four Apple domains. The carboxy-terminal catalytic domain, comprising the light chain of factor XIa, is encoded by exons XI–XV.

C. Molecular and Cellular Interactions

Factor XI is activated as a consequence of a complex interaction of four coagulation proteins, including factor XII, prekallikrein, and high-molecular-weight (M_r) kininogen.[10-18] Factor XI circulates in plasma as a noncovalent complex with high-M_r kininogen[17] and is bound to anionic surfaces such as glass, Celite, or kaolin whereon it can be activated by factor XIIa.[4,5,10-18] Factor XIIa, in turn, arises from the limited reciprocal proteolytic activation of factor XII and prekallikrein, reactions which also occur on negatively charged surfaces. Alternatively, it has recently been demonstrated that factor XI activation can be catalyzed either by thrombin or by factor XIa in the presence of dextran sulfate.[19,20] Because factor XI defi-

[9] R. Asakai, E. W. Davie, and D. W. Chung, *Biochemistry* **26**, 7221 (1987).
[10] O. D. Ratnoff, E. W. Davie, and S. L. Mallet, *J. Clin. Invest.* **40**, 803 (1961).
[11] H. Saito, O. D. Ratnoff, J. S. Marshall, and J. Pensky, *J. Clin. Invest.* **52**, 850 (1973).
[12] S. Schiffman and P. Lee, *Br. J. Haematol.* **27**, 101 (1974).
[13] S. Schiffman and F. J. Markland, Jr., *Thromb Res.* **6**, 273 (1975).
[14] J. H. Griffin and C. G. Cochrane, *Proc. Natl. Acad. Sci. U.S.A.* **73**, 2554 (1976).
[15] C. Y. Liu, C. F. Scott, A. Bagdasarian, J. V. Pierce, A. P. Kaplan, and R. W. Colman, *J. Clin. Invest.* **60**, 7 (1977).
[16] H. L. Meier, C. F. Scott, R. Mandle, M. E. Webster, J. V. Pierce, R. W. Colman, and A. P. Kaplan, *J. Clin. Invest.* **60**, 18 (1977).
[17] R. E. Thompson, R. Mandle, Jr., and A. P. Kaplan, *J. Clin. Invest.* **60**, 1376 (1977).
[18] R. C. Wiggins, B. N. Bouma, C. G. Cochrane, and J. H. Griffin, *Proc. Natl. Acad. Sci. U.S.A.* **74**, 4636 (1977).
[19] K. Naito and K. Fujikawa, *J. Biol. Chem.* **266**, 7353 (1991).
[20] D. Gailani and G. J. Broze, Jr., *Science* **253**, 909 (1991).

ciency is associated with a hemostatic deficiency state in about half the affected patients, it is clear that factor XI is required for normal hemostasis.[21] However, the normal physiological mechanism for activation of factor XI is unknown because deficiencies of factor XII, prekallikrein, and high-M_r kininogen are not associated with abnormal bleeding states, and it is not yet known whether thrombin-catalyzed factor XI activation can occur under physiological conditions.

The normal macromolecular substrate for factor XIa is factor IX, a single-chain 57,000 M_r glycoprotein (17% carbohydrate) found in plasma as a vitamin K-dependent zymogen at a concentration of 70–90 nM.[22-30] The proteolytic activation of factor IX by factor XIa requires the presence of calcium ions and is associated with the formation of a factor IX activation peptide formed when factor XIa cleaves first an internal Arg^{145}-Ala^{146} bond and subsequently an Arg^{180}-Val^{181} bond.[22-30] Because this 10,000 M_r activation peptide contains 50% carbohydrate, the activation of factor IX by factor XIa can be monitored by determining the rate of formation of ^{3}H-labeled activation peptide from zymogen factor IX.[30]

Factor XI has also been shown to interact specifically with activated platelets, which promote the activation of factor XI by factor XIIa in the presence of high-M_r kininogen.[31,32] Additional evidence supports the view that factor XI activation in the presence of stimulated platelets can occur by a factor XII-independent pathway provided kallikrein and high-M_r kininogen are present.[31,32] The specific binding of factor XI to platelets in the presence of high-M_r kininogen and Zn^{2+} ions has been shown to be associated with enhanced rates of factor XI activation on the platelet surface.[33] In addition, endothelial cells possess specific saturable binding

[21] M. V. Ragni, D. Sinha, F. Seaman, J. H. Lewis, J. A. Spero, and P. N. Walsh, *Blood* 65, 719 (1985).

[22] K. Fujikawa, M. E. Legaz, H. Kato, and E. R. Davie, *Biochemistry* 13, 4508 (1974).

[23] R. G. DiScipio, K. Kurachi, and E. W. Davie, *J. Clin. Invest.* 61, 1528 (1978).

[24] B. Østrud, B. N. Bouma, and J. H. Griffin, *J. Biol. Chem.* 253, 5946 (1978).

[25] K. Kurachi and E. W. Davie, *Proc. Natl. Acad. Sci. U.S.A.* 79, 6461 (1982).

[26] M. Jaye, H. DeLaSalle, F. Schamber, A. Balland, V. Kohli, A. Findeli, P. Tolstoshev, and J. P. Lecocq, *Nucleic Acids Res.* 11, 2325 (1983).

[27] D. S. Anson, K. H. Choo, D. J. G. Rees, F. Giannelli, K. Gould, J. A. Huddleston, and G. G. Brownlee, *EMBO. J.* 3, 1053 (1984).

[28] P. Jagadeeswaran, D. E. Lavelle, R. Kaul, T. Mohandas, and S. T. Warren, *Somatic Cell Mol. Genet.* 10, 465 (1984).

[29] S. Yoshitake, B. G. Schach, D. C. Foster, E. W. Davie, and K. Kurachi, *Biochemistry* 24, 3736 (1985).

[30] D. Sinha, F. S. Seaman, and P. N. Walsh, *Biochemistry* 26, 3768 (1987).

[31] P. N. Walsh, *Br. J. Haematol.* 22, 237 (1972).

[32] P. N. Walsh and J. H. Griffin, *Blood* 57, 106 (1981).

[33] J. S. Greengard, M. J. Heeb, E. Ersdal, P. N. Walsh, and J. H. Griffin, *Biochemistry* 25, 3884 (1986).

sites for both factor XI and high-M_r kininogen, and under specific experimental conditions endothelial cells can promote the activation of factor XI by factor XIIa.[34-36] Additional evidence supports the view that factor XIa can bind specifically to activated platelets,[37] whereon it is protected from inactivation by and complex formation with α_1-protease inhibitor.[38] Although the rates of factor IX activation by factor XIa are not increased in the presence of platelets,[39] the platelet surface may provide a protected nidus on which factor IX activation can be localized.[38] Although both α_1-protease inhibitor[40,41] and antithrombin III[42] have been proposed as the major plasma inhibitors of factor XIa, recent evidence suggests that two specific factor XIa inhibitors released from activated platelets, protease nexin II (PNII)[43-45] and a low-molecular-weight platelet inhibitor of factor XIa (PIXI),[46] are major regulators of factor XIa, because they are released by platelets in close proximity to the hemostatic plug, where the local microenvironment favors the formation of factor XIa.

II. Purification and Characterization

A. Assays

1. Coagulation. Factor XI is assayed using factor XI-deficient plasma and minor modifications[47] of the kaolin-activated partial thromboplastin

[34] D. M. Stern, M. Drillings, W. Kisiel, P. Nawroth, H. L. Nossel, and K. LaGamma, *Proc. Natl. Acad. Sci. U.S.A.* **81,** 913 (1984).
[35] D. M. Stern, M. Drillings, H. L. Nossel, A. Hurlet-Jensen, K. S. LaGamma, and J. Owen, *Proc. Natl. Acad. Sci. U.S.A.* **80,** 4119 (1983).
[36] D. M. Stern, P. P. Nawroth, W. Kisiel, D. Handley, M. Drillings, and J. Bartos, *J. Clin. Invest.* **74,** 1910 (1984).
[37] D. Sinha, F. S. Seaman, A. Koshy, L. C. Knight, and P. N. Walsh, *J. Clin. Invest.* **73,** 1550 (1984).
[38] P. N. Walsh, D. Sinha, F. Kueppers, F. S. Seaman, and K. B. Blankstein, *J. Clin. Invest.* **80,** 1578 (1987).
[39] P. N. Walsh, D. Sinha, A. Koshy, F. Seaman, and H. Bradford, *Blood* **68,** 225 (1986).
[40] C. F. Scott, M. Schapira, and R. W. Colman, *Blood* **60,** 940 (1982).
[41] C. F. Scott, M. Schapira, H. L. James, A. B. Cohen, and R. W. Colman, *J. Clin. Invest.* **69,** 844 (1982).
[42] D. L. Beeler, J. A. Marcum, S. Schiffman, and R. D. Rosenberg, *Blood* **67,** 1488 (1986).
[43] W. E. Van Nostrand, A. H. Schmaier, J. S. Farrow, and D. D. Cunningham, *Science* **248,** 745 (1990).
[44] R. P. Smith, D. A. Higuchi, and G. J. Broze, Jr., *Science* **248,** 1126 (1990).
[45] A. I. Bush, R. N. Martins, B. Rumble, R. Moir, S. Fuller, E. Milward, J. Currie, D. Ames, A. Weidemann, P. Fischer, G. Multhaup, K. Beyreuther, and C. L. Masters, *J. Biol. Chem.* **265,** 15977 (1990).
[46] A. L. Cronlund and P. N. Walsh, *Biochemistry* **31,** 1685 (1992).
[47] C. F. Scott, D. Sinha, F. S. Seaman, P. N. Walsh, and R. W. Colman, *Blood* **63,** 42 (1984).

time,[48] and results are quantitated on double logarithmic plots of clotting times versus concentration of normal pooled plasma. Factor XI-deficient plasma (100 μl) is incubated with 100 μl of kaolin (5 mg/ml in saline), 100 μl of 0.2% (w/v) inosithin in 20 mM Tris/saline, pH 7.4, and 10 μl of plasma or sample and 90 μl of the above-mentioned buffer for 5 min at 37°. Then 100 μl of 30 mM CaCl$_2$ is added to initiate clot formation. The observed clotting time is converted to clotting units by comparison to the clotting activities of serial dilutions of a normal pooled plasma.

2. *Radioimmunoassay of Factor XI.* A solid-phase radioimmunoassay (RIA) has been developed to measure factor XI antigen either in plasma samples or as purified protein.[21,47] This is a competitive radioimmunoassay utilizing specific heterologous (rabbit) antifactor XI antibody and the staphylococcal protein A (Staph A) bacterial absorbant as the precipitating agent. Purified factor XI is radiolabeled with ^{125}I either by the technique of Bolton and Hunter or by the Iodogen technique. Monospecific antibody to purified human factor XI is prepared in rabbits as described.[49] The radioimmunoassay is performed by incubating 25 μl of a 1:20,000 dilution of the rabbit antifactor XI antibody with 25 μl of purified factor XI (or test plasma) and 25 μl of ^{125}I-labeled factor XI. After incubation in an Eppendorf microcentrifuge tube (Beckman Instruments, Cedar Grove, NJ) for 1.5 hr at 37°, followed by incubation with 30 μl of a 10% (w/v) suspension of Staph A for 0.5 hr at 23°, the samples are centrifuged, and after amputation, the tips are counted in a gamma counter. Reference plasma consisting of a pool of normal plasma can be used to construct standard curves for the radioimmunoassay. A representative standard curve for the assay shows that the amount of bound ^{125}I-labeled factor XI is inversely proportional to the logarithm of the concentration of normal pooled plasma added. This relationship is linear between 0.005 and 0.5 U/ml. Test materials are diluted as necessary to give values that fall within this range.[47] The coefficient of variation of the assay has been reported as 9.1% with a lower limit of sensitivity for the assay of 0.05 U/ml.[21]

3. *Amidolytic Assay of Factor XIa.* The amidolytic assay of factor XIa using the oligopeptide substrate pyro-Glu-Pro-Arg-pNA has been described previously.[47] Different concentrations of 10 μl of factor XIa (0.6–6.0 μg/ml) are added to 250 μl of 0.1 M sodium phosphate, pH 7.6, containing 0.15 M NaCl, 1 mM EDTA, and 15 μl of the substrate S-2366 (11.5 mM) in a 1-cm cuvette. The rate of hydrolysis is measured using a Gilford System 2600 spectrophotometer (Oberlin, OH).

4. *Factor IX Activation Peptide Release Assay.* The rate of activation of ^{3}H-labeled factor IX can be followed by measurement of the trichloroace-

[48] R. R. Proctor and S. I. Rapaport, *Am. J. Clin. Pathol.* **36,** 212 (1961).
[49] M. S. Lipscomb and P. N. Walsh, *J. Clin. Invest.* **63,** 1006 (1979).

tic acid-soluble activation peptide released during activation by factor XIa, as previously described.[50] For each time point the assay mixture consists of 64 μl of Tris (50 mM), NaCl (100 mM), pH 7.5 (TBS), containing 1 mg/ml bovine serum albumin (BSA), 8 μl of ^{3}H-labeled factor IX [20 μg/ ml, specific radioactivity 0.4×10^6 counts per minute (cpm)/μg], and 8 μl of factor XIa (0.2–0.4 μg/ml). The reaction is stopped by adding 240 μl of an ice-cold mixture containing one part TBS and two parts 50 mM EDTA, pH 7.5. To this mixture is added 160 μl of ice-cold 15% trichloroacetic acid. This is kept on ice and vortexed vigorously and repeatedly for 2 min and then centrifuged at 10,000 g for 3 min at room temperature in a bench-top Brinkman Model 3200 microfuge (Brinkman Instruments Inc., Westbury, NY). Thereafter, 200-μl aliquots of the supernatants are removed into 10 ml of scintillation fluid and counted for ^{3}H in a Beckman LS 8000 scintillation counter (Beckman Instruments Inc., Fullerton, CA). To determine the assay background, 8 μl of enzyme solution is replaced by 8 μl TBS. In each experiment, assay background should be less than 2% of the total number of counts. The initial rates of release of the activation peptide should be determined under conditions wherein less than 20% of the activation peptide has been released.

5. *Proteolytic Cleavage of Factor XI.* The rate of generation of factor XIa can be measured, after incubation of factor XI (containing trace ^{125}I-labeled factor XI) with factor XIIa, by examining the proteolytic cleavage of factor XI by factor XIIa either in the fluid phase or in the presence of high-M_r kininogen and a suitable anionic surface such as kaolin or celite.[51] Factor XI (0.17 μM) containing trace^{125}I-labeled factor XI is incubated in a reaction mixture with BSA (1 mg/ml) and factor XIIa (0.017 mM) at 37°C in a buffer containing phosphate-buffered saline (PBS). At different intervals, aliquots are removed into a buffer containing SDS to examine the cleavage products by SDS-PAGE (8% acrylamide) in the presence of 2-mercaptoethanol followed by autoradiography.

6. *Binding to High-M_r Kininogen.* The binding of factor XI to high-M_r kininogen has been studied[51] using polyvinyl chloride microtiter plates, the wells of which are coated with high-M_r kininogen by incubation with 100 μl of the protein (100 μg/ml) for 2 hr at room temperature. Residual binding sites on the wells are blocked by incubating with 200 μl of 5 mg/ml BSA in PBS (PBS/BSA) at room temperature for 2 hr, after which the excess coating solution containing high-M_r kininogen is aspirated. After washing the wells with PBS/BSA to remove any unbound high-M_r kinino-

[50] P. N. Walsh, H. Bradford, D. Sinha, and J. R. Piperno, *J. Clin. Invest.* **73**, 1392 (1984).
[51] H. Akiyama, D. Sinha, F. Seaman, E. Kirby, and P. N. Walsh, *J. Clin. Invest.* **78**, 1631 (1986).

gen, 100 μl of a mixture of ^{125}I-labeled factor XI (6 μg/ml, with a specific activity of 5×10^6 cpm/μg) and either peptide (50 to 300 μg/ml) or antibody (10^{-13} to 10^{-6} M) or buffer (preincubated for 10 min at room temperature in polypropylene tubes precoated with 5 mg/ml BSA) are added to the wells and incubated for 3–4 hr at room temperature. The same incubation mixtures are also added to additional wells that are coated with BSA (without high-M_r kininogen) to determine background counts arising from binding of ^{125}I-labeled factor XI to BSA. They are thoroughly washed with PBS/BSA, dried, and counted in a gamma counter.

7. *Factor XIa Complex Formation with α_1-Protease Inhibitor.* Because one of the major plasma inhibitors of factor XIa appears to be α_1-protease inhibitor,[41] it may be desirable to study complex formation between factor XIa and α_1-protease inhibitor.[38] Radioiodinated factor XIa (0.8 μg/ml, 2×10^6 cpm/μg) is incubated in polypropylene tubes at 37° in Ca^{2+}-free, HEPES-buffered Tyrode's solution, pH 7.4, in the presence or absence of purified α_1-protease inhibitor (320 μg/ml) for varying time periods. Reactions are stopped by boiling in SDS sample buffer in the presence of 2-mercaptoethanol and samples are analyzed by SDS-polyacrylamide (8%, w/v) gel electrophoresis. The gels are dried and autoradiography is carried out to detect and quantitate complex formation between α_1-protease inhibitor and the light chain of factor XIa. Alternatively, gels can be sliced and counted for ^{125}I.[38]

B. Isolation from Plasma

1. *Conventional Methods.* Factor XI can be purified from human plasma by the method of Walsh and co-workers.[50] Normal human plasma (1 liter) is chromatographed on a DEAE-Sephadex A-50 (Whatman Chemical Separation, Inc., Clifton, NJ) column at pH 8.3. The void fraction containing factor XI is subjected to SP-Sephadex G-50 chromatography at pH 5.3, followed by affinity column chromatography using insolubilized high-M_r kininogen.[52] High-M_r kininogen can be purified by the method of Kerbiriou and Griffin[53] and is insolubilized on cyanogen bromide-activated Sepharose supports using instructions provided by Bio-Rad Laboratories (Richmond, CA). Pooled fractions eluted from the SP-Sephadex G-50 column are dialyzed into 0.04 M Tris and 0.15 M NaCl, pH 7.4, applied to a 1.4×2-cm high-M_r kininogen column, and eluted with 0.2 M sodium acetate and 0.6 M NaCl, pH 5.5. The resulting factor XI appears homogeneous on polyacrylamide gel electrophoresis in the presence of

[52] F. vanderGraaf, J. S. Greengard, B. N. Bouma, D. M. Kerbiriou, and J. H. Griffin, *J. Biol. Chem.* **258**, 9669 (1983).
[53] D. M. Kerbiriou and J. H. Griffin, *J. Biol. Chem.* **254**, 12020 (1979).

SDS with a specific activity of 270 U/ml. It is stored in 0.2 M sodium acetate and 0.6 M NaCl, pH 5.3, at $-70°$.

2. *Immunoaffinity Purification.* The isolation of human factor XI using a monoclonal antibody affinity column was first described by Sinha and co-workers.[54] Fresh frozen plasma (200 ml) is clarified by centrifugation at 10,000 g and applied to an antibody (5F4) affinity column (3.5-ml bed volume) equilibrated in phosphate-buffered saline. Plasma is initially passed through the column at 10 ml/hr, running the column overnight at room temperature. No factor XI clotting activity is detected in the flow-through fraction. The column is then washed with approximately 10 volumes of the same buffer and then with 0.2 M NaHCO$_3$/1 M NaCl, pH 8.0, until the optical density at 280 nm of the wash is <0.1. The column is further washed with 10 volumes of 0.1 M glycine/HCl/1.0 M NaCl, pH 3.0. Factor XI is eluted from the column with 4 M guanidine hydrochloride, pH 4.0, is collected batchwise, and is immediately dialyzed versus 40 mM Tris–succinate buffer, pH 8.3, containing 50 μg/ml Polybrene, 1.0 mM benzamidine, 1.0 mM EDTA and 0.02% NaN$_3$. The dialyzed material is passed through a DEAE A-50 column (5 ml) connected in tandem with a protein A column (2 ml) equilibrated with the same buffer. The column is washed with the same buffer (the volume of the wash being twice the volume of the sample size). The effluent together with the wash is collected batchwise, pooled, and concentrated. Factor XI has been purified 15,000-fold with a final yield of 40–60% by this two-step procedure. The entire purification procedure can be carried out in a total of 2–3 days.

III. Monoclonal Antibodies

A. Production

Murine monoclonal antibodies to purified factor XI can be prepared by immunization of mice and subsequent cellular hybridization.[54,55] BALB/c mice are immunized by three intraperitoneal injections of 50–60 μg of purified factor XI in Freund's complete adjuvant given 1 week apart. At 7–10 days after the last immunization, the mice are boosted with one intravenous injection of the same amount of antigen in PBS. Within 3–5 days after the final injection, the mice are sacrificed, the spleens are removed aseptically, and single-cell suspensions are prepared in minimal essential medium (MEM; GIBCO, Grand Island, NY) containing 20% fetal calf serum (FCS). Cells (approximately 10^8) from a single spleen are

[54] D. Sinha, A. Koshy, F. S. Seaman, and P. N. Walsh, *J. Biol. Chem.* **260**, 10714 (1985).
[55] G. Kohler and C. Milstein, *Nature (London)* **256**, 495 (1975).

mixed with 10^7 Sp 2/0-Ag14 mouse myeloma cells, pelleted, and washed in medium containing no fetal calf serum. Sp 2/0-Ag14 is a hybridoma variant that does not synthesize immunoglobulin components and which was derived from the P3-X63-Ag8 myeloma line by Shulman and co-workers.[56] Cells are fused by a modification of the procedure described by Kennett and McKearn.[57] The medium is removed, the pellet loosened gently, and 1 ml of 35% (v/v) polyethylene glycol (PEG) 1500 in MEM is added. After 7–9 min, the cells are pelleted and then diluted with 5 ml of medium without serum, followed by addition of another 5 ml of medium without serum, followed by addition of another 5 ml of medium with 20% fetal calf serum. The cells in diluted PEG are pelleted and resuspended in 30 ml of HY medium (Dulbecco's MEM with high glucose, 4.5 g/liter), 10% NCTC 109 (Microbiological Assoc., Walkerville, MD), 20% fetal calf serum, 0.15 mg/ml oxaloacetate (Sigma, St. Louis, MO), 0.05 mg/ml pyruvate (Sigma), 0.2 U/ml bovine insulin (Sigma). In addition to the purine and pyrimidine bases present in NCTC 109 medium, thymidine (16 μM) and hypoxanthine (0.1 mM) are added to the medium in which the fused cells are plated. The 30 ml of cells are evenly suspended and distributed into six 96-well microplates (50 μl per well). The next day each well is fed with an equal volume of the same medium with aminopterin (0.8 μM) to make hypoxanthine/aminopterin/thymidine (HAT)-selective medium. The cells are fed twice a week with medium without aminopterin. Colony growth is apparent beginning at day 14.

B. Isolation of Hybridoma Clones with Antifactor XI Activity and Isolation of Antibody

Tissue culture supernatants from hybrids can be tested for antifactor XI activity using the solid-phase radioimmunoassay described previously.[54] Cells from wells that are found to be strongly positive are frozen in fetal bovine serum (90%) with dimethyl sulfoxide (DMSO; 10%) in liquid nitrogen at 5×10^6 cells per ampule. Supernatants positive in the RIA (see Section II,A for details) are also tested for their capacity to neutralize factor XI in the clotting assay. These can be chosen for further cloning. Cloning is done by a limiting dilution method.[57] Cells are grown for freezing and passage into mice as an ascitic tumor. Pristane primed BALB/c mice are injected intraperitoneally with 10^6 cloned hybridoma cells. After 10–12 days, ascitic fluids are collected; 10–12 ml can usually be obtained from

[56] M. Shulman, C. D. Wilde, and G. Kohler, *Nature (London)* **276,** 269 (1978).
[57] R. H. Kennett and T. J. McKearn, *in* "Hybridomas: A New Dimension in Biological Analysis" (R. H. Kennett, T. J. McKearn, and K. B. Bechtol, eds.), p. 365. Plenum, New York, 1980.

each mouse. Antibodies are isolated from the ascitic fluids by precipitation in 40% (w/v) ammonium sulfate followed by gel filtration with BioGel A-1.5m using a column with dimensions of 120×2.5 cm (bed volume 580 ml) in a buffer containing PBS and a fraction size of 5 ml. The purified antibodies are generally greater than 98% pure when characterized by SDS-polyacrylamide gel electrophoresis and quantitated using an extinction coefficient of 14 for a 1% (w/v) solution at 280 nm.[54]

C. Characterization

1. *Immunoglobulin Subtyping.* It is necessary to determine that antibodies are truly monoclonal (not mixed populations) and to define subtypes to use as controls. Therefore, subclasses of the different monoclonal antibodies can be determined using a commercial kit available from Boehringer Mannheim Biochemicals (Indianapolis, IN).[58] Microtiter plates are coated with antigen by adding 100 ml of 0.5 μg/ml of factor XI to each well and incubating for 18 hr at 4°C. After washing, the antibody is then bound. Subclass-specific rabbit antimouse antibody is then added followed by peroxidase-labeled goat antirabbit IgG. The substrate ABTS [2,2′-azinobis(3-ethylbenzthiazoline-6-sulfonic acid)] in citrate buffer, 0.05 M, pH 4.3, at a concentration of 0.2 mg/ml buffer is made before use, and 200 μl is added to each well and the color is read in 5–10 minutes. In control wells, normal rabbit serum is used instead of rabbit antimouse antibody. Positive results are quantitated by measuring the optical density at 415 nm.

2. *Affinity and Stoichiometry of Binding to Factor XI.* To determine the affinity constants in solution of monoclonal antibody complexes with factor XI,[58] an enzyme-linked immunosorbent assay (ELISA) can be used as described by Friguet and co-workers.[59] Briefly, factor XI at various concentrations (5–125 μg/ml) is mixed with a constant amount (1 μg/ml) of monoclonal antibody in 0.01 M Tris, 0.15 M NaCl, pH 7.4, containing 2% (w/v) bovine serum albumin (TBS–BSA), previously determined to be an optimal concentration from a preliminary ELISA calibration. After incubation overnight at 4°, 200 μl of each mixture (factor XI, 5–125 μg/ml; monoclonal antibody, 1 μg/ml) is transferred to the wells of a microtiter plate previously coated with factor XI (200 μl per well, at 1 μg/ml in TBS–BSA, for 24 hr at 4°). After washing with PBS supplemented with 0.5% Tween 20, the bound immunoglobulins are detected

[58] F. A. Baglia, D. Sinha, and P. N. Walsh, *Blood* **74**, 244 (1989).
[59] B. Friguet, A. F. Chaffotte, L. Djavadi-Ohaniance, and M. E. Goldberg, *J. Immunol. Methods* **77**, 305 (1985).

and quantitated by adding goat antimouse IgG coupled with alkaline phosphatase, the activity of which is then measured in each well. The substrate buffer consists of 0.05 M sodium carbonate, pH 9.8, containing 2 mg/ml p-nitrophenyl phosphate disodium and 1 mM MgCl$_2$; 200 μl of substrate solution is added to each well and in 10–30 min the plates are quantitated by measuring the optical density at 405 nm. Dissociation constants (K_D) for binding of each antibody to factor XI are determined as previously described.[58,59]

   3. Determination of Chain Specificity. In order to determine the chain specificity of each antibody, immunoblot analysis can be performed on electrophoretically separated native as well as reduced and alkylated factor XI and factor XIa transferred to nitrocellulose membranes.[54,58] After running an 8% (w/v) SDS-PAGE, the gel is soaked in transfer buffer [containing 25 mM Tris, 192 mM glycine, 20% (v/v) methanol at pH 8.3] for 1 hr at room temperature with gentle shaking to remove the SDS. After soaking, the gel "sandwich" is set up and the proteins transferred to nitrocellulose at 1.5 mA for about 4 hr. After transfer, the nitrocellulose strips are rinsed briefly with distilled water. After putting nitrocellulose strips in a plastic box, 50 ml of nonfat dry milk (NFDM) solution is added to each box. This solution contains 5% NFDM, 10 mM Tris-HCl, 0.15 M NaCl, pH 7.5. The strips are shaken gently for 1–2 hr at room temperature and rinsed briefly with distilled water. Nitrocellulose membranes are soaked in NFDM solution with the first antibody (50 ml of NFDM with an antibody concentration of 20 μg/ml), and gently shaken for 1–2 hr at room temperature, then washed three times with NFDM solution with 0.1% Tween 20 for 10 min at room temperature with shaking (50 ml per box). The fourth wash is carried out with NFDM buffer without Tween 20. Next, the nitrocellulose membranes are soaked in NFDM buffer (no Tween) with the second antibody (50 μl of antimouse IgG alkaline phosphatase conjugated in 50 ml of NFDM solution), shaken for 2 hr at room temperature, and washed four times as described at first antibody addition, and then washed with distilled water. To develop the immunoblot, 50 ml of carbonate buffer (containing 0.1 M NaHCO$_3$, 1.0 mM MgCl$_2 \cdot$6H$_2$O, pH 9.8) is prepared. In two separate tubes, 15 mg of nitro blue tetrazolium (NBT) is dissolved in 0.5 ml of 70% (v/v) N,N-dimethylformamide (DMF), and 7.5 mg of 5-bromo-4-chloro-3-indoyl phosphate (BCIP) is dissolved in 0.5 ml of 100% DMF. Each solution is mixed repeatedly for 15 min. The NBT and BCIP solutions are added to 50 ml of carbonate buffer at room temperature, mixed, and added to the nitrocellulose membranes immediately. When protein bands are adequately visualized the nitrocellulose membranes are rinsed with water to stop the reaction.

D. Structure–Activity Studies

1. Binding of Monoclonal Antibodies to Heavy Chain of Factor XI. We have reported on the production, characterization, and use of murine hybridoma antibodies directed against various epitopes in human coagulation factor XI.[54] One of these antibodies (3C1) was shown to bind the heavy chain of reduced and alkylated factor XIa without affecting the amidolytic activity of intact factor XIa.[51] Another antibody (5F7) also binds to the heavy chain of factor XIa, as demonstrated by the following: (1) 5F7 antibody recognizes the heavy chain of reduced and alkylated factor XIa using an immunoblotting procedure; (2) when ^{125}I-labeled reduced and alkylated factor XIa is passed over a 5F7 immunoaffinity column (3.5-ml bed volume) in phosphate-buffered saline, the light chain passes through during collection of 0.5-ml fractions, whereas the heavy chain is bound and then eluted by application of 4 *M* guanidine hydrochloride.[58] We then used these two antibodies to determine whether they are directed against similar or related sites on the molecule. Figure 1 shows the effects of unlabeled monoclonal antibodies (3C1 or 5F7) on binding of ^{125}I-labeled 5F7 to factor XI bound to microtiter wells. Unlabeled mono-

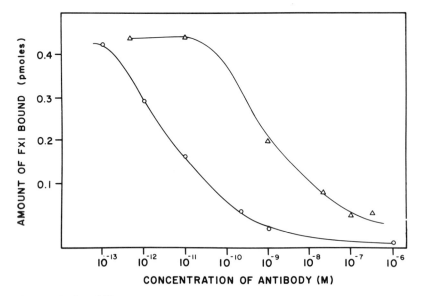

FIG. 1. Binding of ^{125}I-labeled 5F7 monoclonal antibody to factor XI. The conditions for binding are described in the text. Results shown indicate the amount of ^{125}I-labeled 5F7 bound in the presence of unlabeled antibody 5F7 (O) or 3C1 (△). Reproduced with permission from Baglia *et al.*[58]

clonal antibody 3C1 or 5F7 (10 pM to 10 μM) was incubated for 20 min at 20° with 1.3 nM ^{125}I-labeled 5F7 in the presence of BSA (1 mg/ml) in the wells to which purified factor XI had been bound. The concentration of unlabeled 5F7 required to block the binding of 50% of ^{125}I-labeled 5F7 (1.3 nM) to immobilized factor XI was 1.3 nM (as expected), whereas the concentration of unlabeled 3C1 required to inhibit 50% binding was 1.3 μM, or 1000-fold greater than that of 5F7. Because these results might be explained by differing binding affinities of the two antibodies to factor XI, we determined dissociation constants in solution as described above (Section II,C,2). The dissociation constant (K_D) for the binding of 5F7 to factor XI in solution was 6.2×10^{-10} M whereas the corresponding K_D for antibody 3C1 was 9.0×10^{-10} M. Therefore, the differences in concentrations of the two antibodies required to displace ^{125}I-labeled 5F7 from factor XI are not a consequence of different binding affinities. We conclude that antibody 3C1 is directed at an epitope in the heavy chain of factor XI that is distinct from that recognized by the 5F7 monoclonal antibody.

2. Effect of Monoclonal Antibodies on Procoagulant Activity of Factor XI. Because our experiments have implicated the heavy chain in the enzymatic activity of factor XIa[30,54] and in the activation of factor XI,[51] we compared the effects of antibodies 3C1 and 5F7 on factor XI coagulation activity.[58] Factor XI (2.5 nM) was incubated with either buffer or different antibody concentrations (2 to 42.5 nM) of each antibody for 20 min at 37° before measuring the coagulant activity. As shown in Fig. 2A, the heavy-chain-specific monoclonal antibody 5F7 inhibited 100% of factor XI procoagulant activity at a concentration of 30 nM, whereas the 3C1 heavy-chain-specific monoclonal antibody resulted in 75% inhibition at a similar concentration.

3. Effects of Monoclonal Antibodies on Procoagulant Activity of Factor XIa. The procoagulant activity of factor XIa, determined both in a factor XIa clotting assay and by its capacity to activate factor IX, was examined in the presence of various concentrations of heavy-chain-specific antibodies. Factor XIa (2.5 nM) was incubated with either buffer or different concentrations (2 to 40 nM) of each antibody for 20 min at 37°. As shown in Fig. 2B, heavy-chain-specific antibody 3C1 (10 nM), in the absence of kaolin, inhibited 75% of factor XIa procoagulant activity in a coagulant assay, while heavy-chain-specific 5F7 had no effect. The effects of the antibody Fab' fragments on the rate of activation of ^{3}H-labeled factor IX by factor XIa are shown by Fig. 3. Antibody 3C1 Fab' fragments (20 nM) inhibited the initial rate of ^{3}H-labeled activation peptide release from factor IX by nearly 100%, whereas the 5F7 Fab' fragments had no effect. Normal mouse IgG did not inhibit factor XIa either in the clotting assay or in the ^{3}H-labeled factor IX activation peptide-release assay.

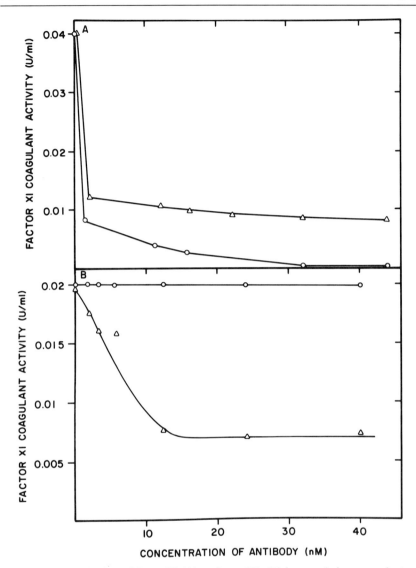

Fig. 2. Neutralization of factor XI (A) or factor XIa (B) in coagulation assays by two different monoclonal antibodies, 3C1 (△) and 5F7 (○). (A) Factor XI (2.5 nM) was incubated with either buffer or different antibody concentrations (2 to 42.5 nM) of each antibody for 20 min at 37° before measuring the coagulant activity. (B) Factor XIa (2.5 nM) was incubated with either buffer or different antibody concentrations (2 to 40 nM) of each antibody for 20 min at 37° before measuring the clotting time in factor XI-deficient plasma in the absence of kaolin. Reproduced with permission from Baglia et al.[58]

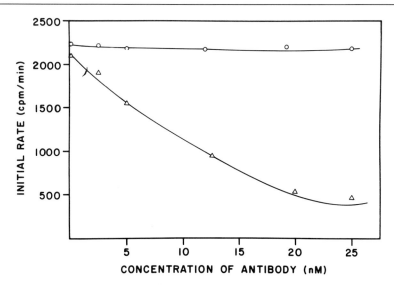

FIG. 3. Effect of Fab' fragments of monoclonal antibodies, 5F7 (O) and 3C1 (△), on the activation of ³H-labeled factor IX by factor XIa. Factor XIa (2.5 nM) was incubated with various concentrations of antibody solutions (2.5 to 25 nM) for 20 min at 20° prior to its use in the ³H-labeled factor IX activation peptide release assay. The final concentrations of factor XIa and ³H-labeled factor IX in the ³H-labeled peptide release assay were 0.04 and 0.2 μg/ml, respectively. Reproduced with permission from Baglia et al.[58]

4. Effects of Monoclonal Antibodies on Binding of Factor XI to High-M_r Kininogen. It is well documented that the heavy chain region of factor XI contains binding sites for high-M_r kininogen.[17,51] Because the heavy-chain-specific antibodies (3C1 and 5F7) inhibit the rate of factor XI activation by factor XIIa in the presence of high-M_r kininogen and kaolin, we determined whether these antibodies might inhibit the binding of factor XI to high-M_r kininogen, thus preventing the complex formation necessary for efficient, surface-mediated activation by factor XIIa. Figure 4 shows the effects of intact IgG and Fab' fragments of the monoclonal antibodies on binding of ¹²⁵I-labeled factor XI to high-M_r kininogen (see Section II,A for details of assay of factor XI binding to high-M_r kininogen). In this experiment, ¹²⁵I-labeled factor XI (0.56 nM) was incubated at 20° for 20 min with high-M_r kininogen bound to microtiter wells in the presence of various concentrations of the 5F7 or 3C1 antibodies (10^{-13} to 10^{-6} M) or their Fab' fragments. It is apparent from these results that monoclonal antibody 5F7 can completely block factor XI binding to high-M_r kininogen and that 50% inhibition of binding occurred at approximately 5×10^{-10} M 5F7 (i.e., close to the K_D for binding 5F7 to factor XI). However, 100-fold less Fab' fragment of 5F7 (approximately 5×10^{-12} M) was required to

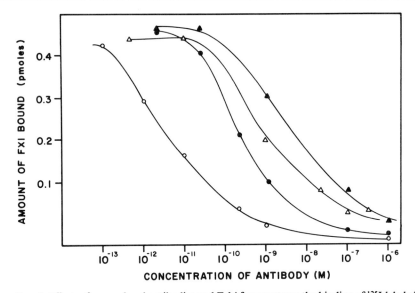

FIG. 4. Effects of monoclonal antibodies and Fab' fragments on the binding of ^{125}I-labeled factor XI to high-M_r kininogen. ^{125}I-Labeled factor XI (0.56 nM) was incubated with Fab' fragments of antibodies at various concentrations (10^{-13} to 10^{-6} M) at 20° for 20 min in microtiter wells with bound high-M_r kininogen. Binding of ^{125}I-labeled factor XI to the wells of microtiter plates not containing high M_r kininogen was <2% of the control value. The maximum variation of the cpm bound for each experimental observation was <2% of total cpm bound. Data shown are those obtained with 5F7 antibody (●) or Fab' fragments (○) or with 3C1 antibody (▲) or Fab' fragments (△). Reproduced with permission from Baglia *et al.*[58]

achieve 50% inhibition of binding of factor XI to high-M_r kininogen than the intact 5F7 monoclonal antibody. The concentration of 3C1 Fab' fragment required to inhibit factor XI binding to high-M_r kininogen was 1000-fold greater (5×10^{-9} M) than that of the 5F7 Fab' fragments (5×10^{-12} M).

5. *Summary of Structure–Function Studies with Monoclonal Antibodies.* These studies suggest that the epitopes recognized by the two heavy-chain-specific monoclonal antibodies are separate and distinct structural sites exposed on the surface of factor XI (XIa). These epitopes are likely to be in close proximity to functional domains within the heavy chain region of factor XI (XIa) that are important for the activation of the zymogen and for the expression of the enzymatic activity of factor XIa. It is possible, on the basis of these studies, to formulate the hypothesis that one functional domain, the cofactor (high-M_r kininogen) binding site, is near the epitope recognized by the 5F7 antibody, whereas another functional domain, the

substrate (factor IX) binding site, is near the epitope recognized by the 3C1 antibody. The precise location and characteristics of the structural sites representing these functional sites are described in detail below.

IV. Protein Modification

A. Chemical and Enzymatic Fragmentation

1. *CNBr Digestion of Factor XI.* Because of the known positions of the five methionine residues[7] and the intrachain disulfide bonds in each identical chain of the homodimer,[8] unreduced factor XI is expected to yield four peptides after CNBr digestion whereas reduced and alkylated factor XI should yield six peptides.

We have developed a procedure for CNBr digestion of factor XI.[60] Briefly, factor XI is dissolved in 70% (v/v) formic acid (1 mg of protein per ml), and solid CNBr is added to a final concentration of approximately 100 mg/ml. After a reaction time of 20 to 24 hr in the dark at room temperature, the solution is dialyzed against distilled water for 18 hr and subsequently concentrated to dryness.

2. *Cleavage of Tryptophanyl Peptide Bonds.* An established method has been used to cleave tryptophanyl peptide bonds[61] in factor XI. Approximately 0.2 mg of factor XI peptide is dissolved in 0.5 ml of 12 mM potassium phosphate in 0.4 M guanidine hydrochloride at pH 6.0. After 30 min of exposure to 1 mg of o-iodosobenzoic acid (Pierce Chemical Co., Rockford, IL) in this buffer, the solution is made 80% (v/v) in acetic acid and allowed to stand for 24 hr. Subsequently, the solution is dialyzed against distilled water for 18 hr and concentrated to dryness.

B. Purification and Characterization of Peptides

1. *HPLC.* The HPLC system employed in our laboratory is from Waters (Milford, MA; Waters 600 gradient module, Model 740 data module, Model 46K universal injector and Lambda-Max Model 481 detector). Reverse-phase chromatography is performed using a Waters C$_8$ μBondapak column which is equilibrated with 0.1% (v/v) trifluoroacetic acid and is then developed with a linear gradient of aqueous 70% (v/v) acetonitrile containing 0.1% (v/v) trifluoroacetic acid with the detector set at a wavelength of 220 nm.

2. *Amino Acid Sequencing.* Automated NH$_2$-terminal sequencing is performed on a gas-phase sequenator (Applied Biosystems, Foster City, CA Model 470A) coupled to an on-line PTH analyzer (Applied Biosys-

---

[60] F. A. Baglia, B. A. Jameson, and P. N. Walsh, *J. Biol. Chem.* **265**, 4149 (1990).
[61] W. C. Mahoney and M. A. Hermodson, *Biochemistry* **18**, 3810 (1979).

tems, Model 120A). Standard protocols of the manufacturer are followed both with regard to Edman degradation and separation of PTH derivatives by HPLC.[62] This sequentator uses gas-phase reagents at the coupling and cleavage steps of the Edman degradation. The HPLC system consists of two Waters Model 510 pumps, a Waters Model 680 gradient controller, an LKB Unicord S detector with 206-nm filter, and a Waters Model 760 integrator. A total of 200 pmol of peptide is used to define 6–10 residues.

3. Compositional Analysis. Hydrolysis in 6 M HCl (at 110°C for 24 hr under nitrogen plus gaseous HCl) and derivatization are carried out in 0.3 ml crimp-top microvials. Precolumn derivatization is performed essentially as described by Koop and co-workers.[63] At the end of the reaction period the samples are dried and stored at −20°. Phenylthiocarbamyl-amino acid analysis after vapor-phase hydrolysis is performed using the methods of Ebert.[64] A Beckman Model 421A HPLC is used which is equipped with a Waters WISP autoinjector with computerized data acquisition utilizing Nelson analytical data acquisition software (P. E. Nelson Co., Cupertino, CA) and hardware with an IBM-AT-compatible computer.

4. Capillary Electrophoresis. One of the first well-characterized comparisons of the use of capillary electrophoresis as an adjunctive technique to HPLC peptide mapping was reported by Cobb and Novotny.[65] With regard to HPLC, capillary electrophoresis offers an alternative and often complementary means of analyzing peptide fragments generated from a proteolytic digest. The theoretical basis of separation is qualitatively different from that of reverse-phase HPLC and capillary electrophoretic systems are able to detect picogram quantities of proteins.

Capillary electrophoresis (CE) is a relatively new and emerging technique for fine analysis of peptide mapping, peptide purity, and conformation. Furthermore, this technique is well-suited for automation with real-time data analysis. Sample volumes in the nanoliter range are used in an analytical run and only microliters of buffer are consumed. Typical sample run times vary from 5 to 20 min as compared with the 40- to 100-min run times used in an HPLC analysis.

In order to utilize CE effectively as a technique for the analysis of peptides, it is important to understand some of the basic principles of electrophoretic separation using narrow-bore (approximately 50 μm i.d.) capillaries. The migration of a sample, defined as its electrophoretic mobility (μ_m), in a given buffer system is an absolute characteristic of the sample,

[62] R. M. Hewick, M. W. Hunkapiller, L. E. Hood, and W. J. Dreyer, *J. Biol. Chem.* **256,** 7990 (1981).
[63] D. R. Koop, E. T. Morgan, G. E. Tarr, and M. J. Coon, *J. Biol. Chem.* **257,** 8472 (1982).
[64] R. F. Ebert, *Anal. Biochem.* **154,** 431 (1986).
[65] K. A. Coob and M. Novotny, *Anal. Chem.* **61,** 2226 (1989).

unlike the relative mobility observed in an SDS-PAGE system. In free solution, the electrophoretic mobility of a solute is simply defined as the electrophoretic velocity of the solute (V_e; measured in cm/sec) per unit electrical field strength (E, measured in V/cm).

$$\mu_m = V_e/E$$

Therefore, increasing the electrical field strength has the added advantage of increasing the electrophoretic velocity of the solutes, leading to fast separations. However, as one increases the electric field, one concomitantly increases local heating, leading to convection currents in the buffer system that disrupt the tight banding patterns of a sample as it moves through the capillary tube. The narrow-bore, fused-silica capillaries typically employed in these analyses take advantage of the "wall effects" described by Jorgenson and Lukacs.[66] Basically, it was observed that diffusion effects (leading to severe peak broadening) caused by joule heating in high electrical fields can be minimized by decreasing the ratio of cross-sectional area of the separation tube to its surface area. Thus, the small diameter of the capillary and the resulting large surface-to-volume ratio, allows for efficient dissipation of resistive heat.

Separation of peptide products in free-solution CE occurs as a function of the differences between the μ_m values of each product. That is to say, the selectivity (α) of a given separation can be expressed in terms of the μ_m:

$$\alpha = \Delta\mu_{1/2}/\mathrm{avg}(\mu_{1,2})$$

where $\Delta\mu_{1,2}$ is the difference in mobility between product 1 and product 2; and $\mathrm{avg}(\mu_{1,2})$ is the average mobility of the two peptide products. Differences in μ_m arise from structural differences in the peptide products such as size, conformation, and charge. Additionally, the electrophoretic velocity (V_m) of a peptide is related to both its net charge and Stokes radius. Therefore, selectivity (α) can be enhanced by manipulating the properties of the solvent, e.g., pH, ionic strength, viscosity, and dielectric constant. It is, consequently, important to consider the physical properties of the peptide in determining the exact buffer conditions for a CE analysis. It is often necessary to try several different buffer conditions before deciding on an optimal set of conditions.

For general analysis of linear synthetic peptides we usually use a 100 mM phosphate buffer, pH 2.4. For cysteine-cyclized peptides, a 50 mM sodium phosphate buffer at pH 9.0 (with NaOH) appears to give good resolution of the resultant products.

---

[66] J. W. Jorgenson and K. D. Lukacs, *J. Chromatogr.* **218**, 209 (1981).

C. Structure–Activity Studies

1. CNBr Digest of Factor XI. Based on the positions of methionine residues in the sequence of each polypeptide chain of factor XI, a CNBr digest (nonreduced) should be composed of four domains having masses of 11,320, 10,745, 17,125, and, 27,425 D. In practice, after digestion of native factor XI with CNBr, bands of apparent M_r 11,000, 17,000, and 29,000 were demonstrated by SDS-polyacrylamide gel electrophoresis.[60] The details of CNBr digestion are factor XI are given in Section IV,A.

2. Isolation of Peptides of Factor XI Containing High-M_r Kininogen Binding Site. In order to identify the structural domain of factor XI that binds high-M_r kininogen, CNBr-digested factor XI is passed over a 5F7 monoclonal antibody affinity column. The following procedure can be used to isolate CNBr peptides of factor XI. A monoclonal antibody affinity column is prepared by incubating purified 5F7 IgG (1 mg/ml) with Affi-Gel 10 (Bio-Rad) according to the conditions described by the manufacturer. CNBr-digested factor XI (1.0 mg/ml) in HPLC-grade water is applied to the antibody affinity column (3.5 ml) and is equilibrated in HPLC-grade water. The column is then washed with approximately 10 volumes of the same solution. The factor XI-derived peptides are eluted from the column with 4 M guanidine hydrochloride at pH 4.0 (20 ml), and immediately dialyzed with 3500 M_r cutoff tubing (Spectropore, Spectrum Medical Industries, Inc., Los Angeles, CA) versus HPLC-grade waters. The dialyzed material is finally dried, rehydrated, and examined by HPLC. The peptides that pass through this column do not inhibit ^{125}I-labeled factor binding to high-M_r kininogen, whereas peptides bound to the affinity column and eluted with 4 M guanidine-hydrochloride are capable of inhibiting ^{125}I-labeled factor XI binding to high-M_r kininogen.

To examine the purity of the material eluted from the antibody affinity column, SDS-PAGE is carried out. This revealed a species with an apparent M_r of 10,000–15,000 (Fig. 5, inset B). However, when this fraction was examined by HPLC, two separate and distinguishable peaks were demonstrated (Fig. 5). When the peptides in these peaks were isolated by HPLC and reinjected, two peaks were obtained with retention times identical to the original mixture (Fig. 5, inset A). When these fractions were tested to determine whether they contained the high-M_r kininogen binding site, only peptide I was able to inhibit ^{125}I-labeled factor XI binding to high-M_r kininogen. Peptide II had no effect in the assay, and there was no additive effect of peptide I and II tested together at equimolar concentrations. We conclude that peptide I contains the high-M_r kininogen binding site.

To place peptide I containing the high-M_r kininogen binding fragment within the structure of the factor XI molecule, its NH_2-terminal sequence was determined as follows: X-X-Val-Thr-Glu-Leu-Leu-Lys-Asp-Thr. With

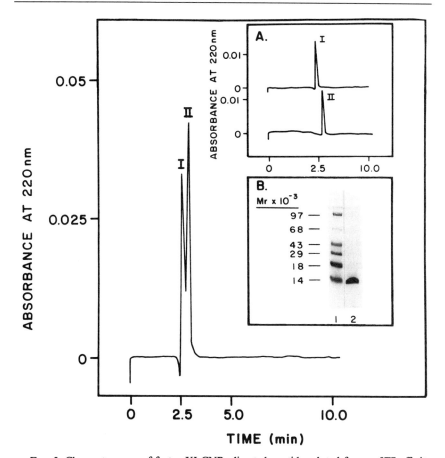

FIG. 5. Chromatograms of factor XI CNBr-digested peptides eluted from a 5F7 affinity column. The chromatography conditions are described in the text. (Inset A) Chromatograms of peak I and peak II, separated by HPLC and reinjected. Time 0 is the point of injection of the sample. The absorbance was measured at 220 nm. (Inset B) SDS-polyacrylamide gel electrophoresis of purified factor XI high-M_r kininogen binding fragment. Lane 1 shows molecular weight standards; lane 2 shows the isolated high-M_r kininogen binding fragment. Silver-stained 15% SDS gel without added reducing agent is shown. Reproduced with permission from Baglia et al.[60]

the exception of the first two residues (not determined), this sequence is identical to that of the 10 NH$_2$-terminal amino acids in the heavy chain of factor XI. Because a methionine residue is present at position 102 where CNBr would be expected to effect chemical cleavage of the molecule and because the apparent M_r of the inhibitory peptide by SDS-PAGE is 10,000–15,000, it is reasonable to conclude that the inhibitory fragment is identical to the amino-terminal domain of the factor XI heavy chain region, Glu-1 through Met-102.

3. Cleavage of Peptide I with o-Iodosobenzoic Acid and Identification of Peptide Fragment that Binds High-M_r Kininogen. In order to map more finely the high-M_r kininogen binding site on peptide I, this CNBr fragment was cleaved with o-iodosobenzoic acid, which should produce two peptides with M_r approximately 5000 because a tryptophan is present at position 55. After treating peptide **I** (0.2 mg/ml) with o-iodosobenzoic acid (1 mg/ml in 12 mM potassium phosphate in 0.4 M guanidine hydrochloride, pH 6.0), the cleavage products were applied to an affinity column to which high-M_r kininogen was bound. A column was prepared by incubation of purified high-M_r kininogen (1 mg/ml) with Affi-Gel 10 (Bio-Rad) according to the conditions described by the manufacturer. Iodosobenzoic acid-treated peptides (1.0 mg/ml) in a buffer containing Tris (40 mM), NaCl (0.15 M), and NaN$_3$ (0.02%), pH 7.4, were applied to a high-M_r kininogen column (3.0 ml) equilibrated in the same buffer. The column was washed with approximately 10 volumes of the same buffer. The peptide fragment **(Ib)** that bound to this column was eluted with a buffer containing sodium acetate (0.2 M), NaCl (0.6 M), and NaN$_3$ (0.02%), pH 5.3. It was analyzed for its NH$_2$-terminal sequence. This sequence is identical to that of the first six amino acids (i.e., residues 55–60) after the Arg-Trp o-iodosobenzoic acid cleavage site of this CNBr peptide of factor XI. Peptide **Ib** also inhibited the binding of factor XI to high-M_r kininogen, whereas the peptide that passed through the column (Glu-1 through Arg-54) did not inhibit factor XI binding to high-M_r kininogen. The identification of peptide **Ib** as comprising residues 55–102 of the factor XI heavy chain was also confirmed by amino acid composition (Table I). These experiments revealed that a high-M_r kininogen binding site in the heavy chain of factor XI is contained within residues Trp-55 through Met-102.

V. Computer Modeling of Domain Structure

A. Rationale

The overall goal of the molecular modeling of a protein domain is to be able to use this information in the design and execution of structure–function studies. The approach we have chosen is to use the resultant structural model as a design template for building conformationally restricted synthetic analogs. These analogs are intended to mimic both the sequence and shape of the parent protein. In an optimal situation, one can use an experimentally derived structure (either through NMR or crystallography) as a template in the design of conformationally restrained syn-

TABLE 1
AMINO ACID COMPOSITION OF FACTOR XI FRAGMENTS[a]

	Peptide I		Peptide Ib	
Amino acids	Observed residues[b]	Expected residues $1-102^c$	Observed residues[b]	Expected residues $55-102^c$
Asx	6.2	9	5.4	6
Glx	ND	9	2.0	3
Ser	8.02	8	5.9	5
Gly	3.3	3	1.8	1
His	1.93	2	1.4	1
Arg	3.2	4	1.5	2
Thr	7.0	13	4.7	4
Ala	5.0	5	4.1	3
Pro	5.66	5	1.0	1
Tyr	3.0	4	1.2	2
Val	7.5	8	4.5	4
Met	1.2	1	0.6	1
Ile	3.8	4	3.0	3
Leu	6.0	7	3.0	3
Phe	5.3	6	1.9	2
Lys	4.4	5	3.0	3
Carboxymethyl-Cys	8.13	8	3.2	3
Trp	ND	1	ND	1

[a] ND, Not determined. Reprinted with permission from Ref. 60.
[b] Observed number of residues determined by amino acid analysis for the indicated peptide.
[c] Expected number of residues for the fragment indicated according to the sequence data of Fujikawa.[7]

thetic analogs. Although the number of protein structures being deposited in the Brookhaven data base is increasing at an astounding rate, the total number of available structures is still limited. Alternatively, molecular modeling, based on homology, is a powerful means of extending this structural data base. The value and accuracy of a molecular model, not surprisingly, depends on the information available for constructing the model. The most accurate molecular models reported to date[67] have been performed with proteins that are evolutionarily related to a protein for which a structure has been solved.

In the case of the factor XI heavy chain, no known structure exists, nor is it closely related to a protein of known structure. Although the accuracy

[67] L. Pearl and W. Taylor, *Nature (London)* **329,** 351 (1987).

of applying the techniques of molecular modeling for such a protein is severely limited, it is, nevertheless, possible. The strategy in this case is not to utilize all of the existing protein but rather to break it into discrete domains and utilize all of the existing information, such as known disulfide bridges, in the construction of the model. The value of a model built in this manner is that it can often provide a hypothetical structure that, in turn, can be used as a template for experimental design. In our experience we have found that, although a model such as this (based on a limited information base) is seldom, if ever, completely correct, it is seldom completely wrong. We have, therefore, used the modeled structure to spawn experimental approaches that would not otherwise be possible.

B. Computational Chemistry

The calculations used in creating a molecular model must treat all atoms within a given protein domain, and even within a given amino acid, individually. As one can well imagine, this creates a very complicated array of algorithms and tables of information that must be utilized in constructing a theoretical structure. Additionally, assumptions must be employed in order to keep the vast number of calculations in the realm of feasibility for computer processing. Unfortunately, this generally means that many quantum mechanical considerations must be simplified or ignored. In spite of these caveats, computational chemistry has proved to be an extraordinarily useful tool in the field of structural biology (for a good review on this subject, see Refs. 68 and 69).

The first component of a modeling package, the energy field, involves a set of parameter tables that are used to create mathematical descriptions of atoms and their potential interactions with one another based on averaged values determined from known structures. One of these parameter tables is an atom data file. This file contains the descriptions of individual atoms, i.e., partial charge, mass, and type of each atom. Such tables provide the basic information necessary to appropriately represent the stereochemical properties of each atom. For example, carbon–carbon interactions of sp^3 and sp^2 hybridizations need to be distinguished (single bond versus double bond) in terms of geometry, bond strength, ideal bond length, etc. A bond description file is used in depicting the covalent structure of the molecule on the basis of the number and type of covalent bonds occurring between atom pairs. This type of table also includes descriptions of the "springlike"

[68] K. B. Lipkowitz and D. B. Boyd, "Reviews in Computational Chemistry," Vol. 1. VCH, New York, 1990.
[69] K. B. Lipkowitz and D. B. Boyd, "Reviews in Computational Chemistry," Vol. 2. VCH, New York, 1991.

properties of various bonds and favorable and unfavorable torsional rotations. Finally, there is a table used in the consideration of nonbonded interactions such as vanderWaals and electrostatic interactions.

The second and third components of the modeling package involve energy minimization functions, molecular mechanics, and energy-dependent molecular motion simulations, molecular dynamics. The theoretical goal of molecular mechanics is to calculate a local energy minimum for a given structure. Due to the number of independent variables and quantum mechanical considerations, such calculations are not truly feasible and must be simplified. The algorithms used in treating potential energy functions of structure through molecular mechanics serve to reduce or eliminate steric conflicts and adjust bond lengths and bond angles to approximate their ideal values. Generally this technique will not produce structures grossly different from the initial structure. Molecular dynamic calculations are used to alter backbone structures. These programs rely on many of the same basic parameters as mechanics except that temperature, pressure, and atomic velocities are included in the calculations as a function of time. Typically these calculations are performed repetitively in 0.002-psec stepwise intervals.

There are quite a few practical restrictions placed on the modeling algorithms. One of the greatest limitations is that most of the spontaneous protein folding events in nature, such as the transition of a random coil to a β strand, occur on a nanosecond time scale. Outside of calculations performed on CRAY supercomputers (which are also time restricted), most computational times are limited to under 100 psec. Thus, it can be underscored that molecular modeling, even under the best of circumstances, is not an exact science. We have found, empirically, that for a globular protein for which at least some basic information is known, the modeling algorithms are fairly accurate at predicting the bends and turns of a protein.

The A1 and A2 domains of factor XI were modeled using the computational chemistry package supplied by Molecular Simulations Inc./Polygen Pasadena, CA on a Silicon Graphics 4D 280 parallel processing computer. Each Apple-like domain was truncated and included the amino acids extending from the first cysteine residue of the domain to the last cysteine residue of the domain. Information concerning cysteine disulfide constraints[70] was used to initiate model building, after which extended energy minimization calculations were carried out. High-energy (900 K) dynamic runs for 10 psec were used to dislodge inappropriate amino acid

---

[70] I. T. Weber, M. Miller, M. Jaskolski, J. Leis, A. M. Skalka, and A. Weodawer, *Science* **243**, 928 (1989).

contacts. The structure was allowed to cool to 300 K over a 100-psec dynamics calculation, followed by minimization of the resulting structure. A trajectory file, recorded over the entire dynamics run, indicated that after approximately 55 psec of dynamics, the calculated backbone structure had stabilized, i.e., reached a low energy well.

C. Rational Peptide Design

Our design goals, with respect to factor XI-derived peptides, have been to stabilize predicted loops and β turns. Excluding helix interactions, most protein–protein interaction sites occur in and around regions of a loop or β turn. Technically, these structures are also the easiest of the various forms of secondary structures to design.

Our principal means of stabilizing loops and turns is through the introduction of conformational restraints into the synthesized peptide. The restraints are designed to limit the conformational freedom of the peptide such that the formation of the desired turn or loop is statistically preferred. Although there are many different types of restraints that can be used to accomplish these objectives, we have primarily relied on the artificial introduction of disulfide bridges. The natural disulfide bridge has an average cross-linking distance of approximately 6.0 Å. In general, a disulfide restraint provides an effective means of stabilizing a loop and turn when using short stretches of the ascending and descending strands. If, however, longer stretches of the stranded portion of a stem–loop structure are included, one must be aware that the hydrogen bonding pattern required to maintain stable antiparallel β strands is extraordinarily difficult because of the propensity of both peptide side chains as well as main chain hydrogen bonding groups to associate preferentially with water molecules in solution.

We have targeted subdomains of factor XI that exhibit stem–loop type configurations for our peptide studies, i.e., regions that contain antiparallel β strands connected by either a β turn or random coil. Because the proximal distance from amino acid to amino acid varies across the "stem" portion of these structures, we look for a region where the natural α-carbon–α-carbon distances approximate 6 Å. In constraining the peptide analogs it is sometimes necessary to compensate for the orientation of amino acid side chains such that torsional stress does not misalign the peptide structure. Thus, in some instances, we will employ D-Cys analogs or appropriate combinations of D- and L-cysteines to mimic the correct stereochemistry. In general, these peptides are then synthesized according to the standard chemistry described below.

After selecting the appropriate placement of the cysteines, the new peptide is modeled as an independent entity. Basically, the peptide se-

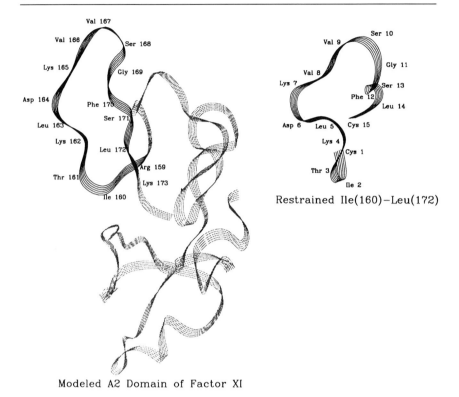

Modeled A2 Domain of Factor XI

Restrained Ile(160)–Leu(172)

FIG. 6. Molecular model of the A2 domain of factor XI compared with the conformationally restrained peptide Ile[160]-Leu[172]. See text for description.

quence is entered into the modeling program. Bulk aqueous solvent is used to form a solvation shell around the peptide. The peptide is minimized and allowed to undergo extended molecular dynamics. The disulfide bond is added to the peptide by extracting a conformation from the trajectory file where the cysteines are naturally at their closest points. The peptide then goes through additional cycles of dynamics and mechanics until equilibrium structures have been obtained. These structures are then compared against the parent structure in an attempt to determine how well the conformation of the new peptide resembles that of the parent domain. An example of this is shown in Fig. 6. The modeled structure of the A2 domain of factor XI is depicted, in which the stem–loop region Arg[159]-Lys[173] is highlighted. Also shown is the modeled synthetic peptide, comprising residues Ile[160]-Leu[172] with restraining cysteines. Although there is expected flexibility in the loop portion of the peptide, one can see that the overall structure is very similar to that of the parent protein.

VI. Synthetic Peptides

A. Solid-Phase Synthesis and Internal Conformational Restraints

The peptides described here were all synthesized on an Applied Biosystems 430 fully automated peptide synthesizer using *t*-butoxycarbonyl (*t*-BOC) protection strategies. The same syntheses can also be performed using fluorenylmethoxycarbonyl (FMOC)-based procedures (for a good review on the FMOC chemistry and practical procedures, see Ref. 71). The chemistries employed in our syntheses used a modification of the procedure of Kent and Clark-Lewis[72] where dimethylformamide (DMF) was used to replace dichloromethane in routine wash cycles. The syntheses were performed using a *p*-methylbenzhydrylamine resin (United States Biochemical Corp., Cleveland, OH). The solvents and protected amino acids used in the generation of these peptides were synthesis-grade biotechnology products purchased from Fisher Scientific (Springfield, NJ). In addition to the routine types of syntheses outlined above, there are several strategies that can be utilized in the covalent closure of the peptides. Two of these strategies are described below.

The peptide can be internally cross-linked via the side chains of a lysine (ε-amino group) and the carboxylic acid function of a glutamic acid side chain, thus creating an amide bond. The peptide is synthesized according to standard procedures on a low-substitution (0.2 mmol/g or less) *p*-methylbenzhydrylamine resin. The first residue added to the resin is an *N*-α-*t*-BOC, ε-FMOC-lysine. The rest of the peptide is continued normally using *t*-BOC chemistry until the final residue is added. The last residue to be added is a Z-protected glutamic acid, in which the carboxylic acid moiety is protected with a *t*-butyl group. Treatment of the peptide–resin with piperidine/DMF removes the FMOC group from the ε-amino group of the initial lysine without affecting any other protection groups and subsequent treatment with trifluoroacetic acid (TFA) removes the protection of the carboxylic acid group of the glutamic acid. Following neutralization, the peptide is covalently closed using a standard diimide-mediated coupling reaction. It should be emphasized that this is only one of the ways in which the synthetic peptide can be covalently closed. Other FMOC/*t*-BOC strategies include covalent closure of the peptide between two free amino groups utilizing toluene 2,4-diisocyanate (TDI), a heterobifunctional cross-linker. The methyl group of the aromatic ring of TDI prevents the isocyanate

[71] E. Atherton and R. C. Sheppard, "Solid Phase Peptide Synthesis: A Practical Approach." IRL Press, Oxford, 1989.

[72] S. B. H. Kent and I. Clark-Lewis, *in* "Synthetic Peptides in Biology and Medicine" (K. Alitalo, P. Partanen, and A. Vaheri, eds.), p. 29. Elsevier, Amsterdam, 1985.

group in the 2-position from reacting at pH 7.5 or below, whereas the isocyanate group in the para position is highly reactive. A shift in pH to greater than pH 9.0 will initiate a reaction with the isocyanate group in the 2-position, thus enabling highly specific and controlled conditions for covalent closure of the peptide. By utilizing a variety of different strategies for restricting the conformation of these peptides, we can control for distance geometries and orientation of the folded peptide.

B. Peptide Folding

In order to refold peptides containing cysteine residues, the peptide is dissolved in deionized water as a 0.1 mg/ml solution in a flask containing a stir bar. The pH is adjusted to pH 8.5 with NH_4OH and the solution is allowed to stir at 5° for at least 3 days. The resulting solution is lyophilized. Alternatively, peptides are reduced with dithiothreitol (DTT) and alkylated with iodoacetamide as previously described.[54]

C. Characterization

1. High-Performance Liquid Chromatography. To determine the homogeneity and purity as well as the monomeric state of synthetic peptides, they are subjected to HPLC utilizing both reversed-phase chromatography and gel filtration. The HPLC system and reversed-phase chromatography are described above (Section II,B,1). Gel filtration is carried out utilizing a Waters Protein-Pak 60 column that is run isocratically with 0.1% (v/v) trifluoroacetic acid in 20% (v/v) acetonitrile.

2. Capillary Electrophoresis. To assess the homogeneity of synthetic peptides and to confirm their monomeric state after refolding, CE is carried out as discussed in detail in Section IV,B,4.

3. Mass Spectrometry (Fast Atom Bombardment). Fast atom bombardment (FAB) mass spectrometry can be utilized to determine the molecular mass and assess the homogeneity of synthesized peptides. Positive-ion FAB mass spectra are obtained in our laboratory utilizing a Finnigan-MAT 4610B Quadrupole Mass Spectrometer, with a mass range of 1800 m/z, equipped with an FAB 11 NF Saddle-field ion source (Finnigan Corporation, San Jose, CA) operating at an ion-accelerating voltage of 8 kV and a mass resolution of approximately 1000. Xenon is used as the primary beam. Solid peptide samples are dissolved directly in the thioglycerol matrix layered on the probe tip. Analysis of synthetic peptides utilized in our studies[64,73] has revealed a single ion species, suggesting homogeneity of the synthetic peptide preparation with masses almost identical to calculated values.

[73] F. A. Baglia, B. A. Jameson, and P. N. Walsh, *J. Biol. Chem.* **266,** 24190 (1991).

TABLE II
EFFECTS OF SYNTHETIC PEPTIDES ON FACTOR IX
ACTIVATION BY FACTOR XIa[a]

Protein with peptide of heavy chain domains		Peptide concentration $(M)^b$
Factor XI		
Phe^{56}-Ser^{86}	(A1)	2.9×10^{-7}
Val^{64}-Tyr^{80}	(A1)	N/E (1.0×10^{-3})
Asn^{145}-Ala^{176}	(A2)	3.5×10^{-8}
Ala^{134}-Ala^{176}	(A2)	1.0×10^{-8}
Asn^{235}-Arg^{266}	(A3)	3.1×10^{-5}
Gly^{326}-Lys^{357}	(A4)	3.0×10^{-5}
Ala^{134}-Ile^{146}	(A2)	8.0×10^{-5}
Leu^{148}-Arg^{159}	(A2)	5.0×10^{-5}
Ile^{160}-Leu^{172}	(A2)	2.5×10^{-7}
Prekallikrein		
Tyr^{143}-Ala^{176}	(A2)	N/E (7.5×10^{-3})

[a] N/E, No effect (highest concentration tested given in parentheses). Reprinted with permission from Baglia et al.[73]

[b] Concentration required to inhibit factor IX activation by factor XIa 50%.

4. Free Thiol Analysis. In order to determine that refolded peptides as well as reduced and alkylated peptides contain no free thiol groups, synthetic peptides are examined[74] for free SH groups using Ellman's reagent [5,5'-dithiobis(2-nitrobenzoic acid)]. Our previously reported studies[64,73] have verified that refolded peptides are homogeneous preparations consisting of intramolecularly disulfide-bonded peptides.

D. Structure–Activity Studies

To define the domain within the heavy chain of factor XIa that binds factor IX, a panel of synthetic peptides containing sequences within the heavy chain domain have been examined for their capacity to inhibit the formation of an activation peptide reflecting factor IX activation by factor XIa.[73] These studies (Table II) demonstrate that peptide Asn^{145}-Ala^{176}, which is located in the second Apple domain of the factor XIa heavy chain, inhibits factor IX activation by factor XIa with a K_i of 30 nM, whereas structurally similar peptides in the first, third, and fourth Apple domains

[74] A. F. S. A. Habeeb, this series, Vol. 25, p. 457.

were required at 10- to 1000-fold higher concentrations for similar effects (2.9×10^{-7}, 3.1×10^{-5}, and 3.0×10^{-5} M concentrations of A1, A3, and A4 peptides). Furthermore, a synthetic peptide identical with a highly homologous region of the heavy chain A2 domain of prekallikrein (Tyr[143]-Ala[176]) had no effect on factor XIa-catalyzed factor IX activation at concentrations of 7.5×10^{-3} M. Further studies[73] demonstrate that the A2 domain peptide is a competitive inhibitor of factor IX activation by factor XIa, indicating that the sequence of amino acids comprising Asn[145]-Ala[176] within the native factor XIa protein contains a substrate binding site for factor IX.

Because of the lack of specific detailed solution or crystal structure for factor XI, we have calculated a potential three-dimensional structure for the factor XI A2 domain as described above. The resulting computer model (Fig. 6) depicted three juxtaposed β-stranded stem–loops comprising residues Ala[134]-Ile[146], Leu[148]-Arg[159], and Ile[160]-Leu[172].[73] When the first and second or the first and third peptides were examined at equimolar concentrations for their effects on factor XIa-catalyzed factor IX activation, a synergistic inhibitory effect was observed compared with each peptide added individually, whereas the second and third stem–loop structures showed additive effects. Thus, all three peptides appear to contain portions of the substrate binding site for factor IX. In addition, a single peptide (Ala[134]–Ala[176]) in the refolded conformation demonstrates striking inhibitory effects with a K_i of 10 nM. Because this concentration is about one-sixth of the concentration of factor XI in plasma, it would appear that the sequence of amino acids comprising these residues in the A2 domain of the heavy chain of factor XI contains three antiparallel β strands connected by β turns that together comprise a continuous surface comprising a substrate binding site for factor IX.

VII. Conclusions

To elucidate structure–activity relationships within the factor XI molecule we have utilized a combined approach employing monoclonal antibodies, protein modification, computational chemistry, and rational synthetic peptide design. Utilizing conformationally constrained synthetic peptides and sensitive and specific assays reflecting binding of factor XI (XIa) to either high-M_r kininogen or factor IX, we have identified two potential heavy-chain-related surfaces utilized for binding high-M_r kininogen and factor IX. The data are consistent with the view that a sequence of amino acids comprising Val[59]-Lys[83] within the A1 domain of factor XI contains two stem–loop structures that comprise a continuous surface utilized for binding high-M_r kininogen.[64] The substrate binding site appears to consist of a sequence of amino acids from Ala-134 through Leu-172 of the A2 domain of the factor XI heavy chain that contains three stem–loop structures that together comprise a continuous surface utilized for binding factor IX.[73]

Acknowledgments

We are grateful to Patricia Pileggi for excellent secretarial assistance. The studies were supported by research grants from the National Institutes of Health (HL36579, HL46213, HL45486, and HL25661) and from the W. W. Smith Charitable Trust (P.N.W.).

[6] Human Factor IX and Factor IXa

By S. Paul Bajaj and Jens J. Birktoft

Introduction

Human factor IX is a vitamin K-dependent multidomain glycoprotein and is synthesized in the liver as a precursor molecule of 461 amino acids.[1] The gene for factor IX consists of eight exons and seven introns, is approximately 34 kilobases (kb) long, and is located on the long arm of the X chromosome at band Xq 27.1[1,2] (Fig. 1). The first exon encodes for the 28-residue hydrophobic signal peptide and the second exon encodes for the 18-residue hydrophilic propeptide necessary for γ-carboxylation of 12 glutamate residues in the γ-carboxyglutamic acid (Gla)-rich domain. A part of the second exon also encodes for the amino-terminal segment (residues 1–38) of the Gla domain. Exon III encodes for residues 39 and 40 of the Gla domain and also for a short hydrophobic segment, residues 41–46, rich in aromatic amino acids. This segment is generally referred to as the aromatic amino acid stack and is thought to favor heterodimerization of Gla proteins.[3] Next, exon IV encodes for the first epidermal growth factor (EGF)-like domain (residues 47–84) and exon V encodes for the second EGF-like domain (residues 85–127). Exon VI encodes for the activation peptide region and a part (residues 181–195) of the protease domain. The balance of the protease domain is encoded by exons VII and VIII.

[1] S. Yoshitake, B. G. Schach, D. C. Foster, E. W. Davie, and K. Kurachi, *Biochemistry* **24,** 3736 (1985).
[2] C. Schwartz, N. Fitch, M. C. Phelan, C. L. Richer, and R. Stevenson, *Hum. Genet.* **76,** 54 (1987).
[3] K. Harlos, S. K. Holland, C. W. G. Boys, A. I. Burgess, M. P. Esnouf, and C. C. F. Blake, *Nature (London)* **330,** 82 (1987).

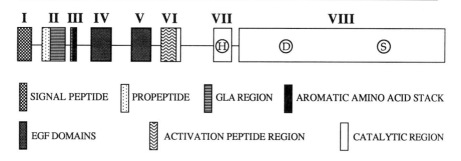

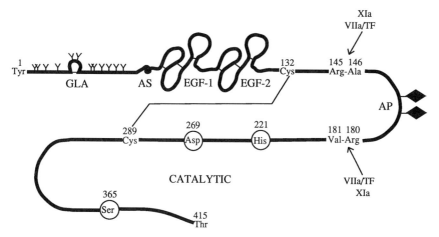

FIG. 1. Gene organization and protein structure of human factor IX. *(Top)* Exon–intron locations in the gene. *(Bottom)* Partial structure of the protein. GLA, γ-Carboxyglutamic acid; AS, aromatic amino acid stack; EGF-1, epidermal growth factor-like domain 1; EGF-2, epidermal growth factor-like domain 2; AP, activation peptide; TF, tissue factor. The diamond-shaped structures represent potential N-linked carbohydrate side chains in the activation peptide.

Several posttranslational modifications occur prior to the secretion of factor IX into blood. These include cleavage and removal of the leader sequence of 46 amino acids, γ-carboxylation of first 12 glutamic acid residues, partial β-hydroxylation of aspartic acid residue 64, and glycosylation of serine residue 53 in the EGF-1 domain and of asparagine residues 157 and 167 in the activation peptide region. The resultant mature protein of 415 amino acids circulates in blood as a zymogen of M_r 56,000[4] and

[4] E. W. Davie, K. Fujikawa, and W. Kisiel, *Biochemistry* **30**, 10363 (1991).

contains approximately 17% carbohydrate by weight[5] (Fig. 1). During hemostasis, factor IX may be activated by factor XIa, requiring calcium, and by factor VIIa, requiring calcium and tissue factor.[4] This activation occurs in two steps. In the first step, the Arg^{145}-Ala^{146} bond is cleaved, which results in the formation of a two-chain inactive intermediate, termed IXα, containing a light chain of residues 1–145 and a heavy chain of residues 146–415. In the second step, which is also the rate-limiting step,[5-7] the heavy chain of factor IXα is cleaved at the Arg^{180}-Val^{181} bond to yield a 35-residue activation peptide rich in carbohydrate, and an active serine protease, factor IXa. Factor IXa thus formed converts factor X to factor Xa in a reaction requiring calcium, phospholipid, and factor VIIIa.[4] The generated factor Xa may also participate in the formation of factor IXα by cleaving the Arg^{145}-Ala^{146} bond in factor IX, thereby augmenting the formation of active factor IXa.[7] The primary known physiologic inhibitor of factor IXa is the serine protease inhibitor, antithrombin III.[4]

The light chain of factor IXa contains the Gla domain and the two EGF-like domains, whereas the heavy chain contains the protease domain. Existing evidence suggests that the Gla domain participates in phospholipid binding whereas the Gla domain, the EGF-1 domain, and the protease domain all participate in calcium binding.[8-10] The protease domain and the EGF-1 domain are also thought to be involved in factor VIIIa binding.[11,12] The function of the second EGF-like domain is not as yet clear.

The purpose of this article is not to provide a comprehensive review of factor IX but to provide methods for its isolation from human plasma and recombinant cultures, preparation of [^{3}H]sialyl and of ^{125}I-labeled tyrosyl factor IX and their use in investigations of the structure–function relationships in the protein. A comparative molecular modeling approach will also be presented to build a putative structure of the protease domain of

[5] R. G. DiScipio, K. Kurachi, and E. W. Davie, *J. Clin. Invest.* **61,** 1528 (1978).

[6] S. P. Bajaj, S. I. Rapaport, and W. A. Russell, *Biochemistry* **22,** 4047 (1983).

[7] J. H. Lawson and K. G. Mann, *J. Biol. Chem.* **266,** 11317 (1991).

[8] R. A. Schwalbe, J. Ryan, D. M. Stern, W. Kisiel, B. Dahlbäck, and G. L. Nelsestuen, *J. Biol. Chem.* **264,** 20888 (1989).

[9] P. A. Handford, M. Mayhew, M. Baron, P. R. Winship, I. D. Campbell, and G. G. Brownlee, *Nature (London)* **351,** 164 (1991).

[10] S. P. Bajaj, A. K. Sabharwal, J. Gorka, and J. J. Birktoft, *Proc. Natl. Acad. Sci. U.S.A.* **89,** 152 (1992).

[11] S. P. Bajaj, S. I. Rapaport, and S. L. Maki, *J. Biol. Chem.* **260,** 11574 (1985).

[12] D. J. G. Rees, I. M. Jones, P. A. Handford, S. J. Walter, M. P. Esnouf, K. J. Smith, and G. G. Brownlee, *EMBO J.* **7,** 2053 (1988).

human factor IX. For a detailed review on factor IX as it relates to blood coagulation, the reader is referred to several recent excellent reviews.[4,13-15]

Assay Methods

Coagulant Assay

A clotting assay for factor IX is carried out by incubating 50 μl of hereditary factor IX-deficient plasma (George King Biomedical), 50 μl of automated activated partial thromboplastin time (aPTT) reagent (General Diagnostics), and 50 μl of a test sample in a 12 × 75-mm glass tube for 5 min in a heating block maintained at 37°. Then, 50 μl of 25 mM $CaCl_2$ kept at 37° is added and the clotting time is noted from the point of addition of $CaCl_2$. Various dilutions of the test sample are made with 0.05 M Tris·HCl, 0.15 M NaCl, pH 7.5 (Tris/NaCl), containing 1 mg/ml of bovine serum albumin (Tris/NaCl/BSA). Citrated human plasma pooled from 20 healthy donors is used as a reference standard and is defined to contain 1 unit of factor IX per milliliter. A standard curve is constructed by using various dilutions between 1:5 and 1:80 of pooled plasma; a linear calibration curve is obtained when the log of plasma dilutions is plotted against the log of clotting times. Samples of purified factor IX are diluted to obtain clotting times between 50 and 65 sec, which represent the midrange of the standard curve.

Electroimmunoassay

Rabbit antiserum against normal factor IX (IX_N) can be obtained from Diagnostica Stago for this purpose. Electroimmunoassay is carried out in 1% Litex agarose gels according to Laurell[16] using 3-mm-thick plates from GelBond Film (FPC Products). The antiserum is used at a final concentration of 0.4 to 0.5% and the gel buffer is 0.08 M Tris/Tricine, pH 8.8, containing 2.5% polyethylene glycol (8000 MW). The reservoir buffer is the same Tris/Tricine buffer but without polyethylene glycol. Samples of 20-μl aliquots are applied in 4-mm-diameter wells and electrophoresis is carried out for 20 hr at 5 V/cm at 10°. The gels are washed in 0.9% NaCl and distilled water and are stained with Coomassie brilliant blue. A stan-

[13] B. Furie and B. C. Furie, Cell (Cambridge, Mass.) 53, 505 (1988).
[14] K. G. Mann, M. E. Nesheim, W. R. Church, P. Haley, and S. Krishnaswamy, Blood 76, 1 (1990).
[15] J. Stenflo, Blood 78, 1637 (1991).
[16] C. B. Laurell, Scan. J. Clin. Lab. Invest. 29, 21 (1972) Suppl. 24.

dard curve is prepared by using 3, 6, 12, 25, 50, and 100% of normal pooled plasma. One unit of factor IX antigen is defined as the amount present in 1 ml of pooled plasma. Electroimmunoassay is used to monitor column fractions during purification of the nonfunctional variants of factor IX from either plasma or the mammalian expression systems.

Purification Procedures

Several procedures for the isolation of human factor IX have been described.[17-19] The procedures described below are routinely used in this laboratory to isolate normal and variant nonfunctional factor IX proteins from human plasma and mammalian cell expression systems.

Conventional Method

This procedure can be carried out with 7 to 10 liters of normal plasma or a variant factor IX plasma obtained from a hemophilia B patient. The purification described below is that employed for the isolation of factor IX_{ER} (Gly363Val)[19] from 8 liters of hemophilia B plasma. Throughout the purification steps, plastic beakers and tubes and siliconized columns are employed. All steps are performed at 4° unless noted otherwise.

1. *Starting Material.* If the laboratory has the facility for collection of blood, it should be collected into a citrate–benzamidine-soybean trypsin inhibitor (STI) (Sigma St. Louis, MO, product T-9003) cocktail in plastic bottles. Use 50 ml of citrate anticoagulant (prepared by mixing 2 parts of 0.1 M citric acid and 3 parts of 0.1 M trisodium citrate), 10 ml of 0.5 M benzamidine hydrochloride, and 1 ml of 10 mg/ml of STI for each unit (450 ml) of blood. The plasma should be obtained by double centrifugation at 6000 g for 10 min. The plasma in 500-ml portions in plastic bottles can be stored frozen at −70° till used. Thaw 8 liters of normal or variant factor IX citrated plasma at 37° and transfer to the cold room in a 10-liter plastic bucket. If inhibitors are not already present in the starting plasma, 160 ml of 0.5 M benzamidine hydrochloride and 40 ml of 10 mg/ml STI are added immediately and the plasma is stirred for 15 min prior to performing step 2.

2. *Barium Citrate Adsorption.* To the gently stirring plasma at 4°, 1 M $BaCl_2$ (80 ml/liter of plasma, or 640 ml for 8 liters of plasma) is

[17] R. G. Discipio, M. A. Harmodson, S. G. Yates, and E. W. Davie, *Biochemistry* **16**, 698 (1977).
[18] J. P. Miletich, G. J. Broze, and P. W. Majerus, *Anal. Biochem.* **105**, 304 (1980).
[19] S. P. Bajaj, S. G. Spitzer, W. J. Welsh, B. J. Warn-Cramer, C. K. Kasper, and J. J. Birktoft, *J. Biol. Chem.* **265**, 2956 (1990).

added dropwise and the stirring is continued for an additional 20 min after all the $BaCl_2$ has been added. The suspension is centrifuged for 20 min at 6000 g in 500-ml plastic centrifuge bottles. The barium citrate pellet in each bottle is mashed extensively with a plastic spatula and suspended in 2.7 liters (one-third volume of starting plasma) of citrate–saline (0.02 M trisodium citrate/0.15 M NaCl) containing benzamidine (0.01 M) and STI (50 μg/ml). The suspension is stirred for 20 min in a 4-liter beaker and the resuspended vitamin K-dependent proteins are reprecipitated by the dropwise addition of 640 ml of 1 M $BaCl_2$. Stirring is continued for an additional 20 min following the addition of $BaCl_2$ and the suspension is then centrifuged for 20 min at 6000 g. The precipitate is again suspended in the same volume of citrate–saline–inhibitor solution and precipitation is repeated with the same volume of 1 M $BaCl_2$ as above.

3. *Ammonium Sulfate Fractionation.* The third barium citrate precipitate obtained above is mashed extensively and suspended in 270 ml (one-thirtieth volume of starting plasma) of cold 35% saturated ammonium sulfate (187 g/liter) solution containing 50 mM benzamidine. The suspension is adjusted to pH 6.0 by the addition of 3.5 M citric acid and stirred for 30 min and centrifuged at 6000 g for 10 min. The supernatant containing the vitamin K-dependent proteins is saved. The $BaSO_4$ precipitate obtained above is mashed extensively and resuspended in 270 ml of 35% saturated ammonium sulfate/50 mM benzamidine solution. The suspension is stirred for 30 min and then centrifuged. The precipitate is discarded. Both supernatants are pooled and solid ammonium sulfate (0.27 g/ml) is added slowly to the stirring solution. The resulting suspension is stirred for an additional 20 min, allowed to stand for another 20 min, and then centrifuged for 30 min at 9000 g. The precipitate is saved.

4. *DEAE-Sephadex Chromatography.* The precipitate obtained in step 3 is dissolved in approximately 60 ml of 0.025 M citrate, 0.1 M NaCl, 1 mM benzamidine, pH 6.0, buffer containing 50 μg/ml of STI (this buffer is made by titration of citric acid with $\approx$5 M NaOH). The protein solution is dialyzed versus 4 liters of 0.025 M citrate, 0.1 M NaCl, 1 mM benzamidine, pH 6.0, overnight (approximately 10 hr). Next day, the dialysis buffer is changed and dialysis continued for an additional 6-hr period. The dialyzed protein is centrifuged at 9000 g for 20 min to remove insoluble material and then applied to a DEAE-Sephadex column (5 $\times$ 95 cm) prepared as described below in step 6. The column is washed with the

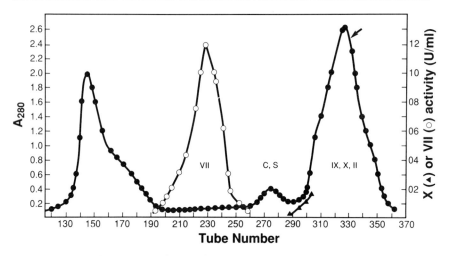

FIG. 2. DEAE-Sephadex gradient elution profile for separation of the human vitamin K-dependent proteins. Factor IX_{ER} (Gly363Val)[19] plasma (8 liters) was used as the starting material. The vitamin K-dependent proteins were adsorbed onto barium citrate, fractionated with ammonium sulfate, and dialyzed into buffer of 0.025 M citric acid, 0.1 M NaCl, 1 mM benzamidine, pH 6.0. The dialyzed protein sample containing approximately 6000 factor IX antigen units in 80 ml of buffer was applied to the DEAE-Sephadex column (5 × 95 cm) at 4°. The column was washed with the initial buffer until the A_{280} of the effluent was ≈0.2. A linear gradient in NaCl (0.1–0.6 M; 2.2 liter/chamber) was then applied and 13.5-ml fractions were collected at a flow rate of 70 ml/hr. The arrow on the descending limb of the last protein peak indicates where the gradient buffer was depleted. At this point, the buffer containing 0.6 M NaCl was applied to the column to complete the elution of prothrombin, factor X, and factor IX. A_{280} (●), factor VII clotting activity[25] (○), and factor X activity[23] (▲) are plotted versus fraction number.

starting buffer until the A_{280} of the effluent is less than 0.2. This usually takes a day and 2 liters of buffer. At this time a linear gradient of NaCl formed by 2200 ml of 0.025 M citrate, 0.1 M NaCl, 1 mM benzamidine, pH 6.0, in the mixing chamber and 2200 ml of the 0.025 M citrate, 0.6 M NaCl, 1 mM benzamidine, pH 6.0, in the reservoir is applied. At the end of the gradient, the column is further developed with 0.025 M citrate, 0.6 M NaCl, 1 mM benzamidine, pH 6.0. The protein and activity elution profile from this column are shown in Fig. 2. In this column, factor VII elutes first and is followed by protein C and protein S. Prothrombin, factor X, and factor IX elute later in the gradient; their elution is never complete at the end of the gradient and is achieved by applying the elution buffer containing 0.6 M NaCl. Further purification of factor VII[20] and of protein C and protein S[21] is reported elsewhere.

5. *Heparin–Agarose Chromatography.* Heparin–agarose for this purpose is prepared as follows:

A. Dissolve 15 g of heparin (Sigma, H-3125) in 150 ml of 0.2 M NaHCO$_3$.

B. Dissolve 90 g of CNBr in 45 ml of acetonitrile under the hood. Keep the beaker covered with a watch glass.

C. Wash 450 ml of agarose beads (BioGel A-15m, 200–400 mesh) with about 6 liters of H$_2$O using a Büchner funnel.

D. Mix 450 ml of agarose beads and 450 ml of cold H$_2$O in a 4-liter glass beaker.

E. Add 900 ml of 2 M Na$_2$CO$_3$ at room temperature to the agarose beads and stir the slurry with the beaker placed in an ice bucket under the hood.

F. Add the dissolved CNBr to the stirring agarose and stir vigorously for 2 min. Add pieces of ice to keep the mixture cold. Some crystallization of Na$_2$CO$_3$ will occur.

G. Immediately filter the mixture through a Büchner funnel. Add pieces of ice to the funnel at all times.

H. Wash the agarose with 3 liters of cold 0.2 M NaHCO$_3$, pH 9.5. Continue to add pieces of ice.

I. Transfer agarose beads to a 2-liter plastic beaker containing 900 ml of cold 0.2 M NaHCO$_3$, pH 9.5. Add dissolved heparin and stir the slurry, very gently, overnight in the cold room.

J. Next day, wash the agarose beads with 2 liters of cold 0.2 M NaHCO$_3$, pH 9.5, followed with 4 liters of 0.1 M sodium acetate, 0.5 M NaCl, pH 4.0, and then with 2 liters of 0.3 M NaCl.

K. Finally, wash the beads with 4 liters of 0.1 M Tris, 0.5 M NaCl, pH 9.0, and finish with 2 liters of 0.3 M NaCl.

For a preparation starting with 8 to 10 liters of plasma, another batch of heparin–agarose should be prepared and both batches combined for a column size of 5 × 40 cm. The heparin–agarose beads should be washed with at least two bed volumes of 0.02 M citric acid, 2 M NaCl, 1 mM benzamidine, pH 7.5, prior to equilibrating with 0.02 M citric acid, 1 mM benzamidine, pH 7.5. Again, at least two bed volumes of buffer should be used for equilibration.

DEAE-Sephadex fractions (from step 4) that contain factor IX,

[20] L. V. M. Rao and S. P. Bajaj, *Anal. Biochem.* **136,** 357 (1984).
[21] S. P. Bajaj, S. I. Rapaport, S. L. Maki, and S. F. Brown, *Prep. Biochem.* **13(3),** 191 (1983).

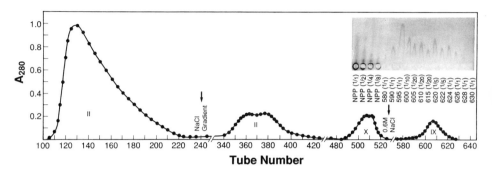

Fig. 3. Purification of factor IX$_{ER}$ (Gly363Val) by heparin–agarose column chromatography. Approximately 650 mg of protein containing prothrombin, factor X, and factor IX$_{ER}$ in 40 ml of 0.02 M citric acid, 1 mM benzamidine, pH 7.5, was applied to the column (5 × 40 cm) and 8-ml fractions were collected at a flow rate of 60 ml/hr. The arrow on the left represents the start of the gradient (1200 ml/chamber, 0.1 M NaCl → 0.6 M NaCl), and the arrow on the right represents the completion of the gradient and application of a buffer of 0.02 M citrate, 0.6 M NaCl, 1 mM benzamidine, pH 7.5. Inset: Electroimmunoassay for factor IX of different fractions at indicated dilutions. Different dilutions of normal pooled plasma (NPP) were used to calculate the amount of factor IX$_{ER}$ present in each fraction.

factor X, and prothrombin are pooled and the protein is precipitated by the slow addition of solid ammonium sulfate (0.532 g/ml). The suspension is centrifuged at 8000 g for 30 min and the proteins are dissolved in approximately 40 ml of 0.02 M citric acid, 1 mM benzamidine, pH 7.5, and stored frozen at −70° until further purification on the heparin–agarose column chromatography. For this step, the protein sample is thawed and dialyzed versus 2 liters of 0.02 M citric acid, 1 mM benzamidine, pH 7.5, for 6–8 hr at 4°. The dialysis is continued overnight with 2 liters of fresh buffer. The dialyzed sample is centrifuged at 8000 g for 10 min to remove insoluble material and loaded onto the heparin–agarose column equilibrated with the same buffer. The column is washed with the equilibration buffer until a majority of the prothrombin is eluted and the absorbance of the effluent fractions is less than 0.05. A linear gradient consisting of 1200 ml of 0.02 M citric acid, 1 mM benzamidine, pH 7.5, and 1200 ml of 0.02 M citric acid, 0.6 M NaCl, 1 mM benzamidine, pH 7.5, is then applied. A second peak containing prothrombin with a slightly lower number of sialic acid residues is eluted early in the gradient[22] (Fig. 3). The protein in the prothrombin fractions can be precipitated by the addition of solid ammonium sulfate to 80% saturation (0.532 g/ml) and stored in

[22] S. P. Bajaj, S. I. Rapaport, C. Prodonos, and W. A. Russell, *Blood* **58,** 886 (1981).

50% glycerol at $-20°$. Factor X is eluted just prior to completion of the gradient (Fig. 3). Factor X can be concentrated by using Diaflo ultrafiltration PM30 membrane (Amicon, Danvers, MA) and stored frozen at $-70°$. Both factor X and prothrombin are effectively homogeneous when analyzed by several electrophoretic systems.[23] At the end of the gradient, additional 0.02 M citric acid, 0.6 M NaCl, 1 mM benzamidine, pH 7.5, is applied to elute factor IX (Fig. 3). Elution of normal factor IX is monitored by activity assays and elution of abnormal nonfunctional variants is monitored by an electroimmunoassay (Fig. 3). The yield of factor IX is usually 10–12 mg ($\approx$ 35% yield) of protein from 8 liters of starting plasma. The protein can be concentrated by using a Diaflo PM30 membrane and stored frozen at $-70°$. When stored in this manner, factor IX is completely stable for at least 2 years. SDS-gel electrophoretic analysis[24] of IX_N and IX_{ER} (Gly363Val) isolated by heparin–agarose chromatography is shown in Fig. 4.

6. *Comments on the Conventional Method of Factor IX Purification.* The length of the DEAE-Sephadex column is crucial for separation of the vitamin K-dependent proteins into individual peaks, i.e., factor VII, protein C, and protein S peaks, and factor IX, factor X, and prothrombin peaks.[25] If isolation involves 8–10 liters of starting plasma, then the preferred size of the column is 5 $\times$ 95 cm and the volume of each gradient buffer is 2200 ml. The preparation of the column is also crucial for obtaining a flow rate of approximately 70 ml/hr. The column should be made ready for this purification as follows: A week before the isolation is planned, approximately 120 g of DEAE-Sephadex A-50 is swelled in 4 liters of 0.025 M citric acid, 0.1 M NaCl, 1 mM benzamidine, pH 6.0, overnight at room temperature. The buffer from the top is siphoned off and fresh buffer is added. The beaker is transferred to the cold room and two fresh changes of buffer are made in the next 2 days. The fine particles are removed by decanting and the gel is degassed to remove trapped air. Failure to remove the fine particles will result in poor resolution and a decreased flow rate. The slurry is poured into a 5 $\times$ 100-cm siliconized column containing a small volume of buffer. The bottom flow valve is to remain closed at all times and the resin is to be packed by gravity. The buffer from the top of the resin is periodically siphoned off prior to adding additional slurry. The column is packed

[23] S. P. Bajaj, S. I. Rapaport, and C. Prodonos, *Prep. Biochem.* **11(4),** 397 (1981).
[24] U. K. Laemmli, *Nature (London)* **227,** 680 (1970).
[25] S. P. Bajaj, S. I. Rapaport, and S. F. Brown, *J. Biol. Chem.* **256,** 253 (1981).

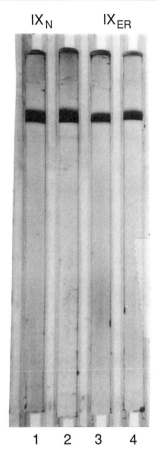

FIG. 4. SDS-gel electrophoretic analysis of IX_N and IX_{ER} isolated by the heparin–agarose chromatography method. Protein (15 μg) was applied to each gel and the concentration of acrylamide was 12%. Gel 1 is unreduced and gel 2 is reduced IX_N (apparent $M_r \approx 62,000$). Similarly, gel 3 is unreduced and gel 4 is reduced IX_{ER} (apparent $M_r \approx 62,000$). Modified with permission from Bajaj et al.[19]

to 95 cm using the above technique. The column is to be run (70 ml/hr) only for approximately 6 hr prior to loading of the sample.

Occasionally, at the second ammonium sulfate precipitation step, a loose protein pellet along with the floating protein aggregates may be seen. If this is the case, a tygon tubing attached to a 50-ml plastic syringe should be lowered close to the bottom of the centrifuge bottle and as much of the liquid removed as possible without disturbing the pellet and the floating protein aggregates. The remaining suspension can be dialyzed directly into 0.025 M citrate, 0.1 M NaCl, 1 mM benzamidine, pH 6.0, buffer in preparation for the DEAE-Sephadex chromatography step.

The heparin–agarose column can be stored at 4° in 0.02 M citric acid, 2 M NaCl, 1 mM benzamidine, pH 7.5, containing 0.05% sodium azide. After three to four runs, the column should be washed with 4 M urea in 0.02 M citric acid, pH 7.5, buffer. Using these precautions, we have made eleven preparations of factor IX using the same heparin–agarose resin over a period of 5 years.

Immunoaffinity Method

1. *Purification from Plasma.* Barium citrate adsorption and the ammonium sulfate fractionation steps are performed as outlined above for the conventional method of purification. The protein eluate of approximately 20 ml (starting with 500 ml of plasma) is dialyzed versus 0.05 M Tris, 0.15 M NaCl, pH 7.5, containing 0.02 M MgCl$_2$ and applied to the metal ion-dependent monoclonal A-7 immunoaffinity column (15-ml size) prepared as described by Smith.[26] Factor IX is eluted with 0.05 M Tris, 0.15 M NaCl, pH 7.5, containing 0.02 M EDTA. Using this method, McCord *et al.*[27] isolated both normal and a variant factor IX (Asp47Gly) molecule, and Ludwig *et al.*[28] isolated normal and two variant molecules (Glu245Val and Arg248Gln). The SDS-gel electrophoretic analysis of normal and the two variant factor IX molecules purified by Ludwig *et al.*[28] is shown in Fig. 5.

2. *Purification from Mammalian Expression Systems.* Active factor IX from the conditioned media of a suitable mammalian expression system can also be isolated by use of the monoclonal A-7 antibody.[12,29] Four parts of conditioned media should be mixed with one part of 3.85% trisodium citrate containing 50 mM benzamidine. Barium citrate adsorption and ammonium sulfate fractionation are carried out as described above. The protein eluate obtained at this stage is dialyzed against 0.05 M Tris, 0.15 M NaCl, pH 7.5, containing 0.02 M MgCl$_2$ and is applied to the monoclonal A-7 immunoaffinity column. Factor IX is then eluted with 0.05 M Tris, 0.15 M

[26] K. J. Smith, *Blood* **72,** 1269 (1988).
[27] D. M. McCord, D. M. Monroe, K. J. Smith, and H. R. Roberts, *J. Biol. Chem.* **265,** 10250 (1990).
[28] M. Ludwig, A. K. Sabharwal, H. H. Brackmann, K. Olek, K. J. Smith, J. J. Birktoft, and S. P. Bajaj, *Blood* **79,** 1225 (1992).
[29] S.-W. Lin, K. J. Smith, D. Welsch, and D. W. Stafford, *J. Biol. Chem.* **265,** 144, 1990.

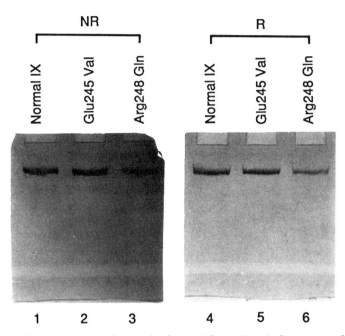

FIG. 5. SDS-gel electrophoretic analysis of normal factor IX and of two mutant factor IX proteins, Glu245Val and Arg248Gln, purified by the monoclonal A-7 antibody affinity column. The amount of protein applied to lanes 1, 2, 4, and 5 was 3 μg each; it was 1.4 μg each for lanes 3 and 6. NR, Nonreduced, R, reduced. From Ludwig et al.[28] with permission.

NaCl, pH 7.5, containing 0.02 M EDTA. Using this technique, we are able to isolate to homogeneity functional factor IX (170 U/mg) and mutant factor IX proteins such as Glu235Lys and Glu245Val from an expression system that utilizes 293 kidney cells and pRc/CMV mammalian expression vector.[30]

3. *Comments on Immunoaffinity Method of Factor IX Purification.* This method is rapid and can be used to isolate factor IX with good yield ($\approx 40\%$ recovery) from small volumes of plasma containing reduced levels of antigen.[28] The method also allows for the purification of active factor IX from *in vitro* expression systems. This method is not suitable for isolation of factor IX proteins that have impaired calcium binding in the Gla domain.

[30] D. Zhong, K. J. Smith, and S. P. Bajaj, unpublished data.

Preparation of [^{3}H]Sialyl Factor IX

^{3}H-Labeled factor IX is prepared by several modifications of the general technique of Van Lenten and Ashwell[31] for preparing [^{3}H]sialyl proteins. A method for preparing bovine ^{3}H-labeled factor IX was described earlier[32]; this method, however, does not appear to be satisfactory for labeling human factor IX.[7,33] Below is a description of the procedure used in our laboratory for labeling human factor IX.

1. Human factor IX (1 ml; ≈ 2 mg/ml) in 0.02 M citric acid, 0.6 M NaCl, 1 mM benzamidine, pH 7.5, stored at $-70°$, is thawed and dialyzed against 4 liters of cold 0.02 M citric acid, 0.1 M NaCl, pH 7.5, buffer for 4 hr. For each milliliter of dialyzed sample, 5 μl of 1 M diisopropyl fluorophosphate (in anhydrous 2-propanol) and 30 μl of 0.3 M EDTA, pH 7.5, are added under the hood and the sample is kept at room temperature for 15–20 min. The sample is then dialyzed against 2 liters of 0.1 M sodium acetate, 0.1 M NaCl, pH 5.8, for 4–6 hr. The dialysis is continued overnight with 2 liters of fresh buffer.

2. The next day, the sample is centrifuged at 4° in an Eppendorf tube at 10,000 rpm for 5 min. The volume of the supernatant is measured and transferred into a new 2-ml Eppendorf tube kept on ice. A 50-μl aliquot is diluted 15-fold to measure A_{280}. The diluted sample can be frozen for analysis by SDS-PAGE and for factor IX activity. Total micromoles of factor IX in the remaining sample are calculated by assuming a molecular weight of 56,000[4] and $E_{280}^{10\%}$ of 13.2. The total number of sialic acid residues is then calculated by assuming 10 mol of sialic acid/mol of protein; accordingly, 2 mg of factor IX is calculated to contain 0.38 μmol of sialic acid.

3. Freshly prepared sodium periodate (Sigma, S-1878) solution is added to the Eppendorf tube containing factor IX kept on ice. Sodium periodate solution should be made by dissolving 0.26 g (taken from a previously unopened container) in 100 ml of H_2O. The amount of sodium periodate to be added is twofold over the sialic acid content. Thus, for 2 mg of factor IX, 63 μl of 12 mM sodium periodate (0.26 g/100 ml) is needed. The Eppendorf tube is gently inverted twice and placed on ice for 10 min. Ethylene glycol

[31] L. Van Lenten and G. Ashwell, *J. Biol. Chem.* **246**, 1889 (1971).
[32] M. Zur and Y. Nemerson, this series, Vol. 80, p. 237.
[33] P. N. Walsh, H. Bradford, D. Sinha, J. R. Piperno, and G. P. Tuszynski, *J. Clin. Invest.* **73**, 1392 (1984).

(5 μl) is added to reduce the excess sodium periodate. After 5 min on ice, the sample is dialyzed against 2 liters of cold 0.1 M sodium borate, pH 8.5, for 3 hr. The dialysis is continued for another 3 hr with fresh cold buffer. The sample is centrifuged for 5 min at 10,000 rpm and then transferred to a 2-ml Eppendorf tube. A small aliquot of the sample is removed for protein measurement, activity assay, and SDS-PAGE analysis.

4. The contents of one vial of sodium boro[^{3}H]hydride (100 mCi, 5–15 Ci/mmol, NEN Research Products, NET-023H) are dissolved in 100 μl of 10 mM NaOH under a hood and added to factor IX from step 3. The Eppendorf tube is kept on ice for 20 min with gentle shaking every 4 min. Unlabeled sodium borohydride (10 μl, 0.2 M) for each milliliter of factor IX solution is then added to reduce the unreacted groups. The sample is allowed to sit at room temperature for 5 min and then placed on ice. Solid ammonium sulfate (0.532 g/ml of reaction mixture) is added slowly and the suspension is allowed to sit at 4° for an additional 20 min after all of the ammonium sulfate has been dissolved. The suspension is centrifuged for 10 min at 10,000 rpm and the supernatant is carefully removed. The protein pellet is washed with 1 ml of 80% saturated cold ammonium sulfate solution prepared by adding 0.532 g of ammonium sulfate to 0.025 M citric acid, 0.1 M NaCl, pH 6.0, buffer. The protein pellet is then dissolved in 1 ml of 0.025 M citric acid, 0.1 M NaCl, pH 6.0, and dialyzed against the same buffer till the counts in the dialyzate are about 50 cpm/100 μl. This usually takes two changes each with 1 liter of dialysis buffer. ^{3}H-Labeled factor IX obtained at this point is centrifuged in an Eppendorf tube to remove any insoluble material, aliquoted in 50-μl portions, and stored frozen at −70°.

Using the above protocol, we have consistently obtained ^{3}H-labeled factor IX of radiospecific activity ranging from 2–5 × 10^8 cpm/mg. Less than 2% ($\approx$ 1.5%) of the radioactive counts are present in the 3% trichloroacetic acid (TCA)-soluble fraction. We routinely count 100 μl of sample in 4 ml of Ready Protein$^+$ liquid scintillation cocktail obtained from Beckman. The clotting activity of ^{3}H-labeled factor IX ranges from 180 to 190 U/mg ($\approx$ 90–95% of control value). On nonreduced and reduced SDS-PAGE, a single radioactivity peak corresponding to factor IX Commassie blue-stained band is observed.[6] After about 1 year of storage, the 3% TCA-soluble cpm value increase to about 2.5% of total counts, at which time it is advisable to prepare a new batch of ^{3}H-labeled factor IX.

Preparation of ^{125}I-Labeled Tyrosyl Factor IX

The method described below uses Iodo-Beads iodination reagent obtained from Pierce (Rockford, IL). Use of Bio-Rad (Richmond, CA) Enzymobead reagent or the Pierce Iodogen reagent also results in a satisfactory labeling with retention of $\approx 90\%$ coagulant activity.[6,34]

1. In an Eppendorf tube, mix 100 μl of 0.1 M sodium phosphate, pH 6.5, with 10 to 20 μl of carrier and reductant-free Na^{125}I (≈ 1 mCi). Add three washed beads prepared as outlined in Pierce Products Bulletin and let stand for 5 min with occasional shaking. Place the Eppendorf tube on ice.
2. Add 100 μl (≈ 1 mg/ml) of factor IX in 0.1 M phosphate, pH 6.5, to the Eppendorf tube containing the beads. Mix gently and keep on ice for a period of no more than 5 min.
3. Transfer 200 μl of the reaction mixture to a fresh 1-ml Eppendorf tube kept on ice. Precipitate ^{125}I-labeled factor IX by adding slowly 106 mg of ammonium sulfate. Wait for 10 min after all of the ammonium sulfate has been dissolved and centrifuge for 10 min at 10,000 rpm in the cold room.
4. Carefully remove the supernatant leaving behind the small protein pellet of ^{125}I-labeled factor IX. Dissolve ^{125}I-labeled factor IX in ≈ 300 μl of 0.025 M citric acid, 0.1 M NaCl, pH 6.0, and dialyze against this buffer to remove remaining free ^{125}I.

Using the above procedure, one can obtain ^{125}I-labeled factor IX with $\approx 1 \times 10^9$ cpm/mg of protein, which roughly corresponds to 1 atom of iodine incorporated per molecule of factor IX. Such preparations retain $> 90\%$ of the biologic activity of the unlabeled control. On both nonreduced and reduced SDS-PAGE, a single radioactivity peak corresponding to factor IX Commassie blue-stained band is observed. The protein begins to lose bioactivity after about 1 month of storage at $-70°$. Thus, it is recommended that bioactivity, radioactivity, and the stability of the protein be checked periodically.

Comparison of Radioactivity Profiles of Activation Products of [125]I- and ^{3}H-Labeled Factor IX

Reduced SDS-PAGE radioactivity profiles obtained on partial activation of ^{125}I-labeled factor IX and of ^{3}H-labeled factor IX by factor XIa are shown in Fig. 6. Two notable differences are obvious. One, a ^{125}I radioac-

---

[34] W.-F. Cheung, D. L. Straight, K. J. Smith, S.-W. Lin, H. R. Roberts, and D. W. Stafford, *J. Biol. Chem.* **266**, 8797 (1991).

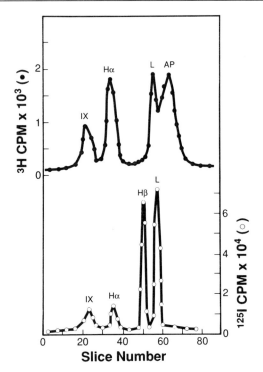

FIG. 6. Comparison of the radioactivity profiles of the activation of ^{3}H-labeled factor IX and ^{125}I-labeled factor IX. The reaction mixture contained 10 μg/ml of [3]H- or ^{125}I-labeled factor IX in Tris/NaCl buffer containing 1 mg/ml BSA, 5 mM Ca^{2+}, and 0.4 μg/ml factor XIa. After 20 min at 37° the reaction mixture was inactivated by the addition of an equal volume of SDS sample buffer containing 10% 2-mercaptoethanol and subjected to SDS gel electrophoresis. The gels were sliced and counted for [3]H (●) or [125]I (○) radioactivity. Reprinted with permission from Ref. 6. Copyright [1983] American Chemical Society.

tivity peak is absent, corresponding to the activation peptide (AP), and second, a ^{3}H radioactivity peak is absent, corresponding to the heavy chain (H$_\beta$) containing the protease domain. These radioactivity profiles have been observed with several hemophilia B mutants of factor IX studied in our laboratory.[19,35] From these data, it would appear that the protease domain of human factor IX lacks a carbohydrate chain containing sialic acid residues. The absence of an ^{125}I radioactivity peak corresponding to the activation peptide is consistent with the known presence of only one tyrosine residue in this region, out of a total of 15 present in the entire protein.[1]

[35] P. Usharani, B. J. Warn-Cramer, C. K. Kasper, and S. P. Bajaj, *J. Clin. Invest.* **75,** 76 (1985).

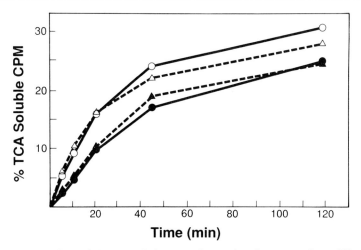

Fig. 7. Comparison of the rates of cleavage of normal and a mutant factor IX protein (Gly363Val) by factor XIa. The reaction mixture contained 10 μg/ml of ^{3}H-labeled normal or the mutant factor IX in Tris/NaCl buffer containing 1 mg/ml BSA, 5 mM Ca^{2+}, and factor XIa at 30 or 60 ng/ml. Mutant factor IX at (▲) 30 and (△) 60 ng/ml of factor XIa; normal factor IX at (●) 30 and (○) 60 ng/ml of factor XIa. From Bajaj et al.[19] with permission.

Use of ^{3}H-Labeled Activation Peptide Release Assay to Measure Activation of Factor IX in Complex Systems

[3]H-Labeled activation peptide release assays can be used to monitor the rates of activation of normal and of nonfunctional variant factor IX molecules.[19,32,35,36] An example of a study of the activation of normal factor IX and of a variant factor IX (Gly363Val) is shown in Fig. 7. The method is also useful in studies of the activation of factor IX in a plasma milieu.[37] The inhibition of release of ^{3}H-labeled activation peptide from ^{3}H-labeled factor IX has also been used to measure the levels of tissue factor pathway inhibitor in complex samples.[38,39] The ^{3}H-labeled peptide release assay, however, is not suitable for measuring the cleavage of only one of the two peptide bonds (Arg^{145}-Ala^{146} or Arg^{180}-Val^{181}) involved in the activation of factor IX, e.g., conversion of factor IX to factor IXα by factor Xa[7] or activation of factor IX mutants in which the amino acids involving the cleavage sites have been altered.[40]

[36] M. Zur and Y. Nemerson, J. Biol. Chem. 255, 5703 (1980).
[37] R. M. Birkowitz and Y. Nemerson, Blood 55, 528 (1981).
[38] M. S. Bajaj, S. V. Rana, R. B. Wysolmerski, and S. P. Bajaj, J. Clin. Invest. 79, 1874 (1987).
[39] T. A. Warr, B. J. Warn-Cramer, L. V. M. Rao, and S. I. Rapaport, Blood 74, 201 (1989).
[40] A. R. Thompson, Prog. Hemostasis Thromb. 10, 175 (1991).

Use of ^{125}I-Labeled Factor IX to Study Binding to Monoclonal
  Antibody

The following method is described to study the binding of ^{125}I-labeled
factor IX or ^{125}I-labeled factor IXa to a mouse monoclonal antibody
(MAb), which has been shown to inhibit the interaction of factor IXa with
factor VIIIa[11] and is specific for the protease domain of factor IX. The
method can be used to study the binding of any ^{125}I-labeled protein or
peptide to a given antibody, and it does not require any specialized equip-
ment. The data can be analyzed to obtain the on rate (k_{on}) and the off rate
(k_{off}) for the ligand interaction with the MAb under a variety of experimental conditions. The following protocol is used to study the effects of
calcium on the binding of factor IX or factor IXa to the MAb.

[125]I-Labeled factor IX (or ^{125}I-labeled factor IXa) in 0.05 M Tris-HCl,
0.15 M NaCl, 1 mg/ml bovine serum albumin, pH 7.5 (Tris/NaCl/BSA), is
incubated with 5- to 15-fold molar excess of the antibody binding sites at
room temperature. The exact concentrations of factor IX (or factor IXa)
and the MAb are given in Figs. 8 and 9. At different times, 20-μl aliquots
are removed and added to 3 ml of cold Tris/NaCl/BSA buffer containing
0.8% *Staphylococcus* A coated with rabbit antimouse immunoglobulin
prepared as follows: Rabbit antimouse gamma globulin (0.1 mg/ml) obtained from Cappel Laboratories is incubated with a 10% suspension of
Staphylococcus A (obtained from Sigma) in 0.05 M Tris-HCl, 0.15 M
NaCl, pH 7.5, buffer for 30 min at room temperature. The suspension is
centrifuged and the pellet is washed three times with copious amounts of
the above Tris/NaCl buffer. The final pellet is suspended to yield a 25%
suspension and 100 μl of it is used for 3 ml of the above Tris/NaCl/BSA
buffer. After incubation for 10 min, the suspensions are centrifuged at
5000 g for 5 min at 4°. The supernatants are decanted and drops remaining on the edges of tubes are blotted by cotton applicators. The background
[125]I counts in the pellets of tubes prepared from reaction mixtures of
$\approx$ 5-sec incubation times are usually between 5 and 7% of the total counts
added to these tubes. This suggests that the binding of ^{125}I-labeled factor IX
(or IXa) to the MAb is effectively stopped on 150-fold dilution of the
reaction mixture. One should note that dilution of 20 μl of reaction mixture at 25° to a 3-ml volume at 4° will slow down the rate of a bimolecular
reaction by a factor of $\approx$ 90,000. Thus, incubation of the diluted sample for
10 min prior to the measurement of the extent of the reaction will result in
an error of less than 1% in all of the samples incubated for up to 5 min. The
counts present in the $\approx$ 5-sec incubation tubes are subtracted from the
counts in the pellets of tubes prepared from the longer incubation mix-

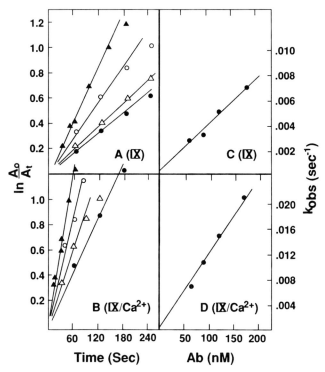

FIG. 8. Effect of saturating concentrations of Ca^{2+} on the association (k_{on}) and dissociation (k_{off}) rate constants for the interaction of factor IX with the MAb FX008.[11] First-order kinetic plots are obtained in the presence of 1 mM EDTA (A) and in the presence of 5 mM Ca^{2+} (B). For both A and B, the concentration of ^{125}I-labeled factor IX used was 11.4 nM. The concentration of antibody was as follows: (●) 57 nM, (△) 85.5 nM, (○) 114 nM, and (▲) 171 nM. A_0 represents the initial concentration of ^{125}I-labeled factor IX and A_t represents the unreacted factor IX at a given time. (C) Observed first-order rate constants plotted against antibody concentration for the reactions in 1 mM EDTA. (D) Observed first-order rate constants plotted against antibody concentration for the reactions in 5 mM Ca^{2+}. Note the difference in k_{obs} scale for plots in C and D. From Bajaj et al.[10] with permission.

tures; the resulting corrected counts are then used to calculate the percentage of factor IX (or IXa) bound to the MAb at a given time.

Apparent first-order rate constants are obtained from plots of ln A_0/A_t versus time, where A_0 represents initial factor IX (or IXa) concentration and A_t represents unreacted factor IX (or IXa) remaining at a given time. The second-order rate constants are obtained from the slopes of observed first-order rate constants versus the MAb concentrations. The data obtained in this manner are plotted in Fig. 8 for factor IX and in Fig. 9 for

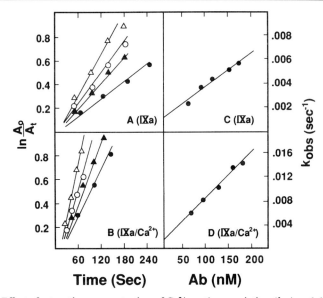

FIG. 9. Effect of saturating concentration of Ca^{2+} on the association (k_{on}) and dissociation (k_{off}) rate constants for the interaction of factor IXa with the MAb FX008.[11] First-order kinetic plots are obtained in the presence of 1 mM EDTA (A) and in the presence of 5 mM Ca^{2+} (B). For both A and B, concentration of ^{125}I-labeled factor IX used was 11.4 nM. The concentration of antibody was as follows: (●) 57 nM, (▲) 85.5 nM, (○) 114 nM, and (△) 171 nM. An antibody concentration of 150 nM was also used; however, because of space limitation, the plot using this antibody concentration is not shown. A_0 represents the initial concentration of ^{125}I-labeled factor IXa and A_t represents the unreacted factor IXa at a given time. (C) Observed first-order rate constants plotted against antibody concentration for the reactions in 1 mM EDTA. (D) Observed first-order rate constants plotted against Ab concentration for the reactions in 5 mM Ca^{2+}. Note the difference in k_{obs} scale for plots in C and D. From Bajaj et al.[10] with permission.

factor IXa. Calculated rate constants are listed in Table I. The k_{on} for both factor IX and factor IXa is ≈ 3-fold higher in the presence of 5 mM Ca^{2+} than that observed in the presence of 1 mM EDTA. Also, the value of k_{on} does not change on conversion of factor IX to factor IXa. However, the k_{off} is ≈ 10-fold higher for factor IXa than for factor IX both in the absence or presence of Ca^{2+}. Because the rate of collision (k_{on}) is determined primarily by the steric factors and the rate of dissociation (k_{off}) is determined primarily by the bonding strength, the rate data (Table I) suggest that the protease domain goes through one conformational change on binding of calcium and another on conversion to factor IXa. Thus, measurements of a specific MAb binding can be used to detect subtle and distinct conformational changes in the target protein.

TABLE I

ASSOCIATION (k_{on}) AND DISSOCIATION (k_{off}) RATE
CONSTANTS FOR BINDING OF FACTOR IX OR IXa[a]

Protein	k_{on} ($M^{-1} s^{-1}$)	k_{off} (s^{-1})	Binding constant (M^{-1})
IX/EDTA	4×10^4	1.9×10^{-4}	2×10^8
IX/Ca^{2+}	1.2×10^5	1.7×10^{-4}	7×10^8
IXa/EDTA	3.1×10^4	1.5×10^{-3}	2×10^7
IXa/Ca^{2+}	1.0×10^5	2×10^{-3}	5×10^7

[a] Binding to the MAb ± 5 mM Ca^{2+}.

To obtain a $K_{1/2}$ value for the Ca^{2+}-induced conformational change, first-order rate constants for the binding of factor IX (or IXa) to the MAb should be obtained at several increasing concentrations of calcium in the reaction mixture. The data can then be plotted as k_{obs} versus concentration of calcium (Fig. 10). The concentration of calcium required for the half-maximal increase in the k_{obs} reflects the affinity for Ca^{2+} binding. In the present study, it was shown to be 300 μM for factor IX and 250 μM for factor IXa. Using such an approach (including the molecular modeling approach described below), a hitherto unrecognized Ca^{2+} binding site in the protease domain of factor IX was identified.[10]

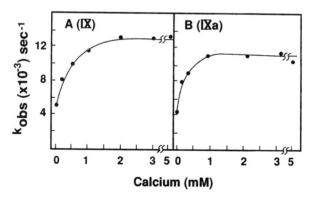

FIG. 10. Reaction of factor IX or factor IXa with the MAb at different concentrations of Ca^{2+}. Observed first-order rate constants for factor IX (A) and for factor IXa (B) are plotted against varying concentrations of Ca^{2+}. ^{125}I-Labeled factor IX (or ^{125}I-labeled factor IXa) concentration was 11.4 nM and the MAb binding site concentration was 114 nM. Ca^{2+} concentrations were as indicated. From Bajaj et al.[10] with permission.

Building of Putative Model Structure of Protease Domain of Factor IX

Full appreciation of the relationship between amino acid sequence and biologic function of a protein is only possible when the three-dimensional structure of the protein is known. Only two experimental approaches are capable of providing structural information of sufficient detail and accuracy: single crystal X-ray diffraction and multidimensional NMR. However, both methods suffer from experimental limitations. X-Ray diffraction analysis is critically dependent on obtaining crystals that diffract to high resolution, a hurdle that has yet to be overcome for all coagulation proteases, except thrombin.[41] A detailed description of this technique is presented in two volumes in this series.[42] Two- and three-dimensional NMR-based structure determinations are limited by the size of the molecules to which this method can be applied, and the overall accuracy of NMR-derived structures is also not as high as that obtainable by high-resolution diffraction methods. Details of NMR techniques as applied to the proteins are presented in two volumes in this series.[43] The upper limit is molecules with molecular mass of up to about 12,000–15,000 Da, precluding analysis of the entire factor IX molecule. The method has, however, been successfully applied to the structure determination of fragments of factor IX, including the first EGF-like domain.[9,44]

In the absence of an experimentally determined three-dimensional structure of factor IX, the technique of comparative molecular model building can be employed in order to gain some insight into the structure–function relationships of factor IX. While molecular models derived by this approach in no way can replace an experimentally determined structure in terms of accuracy and level of confidence, such models can nevertheless serve as valuable tools that facilitate the understanding of the biologic properties of factor IX.

The underlying principle for the knowledge-based model-building procedure described here is that the experimentally determined three-dimensional structure of one protein can serve as a template for the generation of a model of another, provided it is homologous to the first one.[45] This approach was first applied to the serine proteases by Hartley,[46]

---

[41] W. Bode, I. Mayr, U. Baumann, R. Huber, S. R. Stone, and J. Hofsteenge, *EMBO J.* **8**, 3467 (1989).

[42] This series, Vols. 114 and 115.

[43] This series, Vols. 176 and 177.

[44] L. H. Huang, H. Cheng, A. Pardi, J. P. Tam, and W. V. Sweeney, *Biochemistry* **30**, 7402 (1991).

[45] W. J. Browne, A. C. T. North, D. C. Phillips, K. Brew, T. C. Vanaman, and R. L. Hill, *J. Mol. Biol.* **42**, 65 (1969).

[46] B. S. Hartley, *Philos. Trans. R. Soc. London, Ser.* **B 257**, 77 (1970).

who derived the structure of trypsin from that of tosyl-α-chymotrypsin.[47] Since this seminal effort, several papers describing the construction of putative serine protease structures have been published.[48-52] The different approaches are similar in principle and as far as can be appraised yield comparable results. While a direct comparison of the structures generated by the different model-building procedures has not been published, it is reasonable to assume that differences in the resulting structures are most likely a reflection of the overall inaccuracies in the different approaches and that no single approach is significantly superior to the others. Detailed presentation of the general principles and methods in protein modeling can be found in two volumes of this series.[42] The procedure described here is the one we have employed for the generation of putative models of the protease domain of factor IX and related coagulation pro-teases.[10,19,28,51,53,54] The approach is a general one and is applicable for the generation of a putative protein model based on the crystal structure of a homologous protein. The brief description outlined here contains details that pertain to the serine proteases and specifically to human factor IX.

Programs and Computer Hardware

In our own studies we have employed the IRIS graphics systems from Silicon Graphics and the software programs TOM[55,56] and TURBO.[57] The procedure is, however, hardware and software independent and program packages such as SYBYL,[58] QUANTA,[59] and INSIGHT,[60] which all run on a number of different platforms, are equally applicable. Suitable programs for the alignment of amino acid sequences include the PRI[61] and

[47] J. J. Birktoft and D. M. Blow, *J. Mol. Biol.* **68**, 187 (1972).
[48] B. Furie, D. H. Bing, R. J. Feldman, D. J. Robison, J. P. Burnier, and B. C. Furie, *J. Biol. Chem.* **257**, 3875 (1982).
[49] J. Greer, *J. Mol. Biol.* **153**, 1027 (1981).
[50] R. J. Read, G. D. Brayer, L. Juršek, and M. N. G. James, *Biochemistry* **23**, 6570 (1984).
[51] L. J. Berliner, J. J. Birktoft, T. L. Miller, G. Musci, J. E. Scheffler, Y. Y. Shen, and Y. Sugawara, *Ann. N.Y. Acad. Sci.* **485**, 80 (1986).
[52] A. Sali and T. L. Blundell, *J. Mol. Biol.* **212**, 403 (1990).
[53] Y. Sugawara, J. J. Birktoft, and L. J. Berliner, *Semin. Thromb. Hemostasis* **12**, 209 (1986).
[54] V. L. Nienaber, S. L. Young, J. J. Birktoft, D. L. Higgins, and L. J. Berliner, *Biochemistry* **31**, 3852 (1992).
[55] T. A. Jones, this series, Vol. 115, p 157.
[56] C. Cambillau and E. Horales, *J. Mol. Graphics* **5**, 174 (1987).
[57] A. Roussel and C. Cambillau, "TURBO," LCCMB-CNRS, Faculté de Médicine Nord, F-13326 Marseille Cedex 15, France.
[58] "SYBYL Molecular Modeling System." Tripos Associates Inc., St. Louis, MO.
[59] "QUANTA." Polygen Corporation, Boston, MA.
[60] "INSIGHT." Biosym Technologies Inc., San Diego, CA.
[61] "PIR, Protein Identification Resource." National Biomedical Research Foundation, 3900 Reservoir Road NW, Washington, DC 20007.

GCG[62] packages; the latter was employed in our studies. A detailed description of principles and programs for sequence analysis can be found in a volume in this series.[63] Atomic protein coordinates can be obtained from the Protein Data Bank (PDB) at Brookhaven National Laboratory (Upton, NY).[64]

Data Preparation

Prior to the actual model construction, the following preliminary steps should be completed: (a) identify amino acid sequences of homologous proteins, (b) identify experimentally determined crystal structures of homologous proteins, (c) create a structural data base by structurally aligning the homologous structures, (d) align the amino acid sequences, and (e) assign insertions/deletions in relation to the elements of secondary structure. The protease domain of factor IX is homologous to the members of the trypsin family of serine proteases and the structural data base used for the modeling of factor IX is generated by superposition of the atomic coordinates of all known mammalian serine proteases. Segments that form the core of the protease domain structure, the β strands and the α helices, have essentially the same structure in all serine proteases, while segments that connect the strands and helices show a high degree of variation from one protease to the other. These two types of structural regions are sometimes referred to as structural conserved regions (SCRs) and variable regions (VRs).[49] The alignment of the corresponding amino acid sequences is based *entirely* cn the structural superpositions. The amino acid sequence of the unknown structure is then incorporated into this structurally based amino acid sequence alignment. The main consideration in this step is the matching of sequence patterns of residues located in the β strands and α helices. Attention is given to both sequence identities and to the polarity and electrostatic charges of the side chains. In order to obtain an optimal sequence alignment it is generally necessary to incorporate insertions and deletions into the amino acid sequences. Among the serine proteases such variations in polypeptide length always occur in the loops that connect the elements of secondary structure, and in particular the loops joining the β-sheet strands.[65] Consequently insertions and deletions are always placed either at the end or between the elements of secondary structure. A sequence alignment of mammalian serine proteases (domains) derived from these considerations is shown in Table II.

[62] J. Deveraux, P. Haeberli, and O. Smithies, *Nucleic Acids Res.* **12,** 387 (1984).
[63] This series, Vol. 183.
[64] F. C. Bernstein, T. F. Koetzle, G. J. B. Williams, E. F. Meyer, M. D. Brice, J. R. Rodgers, O. Kennard, T. Shimanouchi, and M. Tasumi, *J. Mol. Biol.* **112,** 535 (1984); address: F. C. Bernstein, Chemistry Department, Brookhaven National Laboratory, Upton, NY 11973.
[65] C. S. Craick, W. J. Rutter, and R. J. Fletterick, *Science* **220,** 1125 (1983).

TABLE II
AMINO ACID SEQUENCE ALIGNMENT OF SEVERAL SERINE PROTEASE DOMAINS

```
Fac-9#  181 185   190    195 200        205    210    215 220     225
CHT#:   16 20    25     30 35           40     45     50  55 57   60

Fac-9-Hum   VVGGEDAKPGQF PWQVVLNG K----VDAF CGGSIV NE KWIVTAA H CVET-------GV
Fac-9-Bov   VVGGEDAERGQF PWQVLLHG E----IAAF CGGSIV NE KWVVTAA H CIKP-------GV
Fac-9-Pig   IVGGEDAKPGQF PWQVLLNG K----IDAF CGGSII NE KWVVTAA H CIEP-------GV
Fac-9-Dog   VVGGKDAKPGQF PWQVLLNG K----VDAF CGGSII NE KWVVTAA H CIEP-------DV
Fac-9-Rab   IVGGENAKPGQF PWQVLLNG K----VEAF CGGSII NE KWVVTAA H CIKP-------DD
Fac-9-She   VVGGEDAARGQF PWQVLLHG E----IAAF CGGSIV NE KWVVTAA H CIKP-------GV
Fac-9-Gui   VVGGEDAKPGQF PWQVLLNG E----TEAF CGGSIV NE KWIVTAA H CILP-------GI
Fac-9-Mou   VVGGENAKPGQI PWQVILNG E----IEAF CGGAII NE KWIVTAA H CLKP-------GD
Fac-9-Rat   VVGGENAKPGQI PWQVILNG E----IEAF CGGAII NE KWIVTAA H CLKP-------GD
Cht-A-Bov   IVNGEEAVPGSW PWQVSLQD K---TGFHF CGGSLI NE NWVVTAA H CGV----------TT
RMCPII-Rat  IIGGVESIPHSR PYMAHLDI VTEKGLRVI CGGFLI SR QFVLTAA H CKG-----------R
Elast-Pig   VVGGTEAQRNSW PSQISLQY RSGSSWAHT CGGTLI RQ NWVMTAA H CVDR-------EL
Elast-Neu   IVGGRRARPHAW PFMVSLQL R----GGHF CGATLI AP NFVMSAA H CVAN------VNVR
Tonin-Rat   IVGGYKCEKNSQ PWQVAVIN ------EYL CGGVLI DP SWVITAA H CYS----------N
Kalli-Pig   IIGGRECEKNSH PWQVAIYH Y----SSFQ CGGVLV NP KWVLTAA H CKN----------D
Tryp-Bov    IVGGYTCGANTV PYQVSLNS G-----YHF CGGSLI NS QWVVSAA H CYK----------S
Throm-Hum   IVEGSDAEIGMS PWQVMLFR KS--PQELL CGASLI SD RWVLTAA H CLLYPPWDKNFTEN
Fac-7-Hum   IVGGKVCPKGEC PWQVLLLV N----GAQL CGGTLI NT IWVVSAA H CFDK-----IKNWR
Fac-10-Hum  IVGGQECKDGEC PWQALLIN E---ENEGF CGGTIL SE FYILTAA H CLYQ--------AK
Fac-11-Hum  IVGGTASVRGEW PWQVTLHT TS-PTQRHL CGGSII GN QWILTAA H CFYG-----VESPK
Fac-12-Hum  VVGGLVGLRGAH PYIAALYW G-----HSF CAGSLI AP CWVLTAA H CLQD-----RPAPE
Pro-C-Hum   LIDGKMTRRGDS PWQVVLLD S---KKKLA CGAVLI HP SWVLTAA H CMDE-------SK
Plasm-Hum   VVGGCVAHPHSW PWQVSLRT R---FGMHF CGGTLI SP EWVLTAA H CLEK-----SPRPS

            β-1           β-2      β-3
```

```
Fac-9#  230  235 240  245       250 255   260              270   270 275
CHT#:   65  70 75  80           85 90   95               100 102 105   110

Fac-9-Hum   KITVVAG EHNIEETE-HTE QKRNVIRIIPHH NYNAAI---------NKYN H D IALLELDE
Fac-9-Bov   KITVVAG EHNTEKPE-PTE QKRNVIRAIPHH SYNASI---------NKYS H D IALLELDE
Fac-9-Pig   KITVVAG EYNTEETE-PTE QRRNVIRAIPHH SYNATV---------NKYS H D IALLELDE
Fac-9-Dog   KITIVAG EHNTEKRE-HTE QKRNVIRTILHH SYNATI---------NKYN H D IALLELDE
Fac-9-Rab   NITVVAG EYNIQETE-NTE QKRNVIRIIPYH KYNATI---------NKYN H D IALLELDK
Fac-9-She   KITVVAG EHNTEKPE-PTE QKRNVIRAIPHH SYNASI---------NKYS H D IALLELDE
Fac-9-Gui   KIEVVAG KHNIEKKE-DTE QRRNVTQIILHH SYNASF---------NKYS H D IALLELDK
Fac-9-Mou   KIEVVAG EYNIDKKE-DTE QRRNVIRTIPHH QYNATI---------NKYS H D IALLELDK
Fac-Rat     KIEVVAG EHNIDEKE-DTE QRRNVIRTIPHH QYNATI---------NKYS H D IALLELDK
Cht-A-Bov   SDVVVAG EFDQGSSS-EKI QKLKIAKVFKNS KYNSLT-----------IN N D ITLLKLST
RMCPII-Rat  EITVILG AHDVRKRE-STQ QKIKVEKQIIHE SYNSVP-----------NL H D IMLLKLEK
Elast-Pig   TFRVVVG EHNLNQNN-GTE QYVGVQKIVVHP YWNTDD---------VAAG Y D IALLRLAQ
Elast-Neu   AVRVVLG AHNLSRRE-PTR QVFAVQRIFEDG YDPVN-------------LL N D IVILQLNG
Tonin-Rat   NYQVLLG RNNLFKDE-PFA QRRLVRQSFRHP DYIPLIVTNDTEQPVHDHS N D LMLLHLSE
Kalli-Pig   NYEVWLG RHNLFENE-NTA QFFGVTADFPHP GFNLSADGK-------DYS H D LMLLRLQS
Tryp-Bov    GIQVRLG EDNINVVE-GNE QFISASKSIVHP SYNSNT-----------LN N D IMLIKLKS
Throm-Hum   DLLVRIG KHSRTRYERNIE KISMLEKIYIHP RYNWREN----------LD R D IALMKLKK
Fac-7-Hum   NLIAVLG EHDLSEHD-GDE QSRRVAQVIIPS TYVPGT-----------TN H D IALLRLHQ
Fac-10-Hum  RFKVRVG DRNTEQEE-GGE AVHEVEVVIKHN RFTKE-----------TYD F D IAVLRLKT
Fac-11-Hum  ILRVYSG ILNQSEIK-EDT SFFGVQEIIIHD QYKMA-----------ESG Y D IALLKLET
Fac-12-Hum  DLTVVLG QERRNHS---CE PCQTLAVRSYRL HEAFSPV----------SYQ H D LALLRLQE
Pro-C-Hum   KLLVRLG EYDLRRWE-KWE LDLDIKEVFVHP NYSKST-----------TD N D IALLHLAQ
Plasm-Hum   SYKVILG AHQEVNLE-PHV QEIEVSRLFLEP TR---------------- K D IALLKLSS

            β-4             β-5       β-6
```

(continued)

TABLE II (continued)

```
Fac-9#      280   285   290   295   300   305 310   315       320   325
CHT#         115   120   125       130   135 140   145       150   155 160
Fac-9-Hum  P-----LVLNSYVTPICIADKEYTNI--FLK FGSGYVSGW GRVFHKGR------SALV LQYLRV
Fac-9-Bov  P-----LELNSYVTPICIADRDYTNI--FSK FGYGYVSGW GKVFNRGR------SASI LQYLKV
Fac-9-Pig  P-----LTLNSYVTPICIADKEYTNI--FLK FGSGYVSGW GRVFNRGR------SATI LQYLKV
Fac-9-Dog  P-----LTLNSYVTPICIADREYSNI--FLK FGSGYVSGW GRVFNKGR------SASI LQYLKV
Fac-9-Rab  P-----LTLNSYVTPICIANREYTNI--FLN FGSGYVSGW GRVFNRGR------QASI LQYLRV
Fac-9-Shp  P-----LELNSYVTPICIADREYTNI--FLK FGYGYVSGW GRVFNRGR------SASI LQYLRV
Fac-9-Gui  P-----LSLNSYVTPICIANREYTNI--FLK FGAGYVSGW GKLFSQGR------TASI LQYLRV
Fac-9-Mou  P-----LILNSYVTPICVANREYTNI--FLK FGSGYVSGW GKVFNKGR------QASI LQYLRV
Fac-9-Rat  P-----LILNSYVTPICVANKEYTNI--FLK FGSGYVSGW GKVFNKGR------QASI LQYLRV
CHT-A-Bov  A-----ASFSQTVSAVCLPSASDD----FAA GTTCVTTGW GLTRYTNA------NTPDR LQQASL
RMCPII-Rat K-----VELTPAVNVVPLPSPSDF----IHP GAMCWAAGW GKTGVRDP------TSYT LREVEL
Elast-Pig  S-----VTLNSYVQLGVLPRAGTI----LAN NSPCYITGW GLTRTNGQ------LAQT LQQAYL
Elast-Neu  S-----ATINANVQVAQLPAQGRR----LGN GVQCLAMGW GLLGRNRG------IASV LQELNV
Tonin-Rat  P-----ADITGGVKVIDLPTKE------PKV GSTCLASGW GSTNPSEM-----VVSHD LQCVNI
Kalli-Pig  P-----AKITDAVKVLELPTQE------PEL GSTCEASGW GSIEPGPD---DFEFPDE IQCVQL
Tryp-Bov   A-----ASLNSRVAS-SLPTSC------ASA GTQCLISGW GNTKSSGT-----SYPDV LKCLKA
Throm-Hum  P-----VAFSDYIHPVCLPDRETAA-SLLQA GYKGRVTGW GNLKETWTANVGKGQPSV LQVVNL
Fac-7-Hum  P-----VVLTDHVVPLCLPERTFSE-RTLAF VRFSLVSGW GQLLDRGA------TALE LMVLNV
Fac-10-Hum P-----ITFRMNVAPACLPERDWAE-STLMT QKTGIVSGF GRTHEKGR------QSTR LKMLEV
Fac-11-Hum T-----VNYTDSQRPICLPSKGDR----NVI YTDCWVTGW GYRKLRDK------IQNT LQKAKI
Fac-12-Hum DADGSCALLSPYVQPVCLPSGAAR----PSE TTLCQVAGW GHQFEGAE-----EYASF LQEAQV
Pro-C-Hum  P-----ATLSQTIVPICLPDSGLAERELNQA GQETLVTGW GYHSSREKEA-KRNRTFV LNFIKI
Plasm-Hum  P-----AVITDKVIPACLPSPNYV----VAD RTECFITGW GETQGT------FGAGL LKEAQL
                                       β-7                           β-8
```

```
Fac-9#      330   335       340   345   350   355     360   365   370
CHT#         165   170       175   180   185       190   195   200
Fac-9-Hum  PL V DRATCLR-----S TKFTIYN NMFCAG FHEG---GR D SCQGD S G GPHVT EVE--
Fac-9-Bov  PL V DRATCLR-----S TKFSIYS HMFCAG YHEG---GK D SCQGD S G GPHVT EVE--
Fac-9-Pig  PL V DRATCLR-----S TKVTIYS NMFCAG FHEG---GK D SCQGD S G GPHVT EVE--
Fac-9-Dog  PL V DRATCLR-----S TKFTIYN NMFCAG FHEG---GK D SCQGD S G GPHVT EVE--
Fac-9-Rab  PF V DRATCLR-----S TKFTIYN NMFCAG FDVG---GK D SCEGD S G GPHVT EVE--
Fac-9-Shp  PL V DRATCLR-----S TKFTIYN HMFCAG YHEG---GK D SCQGD S G GPHVT EVE--
Fac-9-Gui  PL V DRATCLR-----S TKFTIYN NMFCAG FHEG---GR D SCQGD S G GPHVT EVE--
Fac-9-Mou  PL V DRATCLR-----S TTFTIYN NMFCAG YREG---GK D SCEGD S G GPHVT EVE--
Fac-9-Rat  PL V DRATCLR-----S TKFSIYN HMFCAG YREG---GK D SCEGD S G GPHVT EVE--
Cht-A-Bov  PL L SNTNCKK-----Y WGTKIKD AMICAG -AS---GV  S SCMGD S G GPLVC KKN--
RMCPII-Rat RI M DEKACVD------ YRYYEYK FQVCVG SPTT---LR A AFMGD S G GPLLC --A--
Elast-Pig  PT V DYAICSSS---SY WGSTVKN SMVCAG -GNR---GV S GCQGD S G GPLHC LVN--
Elast-Neu  TV V TS-LCRR------ ------- SNVCTL VRGR---QA G VCFGD S G GPLVC --N--
Tonin-Rat  HL L SNEKCIE-----T YKDNVTD VMLCAG EMEG---GK D TCAGD S G GPLIC --D--
Kalli-Pig  TL L QNTFCAH-----B HBDKVTE SMLCAG YLPG---GK D TCMGD S G GPLIC --N--
Tryp-Bov   PI L DNSSCKS-----A YPGQITS NMFCAG YLEG---GK D SCQGD S G GPVVC --S--
Throm-Hum  PI V ERPVCKD-----S TRIRITD NMFCAG YKPDEGKRG D ACEGD S G GPFVM KSPFN
Fac-7-Hum  PR L MTQDCLQQSRKVG DSPNITE YMFCAG YSDG---SK D SCKGD S G GPHAT HYR--
Fac-10-Hum PY V DRNSCKL-----S SSFIITQ NMFCAG YDTK---QE D ACQGD S G GPHVT RFK--
Fac-11-Hum PI L TNEECQV---RY  RGHKITH KMICAG YREG---GK D ACKGD S G GPLSC KHN--
Fac-12-Hum PF L SLERCSA---PDV HGSSILP GMLCAG FLEG---GT D ACQGD S G GPLVC EDQAA
Pro-C-Hum  PV V PHNECSE-----V MSNMVSE NMLCAG ILGD---RQ D ACEGD S G GPMVA SFH--
Plasm-Hum  PV I ENKVCNR---YEF LNGRVQS TELCAG HLAG---GT D SCQGD S G GPLVC FEK--
           β-8    α-1              β-9                          β-10
```

TABLE II (continued)

Fac-9#	375	380	385			390	395		400	405	410	415
CHT#	205	210	215			220	225		230	235	240	245
Fac-9-Hum	-GTSFLTG	IISW	G	E--ECAMK	GKY	G	IYT	KVSRY	VNWIKEKTKLT			
Fac-9-Bov	-GTSFLTG	IISW	G	E--ECAMK	GKY	G	IYT	KVSRY	VNWIKEKTKLT			
Fac-9-Pig	-GTSFLTG	IISW	G	E--ECAVK	GKY	G	IYT	KVSRY	VNW????????			
Fac-9-Dog	-GISFLTG	IISW	G	E--ECAMK	GKY	G	IYT	KVSRY	VNWIKEKTKLT			
Fac-9-Rab	-GTSFLTG	IISW	G	E--ECAIK	GKY	G	VYT	RVSWY	VNWIKEKTKLT			
Fac-9-She	-GTSFLTG	IISW	G	E--ECAMK	GKY	G	IYT	KVSRY	????????????			
Fac-9-Gui	-GTNFLTG	IISW	G	E--ECAMK	GKY	G	IYT	KVSRY	VNW????????			
Fac-9-Mou	-GTSFLTG	IISW	G	E--ECAMK	GKY	G	IYT	KVSRY	VNWIKEKTKLT			
Fac-9-Rat	-GTSFLTG	IISW	G	E--ECAMK	GKY	G	IYT	KVSRY	VNW????????			
Chy-A-Bov	-GAWTLVG	IVSW	G	SS-TCS-T	STP	G	VYA	RVTAL	VNWVQQTLAAN			
RMCPII-Rat	-G--VAHG	IVSY	G	----HPDA	KPP	A	IFT	RVSTY	VPTINAVIN--			
Elast-Pig	-GQYAVHG	VTSF	V	SRLGCNVT	RKP	T	VFT	RVSAY	ISWINNVIASN			
Elast-Neu	-G--LIHG	IASF	V	RG-GCASG	LYP	D	AFA	PVAQF	VNWIDSIIQ--			
Tonin-Rat	-G--VLQG	ITSG	G	AT-PCAKP	KTP	A	IYA	KLIKF	TSWIKKVMKEN			
Kalli-Pig	-G--MWQG	ITSW	G	HT-PCGSA	NKP	S	IYT	KLIFY	LDWINbTITEN	DGK		
Tryp-Bov	-G--KLQG	IVSW	G	S--GCAQK	NKP	G	VYT	KVCNY	VSWIKQTIASN			
Throm-Hum	-NRWYQMG	IVSW	G	E--GCDRD	GKY	G	FYT	HVFRL	KKWIQKVIDQF	GE		
Fac-7-Hum	-GTWYLTG	IVSW	G	Q--GCATV	GHF	G	VYT	RVSQY	IEWLQKLMRSE	PRPGVLLRAPFP		
Fac-10-Hum	-DTYFVTG	IVSW	G	E--SCARK	GKY	G	IYT	KVTAF	LKWIDRSMKTR	GLPKAKSHAPEVITSSPL		
Fac-11-Hum	-EVWHLVG	ITSW	G	E--GCAQR	ERP	G	VYT	NVVEY	VDWILEKTQAV			
Fac-12-Hum	ERRLTLQG	IISW	G	S--GCGDR	NKP	G	VYT	DVAYY	LAWIREHTVS-			
Pro-C-Hum	-GTWFLVG	LVSW	G	E--GCGLL	HNY	G	VYT	KVSRY	LDWIHGHIRDK	EAPQKSWAP		
Plasm-Hum	-DKYILQG	VTSW	G	L--GCARP	NKP	G	VYV	RVSRF	VTWIEGVMRNN			

β-11   β-12   α-2

[a] Boxed residues are located in β-strands or α-helices. The numbering systems based on the bovine chymotrypsinogen and human factor IX are given. The last digit in each number indicates the location of the residue. CHT, Chymotrypsin; RMCP, rat mast cell protease; Hum, human; Bov, bovine; Rab, rabbit; She, sheep; Gui, guinea pig; Mou, mouse; Elast–Neu, elastase neutrophil. A question mark indicates that the sequence at that position is unknown. The one-letter amino acid code shown in bold indicates identity with human factor IX sequence. The catalytic triad residues (H57, D102, and S195) and the substrate specificity pocket residues (189, 216, and 226) are also boxed.

Selection of Template Structure for Model Construction

Although the substrate specificities of factor IX and most other coagulation proteases resemble that of trypsin, chymotrypsin was chosen at the starting point in our model construction. While the amino acid sequence of the protease domain of factor IX and chymotrypsin are about 40% identical (versus 44% identity with trypsin), the distribution of disulfide bridges is most similar to that in chymotrypsin and fewer insertions are necessary to align the sequences (Table II). Therefore, our starting template for the construction of the putative factor IX model was the structure of α-chymotrypsin.[47,66] However, in regions where the amino acid sequence of factor IX more closely resembles that of another protease, that structure was used as a template(s) in those regions. Examples of this are

[66] R. A. Blevins and A. Tulinsky, J. Biol. Chem. 260, 4264 (1985).

the calcium binding region,[10] consisting of residues 235–245, and the substrate binding pocket residues 352–360 of factor IX which display the greatest similarity to the corresponding sequences in trypsin and elastase and to those in trypsin and kallikrein, respectively.

Replacements of Amino Acid Residues

When replacing an amino acid residue, the sizes of the two side chains have to be taken into consideration; this is particularly important for internal residues. In order to minimize the introduction of errors it is important that the side-chain replacement is performed in the following sequence: (a) larger → smaller, (b) same size, (c) smaller → larger, (d) glycine → any nonglycine, and (e) any nonproline → proline. The reason is that the interior of the proteins are tightly packed. When a side chain in a homologous protein is replaced with a larger residue, adjacent side-chain residues in general are concomitantly decreased in size.

Specifically, replacements that involve glycines and prolines require special attention. The main-chain conformation of glycine residues can occupy most of the Ramachandran space while nonglycine residues as a general rule only occupy the left half of the Ramachandran space. Thus, when replacing a glycine with a nonglycine residue, consideration has to be given to the (ϕ, ψ) values. If these values suggest a conformation with nonpermissible (ϕ, ψ) values, it is likely that an alteration of the main-chain conformation is required. A rotation of 180° of the peptide preceding the glycine residue will frequently generate an acceptable conformation. An example of this is the replacement of residue 243 in the calcium binding loop from glycine in trypsin to histidine in factor IX.[10] Because of the cyclic structure of proline, replacement to a proline can take place without an adjustment in the main chain only if the ϕ value is near $-60°$.

In our study, the side-chain replacements were performed with the program DELPHI,[67] which permits the automatic replacement of a specified residue by another, followed by torsional readjustments to relieve bad contacts caused by the replacement. The polypeptide main chain is left unaltered. Analysis of the known homologous structures can frequently suggest the conformation of the replaced residues, and advantage is also taken of the side-chain rotamer library.[68] For residues containing polar atoms and which are located in internal positions, care has to be taken that the polar atoms participate in hydrogen bond interactions, either with other protein atoms or with solvent molecules. The serine proteases con-

---

[67] R. Kimmel, Masters Degree Thesis, Washington University, St. Louis, MO (1985).
[68] J. W. Ponder and F. M. Richards, J. Mol. Biol. 193, 775 (1987).

tain a number of solvent molecules located in internal positions, and several of these are conserved in the various protease structures.

In regions of insertions and deletions where the length of the polypeptide chain of the altered structure matches that of a protease structure different from that of chymotrypsin, a fragment-fitting method[69] was used. The structural data base for this purpose was assembled from all known serine protease structures, as discussed above. In cases in which the length of the insertion loops did not have any counterparts in other protease structures, a secondary data base formed from all known crystal structures composed predominantly of antiparallel β sheets served as the source of fragments.

Areas of unreasonable stereochemistry consisting mainly of close non-bonded contacts were identified and corrected using a molecular graphics system. Simulated annealing molecular dynamics procedures[70] were used to improve the overall stereochemistry of the factor IX model. In our work we have used XPLOR,[70] but the minimization of the empirical energy of the putative models can also be achieved by programs such as AMBER, CHARMM, and SYBYL. Care should be taken during energy minimization of the putative models; extensive structural changes suggest that the starting model is likely to contain significant errors.

The validity of our approach is supported indirectly by a number of observations: (a) The overall amino acid sequence of the factor IX protease domain can be incorporated into the chymotrypsin framework with little difficulty. (b) Residues located in the interior of the protease domain, i.e., not in direct solvent contact, are either largely unchanged or replaced by similar residues. Importantly, such replacements cause few additional polar side chains and no charged residues to be positioned in internal locations. (c) Insertions can easily be accommodated in loop regions that connect elements of the secondary structure, and which are located on the molecular surface. However, in general there is little guidance in modeling such insertion loops and several equally likely conformations are possible. A putative model of the protease domain of human factor IX based on these considerations is shown in Fig. 11.

The overall conclusions that can be drawn from analysis of the putative protein models are that the conformation of the protease domain of factor IX (and related enzymes) is likely to be quite similar to that of chymotrypsin and that the putative models can serve as acceptable working models. It should, however, be borne in mind that, in general, those parts of the protease domain that are involved in unique biologic functions are located

[69] T. A. Jones and S. Thirup, *EMBO J.* **5,** 819 (1986).
[70] A. T. Brunger, *J. Mol. Biol.* **203,** 803 (1988).

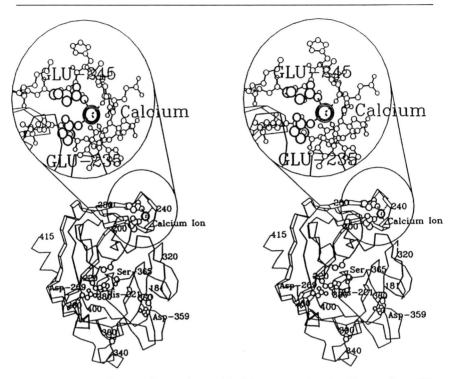

FIG. 11. Stereo diagram of a putative model of the protease domain of human factor IXa showing a C_α model and a few selected side chains. These include the catalytic triad residues: Asp-269[102], His-221[57], and Ser-365[195], and the Asp-359[189] residue, which is located in the substrate binding pocket. Also included are the calcium ion and the two glutamic acid residues that are proposed to act as calcium ligands. The part of the C_α model that includes the two peptide segments (residues 231–265 and 276–307), which were tested for their binding to MAb (FX008) and calcium,[10] is shown as a ribbon. The remaining part of the 181[15]–310[141] segment of the protease domain is shown in heavy tracing. The last half of the protease domain is shown in thin tracing. The three cystine bridges in the protease domain are shown as two balls connected with thin lines to the C_α model. The N and C termini and every 20th residue are labeled at the C_α positions. The selected side chains mentioned above are also labeled. The enlarged circular area shows the environment around the calcium ion. All atoms in residues 232[67]–248[83] are shown as ball and stick models. Ligand bonds between the calcium ion and the carbonyl oxygens of residues 237[72] and 240[75], and the carboxylates of Glu-235[70] and Glu-245[80], are shown as stippled lines. The view of the model is the one commonly used for presentation of α-chymotrypsin.[27] The numbers in brackets represent the chymotrypsinogen numbering system. From Bajaj et al.[10] with permission.

in the loop regions. These are also those segments of the putative structures that are least likely to be predicted correctly using comparative model-building methods. Any conclusions drawn from the use of such models therefore have to be verified by other means.[10]

Use of Putative Models to Study Protease Domain Mutations in Factor IX

Using the structure-based models, several investigators have analyzed the effect of naturally occurring mutations in the protease domain of factor IX that result in a bleeding disorder commonly known as hemophilia B. A mutant factor IX protease domain model suggests that replacement of isoleucine at position 397 by threonine will result in the formation of a hydrogen bond between the OH group of Thr-397 and the carbonyl oxygen of Trp-385; as a result, factor X binds to the activated Ile[397]-Thr mutant in a configuration inefficient for rapid catalysis.[71,72] Mutant Gly363Val is inactive because this substitution prevents the development of active-site conformation in factor IXa such that the substrate binding and the oxyanion hole are not formed in the mutated enzyme.[19] Three mutants of factor IX in which Asp-364 is replaced by either His, Arg, or Val are inactive because the mutant enzymes cannot form the critical ion pair between the carboxylate of Asp-364 and the α-amino group of Val-181. As a result, each of these three mutants does not achieve the proper conformation in the active-site region and has an incompletely formed substrate binding pocket. In addition, these mutants are incapable of making the hydrogen bonds between the main-chain amide groups of residues 363 and 365 and the carbonyl oxygen of the substrate, which is essential for stabilization of the tetrahedral transition state,[28] a situation analogous to that described for Gly363Val mutant.[19] Two other factor IX mutants, Gly309Val and Gly311   Asp-364 and Val-181 could be prevented, have also been described.[40,73] In the putative structural models, it has been observed that Gly-309 and Gly-311 are spatially close to the Asp[364]-Val[181] ion pair and the bulkier side chain of valine or glutamic acid could physically displace the atoms necessary for the ion-pair formation.[40,73] A mutant in which Glu-245 is replaced by Val

[71] V. A. Geddes, B. F. Le Bonniec, G. V. Louie, G. D. Brayer, A. R. Thomson, and R. T. A. MacGillivray, *J. Biol. Chem.* **264**, 4689 (1989).

[72] N. Hamaguchi, P. S. Charifson, L. Pedersen, G. D. Brayer, K. J. Smith, and D. W. Stafford, *J. Biol. Chem.* **266**, 15213 (1991).

[73] T. Miyata, T. Sakai, M. Sugimoto, H. Naka, K. Yamamoto, A. Yoshioka, H. Fukui, K. Mitsui, K. Kamiya, H. Umeyama, and K. Yamamoto, *Biochemistry* **30**, 11286 (1991).

has also been described.[28] This mutant lacks the carboxyl group of Glu-245, which has been proposed to be one of the ligands for the high-affinity Ca^{2+} binding site in the protease domain of factor IX.[10] Thus, this mutant may be impaired in its binding to Ca^{2+} at this site. One should note, however, that analysis of the above mutants is based on putative structural models and caution must be exercised in interpreting them in the absence of an experimentally determined structure.

[7] Factor VIII and Factor VIIIa

By PETE LOLLAR, PHILIP J. FAY, and DAVID N. FASS

Overview: Intrinsic Pathway Factor X Activation

The activation of factor X (fX) in the intrinsic pathway of blood coagulation occurs by means of hydrolysis of a single bond at Arg^{194}-Ile^{195} catalyzed by a trypsinlike serine protease, factor IXa. In the absence of additional cofactors, the catalytic efficiency of factor IXa toward factor X is very low. Additionally, factor IXa is generally poorly reactive toward other substrates and inhibitors, including diisopropyl fluorophosphate, p-nitrophenyl-p'-guanidinobenzoate, N^{α}-benzoyl-L-arginine ethyl ester, and thiobenzyl benzyloxycarbonyl-L-lysinate, which usually are very reactive toward other proteases in this family. However, in the presence of three cofactors, activated factor VIII (fVIIIa), acidic phospholipid, and Ca^{2+} (Scheme 1), the catalytic efficiency of factor IXa toward factor X is increased by several orders of magnitude.[1]

$$\text{factor X} \xrightarrow[\substack{\text{calcium ions}}]{\substack{\text{factor IXa} \\ \text{factor VIIIa} \\ \text{membrane surface}}} \text{factor Xa}$$

SCHEME 1

The catalytic advantage conferred by cofactors in this reaction involves binding of factors VIIIa, IXa, and X to the phospholipid membrane and appears to produce a rate enhancement in part through an increase in collisional frequency by restricting the reactants to a two-dimensional surface, i.e., by a reduction in dimensionality, which has been observed in several biological systems.[2] Additionally, factor VIIIa apparently alters the

---

[1] G. van Dieijen, G. Tans, J. Rosing, and H. C. Hemker, *J. Biol. Chem.* **256,** 3433 (1981).
[2] P. H. von Hippel and O. G. Berg, *J. Biol. Chem.* **264,** 675 (1989).

active site structure of factor IXa to make it better recognize a rate-limiting transition state for factor X activation, because its dominant kinetic effect is to increase the k_{cat}.[1] A k_{cat} effect has also been observed in the homologous prothrombinase complex,[3] and suggests that unlike trypsin and other solution-phase serine proteases, the active sites of factor IXa and factor Xa are not fully formed in the absence of cofactors.

Factor VIII and Factor VIIIa

Factor VIII (fVIII) is defined as that substance that is missing or deficient in hemophilia A; it is a large heterodimeric plasma glycoprotein that circulates polydisperse ($M_r \approx 160,000 - 280,000$). Because of its clinical importance, it has been the subject of intensive investigation since first being defined in 1936.[4] When factor VIII was last reviewed in this series in 1976, it was concluded correctly that available purification procedures for factor VIII had led to the characterization not of factor VIII, but of a large carrier protein for factor VIII.[5] This carrier protein has been identified as von Willebrand factor (vWf), a huge polydisperse, homopolymeric plasma glycoprotein (M_r 0.5 to greater than 20×10^6) that binds to and copurifies with factor VIII unless specific steps are taken to dissociate the complex. The distinction is an important one because vWf still is occasionally erroneously referred to as factor VIII and some commercially available antibodies to vWf (used commonly to identify endothelial cells, which contain vWf but not detectable factor VIII) are sold as "anti-factor VIII."

Structure of Factor VIII

Factor VIII is synthesized as a single chain with internal sequence homology defining three A domains that are homologous to regions in factor V and ceruloplasmin, two C domains that are homologous to lipid-binding regions of a cellular agglutinin from *Dictyostelium discoidium,* and a large B domain that is not homologous to other known structures[6-8]

[3] J. Rosing, G. Tans, J. W. P. Govers-Riemslag, R. F. A. Zwaal, and H. C. Hemker, *J. Biol. Chem.* **255**, 274 (1980).

[4] A. J. Patek, Jr. and R. H. Stetson, *J. Clin. Invest.* **15**, 531 (1936).

[5] M. E. Legaz and E. W. Davie, this series, Vol. 45, p. 83.

[6] G. A. Vehar, B. Keyt, D. Eaton, H. Rodriguez, D. P. O'Brien, F. Rotblat, H. Oppermann, R. Keck, W. I. Wood, R. N. Harkins, E. G. D. Tuddenham, R. M. Lawn, and D. J. Capon, *Nature (London)* **312**, 337 (1984).

[7] J. J. Toole, J. L. Knopf, J. M. Wozney, L. A. Sultzman, J. L. Buecker, D. D. Pittman, R. J. Kaufman, E. Brown, C. Shoemaker, E. C. Orr, G. W. Amphlett, W. B. Foster, M. L. Coe, G. J. Knutson, D. N. Fass, and R. M. Hewick, *Nature (London)* **312**, 342 (1984).

[8] W. R. Church, R. L. Jernigan, J. J. Toole, R. M. Hewick, J. Knopf, G. J. Knutson, M. E. Nesheim, K. G. Mann, and D. N. Fass, *Proc. Natl. Acad. Sci. U.S.A.* **81**, 6934 (1984).

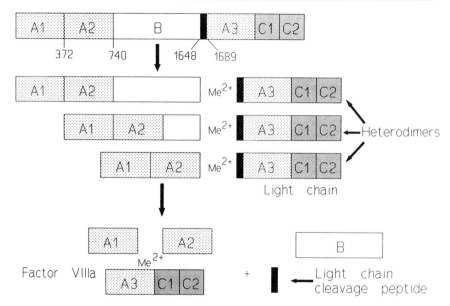

FIG. 1. Domain structure of factor VIII and thrombin-activated factor VIII. Amino acid numbering is based on the human sequence.

(Fig. 1). Although the boundaries between the domains were originally defined based on intron–exon junctions,[6] A domains now are commonly designated at thrombin cleavage sites: at position 372 between the A1 and A2 domains, at position 740 between the A2 and B domains, and at position 1689 to define the NH$_2$ terminus of the A3 domain. Due to variable proteolysis in the B domain, perhaps intracellularly prior to secretion, human and porcine factor VIII are isolated as a heterogeneous mixture of heterodimers that contain a common light chain, containing residues 1649–2332 with respect to the human sequence, and a variably sized heavy chain. The dimers are held together noncovalently by a linkage that requires Ca^{2+}, Mg^{2+}, or Mn^{2+}.[9,10] It is convenient to name proteolytic fragments of factor VIII relative to their domains to avoid the variable nomenclature that results from naming fragments by their apparent molecular masses by SDS-PAGE. Noncovalent fragment associations are denoted by a "/" whereas covalent associations between domains are denoted by a "-". For example, the dimer resulting from the association of the A1 fragment and the thrombin-cleaved light chain, A3–C1–C2, is designated fVIII$_{A1/A3-C1-C2}$.

[9] D. N. Fass, G. J. Knutson, and J. A. Katzmann, *Blood* **59**, 594 (1982).
[10] P. J. Fay, *Arch. Biochem. Biophys.* **262**, 525 (1988).

Factor VIII has no detectable activity prior to proteolytic activation. Thrombin and factor Xa are the only two mammalian enzymes that have been shown to activate highly purified factor VIII. The activation of fVIII is complex because several proteolytic cleavages are involved. Additionally, analysis of fVIII activation is complicated by the fact that activated fVIII undergoes nonproteolytic inactivation under certain conditions, including conditions of its plasma concentration (≈ 1 nM) at physiological pH. Although there have been conflicting reports regarding the functional significance of cleavages in fVIII catalyzed by thrombin, a consistent model has emerged that describes the activation of human and porcine fVIII. Human and porcine thrombin-activated fVIII are heterotrimers that contain the A1 fragment, A2 fragment, and the thrombin-cleaved light chain as subunits.[11-13] The nonproteolytic inactivation of fVIIIa is accompanied by dissociation of the A2 subunit.[14] The A2 subunit of human fVIIIa is more weakly associated than in the porcine homolog,[12] which may explain the observation that human fVIIIa is more labile than porcine fVIIIa. Cleavage of fVIII at position 1689 to release a 41-residue peptide from the NH$_2$ terminus of the fVIII light chain is necessary for dissociation of fVIII from vWf (Fig. 1). This cleavage does not appear necessary to activate the procoagulant function of fVIII per se (i.e., that function of fVIIIa which enhances the catalytic efficiency of factor IXa),[15,16] although this issue is controversial.[17,18]

The fVIII B domain, which is heavily glycosylated and contains about 50% of the mass of fVIII, has no known function. Recombinant mutants[19,20] or naturally occurring dimers[21-23] lacking nearly the entire B domain have the same or slightly increased coagulant activity as B do-

[11] P. Lollar and C. G. Parker, *Biochemistry* **28**, 666 (1989).

[12] P. Lollar and E. T. Parker, *J. Biol. Chem.* **266**, 12481 (1991).

[13] P. J. Fay, P. J. Haidaris, and T. M. Smudzin, *J. Biol. Chem.* **266**, 8957 (1991).

[14] P. Lollar and C. G. Parker, *J. Biol. Chem.* **265**, 1688 (1990).

[15] D. C. Hill-Eubanks, C. G. Parker, and P. Lollar, *Proc. Natl. Acad. Sci. U.S.A.* **86**, 6508 (1989).

[16] A. M. Aly and L. W. Hoyer, *Thromb. Haemostasis* **65**, 942a (1991).

[17] D. D. Pittman and R. J. Kaufman, *Proc. Natl. Acad. Sci. U.S.A.* **85**, 2429 (1988).

[18] D. P. O'Brien and E. G. Tuddenham, *Blood* **73**, 2117 (1989).

[19] J. J. Toole, D. D. Pittman, E. C. Orr, P. Murtha, L. C. Wasley, and R. J. Kaufman, *Proc. Natl. Acad. Sci. U.S.A.* **83**, 5939 (1986).

[20] D. L. Eaton, W. I. Wood, D. Eaton, P. E. Hass, P. Hollingshead, K. Wion, J. Mather, R. M. Lawn, G. A. Vehar, and C. Gorman, *Biochemistry* **25**, 8343 (1986).

[21] L. O. Andersson, N. Forsman, K. Huang, K. Larsen, A. Lundin, B. Pavlu, J. Sandberg, K. Sewerin, and J. Smart, *Proc. Natl. Acad. Sci. U.S.A.* **83**, 2979 (1986).

[22] P. J. Fay, M. T. Anderson, S. I. Chavin, and V. J. Marder, *Biochim. Biophys. Acta* **871**, 268 (1986).

[23] P. Lollar, C. G. Parker, and R. P. Tracy, *Blood* **71**, 137 (1988).

main-rich molecules. Scanning transmission electron microscopy[24] and rotary shadowing electron microscopy[25] have revealed that the B domain projects as a long, thin stalk from the core of the fVIII molecule.

Isolation of Factor VIII: General Procedures

Factor VIII was first purified from bovine blood.[26] However, low yields rendered this procedure impractical for routine isolation of quantities suitable for structure–function studies. Subsequently, successful procedures were developed for the isolation of fVIII from porcine and human blood, which led to the cloning of the 9-kilobase (kb) human cDNA and the expression of active, recombinant fVIII.[6,7,27,28] Factor VIII is present at low concentrations in plasma. Additionally, it is extraordinarily sensitive to activation by thrombin, which can form unintentionally during the purification process. As discussed below, thrombin-activated fVIII (fVIIIa) is unstable under many conditions. This appears to account for the difficulties associated with fVIII purification. Nonetheless, several research laboratories now isolate several milligrams of plasma-derived human or porcine fVIII per year for structure–function analysis. Additionally, recombinant fVIII, which appears functionally identical to plasma-derived fVIII, has been produced in heterologous mammalian expression systems. The efficiency of recombinant fVIII expression, currently the purview of large biotechnology groups, is limited by poorly understood factors. Thus, in contrast to other large glycoproteins such as tissue plasminogen activator, fVIII that is purified from plasma remains an important source of study material.

Several procedures to isolate human[21,22,27,29-31] and porcine fVIII[9,11] from plasma have been published. These methods are specialized in the sense that they rely on access to or preparation of human or porcine

[24] M. W. Mosesson, D. N. Fass, P. Lollar, J. P. DiOrio, C. G. Parker, G. J. Knutson, J. F. Hainfeld, and J. S. Wall, *J. Clin. Invest.* **145**, 1310 (1990).

[25] W. E. Fowler, P. J. Fay, D. S. Arvan, and V. J. Marder, *Proc. Natl. Acad. Sci. U.S.A.* **87**, 7648 (1990).

[26] G. A. Vehar and E. W. Davie, *Biochemistry* **19**, 401 (1980).

[27] J. Gitschier, W. I. Wood, T. M. Goralka, K. L. Wion, E. Y. Chen, D. H. Eaton, G. A. Vehar, D. J. Capon, and R. M. Lawn, *Nature (London)* **312**, 326 (1984).

[28] W. I. Wood, D. J. Capon, C. C. Simonsen, D. L. Eaton, J. Gitschier, B. Keyt, P. H. Seeburg, D. H. Smith, P. Hollingshead, K. L. Wion, E. Delwart, E. G. D. Tuddenham, G. A. Vehar, and R. M. Lawn, *Nature (London)* **312**, 330 (1984).

[29] C. A. Fulcher and T. S. Zimmerman, *Proc. Natl. Acad. Sci. U.S.A.* **79**, 1648 (1982).

[30] F. Rotblat, D. P. O'Brien, F. J. O'Brien, A. H. Goodall, and E. G. D. Tuddenham, *Biochemistry* **24**, 4294 (1985).

[31] R. J. Hamer, J. A. Koedam, N. H. Beeser-Visser, and J. J. Sixma, *Biochim. Biophys. Acta* **873**, 356 (1986).

plasma concentrates derived from large volumes of blood or access to cell lines producing recombinant fVIII. The concentrations of fVIII in human and porcine plasma are approximately 0.3 and 2 μg/ml, respectively. This, combined with the fact that the overall yield of fVIII from starting plasma is frequently in the range of 5–10%, requires large starting volumes to produce a milligram of highly purified fVIII (e.g., approximately 70 liters of human plasma if the yield is 5%). Factor VIII is tightly bound to vWf in plasma. Although each 270-kDa subunit of vWf appears capable of binding one fVIII molecule, most of these binding sites are empty in plasma. Thus, most of the mass of the fVIII–vWf complex is due to vWf. Because of this, early steps during fVIII isolation take advantage of differential solubility (e.g., by using polyethylene glycol) and/or ion-exchange behavior of vWf; fVIII copurifies during these steps. Immunoaffinity chromatography using anti-fVIII antibodies is a key step during most procedures currently in use for the isolation of plasma-derived and recombinant fVIII. Specific measures are taken to dissociate the fVIII–vWf complex either before or during immunoaffinity chromatography. These include dissociation of the complex with 0.25 M CaCl$_2$, 0.25 M MgCl$_2$, or 35 mM 2-mercaptoethanol. 2-Mercaptoethanol irreversibly abolishes vWf binding to fVIII but has no apparent effect on fVIII function. Presented below are procedures used by the authors to isolate human and porcine fVIII.

Isolation of Human Factor VIII

Commercial fVIII concentrates (Koate or Koate-HP, Cutter Biologicals, Berkeley, CA) have been used successfully as starting material. A single-step immunoaffinity purification from Koate-HP,[32] a higher purity preparation consisting primarily of fVIII–vWf complex and albumin, which is added to stabilize the fVIII during freeze drying, is described here. A monoclonal anti-A2 antibody, R8B12,[13] is used to achieve the isolation of fVIII. Purification from Koate, a cruder concentrate, involves several steps which have previously been described in detail.[33,34] The concentrate (6000 units, corresponding to approximately 1.5–2 mg of fVIII), in 0.15 M NaCl, 0.02 M histidine chloride, 1 mM CaCl$_2$, pH 7.2, is adjusted to 0.35 M CaCl$_2$, 0.4 M NaCl, 1 mM benzamidine and incubated for 1 hr at room temperature. The high Ca^{2+} buffer produces dissociation of the fVIII–vWf complex. The material is applied to a 4 ml R8B12 Affi-Gel column (3 mg/ml IgG) at a flow rate of 10 ml/hr. The column is washed with approximately 50 column volumes of the high Ca^{2+}-containing buffer

[32] P. J. Fay, T. M. Smudzin, and F. J. Walker, *J. Biol. Chem.* **266,** 20139 (1991).
[33] P. J. Fay and T. M. Smudzin, *J. Biol. Chem.* **265,** 6197 (1990).
[34] P. J. Fay and T. M. Smudzin, *J. Biol. Chem.* **264,** 14005 (1989).

and then washed with 0.15 M NaCl, 0.01 M histidine, 5 mM CaCl$_2$, and 0.01% Tween 20, pH 6.0. Factor VIII is eluted in this buffer plus 70% ethylene glycol (v/v). Occasionally, low levels of contaminating vWf are present, which can be removed by Mono S HPLC (see below).

Isolation of Porcine Factor VIII

Porcine fVIII can be isolated by using either porcine whole blood[9,35] or a lyophilized commercial concentrate (Hyate:C, Porton Products, Agoura Hills, CA)[11,23] as starting material. In the purification process, both of these procedures use a monoclonal antibody, W3-3[26], to the porcine fVIII light chain. A fVIII concentrate from whole blood (1–10 liters) can be prepared in the following way. Blood (0.5–1 liter) is collected from the femoral artery or vein of normal 70- to 100-kg animals into an evacuated container (w/v) containing 1/10 volume of anticoagulant containing 3.8% sodium citrate, 0.5 M ε-aminocaproic acid, 0.1 M benzamidine, and 50 μg/ml soybean trypsin inhibitor. At 5 min prior to the blood collection, the animals are given an intravenous injection of heparin, 55 U/kg.

Subsequent steps are carried out at room temperature. Cells are removed by centrifugation at 3500–6000 g for 10 min. Vitamin K-dependent proteins are removed by adsorption with 1/10 volume Al(OH)$_3$ for 10 min. The Al(OH)$_3$ precipitate is removed by centrifugation at 3500–6000 g for 10 min. A 50% solution (w/v) of polyethylene glycol (PEG) 6000 is added dropwise to the adsorbed plasma to a final concentration of 5% and stirred 30 min. The precipitate is collected by centrifugation at 3500–6000 g for 10 min.

The PEG precipitate is dissolved in 10 mM histidine, 5 mM CaCl$_2$, 1 mM benzamidine, pH 6.0 (Buffer A), plus 0.15 M NaCl to 1/5 plasma volume. The solution is clarified by centrifugation at 3500–6000 g for 10 min and the supernatant is applied to 200 ml QAE-cellulose packed into a coarse sintered glass funnel. The resin is washed with 2–3 volumes of buffer until the A_{280} is less than 0.5. The proteins are eluted with Buffer A plus 0.8 M NaCl. Fractions (12 ml) are collected; slightly turbid fractions are pooled because they are found to contain fVIII by clotting assay. To 1 volume of the pooled eluate is added 1 volume Buffer A plus 0.075 M NaCl. Then diisopropyl fluorophosphate and aprotinin is added to 1 mM and 10 μg/ml, respectively. The eluate is batch adsorbed to 150 ml dextran sulfate-Sepharose prepared as described.[36] The slurry is poured into a column and the supernatant is removed by siphoning after the resin has settled. The column is washed with Buffer A plus 0.1 M NaCl until the A_{280}

[35] G. J. Knutson and D. N. Fass, *Blood* **59**, 615 (1982).
[36] W. Kisiel, *J. Clin. Invest.* **64**, 761 (1979).

is within 0.1 of the buffer. The proteins are eluted with the same buffer plus 0.255 M CaCl$_2$ at 60 ml/hr. This elutes fVIII that is bound to vWf, which in turn is bound to the column. The A_{280} peak is pooled and assayed for fVIII activity.

The resulting solution is then applied to 2–4 ml W3-3-Sepharose 4B (6 mg IgG/ml) equilibrated in buffer A plus 0.1 M NaCl plus 0.25 M CaCl$_2$ at 20 ml/hr. The column is washed with 2–3 volumes of buffer A plus 0.1 M NaCl, followed by buffer A plus 2 M NaCl until the A_{280} is less than 0.005. Factor VIII is eluted with buffer A plus 2 M NaCl diluted 1 : 1 with ethylene glycol. The resulting preparation contains free fVIII light chain and an unidentified high-molecular-weight protein as minor contaminants. These contaminants are removed by Mono Q HPLC[11] (see below).

Hyate:C can be dissolved and applied directly to W3-3 with similar results. Each bottle of Hyate:C (10–30 bottles total, representing 0.4–1.2 mg of porcine fVIII) is dissolved in 4 ml 0.1 M NaCl, 10 mM histidine, 0.255 M MgCl$_2$; pH 6.0. Mg^{2+} is used instead of Ca^{2+} to avoid the formation of an insoluble calcium salt.[23] Washing and elution steps then proceed as above. Morpholineethanesulfonic acid (MES) buffers can be used instead of histidine buffers during the immunoaffinity purification step.

Characterization of Factor VIII by SDS-Polyacrylamide Gel Electrophoresis

Electrophoretic analysis of fVIII preparations is commonly used to evaluate impurities and to determine whether fVIII has been unintentionally activated. SDS-PAGE (8%) analysis of 200- to 500-ng samples using the Laemmli buffer system in a Hoeffer Mighty Small apparatus Hoeffer Scientific Instruments, San Francisco, CA followed by silver staining using the method of Morrissey[37] is a convenient way to evaluate samples in about 2 hr. The silver-staining method can be accelerated by heating the gels in a microwave oven for 1 min during the methanol fixation, washing steps, and incubation with reducing agent. A good fVIII preparation shows staining due to heavy chains and the light chain and no bands migrating ahead of the light chain (Fig. 2). Additionally, prior to loading the gel, a separate sample is diluted to 10–20 μg/ml into 0.15 M NaCl, 20 mM HEPES, pH 7.4, and reacted with 1 μg/ml (30 nM) thrombin for 2 min at room temperature to identify the appropriate cleavage products known as fragments A1, A2, and A3-C1-C2. All of the stainable bands in good fVIII preparations should disappear after the addition of thrombin. On heavily loaded gels, the B domain is also frequently identified as a thrombin cleavage product. This band disappears with time due to ongoing proteoly-

[37] J. H. Morrissey, *Anal. Biochem.* **117**, 307 (1981).

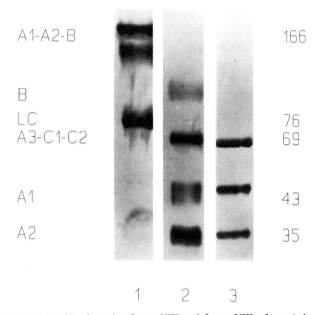

FIG. 2. SDS-PAGE (8%) of porcine factor VIII and factor VIIIa. Lane 1, heterodimeric factor VIII—two heavy chain species are visible along with the light chain (LC); lane 2, products of activation of factor VIII by thrombin—B, A3-C1-C2, A1, and A2 fragments are visible; lane 3, purified factor VIIIa. Apparent molecular masses (kDa) are shown on the right.

sis by thrombin and the fragments associated with this degradation are not identified.

Coagulation Assays for Factor VIII

Assays for fVIII coagulant function continue to play an important role in the assessment of fVIII preparations. These assays are based on the ability of fVIII to shorten the clotting time of plasma derived from a patient with hemophilia A. Two types of assays are commonly employed: the one-stage assay and the two-stage assay.[9,38]

In the one-stage assay, 0.1 ml hemophilia A plasma (e.g., from George King Biomedical, Overland Park, KA), 0.1 ml activated partial thromboplastin (aPTT) reagent (e.g., from Organon Teknika, Durham, NC), and 0.01 ml sample or standard consisting of diluted, citrated normal plasma are incubated for 5 min at 37°. Then 0.1 ml of 20 mM CaCl$_2$ is added and

[38] E. J. W. Bowie and C. A. Owen, *in* "Disorders of Hemostasis" (O. D. Ratnoff and C. D. Forbes, eds.), p. 43. Grune & Stratton, Orlando, FL, 1984.

the time to develop a fibrin clot is measured by visual inspection (tilt-tube assay) or by using a mechanical device (e.g., a Fibrometer, Becton-Dickinson Co., Cockeysville, MD). All dilutions can be made into 0.15 M NaCl, 0.02 M HEPES, pH 7.4. Borosilicate test tubes (10 × 75 mm) are used throughout.

A unit of fVIII is defined as the amount that is present in 1 ml of citrated normal plasma. The standard curve is constructed by making three or four dilutions of plasma (e.g., $\frac{1}{5}$, $\frac{1}{10}$, $\frac{1}{20}$, and $\frac{1}{50}$) and by plotting log clotting time versus log plasma concentration. This produces a linear plot from which the concentration of an unknown can be determined by interpolation. The one-stage assay relies on endogenously formed enzymes (thrombin and perhaps factor Xa) to activate fVIII. The factor IXa which is required for the formation of the fibrin clot results from the activation of plasma factor IX by factor XIa that has been produced during the preincubation of plasma with the aPTT reagent.

Highly purified human fVIII has a specific activity of about 3000–4000 U/mg by using as a standard citrated normal human plasma (George King Biomedical, Inc.) which has been referenced against a World Health Organization reference plasma. Highly purified porcine fVIII has a specific activity of 800–1200 U/mg relative to a standard developed at the Mayo Clinic (Rochester, MN) by using citrated male porcine plasma. The difference between the specific activities of porcine and human plasma occurs at least partly because there is more fVIII in porcine plasma.[23]

The two-stage assay measures the activation of fVIII by thrombin and is usually done simultaneously with the one-stage assay. In the assay, 0.1 ml hemophilia A plasma and 0.1 ml aPTT reagent are incubated for 5 min at 37° in the absence of sample. At 1 min prior to the end of this 5-min preincubation, thrombin is added to a sample of fVIII or partially purified fVIII to a concentration of 1 U/ml (≈ 10 nM). At 5 min, this sample is rapidly diluted and then 0.01 ml is added to the preincubation mixture, immediately followed by addition of 0.1 ml 20 mM CaCl$_2$ and the time to develop a fibrin clot is measured.

An example set of dilutions to compare purified human fVIII by the one- and two-stage assays using the tilt-tube method is as follows. Factor VIII, 0.4 mg/ml, approximately 1400 U/ml, is diluted 1/1500 at room temperature (by two dilutions) to decrease the concentration to less than 1 U/ml (tube 1). At 1 min after 0.01 ml of this sample has been added to hemophilia A plasma/aPTT reagent to start the 5-min incubation of the one-stage assay (tube 2), hemophilia A plasma and aPTT reagent are mixed to start the preincubation of the two-stage assay (tube 3). This produces two assay tubes that are staggered. Just prior to the 5-min point, thrombin is added to fVIII in tube 1 as part of the two-stage assay. At 5 min, CaCl$_2$ is added to tube 2 to measure the clotting time in the one-stage assay. After the clot has formed (at about 50–60 sec), the sample of thrombin-activated fVIII in tube 1 is diluted $\frac{1}{20}$ and 0.01 ml is added immediately to the preincubation mixture (tube 3), followed immediately by addition of $CaCl_2$. The sample is diluted because thrombin-activated fVIII is more potent than nonactivated fVIII because time-consuming, endogenous activation of fVIII must occur in the one-stage assay. The clotting time from the two-stage assay is converted to units/milliliter using the same standard curve as is used for the one-stage assay. The activation quotient is defined as activity in the two-stage assay divided by the activity in the one-stage assay. A hallmark of a good fVIII preparation is a high activation quotient. The activation quotient for human fVIII typically is 15–25. The activation quotient of porcine fVIII is higher, typically 40–60. The mechanism for the higher activation quotient of porcine fVIII is not known. A low activation quotient indicates the presence of unintentionally activated fVIII due to contaminating proteases. Because activated fVIII is not stable at pH 7.4, preparations with a low activation quotient typically lose activity with time in the one-stage assay. In contrast, good fVIII preparations are indefinitely stable in a variety of buffers at 4°.

Plasma-Free Assay of Factor VIII

In its activated form, fVIII can also be measured using purified factors IXa and X in the presence of a source of phospholipid.[39] At sufficiently low concentrations of fVIIIa, the initial rate of activation of factor X by factor IXa is linearly dependent on the concentration of fVIIIa. FVIIIa may be measured directly in this assay; alternatively, fVIII may be measured by adding thrombin to the system. The plasma-free assay is useful to measure the kinetics of activation of fVIII and the kinetics of inactivation of fVIIIa, and to characterize the kinetics of factor X activation.

In a typical experiment in which fVIII is measured, fVIII is added to a final concentration of 0.02–0.2 nM to a solution containing 2 nM factor IXa, 20 μM phosphatidylcholine–phosphatidylserine (75/25, w/w) vesicles in 0.15 M NaCl, 0.02 M HEPES, 5 mM $CaCl_2$, and 0.1% polyethylene glycol 8000 or 0.01% Tween 80 at pH 7.4. The total volume is 0.25 ml. Thrombin is added to a final concentration of 20 nM for 30 sec to activate fVIII. Then factor X is added to 300 nM and 50-μl samples are withdrawn at 15-sec intervals for 1 min and added to 5 μl of 0.5 M EDTA, pH 7.4, in small bullet tubes to quench factor X activation. These samples are stable for several hours or can be frozen. Factor Xa samples (20 μl) are added to 80 μl of 0.4 mM Spectrozyme Xa (American Diagnostica, New York, NY) and the A_{405}/min is measured. This is conveniently done in a V_{max} micro-

[39] P. Lollar, G. J. Knutson, and D. N. Fass, *Biochemistry* **24,** 8056 (1985).

titer kinetic plate reader (Molecular Devices, Menlo Park, CA). Plots of the initial rate of Spectrozyme Xa hydrolysis versus time of aliquot removal should be linear and proportional to the concentration of fVIII. Nonlinearity in the assay can result from either factor X depletion due to high enzyme (factor IXa/fVIIIa/phospholipid) concentration or from high fractional saturation of factor IXa by fVIIIa. Preferably, A_{405}/min is converted to the concentration of factor Xa using a standard curve prepared using purified factor Xa. The necessary proteins may be purified in the laboratory by using published procedures. Human thrombin, factor IXa, factor X, and factor Xa are available commercially. The quality of these preparations should be assessed by SDS-PAGE. Additionally, chromogenic substrate assays of factor IXa and factor X preparations for factor Xa and thrombin should be done, because the presence of low concentrations of these enzymes as contaminants can seriously affect the assay.

Mono Q and Mono S HPLC of Factors VIII and VIIIa

High-performance Liquid Chromatography (HPLC) of fVIII and its derivatives using Mono Q and Mono S columns (Pharmacia-LKB Biotechnology, Piscataway, NJ) is an important tool. These columns are expensive but can be reused many times if all solutions including those containing fVIII are passed through 0.2-μm filters prior to application to the columns. The lines in the chromatography system should be cleaned with 25% (v/v) ethanol/3 M HCl and then washed with H_2O before use to inactivate possible contaminating proteases.

Immunoaffinity-purified fVIII can be fractionated into B domain-rich and B domainless dimers by Mono Q HPLC.[22] Additionally, free light chains in the preparations, which especially accompany preparations that are derived from anti-light-chain immunoaffinity columns, can be removed because the fVIII light chain binds Mono Q less avidly than heavy chains. Mono Q HPLC of porcine fVIII also removes an unidentified, high-molecular-weight contaminant which binds tightly to the W3-3[9] immunoaffinity resin. Mono Q HPLC can be done by using HEPES/Ca^{2+} buffers, pH 7–7.5, and by using NaCl to develop the gradient. Whether to include a detergent such as 0.01% (v/v) Tween 80 or another agent to prevent loss of fVIII by adsorption is always a consideration. The presence of detergents in the preparation can be undesirable in certain instances (e.g., during analytical ultracentrifugation of fVIII). As the amount of fVIII loaded increases, the fractional loss to the column decreases. Application of greater than 300 μg of fVIII leads to a reasonable recovery (about 60% on the average) in the absence of a detergent. On the other hand, in the presence of 0.01% Tween 80 an average recovery of about 70% can be expected when as little as 1 μg of fVIII is applied, whereas in the absence of detergent, little or no recovery occurs. Loss of fVIII and its derivatives at low concentrations because of adsorption is also decreased by collecting fractions in Nunc (Naperville, IL) Mini-Sorb tubes.

Isolation of Subunits of Factor VIII

Factor VIII heavy chains and fVIII light chain are isolated by Mono Q and Mono S ion-exchange chromatography, respectively,[10,34,40] as follows. The subunits of intact, heterodimeric fVIII (0.1–0.3 mg/ml) in 0.6–0.8 M NaCl, 5 mM CaCl$_2$ in either 10 mM histidine chloride, pH 6.0, or 20 mM HEPES, pH 7.4, are dissociated by the addition of a solution of 0.5 M EDTA, pH 7.4, to a final concentration of 0.05 M for 12–24 hr at room temperature. An equal volume of 20 mM MES, 10 mM EDTA, 0.02% Tween 80, pH 6.0, is added, and the sample is applied at 1 ml/min to a Mono S HR5/5 column previously equilibrated in 0.25 M NaCl, 10 mM histidine chloride, 5 mM CaCl$_2$, 0.01% Tween 80, pH 6.0. The column is washed with 10 ml of column equilibration buffer plus 0.25 M NaCl. Factor VIII light chain is eluted by application of a linear, 20-ml gradient of NaCl from 0.25 to 1.0 M in the column equilibration buffer. Factor VIII heavy chains, which do not bind to Mono S under these conditions, are adjusted to pH 7.2–7.4, adsorbed to a Mono Q HR5/5 column equilibrated in 0.25 M NaCl, 10 mM HEPES, 5 mM CaCl$_2$, 0.01% Tween 80, pH 7.4, and eluted by application of a linear, 20-ml gradient of NaCl from 0.25 to 1.0 M in the column equilibration buffer. These procedures work well with both human and porcine fVIII. MES buffer (5 mM) at pH 6.0 can be used instead of histidine chloride.

Factor VIII can be reconstituted from its isolated subunits when supplemented with a divalent cation.[10] Some specificity of cations exists, with Ca^{2+} and Mn^{2+} being most effective in promoting reformation of the fVIII heterodimers.[10] This reaction is also affected by ionic strength with peak activity observed at ≈0.4 M NaCl. The presence von Willebrand factor[10] and of reducing agents such as dithiothreitol (DTT)[41] enhance the reconstitution of fVIII. The reducing agents may prevent disulfide exchange reactions leading to aggregation of subunits, whereas von Willebrand factor may help orient the fVIII light chain to increase the frequency of productive collisions.

[40] P. Lollar, D. C. Hill-Eubanks, and C. G. Parker, J. Biol. Chem. 263, 10451 (1988).
[41] O. Nordfang and M. Ezban, J. Biol. Chem. 263, 1115 (1988).

Isolation of Factor VIIIa

The isolation of thrombin-activated fVIII in stable form has only been achieved in the porcine system,[11] although human fVIIIa has been partially reconstituted from the A2 subunit and the A1/A3-C1-C2 dimer.[13] Both human and porcine fVIIIa are more stable at pH 6 than at pH 7.4. Porcine fVIIIa is indefinitely stable at pH 6 and 4° or room temperature at concentrations greater than $0.2 \mu M$, whereas human fVIIIa loses most of its activity within minutes when the same conditions are used to attempt its isolation.[12] To isolate porcine fVIIIa, HEPES and MES obtained from BDH Chemicals, Ltd (Poole, England) are used; these buffers obtained from several other manufacturers have been observed to inactivate fVIIIa (P. Lollar, unpublished observations, 1989). Fractions from Mono Q HPLC containing fVIII coagulant activity (0.3–1.0 mg) are pooled and diluted $\frac{1}{3}$ with 0.01 M Tris-Cl, 5 mM CaCl$_2$, 0.01% Tween 80, pH 7.4, to lower the ionic strength to ≈ 0.2. Porcine thrombin is added to a concentration of 35 nM and the reaction mixture is assayed by the two-stage assay at 2-min intervals until peak coagulant activity is reached (at about 4 min), followed by the addition of D-phenylalanylprolylarginyl chloromethyl ketone to a final concentration of $0.15 \mu M$ to inactivate thrombin. The reaction mixture is diluted $\frac{1}{3}$ into 0.03 M MES, 5 mM CaCl$_2$, pH 6.0, and applied to a Mono S HR5/5 column equilibrated in 0.1 M NaCl, 5 mM MES, 5 mM CaCl$_2$, pH 6.0, at 2 ml/min. After washing the column with this buffer until the absorbance at 280 nm is less than 0.003, fVIIIa is eluted with a 20-ml linear NaCl gradient from 0.1 to 0.75 M in the same buffer at a flow rate of 1 ml/min. Thrombin elutes at about ≈ 0.4 M NaCl. FVIIIa elutes at about 0.65 M NaCl and is stored in the elution buffer. Tween 80 may be included in the column buffers but has not been documented to improve yields.

Radioiodination of Factor VIII

Porcine FVIII may be labeled with ^{125}I by using a modification[42] of the method of Thorell and Johansson.[43] Factor VIII in 0.65 M NaCl, 0.02 M HEPES, 5 mM CaCl$_2$, 0.01% Tween 80, pH 7.4, from Mono Q HPLC (0.1–0.4 mg/ml) is desalted into 0.2 M sodium acetate, 5 mM calcium nitrate, pH 6.8 (radioiodination buffer), using a Pharmacia LKB fast desalting column. FVIII can be stored in small aliquots at $-80°$, which is preferable because several iodinations, each lasting at least 1 month, can be

[42] D. C. Hill-Eubanks and P. Lollar, *J. Biol. Chem.* **265,** 17854 (1990).
[43] J. I. Thorell and B. G. Johansson, *Biochim. Biophys. Acta* **251,** 363 (1971).

done on a single preparation of fVIII desalted into radioiodination buffer. Prior to iodination, fVIII is diluted to 0.25 mg/ml with radioiodination buffer containing human vWf (0.375 mg/ml). vWf is included during the reaction to protect against labeling the fVIII light chain in the region of the vWf binding site, since fVIII preparations labeled in the absence of vWf have occasionally shown a variable reduction in vWf binding. Factor VIII/vWf complex is labeled in radioiodination buffer by addition of hydrogen peroxide to 0.0025% (v/v) $Na^{125}I$ to 1 mCi/ml, and lactoperoxidase (Calbiochem, La Jolla, CA) to 2 U/ml for 3 min. Lactoperoxidase is diluted to 20 IU/ml and stored in radioiodination buffer in small aliquots at $-80°$. There appears to be lot-to-lot variation in lactoperoxidase in terms of radioiodination of fVIII, which may require the concentration to be altered. The reaction is stopped by the addition of sodium azide to 0.1% (w/v) to inhibit lactoperoxidase. Factor VIII is dissociated from vWf with Mg^{2+} by the addition of 1 part 0.5 M NaCl, 1.3 M $MgCl_2$, 0.5 M MES, pH 6.0, to 4 parts of the reaction mixture and letting the mixture stand 5 min. ^{125}I-Labeled fVIII is purified by immunoaffinity chromatography[9] and stored at $4°$ in the elution buffer, which contains 50% (v/v) ethylene glycol, before further use. The preparation is stable for at least several months in this form. ^{125}I-Labeled fVIII prepared in this fashion contains from 0.04 to 0.2 atoms of ^{125}I per fVIII molecule, resulting in a specific radioactivity of $0.6-3 \times 10^9$ dpm/mg. ^{125}I-Labeled fVIII can be exchanged into HEPES buffer by using Mono Q HPLC. Samples as small as 1 μg can be recovered in good yield when 0.01% Tween 80 is included in the buffers. The clotting activity of ^{125}I-labeled fVIII is not significantly different from unlabeled fVIII.

Fluorescent Labeling of Factor VIII

The ability to modify a protein covalently at unique locations with fluorescence reporter groups can provide information on intra- and intermolecular interactions. FVIII possesses no active site so that labeling at a single, unique position is difficult. However, the three A domains each contain five cysteine residues, of which one likely does not participate in a disulfide bond. Thus, both fVIII light chain and heavy chains[34] have been derivatized with fluorophores possessing maleimide reactive groups. In a typical labeling experiment, fVIII or a factor VIII subunit ($\approx 0.1-0.5$ μM) in 0.2 M NaCl, 0.02 M HEPES, 1 mM $CaCl_2$, 0.01% Tween 80, pH 7.2, is reacted with a 20- to 50-fold molar excess of fluorophore for 2 to 3 hr at room temperature or overnight at $4°$ in the dark. Prior to reaction, the fluorophores are dissolved in a small volume of dry dimethylformamide. Unbound fluorophore is separated from the modified protein following exhaustive dialysis or by a combination of gel-permeation chromatography using Sephadex G-25 and dialysis. The extent of modification can be estimated from the protein concentration and knowledge of the extinction coefficient of the fluorophore. Experience with several fluorophores suggests that a single cysteine in each chain is reactive under these conditions. Sites of sulfhydryl-reactive fluorophore attachment in both light chain and heavy chains have been determined following sequence analysis of fluorescent peptides derived from a tryptic digest of the modified subunit and resolved by reversed-phase HPLC.[34] Both the modified heavy chain and light chain yielded a single predominant fluorescent peak on the chromatogram, suggesting modification at a single cysteine residue in each subunit. Sequence analysis indicated that Cys-528 in the heavy chain (A2 domain) and Cys-1858 in the light chain (A3 domain) were modified. The failure to modify a Cys residue in the A1 domain may indicate that it is inaccessible to the fluorophore due to its location in the folded protein. Modification of these residues does not appear to affect cofactor function because high-specific-activity fVIII can be reconstituted from the modified subunits and these subunits appear to interact normally with vWf.[33]

Acknowledgments

This work was supported by American Heart Association Established Investigator Awards to P.L. and P.J.F; by grants-in-aid HL-40921 (to P.L.), HL-38199 (to P.J.F.), and HL-17430 (to D.N.F.) from the National Institutes of Health; by a grant-in-aid from the Georgia Affiliate of the American Heart Association; and by the Mayo Foundation.

[8] Characterization of Factor IX Defects in Hemophilia B Patients

By ARTHUR R. THOMPSON and SHI-HAN CHEN

Introduction

Hemophilia B results from deficient activity of plasma clotting factor IX. It represents 20–25% of all hemophilia and was distinguished from hemophilia A (factor VIII deficiency) nearly 40 years ago.[1] Both types of hemophilia are X-linked disorders, and the two are clinically indistinguishable but require distinct therapeutic products. In hemophilia B, marked heterogeneity of defects among different families was suggested

---

[1] A. R. Thompson, *Prog. Hemostasis Thromb.* **10,** 175 (1991).

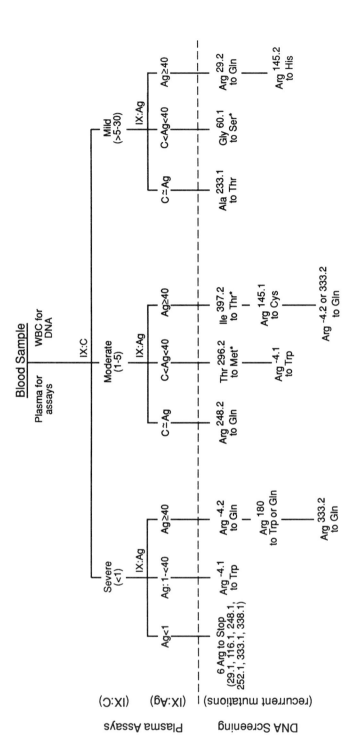

FIG. 1. Recurrent sites of point mutations in hemophilia B patients. Of 19 common recurrent point mutations,[3] 18 are transitions within CG dinucleotides; these account for around half of the mutations in different families with hemophilia B. Specific codons are indicated by superscripts, and decimals reflect the base within the triplet; an asterisk indicates a founder effect that may affect frequencies in different patient populations. General exon screening or sequencing is reserved for the nearly half of patients failing to test positively for a mutation at one of the recurrent sites (see text).

when quantitative immunoassays were developed and a variety of specific activities (i.e., the ratio of clotting activity to antigen level) was found among all subgroups of patients, including those with severe, moderate, or mild bleeding tendencies.[2]

The molecular basis for hemophilia B is diverse,[1,3] but nearly half of the families have point mutations at 1 of 19 recurrent sites, 18 of which are due to transitions within CG dinucleotides.[3] Based on the baseline (nontransfused) factor IX clotting activity (IX : C) and the antigen level (IX : Ag) of a patient, the recurrent mutations can be assigned to different categories as indicated in Fig. 1. Considering all reported cases, there is some overlap; for example, the disease in patients with Arg^{-4} or Arg^{333} to Gln is either severe or moderately severe. Determination of factor IX clotting activity and antigen level is thus important to define mutations in patients with hemophilia B efficiently.

Factor IX Assays

Factor IX Clotting Activity

The procoagulant activity of factor IX is determined by the ability of dilutions of a patient's plasma to correct the prolonged clotting time of a severely factor IX-deficient plasma in a partial thromboplastin-like assay.[4] The preferred normal control is a plasma pool from several (20–30) different normal donors, who should average ~ 100 U/dl (100% activity). Whole blood samples are drawn into one-tenth volume of 3.8% (v/v) sodium citrate. Plasmas are pooled and rendered platelet poor by centrifugation (20 min at 45,000 g, 4°) before freezing at −80° for storage. There is approximately a threefold variation in clotting activities found within 2 SD of the mean (i.e., 50–150 U/dl or percentage of normal).[2] For the assay, partial thromboplastin reagents vary as to the source of contact phase activator (e.g., kaolin, ellagic acid), the phospholipid source (species and tissue extract, concentration), incubation time, instrumentation (siliconized "tilt" tubes and a water bath, fibrometers, or various semi- and fully automated systems), and importantly, the substrate plasma.

As with commercial reagents and normal control plasmas, there is variation in the sensitivity of different deficient (substrate) plasmas. Ci-

[2] A. R. Thompson, J. Clin. Invest. 59, 900 (1977).
[3] F. Giannelli, P. M. Green, K. A. High, S. Sommer, D. P. Lillicrap, M. Ludwig, R. Olek, P. H. Reitsma, M. Goossens, A. Yoshioka, and G. G. Brownlee, Nucleic Acids Res. 20, 2027 (1992).
[4] R. R. Procter and S. I. Rapaport, J. Clin. Pathol. 36, 212 (1961).

trated, platelet-poor substrate plasmas (preferably from a severe hemophilia B patient with undetectable factor IX antigen) can be prepared as above, rapidly frozen in small aliquots and stored at $-70°$ until used. Commercial patient substrate plasmas can also be purchased (e.g., George King, Overland Park, KS; Helena, Beaumont, TX; Pacific Hemostasis, Huntsville, NC). Immunodepleted normal plasmas can be prepared by passing normal plasma over a column with an insolubilized antifactor IX antibody or purchased commercially (e.g., Baxter-Dade, Miami, FL; Organa Technika, Durham, NC) and are usually comparable for normal to moderately factor IX-deficient patient plasma samples; not all give a sufficiently prolonged background time (i.e., with contact activation omitted), as plasmas from a severely deficient patient. With either type of substrate plasma, the reproducibility and sensitivity may be impaired for levels below 5 U/dl, depending on the reagent. Viral agents more likely contaminate severely deficient plasma from a hemophilic patient, making it more hazardous than the immunodepleted type, although precautions in handling apply to all plasma products.

For a typical semiautomated assay, equal volumes (50 μl) of human deficient plasma, human brain cephalin[5] (any of a number of commercial rabbit brain preparations with activator is often substituted), and kaolin (0.5% w/v; Fisher, Fair Lawn, NJ) are preincubated with dilutions of the patient's plasma for 10 min at 37° in disposable plastic cuvettes (FTX-C11-1, AmLabor, Durham, NC). CaCl$_2$ (50 μl, 40 mM) is added and the time for clot formation detected. For a CoaScreener Apparatus (AmLabor), typical clotting times for 1 to 10, 1 to 20, 1 to 40, 1 to 80, and 1 to 160 dilutions of normal plasma into 0.1 M imidazole buffer (pH 7.4) are 95, 110, 125, 140, and 155 sec. A standard curve is constructed from the five dilution points as programmed for log clotting time versus log dilution (1 to 10, equal to 100%). For factor levels near normal, a semilogarithmic plot (clotting time versus log dilution) can be substituted. Dilutions of a patient plasma are then compared to the dilutions of the control (normal) plasma and the two curves should be parallel. The lower level of sensitivity is 1 U/dl when severely deficient substrate plasma is used. The coefficient of variation for repeat determinations is around 10%, but may be higher below 5 U/dl.

Factor IX Antigen Level

The most sensitive assessment of the total amount of factor IX in a plasma sample is by radioimmunometric or ELISA (sandwich) assays. A

---

[5] W. N. Bell and H. G. Alton, *Nature (London)* **174**, 880 (1954).

number of highly specific polyclonal antibodies have been used to detect levels as low as 0.1 to 1.0 U/dl (0.1 to 1% of normal or ~5–50 ng/ml) of antigen in a patient's plasma. Baseline plasmas from patients with severe hemophilia due to gross gene deletions, premature stop codons, or splice junction defects serve as controls for assay *specificity* that should be >99%. *Sensitivity* is apparent as the lowest level of normal plasma (diluted in an albumin-containing buffer to limit loss due to adsorption) readily detectable with three or more dilutions parallel to the normal dose–response curve (e.g., log–linear or log–logit). In electroimmunoassays, levels below 5–20 U/dl are usually undetectable and levels near the detection limit have a greater variability on repeated measures. In contrast, coefficients of variation of "sandwich" immunoassay results remain low (e.g., <5–10%) even at high dilutions of plasma.[6] Among normal individual control plasmas, the correlation coefficient between IX:C and IX:Ag should be high (>0.8).

As with clotting activity results, there is a wide range of factor IX antigen levels among different subjects, suggesting a true biologic variance that is greater than that of autosomally linked clotting factors. Whereas affected individuals from the same family usually have similar factor IX clotting activities and antigen levels,[2] patients from different families with the same genetic defect sometimes vary as much as threefold.[3,7] In these situations, the clinical severity can also vary from a severe to moderately severe or a moderate to mild bleeding tendency. This suggests that whatever is responsible for variability in the normal population may be independently inherited and affect the severity of a given factor IX mutation. Alternatively, polymorphic or variant mutations in the factor IX gene may interact with a given mutation to modify the degree of its effect on expression.

For a typical immunoradiometric assay,[6] a standard curve of a normal plasma pool (with the same anticoagulant as that in the patient's plasma) is diluted serially (e.g., 1- to 50- through 1- to 800- or 1600-fold dilutions). For citrated plasma, there is a 17–18% dilution by the anticoagulant liquid, when compared to EDTA or heparin; citrate is conveniently used, however, as it also represents the anticoagulant for samples run in clotting activity assays. A purified rabbit antifactor IX fraction diluted in 0.1 M $NaCO_3$, pH 9.5 (approximately 20 μg in 100 μl), is used to coat microtiter wells (Immunlon I, Dynatech, Chantilly, VA) at 37° for 1 hr. Nonbound

[6] G. L. Bray, A. F. Weinmann, and A. R. Thompson, *J. Lab. Clin. Med.* **107**, 269 (1986).
[7] S. G. Spitzer, B. J. Warn-Cramer, C. K. Kasper, and S. P. Bajaj, *Biochem. J.* **265**, 219 (1990).

antibody is removed and the wells washed three times with 0.1 M NaCl, 50 mM Tris, 1% (w/v) bovine serum albumin (BSA). Then 100-μl samples of four to five serial dilutions in the Tris-buffered saline with albumin are added and incubated at 37° for 2 hr (alternatively they are incubated overnight at 4°). After removing and rinsing as before, 100 μl of a diluted rabbit ^{125}I-labeled antibody are added and incubated at 37° for 3 hr. The rabbit antibody (50–100 ng) is labeled with Na[125]I (0.3 μCi) in NaOH (Amersham, Arlington Heights, IL) with either a chloramine-T (60 sec) method[2] or with lactoperoxidase beads (Bio-Rad, Richmond, CA).[6] After removing the excess labeled antibody and rinsing as before, wells are counted in a gamma counter. Each dilution point is expressed as the percentage of total counts added that are bound and plotted versus concentration on log–logit paper or as programmed in a RIA-Star 10-crystal gamma counter (Packard, Meridian, CT). Coefficients of variation are <5% for repeat determinations, even at low levels. The assay for total factor IX antigen is sensitive and specific to levels as low as 0.1 U/dl (5 ng/ml) with rabbit polyclonal antibodies.[6]

For variations in the antigen assay, different coating antibodies are used. These include a calcium-dependent rabbit antifactor IX fraction that predominantly binds to epitopes within the Gla domain, or monoclonal antibodies to epitopes in the product of the fourth exon or within the heavy chain.[8]

Methods for Defining Mutations

With cloning and sequencing of the cDNA of factor IX[1] and the entire gene,[9] it is now possible to analyze patient DNA directly for mutations. Altered restriction patterns on Southern blots of DNA digests hybridized to radiolabeled cDNAs or genomic probes have indicated gross gene deletions in several patients with severe hemophilia B.[1,10] Restriction analysis can show loss of one of three normal TaqI sites (T-CGA) in severe patients where a transition in a CG dinucleotide introduces a TGA or Stop codon in place of CGA for Arg[116], Arg[252], or Arg[338]. Although some defects have been identified by amino acid sequencing of peptide fragments, or by nucleotide sequencing of the cloned fragments, the majority have been determined by sequencing amplified DNA fragments. The polymerase chain reaction (PCR) became a convenient and powerful tool to study

[8] S.-H. Chen, A. R. Thompson, M. Zhang, and C. R. Scott, *J. Clin. Invest.* **84,** 113 (1989).
[9] S. Yoshitake, B. G. Schach, D. C. Foster, E. W. Davie, and K. Kurachi, *Biochemistry* **24,** 3736 (1985).
[10] A. R. Thompson, *in* "Molecular Defects in Hemophilia B Patients" (L. W. Hoyer and W. N. Drohan, eds.), p. 115. Plenum, New York, 1991.

mutations once a heat-stable polymerase was obtained.[11] PCR-amplified fragments can either be sequenced directly, or screened by restriction digestion or alternative techniques.

PCR with Direct Sequencing

Several groups have used PCR amplification to prepare fragments containing the factor IX coding regions including intron–exon splice junctions and promoter regions. Nine sets of primers have been used for the Seattle series,[12] and the preferred versions of these are presented in Table I. Several were selected with the aid of a computer program, "Primers" (Scientific and Education Software, State Line, PA). The nine amplified fragments total 3383 nucleotides of the factor IX gene and include all eight exons, the seven pairs of splice junctions, and a regulator region to nucleotide -266 (266 bases centromeric to the proposed transcription initiation site). These sequences encompass all hemophilic point mutations reported to date.[3,13–16] No hemophilic mutations have been found in the transcribed 3' noncoding region of the gene despite its being as long as the coding sequence.[3]

Each amplification reaction is carried out in a final volume of 100 μl with 0.1 to 0.5 μg of genomic DNA, 25 pmol of each primer, and 2.5 units of polymerase enzyme (AmpliTaq, Perkin Elmer-Cetus, Norwalk, CT). The conditions for amplification are 1 min at 94° for denaturation, 30 cycles of 1 min at 55° for annealing, and 1 min at 72° for extension, in 10× buffer (Cetus). A modified program uses 7 min for denaturation followed by 30 cycles of 6 min at 60° for annealing and 1 min at 91° for extension (for most fragments that undergo digestive screening). PCR with restriction digests is the basis of the strategy, presented below, to screen for recurrent mutations.

Prior to sequencing, amplified DNA fragments are further purified by electrophoresis on 1.4% (w/v) agarose gels[17] and electroeluted. Direct DNA

[11] R. K. Saiki, D. H. Gelfand, S. Stoffel, S. J. Scharf, R. Higuchi, G. T. Horn, K. B. Mullis, and H. A. Erlich, *Science* **239**, 487 (1988).

[12] S.-H. Chen, M. Zhang, E. W. Lovrien, C. R. Scott, and A. R. Thompson, *Hum. Genet.* **87**, 177 (1991).

[13] C. D. K. Bottema, M. J. Bottema, R. P. Ketterling, H.-S. Yoon, R. L. Janco, J. A. Phillips, and S. S. Sommer, *Am. J. Hum. Genet.* **49**, 820 (1991).

[14] C. D. K. Bottema, R. P. Ketterling, S. Ii, H.-S. Yoon, J. A. Phillips, and S. S. Sommer, *Am. J. Hum. Genet.* **49**, 839 (1991).

[15] A. R. Thompson, J. M. Schoof, A. F. Weinmann, and S.-H. Chen, *Thromb. Res.* **65**, 289 (1992).

[16] S.-W. Lin and M. C. Shen, *Thromb. Haemostasis* **66**, 459 (1991).

[17] T. Maniatis, E. F. Fritsch, and J. Sambrook, "Molecular Cloning: A Laboratory Manual." Cold Spring Harbor Lab., Cold Spring Harbor, New York, 1982.

TABLE I

OLIGONUCLEOTIDE PRIMERS FOR FACTOR IX FRAGMENT AMPLIFICATIONS

Primer	Nucleotide Numbers[9]	Sequence
Exon amplifications[a]		
5' Exon 1	−266 to −247	5'-TGATGAACTGTGCTGCCACA-3'
3' Exon 1	250 to 231	5'-TGCGTGCTGGCTGTTAGACT-3'
5' Exon 2	5880 to 5999	5'-CAGCTGGACCATAATTAGGC-3'
3' Exon 2	6530 to 6511	5'-CTATGCTCTGCATCTGAAGG-3'
5' Exon 3	6511 to 6530	5'-CCTTCAGATGCAGAGCATAG-3'
3' Exon 3	7073 to 7054	5'-GGCCATGATAATGAGGAGGA-3'
5' Exon 4	10339 to 10363	5'-CGGGCATTCTAAGCAGTTTACGTGC-3'
3' Exon 4	10587 to 10563	5'-TACACCAATATTGCATTTTCCAGTT-3'
5' Exon 5	17573 to 17592	5'-AATGTATATTTGACCCATAC-3'
3' Exon 5	17857 to 17838	5'-GTTAAAATGCTGAAGTTTCA-3'
5' Exon 6	20252 to 20271	5'-GGCCTGCTTCTCAGAAGTGA-3'
3' Exon 6	20617 to 20598	5'-CTGTGTCTTGCCAGCTGAGC-3'
5' Exon 7	29929 to 29948	5'-GCCAGCACCTAGAAGCCAAT-3'
3' Exon 7	30319 to 30300	5'-ATGGATGCTTGGAGGACGGT-3'
5' Exon 8(a)	30741 to 30760	5'-CAGTGGTCCCAAGTAGTCAC-3'
Mid-exon 8(a)	31092 to 31073	5'-TGAAGAACTAAAGCTGATCT-3'
Mid-exon 8(b)	30969 to 30988	5'-GCTACGTTACACCTATTTGC-3'
3' Exon 8(b)[b]	31432 to 31413	5'-AGTTAGTGAGAGGCCCTGTT-3'
Amplification mutagenesis (creation of new or "neo" cleavage sites)[c]		
Codon 29 (exon 2)	6459–6479	5'-AAGTGTAGTTTTGAAGAAGCG-3'
Codon 145 (exon 6)	20392–20412	5'-TCACAAACTTCTAAGCTCACG-3'
Codon 180 (exon 6)	20540–20520	5'-GCATCTTCTCCACCAACAACG-3'
Codon 248 (exon 8)	30841–30861	5'-GACAGAACATACAGAGCAAAC-3'
Codon 333 (exon 8)	31140–31120	5'-GTAGATCGAAGACATGTGGCG-3'
Codon 397 (exon 8)	31289–31309	5'-GCAATGAAAGGCAAATATGGT-3'

[a] The second of each pair of primers represent the "antisense" or reverse sequence.

[b] The normal 3' Stop signal (nucleotides 31367–9) is codon 416.[9]

[c] The underlined nucleotide is substituted for the normal one. Primers are paired with primers of the opposite sense from the appropriate exon (see exon amplifications above).[12] Those listed for codons 180 and 333 screening are in the antisense direction whereas for codons 29, 145, 248, and 397, mutagenesis primers are in the sense direction. The 397 primer creates a new RsaI site (GTAC) when the mutation is present; the others create new BstUI sites (CGCG) when the normal sequence is present.

sequencing can then be performed by the dideoxy method of Sanger *et al.*[18] with [32]P- or ^{35}S-labeled nucleotides and the sequence read from the radioautograph of sequence gels. Most of the mutations can be determined using the amplification primers as sequencing primers, although some internal sequencing primers have also been used.[12,15] In examination of 100 patients in a Seattle series, a mutation responsible for the hemophilia has been determined in 99. The one negative patient has severe hemophilia B and no detectable antigen; no mutation was indicated when all exons and splice junctions were analyzed either by direct sequencing (S.-H. Chen and A. R. Thompson, unpublished, 1990) or by chemical mismatch screening[19] (F. Giannelli, unpublished, 1991). For known mutations, restriction enzymes can be used to confirm and rapidly screen for a defect.[15]

Southern Blot Analysis

For restriction enzyme digests, 5–10 μg of genomic DNA is digested overnight according to the enzyme manufacturer's specifications. Digested DNA samples are subjected to 0.8% (w/v) agarose gel electrophoresis (2V/cm) in a Tris–phosphate–EDTA buffer, pH 8.0, at room temperature overnight. After electrophoresis, gels are stained with ethidium bromide (0.5 μg/ml) for 20 min and photographed to confirm digestion. Electrophoresed DNA fragments are transferred to a nitrocellulose or nylon filter overnight. The filter is baked for 2 hr at 80°, prehybridized for 4 hr at 42° with 1% (w/v) glycine in a prehybridization solution containing 5× SSC (0.3 M sodium chloride and 0.03 M sodium citrate, pH 7.0), 1× Denhardt's solution, 0.3 mg/ml salmon sperm DNA, and 50% (v/v) formamide.[17] Hybridizations are carried out overnight with a random hexamer (BRL, Gaithersburg, MD) [32]P- or ^{35}S-labeled probe, approximately 5 × 10[6] counts per minute (cpm)/filter in the solution. The factor IX cDNA probe[20] contains the sequence coding for entire signal peptide, mature protein, and part of the 3′ noncoding sequence. After hybridization, filters are washed three times for 5 min each with 2× SSC and 0.1% SDS at room temperature and 2 times for 30 min each at 65°C in 0.1× SSC and 0.1% (w/v) SDS. After drying, filters are exposed to Kodak XAR film (Eastman Kodak, Rochester, NY) at −70° for 3–10 days, before the film is developed.

By using the enzymes *Taq*I, *Hin*dIII, and *Eco*RI (New England Biolabs, Beverly, MA), the entire 33.5 kb of the factor IX gene is screened and any alteration of over ~100 nucleotides should be detected.[20] Once an altered

[18] F. Sanger, S. Nicklen, and A. R. Coulson, *Proc. Natl. Acad. Sci. U.S.A.* **74**, 5463 (1977).
[19] A. J. Montandon, P. M. Green, F. Giannelli, and D. R. Bentley, *Nucleic Acids Res.* **17**, 3347 (1989).
[20] S.-H. Chen, S. Yoshitake, P. F. Chance, G. L. Bray, A. R. Thompson, C. R. Scott, and K. Kurachi, *J. Clin. Invest.* **76**, 2161 (1985).

pattern of restriction enzyme fragments is detected, additional restriction enzyme digests are performed, based on the restriction map of Yoshitake *et al.*,[9] often with hybridization to genomic probes, to define the approximate extent of the alteration. Breakpoint analysis can follow, assuming the sequences near either side of a deleted fragment are known.

Breakpoint Analysis of Gross Gene Alterations

An example of determining the actual site where a deletion has occurred is provided in a family with seven affected members who have severe hemophilia B.[21] Based on Southern analysis with cDNA and genomic probes, these patients had a partial gene deletion, including exons 5 and 6.[20] They also had evidence for trace levels of a truncated protein, including epitopes in the calcium-binding or amino-terminal Gla domain, especially in urine samples.[22] From Southern analysis the length of the deletion was estimated as 10.0 ± 0.3 kb. Primers were then designed to flank the deletion site by selecting sequences in introns 4 and 6; the primers were ~11 kb apart. Their sequences were 5'-TGATTGCT-TAACTTCCTGGGACTGT-3' (14088–14112, sense strain), and 5'-TTCAGAGACCCTGCTTCCTTCCCCG-3' (24982–24958, antisense strain). A single fragment of ~0.9 kb was found when patient DNA, but not normal DNA, was amplified in the presence of these primers. In sequencing this fragment, breakpoints were localized to sequences where 13 of 14 bases were identical in intron 4 and in intron 6 (Fig. 2).[21] The breakpoints were 10.0 kb apart.

Strategy for Detecting Recurrent Mutations

In detecting the mutation of a patient with hemophilia B, one approach is to screen or sequence amplified fragments containing all coding regions and splice junctions. When a reproducible change has been found that would readily account for the severity of the patient's hemophilia or represents a mutation known to be a recurrent one associated with the bleeding tendency in other patients, it is reasonable to consider that mutation as causative. Otherwise, complete sequencing of exons, splice junctions, and the 5' region is required to exclude a second mutation that may be more likely to produce the mutant phenotype.[1,3] However, at least half of the mutations are recurrent at 19 sites in 13 codons within 5 different exons. Therefore, a strategy has been devised to screen rapidly for these recurring mutations, reserving the more time consuming, complete screening or sequencing for those patients without mutations at a recurrent site.

[21] S.-H. Chen and C. R. Scott, *Am. J. Hum. Genet.* **47**, 1020 (1990).
[22] G. L. Bray and A. R. Thompson, *J. Clin. Invest.* **77**, 1194 (1986).

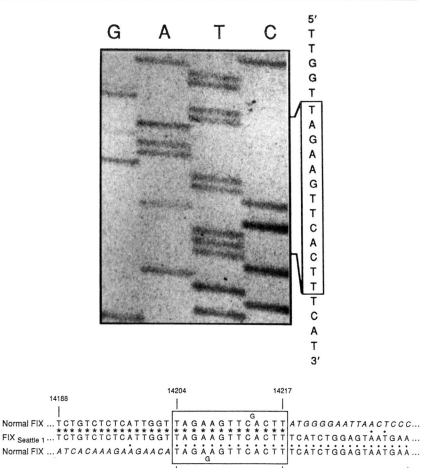

FIG. 2. Breakpoint sequence of a 10.0-kb factor IX gene deletion. A fragment was amplified by using primers near putative breakpoints in exons 4 and 6. The resultant sequence (above) revealed a homologous recombination sequence where, of 14 bases, 13 were identical (as boxed).[21] Patient sequence identity with the exon 4 sequence is indicated by asterisks; with exon 6, by dots. The actual breakpoint and recombination appears to have occurred within the central AGTTC sequence.

Screening for Recurrent Mutations

When patients are divided into nine groups according to their clotting activities and antigen levels (Fig. 1), about half of them in each category will have a mutation indicated by rapid screening of one to four amplified products. Some mutations overlap more than one category. These 19 recurrent mutations can be identified by amplification of the DNA of a patient and restriction enzyme digests as presented in Table II.[23]

PCR amplifications, electrophoresis, and ethidium bromide staining are performed as described above for preparing exon fragments for sequencing. The modified program was used for fragments containing recurrent mutations except for the 296.2 mutation. Restriction enzymes (indicated in Table II; New England Biolabs, except *Mae*III, Boeringer-Mannheim, Indianapolis, IN) are incubated with aliquots of amplified fragments according to manufacturer's directions. For the *Bam*HI polymorphism, primers are 20 bp long: 5'-GACCTAAACATCATACTTTA-3' (−701 through −682) and 5'-GTATGAGTGGTCCAGTTAGT-3' (−346 through −365).[24]

Once any mutation is indicated by screening, its presence should be considered presumptive. The corresponding exon fragment can then be directly sequenced through the restriction recognition site (see sequencing methods, above) to exclude a different mutation within the recognition sequence. Examples of different mutations are the occasional transversions found within CG dinucleotides.[3] Even without confirmation, restriction site screening serves to track a hemophilic gene within a family, providing rapid carrier detection and prenatal diagnosis.

Severe Hemophilia B. Clinically severe hemophilia occurs in patients who have recurrent, spontaneous bleeding episodes into their joints or, less frequently, into their muscles, and these may occur several times a year. Prolonged bleeding or delayed wound healing are also observed after trauma, which need only be relatively mild. This clinical pattern usually correlates with a clotting activity of <1 U/dl, although among different patients with severe hemophilia there may be some who have relatively less frequent spontaneous bleeding episodes than others, and in some patients with 1 to 2 U/dl, occasional spontaneous bleeding episodes occur. Overall, about one-third of patients with hemophilia B are severely affected.

Of patients with severe hemophilia B, about two-thirds will have undetectable (<1 U/dl) factor IX antigen. Lack of detectable antigen is often due to a recurrent nonsense mutation in one of six CGA codons (Fig. 1). These are screened as a new *Nla*III site (CA̲TG) in exon 2, absence of normal *Taq*I sites (T̲CGA) in exon 5 or in two codons in exon 8, or removal of one of two *neo-Bst*UI sites (CGC̲G or C̲GCG) in exon 8 (Table II). Some additional patients without detectable antigen will have gross gene deletions apparent as a failure to amplify certain exon fragments that are included in a partial or complete gene deletion.

---

[23] A. R. Thompson, and S.-H. Chen, *Circulation* **86,** I-686 (1992).
[24] M. Zhang, S.-H. Chen, C. R. Scott, A. R. Thompson, *Hum. Genet.* **82,** 283 (1989).

About 5% of all hemophilia B patients (between 10 and 15% of those who are severely affected) will have antigen levels of $1-40$ U/dl, or reduced cross-reactive material (Crm). About half of these will have an Arg^{-4} to Trp mutation due to a C to T transition within the CG dinucleotide coding for Arg^{-4}. The range of antigen levels observed in patients previously reported with this mutation is $26-36$ (Table II)[3]; either severe or moderately severe subjects with clotting activities ranging from <1 to 4 U/dl have been found. As shown in Table II, an MscI site ($\underline{T}$GGCCA) is created by the C to T mutation, and it is readily appreciated after MscI digesting an amplified exon 2 fragment. A -4 mutation can be suspected also by immunoassays specific for the calcium-dependent conformation of factor IX that indicate somewhat lower levels than the total factor IX antigen level (by antibodies that do not depend on the calcium-dependent conformation). The mutation may also be demonstrated by Western blots of a factor XIa-cleaved, dysfunctional protein as a slightly larger light chain corresponding to the addition of 18 amino acids at the amino-terminal end or by an immunoassay with an antibody specific for the attached propetide.[25]

Another 5% of all hemophilic patients will have severe hemophilia and normal antigen levels ($40-130$ U/dl). Half of these will have Arg^{180} to Trp or Gln or Arg^{-4} or Arg^{333} to Gln mutations; the latter two have also been described in patients with moderately severe hemophilia. Cleavage of the Arg^{180}-Val^{181} bond is required for activity, explaining why patients with either Arg^{180} mutation have a severe bleeding tendency. In amplified exon 6 fragments, the C to T or G to A transition in codon 180 will remove a neo-BstUI site (CG$\underline{C}$G); G to A will also create a new DdeI site (CTN$\underline{A}$G). The $Arg^{-4.2}$ G to A transition removes an HaeIII site ($\underline{G}$GCC) in exon 2 or the protein can be screened immunologically or by Western blots as with the $Arg^{-4.1}$ C to T transition (above). The $Arg^{333.2}$ is apparent as removal of a neo-BstUI site in an exon 8 fragment (as with $Arg^{333.1}$ to Stop, above).

Moderately Severe Hemophilia B. Overall, about one-third of all patients with hemophilia B have baseline factor IX clotting activities between 1 and 5 U/dl. This usually corresponds to a more moderately severe bleeding tendency without (or with occasional) spontaneous bleeding episodes. Relatively minor trauma can induce bleeding, and deep muscle hematomas are often at least as common as hemarthroses. Some of these patients will have a normal "specific activity," that, is a clotting activity which is at about the same level as their total protein by immunoassays. The one recurrent mutation in this group is Arg^{248} to Gln. It was originally detected in patients who, despite their low baseline antigen level, had no reactivity when a monoclonal antibody specific for an epitope within the

[25] J. A. Bristol, B. Furie, and B. C. Furie, *Blood* 78, 58a (1991).

TABLE II

SCREENING FOR RECURRENT MUTATIONS IN PATIENTS WITH HEMOPHILIA B[a]

Codon (exon)	Mutation		Range of factor IX levels (U/dl)		Screening procedure	
	Base	Amino acid	IX:C	IX:Ag	Restriction sites	Immunologic or protein
Severe						
29.1 (2)	C to T	Arg to Stop	<1	<1	neo-*Bst*UI, removed; *Nla*III, created	None detected
116.1 (5)	C to T	Arg to Stop	<1	<1	*Taq*I, removed	None detected
248.1 (8)	C to T	Arg to Stop	<1	<1	neo-*Bst*UI, removed	None detected
252.1 (8)	C to T	Arg to Stop	<1	<1	*Taq*I, removed	None detected
333.1 (8)	C to T	Arg to Stop	<1	<1	neo-*Bst*UI, removed	None detected
338.1 (8)	C to T	Arg to Stop	<1	<1	*Taq*I, removed	None detected
180.1 (6)	C to T	Arg to Trp	<1	74–130	neo-*Bst*UI, removed;	10K ↑ heavy chain (IXa)
180.2 (6)	G to A	Arg to Gln	<1	100–120	*Dde*I, created	
Severe to moderate						
−4.1 (2)	C to T	Arg to Trp	<1–4	26–36	*Msc*I, created	Mild ↓ binding to Ca^{2+} anti-IXs
−4.2 (2)	G to A	Arg to Gln	<1–4	40–98	*Hae*III, removed	and 18aa ↑ light chain (IXa)
333.2 (8)	G to A	Arg to Gln	<1–4	32–135	neo-*Bst*UI, removed	—

Moderate to mild						
248.2 (8)	G to A	Arg to Gln	1–4	1–4	*neo-BstUI*, removed	Nonbinding to m-anti-IX
296.2 (8)	C to T	Thr to Met[b]	2–9	7–20	*NlaIII*, created	—
397.2 (8)	A to C	Ile to Thr[b]	1–5	45–100	*neo-RsaI*, created; *BamHI* (5′E1), linked	↓ Xa activation
145.1 (6)	C to T	Arg to Cys	1–3	30–66	*neo-BstUI*, removed[c];	Mild ↓ binding to m-anti-IX[c]
145.2 (6)	G to A	Arg to His	4–11	60–160	*NlaIII*, created	and 10K ↑ light chain (IXa)
Mild						
233.1 (7)	G to A	Ala to Thr	10–22	11–15	*MaeII*, created	Nonbinding to m-anti-IX
60.1 (4)	G to A	Gly to Ser[b]	10–17	18–54	*BbvI*, created	↓ Binding to m-anti-IX
29.2 (2)	G to A	Arg to Gln	20–30	70, 77	*neo-BstUI*, removed	—

[a] Codons are numbered[9]; decimals reflect the base within the triplet; IXa refers to factor XIa-cleaved factor IX, which normally produces an amino-terminal light chain (17K), an activation peptide (10K), and the carboxy-terminal heavy chain (30K); *neo* refers to a restriction site created by amplification mutagenesis; IX:C is factor IX procoagulant (clotting activity); IX:Ag is circulating antigen level and m is for a monoclonal antibody.

[b] Founder effect.

[c] *BsI* (CCN–GG) is normally present in subjects with the less common dimorphic form Ala[148] (*MnlI* positive) in exon 6, where the first base of codon 148 is G; a 145.1 C to T transition removes this site when present. Decreased binding seen only with the more common Thr[148] (*MnlI* negative) form is present; compared to the normal Thr[148] variant, the Cys[145] and Thr[148] protein shows a mild reduction in binding to the monoclonal antibody used to detect the codon 148 polymorphism.

heavy chain of factor IXa was used in immunoassays.[8] A second monoclonal antibody[26] has similar properties and likewise does not react to the factor IX in patients with this mutation. More than 10 patients have been described in the United States, Canada, and Germany.[3] Of 4 patients in the Seattle series, 3 share a haplotype present in only 10% of Caucasian males,[12] suggesting that a founder effect may be present. As the fourth patient had a distinct haplotype, recurrent mutations also occur. Screening can be accomplished by specific immunoassays or by detection of lack of cleavage at a BstUI site (CGCG), introduced by mutagenesis.

More commonly, patients with moderately severe hemophilia have excess, dysfunctional antigen present. Of those with intermediate elevations of antigen (7–20 U/dl), the Thr^{296} to Met mutation has been found in over 25 families in the United States, United Kingdom, Sweden, and Germany. Of 10 patients in the Seattle series, all share a common haplotype present in 14% of Caucasian males. A founder effect is likely and all but one patient in the Mayo Clinic series shares this haplotype; the shared haplotype is present among several patients of German and Amish decent.[27] The mutation creates a new NlaIII site (CATG) in exon 8 which is readily demonstrated by cleavage of an amplified fragment.

About half of Caucasian patients with moderately severe hemophilia B in North America will have normal factor IX antigen levels and this is largely due to a founder effect that is documented in patients with a non-CG mutation, namely Ile^{397} to Thr.[28,29] These patients have been described in France, eastern Canada, and central and western United States. Of 21 who have been haplotyped, all share a rare (<1%) type including a BamHI site which is only present in 3% of Caucasian factor IX genes and is about 0.5 kb 5' to exon 1. The defect in exon 8 has not been found in hemophilia B patients in Sweden or the United Kingdom.[30,31] Thus, the proportion of patients in this category may vary depending on their ethnic origin. The simplest screen, although indirect, is for the presence of the BamHI polymorphism.[24] More directly, an RsaI site, introduced near the 3' end of the coding region of exon 8, indicates the Thr^{397}

[26] S. P. Bajaj, A. K. Sabharwal, J. Gorka, and J. J. Birktoft, Proc. Natl. Acad. Sci. U.S.A. 89, 152 (1992).
[27] R. P. Ketterling, C. D. K. Bottema, D. D. Koeberl, and S. S. Sommer, Hum. Genet. 87, 333 (1991).
[28] A. R. Thompson, S. P. Bajaj, S.-H. Chen, and R. T. A. MacGillivray, Lancet 1, 418 (1990).
[29] R. P. Ketterling, C. D. K. Bottema, J. A. Phillips, and S. S. Sommer, Genomics 10, 1093 (1991).
[30] P. M. Green, A. J. Montandon, R. Ljung, D. R. Bentley, I.-M. Nilsson, S. Kling, and F. Giannelli, Br. J. Haematol. 78, 390 (1991).
[31] P. M. Green, A. J. Montandon, D. R. Bentley, R. Ljung, I.-M. Nilsson, and F. Giannelli, Nucleic Acids Res. 18, 3227 (1990).

substitution (Table II). Of other patients with moderately severe hemophilia and normal antigen levels, the Arg[145] to Cys mutation has been found in five families and it is usually moderately severe with low-normal antigen levels. Screening can be by Western blots of a factor XIa-cleaved preparation in which the light chain-activation peptide bond is not cleaved, by a nonparallel dose response in a specific immunoassay that recognizes the common Thr[148] dimorphism, or by restriction analysis for either lack of a new *Bst*UI site in exon 6 or (for those with the less common Ala[148] dimorphic type) loss of a *Bsl*I site, CCN_7GG. In considering moderately severe hemophilia, the Arg[−4.1] to Trp mutation is usually associated with mildly reduced factor IX antigen levels and may be either severe or moderately severe. The Arg[−4.2] or Arg[333.2] to Gln mutations may likewise be severe or moderately severe but are associated with normal antigen levels (Table II).

Mild hemophilia B. This group represents just over one-third of all patients, and again there are three categories, with the first being those with normal specific activities. A recurrent mutation in those with normal specific activity is Ala[233] to Thr. This, like the Arg[248] to Gln substitution,[8] results in a protein that is nonreactive to a specific monoclonal antibody,[12,26] suggesting that this antibody's epitope is conformation dependent rather than a linear sequence. The presence can also be confirmed by noting a new *Mae*III site (GTNAC) in an amplified exon 7 fragment. Nine families have been described,[3] and at least two represent independent mutations.[12,15]

The majority of patients with mild hemophilia B and a recurrent mutation (about 10% of all hemophilia B patients) have mildly reduced antigen levels and a Gly[60] to Ser mutation; a founder effect is probably present,[12,15,26,29] although distinct haplotypes have also been noted.[8] The shared haplotype, including the presence of a 50-base insert in intron 1, and a 5′ *Mse*I site are only found in 3% of Caucasian males. This mutation and several others within the fourth exon can also be suggested by mildly reduced binding of the dysfunctional protein to a monoclonal antifactor IX antibody, which binds to an epitope within the first growth factor-like domain.[8] Over 20 families have been reported.[3]

When mild hemophilia occurs with a normal factor IX antigen level, a recurrent mutation has been noted as Arg[29] to Gln. Although this is within the Gla domain, there is normal binding to calcium-dependent antibodies. The mutation is most readily screened for by restriction digest, in this case by absence of a *Bst*UI site (CGCG) inserted by mutagenesis. Three of four families described have a second mutation, a rare variant, T to C at Val[227.3],[3,12,23] and must reflect a single mutation more than four generations ago. The Arg[145] to His mutation is also mild with a normal antigen level. Here, the bond between the activation peptide and the light chain cannot be cleaved. Somewhat higher clotting activities and antigen levels are found with the $Arg^{145.2}$ to His as opposed to the $Arg^{145.1}$ to Cys mutation; either can be screened by protein, immunologic (blot), or restriction digest procedures described above. The $Arg^{145.2}$ to His mutation can be distinguished, as it creates a new NlaIII site (C<u>A</u>TG) in exon 6.

Efficacy of Screening for Recurrent Mutations

In a previous study,[15] a patient with a new $Arg^{29.1}$ to Stop mutation and one each with $Arg^{180.2}$ and $Arg^{-4.2}$ to Gln mutations were initially identified by restriction digest screening leading to the current strategy of screening for recurrent mutations. Plasma and DNA samples from the next 21 unrelated patients whose mutations were not previously reported were screened prospectively (Table III). Fourteen had a recurrent mutation indicated by screening, and all were confirmed by DNA sequencing.[23]

The general strategy of screening should allow rapid detection of at least half of the mutations in patients with hemophilia B; two-thirds of the prospective series were localized. As an alternative to restriction digests, oligonucleotide hybridization with dot blots and probes for both the normal and abnormal sequence, could be substituted. This would involve amplification and then optimizing conditions for hybridization of each probe for each normal and mutant base and use of radioisotopes or other labels. The ligase – polymerase chain reaction variant might be another way to screen and would have the advantage of allowing screening of several sites simultaneously.[32]

Detection of Other Mutations

Direct DNA Sequencing of PCR Fragments

Most of the patients testing negatively in the specific screening will have distinct mutations that are "private," that is, restricted to their own family. Of all patients with severe hemophilia, around half will have no detectable defect localized by screening. For those with undetectable antigen, there are often one of four types of mutations that have occurred anywhere throughout the coding regions: gross gene alterations, frameshift mutations, splice junction defects (that remove either an acceptor or donor consensus sequence, thus altering transcription), or nonsense mutations.[1,3]

---

[32] F. Barany, Proc. Natl. Acad. Sci. U.S.A. **88,** 189 (1991).

TABLE III
SCREENING FOR RECURRENT MUTATIONS
IN 21 UNRELATED PATIENTS WITH HEMOPHILIA B[a]

Patient	Factor IX level (U/dl)		Exons and enzymes screened	Mutation indicated
	IX:C	IX:Ag		
Severe				
A	<1	<1	E2 and 8, *Bst*UI;	None
B	<1	<1	E5 and 8, *Taq*I	None
C	<1	50	E2, *Msc*I and *Hae*III;	Arg$^{180.1}$ to Trp[c]
D	<1	100	E6, *Bst*UI and *Dde*I;	None
E	<1	110	E8, *Bst*UI	Arg$^{180.1}$ to Trp[c]
Moderate				
F	3	3	E8, *Bst*UI	Arg$^{248.2}$ to Gln[b]
G	2	5		None
H	2	5		Thr$^{296.2}$ to Met
I	(17)	6	E8, *Nla*III	Thr$^{296.2}$ to Met
J	(2)	7		Thr$^{296.2}$ to Met
K	3	8		Thr$^{296.2}$ to Met
L	4	10		Thr$^{296.2}$ to Met
M	2	40	E6, *Bst*UI and	Arg$^{145.1}$ Cys[b,c]
N	2	68	*Nla*III;	Ile$^{397.2}$ to Thr
O	2	72	E2 *Hae*III;	None
P	(9)	74	E8, *Bst*UI;	Ile$^{397.2}$ to Thr
Q	2	80	E8, *Rsa*I;	Arg$^{333.2}$ to Gln
R	(2)	110	5' E1, *Bam*HI	None
Mild				
S	6	15	E4, *Bbv*I	None[b]
T	13	30		Gly$^{60.1}$ to Ser[b]
U	10	77	E2, *Bst*UI	Arg$^{29.2}$ to Gln

[a] IX:C is factor IX clotting activity; IX:Ag, antigen level; values in parentheses are from other laboratories, and a citrated sample was not available for confirmation. Families are ordered by severity and antigen level category as in Fig. 2.[23]

[b] Some mutations were also suggested by results of immunoassays with a monoclonal antifactor IX.

[c] Others by altered factor XIa cleavage patterns on Western blots.

"Micro" deletions or insertions often occur near a consensus sequence T-G-A/G-A/G-G/T-A/C[33] and one 4-base deletion is most likely recurrent.[34,35] For the remaining patients with near normal or normal antigen levels, several have had missense mutations at or around the active center Ser[365] such that the majority of defects undefined by the initial screening will be in exon 8. The others will be found throughout the coding regions and at least 98% of patients in most series will have a definable mutation.[30,31] Of hemophilic patients with more mild bleeding tendencies, there are clear examples of less prevalent "founder" mutations as well.

For patients with severe hemophilia B and no detectable antigen and who do not have one of the six recurrent nonsense codons due to CG transitions, gross gene alterations will be found in about one-third. All of the complete or partial gross gene deletions, with the exception of one that deletes exon 4, result in severe hemophilia B without detectable antigen. With the splice junctions of exon 4 and 5 "inframe," one would predict that a protein could be made. Indeed, the family with the exon 4 deletion has about one-third of normal factor IX antigen level and that antigen is smaller than normal due to absence of amino acids from the first growth factor-like region coded for by exon 4.[10]

Of 71 patients with point mutations in this category,[3,13,14] 21 have mutations in exon 8, 15 in exons 2 or 3, 13 in exon 6, 10 in exon 5, 5 each in exons 4 and 7, and 2 in exon 1. The substitutions are most often new nonsense codons, frameshift mutations, or mutations at splice junctions; few are missense mutations. Thus in this category, mutations can be found anywhere throughout the coding region.

Because exon 8 represents the largest coding region, it is most efficient to begin with its sequence. Exon 8 is also where the majority of patients in other categories have defects, with the exceptions of severe hemophilia with reduced or normal factor IX antigen levels, which also has several mutations in exon 2, and those with normal antigen, also found in exon 6, usually associated with cleavage site mutations. The other exception is in mild hemophilia B with low-normal to normal factor IX antigen levels. Half of the patients in this category have defects within the fourth exon and most of these can also be screened as mild to moderately reduced ability to bind to a monoclonal antibody specific for an epitope in the first growth factor-like region, as with the recurrent Gly[60] to Ser mutation.[8]

[33] D. N. Cooper, *Blood Rev.* **5**, 55 (1991).
[34] S.-H. Chen, M. Zhang, A. R. Thompson, G. L. Bray, and C. R. Scott, *Nucleic Acids Res.* **19**, 1172 (1991).
[35] S. R. Poort, E. Briet, R. M. Bertina, and P. H. Reitsma, *Thromb. Haemostasis* **64**, 379 (1990).

Criteria for a Causative Role of a Mutation

When a gross or small gene alteration or a point mutation changes an acceptor or donor splice junction consensus sequence, or causes a frame-shift, or when a point mutation introduces a nonsense codon, it will interfere with normal transcription or translation. For hemophilia B, such changes in the factor IX genes are nearly always associated with severe hemophilia with undetectable circulating antigen.[1,3] Missense mutations are often more subtle and once one is found, a case must be built that the particular mutation is responsible for the hemophilia. Neutral polymorphisms can be distinguished by screening a number of genes from normal, nonhemophilic subjects. Only one common polymorphism, Thr-Ala[148], in the activation peptide region, has been found in the coding region. Rare variants, unrelated to the hemophilic mutation are more problematic and 10 of these have been identified in the factor IX genes of hemophilic patients.[3,13,14] In each case, however, it was possible to distinguish the rare variant from a distinct mutation that was far more likely to cause the hemophilia.

In deciding that a given missense mutation is probably causative of the hemophilia, one must first perform adequate controls to exclude cloning or, more commonly, amplification artifacts. The most rigorous criterion that a missense mutation is responsible for a dysfunctional factor IX is by inserting a given patient's mutation into a normal factor IX gene by site-directed mutagenesis. That gene is then expressed in a mammalian cell line and the mutant factor IX protein is purified from the conditioned media. This "synthetic" protein is compared with expressed wild-type and plasma-derived factor IX preparations. Abnormalities of expression can also be defined by comparing native and mutant mRNA or intracellular processing of a mutant protein. Expression criteria have only been applied to a few point mutations, including an Arg^{-4} to Gln substitution that alters a processing protease site,[36] various mutations in the first epidermal growth factor-like region,[37] and the Ile^{397} to Thr change that alters substrate binding in the trypsin-like domain.[38]

When a given missense mutation is the only one found on screening or sequencing of all eight exons and their splice junctions, it is likely that it is responsible for the hemophilia. The case is strengthened when the particular mutation is found in more than one family, particularly if it can be

---

[36] P. Galeffi and G. G. Brownlee, *Nucleic Acids Res.* **15,** 9595 (1987).
[37] D. J. G. Rees, I. M. Jones, P. A. Hanford, S. J. Walter, M. P. Esnouf, K. J. Smith, and G. G. Brownlee, *EMBO J.* **7,** 2053 (1988).
[38] N. Hamaguchi, P. S. Charifson, L. G. Pedersen, K. J. Smith, and D. W. Stafford, *J. Biol. Chem.* **266,** 15213 (1991).

demonstrated that the mutations were distinct by a different intragenic haplotype pattern or *de novo* occurrence. With some of the more mild or moderately severe mutations, recurrence is often due to a "founder" effect, or common origin within several families.[28,29] Most recurrent mutations have occurred as transitions within 12 of the 20 CG dinucleotides. A recurrent promoter mutation, -6 bp G to A, and a nonhemophilic variant transition (G to A changing the codon for Arg^{-44} to one for His) occur in other CG dinucleotides.[3] Within the coding regions, transitions within the other seven CGs are unlikely to cause hemophilia[10] and several are noted in comparing sequences among different species.[31]

Additional evidence for causation is sometimes provided by protein, immunologic, or enzymologic studies. Alterations of the processing protease cleavage site or of either of the two activation cleavage sites have occurred in several families and can be distinguished on Western blots, particularly following activation by factor XIa of even partially purified, dysfunctional protein. Kinetic studies are more likely to be investigated once a mutation has been defined because they require a large amount of purified, dysfunctional protein. Alteration of various binding properties, however, can be suggested by *in vitro* screening. The most readily assessed are those of the conformation within the calcium-binding Gla domain where both polyclonal antibody fractions and monoclonal antibodies that are dependent on the conformation achieved when both the propeptide is cleaved and/or sufficient carboxylation of the 12 amino-terminal Glu residues has occurred.[8,25,39] Other conformational changes elsewhere in the factor IX molecule can be suggested by altered binding to specific monoclonal antibodies, including epitopes in the first growth factor-like region[8,40] or epitopes within trypsin-like domain.[8,26]

Comparison of the mutation found with sequences both of factor IX genes from different species[41-44] and from homologous serine proteases[14,45,46] can provide indirect evidence that a given mutation, when at a conserved position, may well be the cause of the hemophilia. Within the trypsin-like domain, sequence is known for nine different species and the

[39] A. K. Bentley, D. J. G. Rees, C. Rizza, and G. G. Brownlee, *Cell (Cambridge, Mass.)* **45**, 343 (1986).
[40] P. H. Denton, D. M. Folk, S. T. Lord, and H. M. Reisner, *Blood* **72**, 1407 (1988).
[41] J. P. Evans, H. H. Watzke, J. L. Ware, D. W. Stafford, and K. A. High, *Blood* **74**, 207 (1989).
[42] S.-M. Wu, D. W. Stafford, and J. Ware, *Gene* **86**, 275, 1990.
[43] U. R. Pendurthi, R. H. Tukey, and L. V. M. Rao, *Thromb. Res.* **65**, 177 (1992).
[44] G. Sarkar, D. D. Koeberl, and S. S. Sommer, *Genomics* **6**, 133 (1990).
[45] C. P. Pang, M. Crossley, G. Kent, and G. G. Brownlee, *Nucleic Acids Res.* **18**, 6731 (1990).
[46] B. Furie, D. H. Bing, R. J. Feldmann, D. J. Robison, J.-P. Burnier, and B. C. Furie, *J. Biol. Chem.* **257**, 3875 (1982).

complete sequence of canine, murine, and (for most of the residues) rabbit and bovine factor IXs is also known.[41-44] Within the 5′ regions, where potential promoter or regulator defects might occur, sequence data from five species of factor IX were obtained and compared.[45] When a missense mutation alters a conserved residue known to be essential for the catalytic function of all serine proteases, it is reasonable to assume that it will render the mutant factor IX inactive. At the other extreme, a point mutation that is found in a patient and also occurs in factor IX genes from some nonhuman species is unlikely to affect clotting activity because, for example, only subtle differences can be observed in the reaction kinetics of bovine as opposed to human factor IXs, at least *in vitro*.

From data on homologous proteins, models of the three-dimensional structure of regions of factor IX can be constructed. The propeptide structure has been estimated for prothrombin by NMR studies[47] as has the structure of epidermal growth factor, which is homologous to the domains coded for by exons 4 and 5.[10] The latter has been confirmed on synthetic, oxidized sequences of the first growth factor-like region[48] of factor IX. The Gla domain itself is highly homologous with that of prothrombin, where the tertiary structure has been solved by X-ray crystallography.[49] There is no detail other than what can be predicted from the primary structure for the conformation of the activation peptide region, but this region is highly variable both in terms of the number of Asn-linked carbohydrate chains and even the number of residues present in comparing different species.[41-44] The highly conserved or "constant" sequences within the trypsinlike domain[46] can be three-dimensionally modeled by computer to the known X-ray crystallographic coordinates for trypsin, chymotrypsin, and, more recently, thrombin.[50] The extent to which a variable region differs in factor IX, however, requires assumptions that potentially introduce error in these regions of the models.

Within all of the mutations described in different patients, there remain a few that meet the criterion of being the only mutation found on complete screening or sequencing of coding regions and splice junctions. For some of these, it is proposed that a cryptic or alternative splicing site is introduced when a consensus sequence, 5′-pyrimidine-A-G-3′, has been formed.[15,30] In others, such as a molecular defect leading to severe hemophilia B in a dog

[47] D. G. Stanford, C. Kangy, J. L. Sudmeier, B. C. Furie, and B. Furie, *Biochemistry* **30**, 9835 (1991).
[48] L. H. Huang, H. Cheng, A. Pardi, J. P. Tam, and W. V. Sweeney, *Biochemistry* **30**, 7402 (1991).
[49] M. Soriano-Garcia, K. Padmanabhan, A. M. de Vos, and A. Tulinsky, *Biochemistry* **31**, 2554 (1992).
[50] W. Bode, I. Mayr, U. Baumann, R. Huber, S. R. Stone, and J. Hofsteenge, *EMBO J.* **8**, 3467 (1989).

colony[51] and in a hemophilic patient with an inhibitor (associated with a Gln[191] to Lys substitution),[12] it is possible that the single nucleotide substitution drastically alters the stability of the protein or, more likely, the transcript as has been observed in some other congenital disorders.[52] Studies to define transcript or protein instabilities would require site-specific mutagenesis and an expression system, however.

Other Variants and Mutations

In the 5' region flanking the transcription initiation site, several mutations from base -20 to $+13$ have been found[3] and are associated with a phenotype referred to as "factor IX Leyden." In these variants, factor IX levels are more severe from birth until puberty than they are later in life.[53] Bleeding tendencies can vary from severe, becoming mild hemophilia or a milder form that normalizes in adults.

A final "variant" should also be mentioned. The B_m type was originally described by Hougie and Twomey,[54] and was distinguished by prolongation of the prothrombin time when bovine (but not human or rabbit) thromboplastin was used. However, the degree of prolongation is variable in different assay systems and among different patients with the same mutation.[7] Variants with the B_m type have normal antigen levels and usually severe or moderately severe defects, but the mutations have involved different regions of the trypsinlike domain.[3] The Arg[180] to Trp or Gln mutations and others that involve the β activation cleavage site, Arg[180]-Val[181], are one example. A second type has been found sterically near the ion pair formed after cleavage but involving a Gly[311] to Glu substitution, for example. Finally, mutations of residues near the active-center serine and near the substrate binding sites lead to a B_m type of dysfunctional protein. The substrate binding sites include mutations at residues 390, 396, and 397. Presumably, these dysfunctional factor IXs share the ability to compete in extrinsic system clotting with bovine tissue factor, whereas binding of human VIIa to either human or rabbit tissue factor is not affected.

In any patient, determining a mutation in DNA that is negative to screening for recurrent mutations (regardless of severity) will provide additional information about factor IX. Dysfunctional proteins will extend the

[51] J. P. Evans, K. M. Brinkhous, G. D. Brayer, H. M. Reisner, and K. A. High, *Proc. Natl. Acad. Sci. U.S.A.* **86,** 10095 (1989).

[52] A. R. Thompson, *in* "Molecular Genetics of Hemostatic Proteins" (R. W. Colman, J. Hirsh, V. J. Marder, and E. W. Salzman, eds.), Lippincott, Philadelphia, 1993 (in press).

[53] E. Briet, R. M. Bertina, N. H. van Tilburg, and J. J. Veltkamp, *N. Engl. J. Med.* **306,** 788 (1982).

[54] C. Hougie and J. J. Twomey, *Lancet* **1,** 698 (1965).

understanding of the relationship between structure and function of normal factor IX. Other mutations, such as those with undetectable levels of antigen or normal specific activities, will help identify bases or residues important for mRNA or protein stability. Finally, new mutations within regulator elements, as in the Leyden-like region, will enhance the understanding of control mechanisms in regulating expression of the factor IX gene.

Alternative Approaches

Amino Acid Sequencing of Hemophilic Proteins

Some patients have a dysfunctional factor IX protein present in amounts sufficient for purification. One can then screen peptide digests for altered mobility on two-dimensional electrophoresis.[55-59] Lysyl aminopeptidase produces a characteristic pattern of around 30 different peptides.[57-59] Simpler patterns can be obtained when normal and dysfunctional factor IXs are first cleaved by factor XIa and reduced alkylated light and heavy chains are isolated.

Polymerase Chain Reaction Amplification: Other Applications

PCR with Genomic Amplification and Transcript Sequencing. This method involves attaching a T7 phage promoter onto one of the PCR primers.[60] The DNA fragments amplified by this method are further transcribed to increase the signal and provide a single-stranded template for reverse transcriptase-mediated, dideoxy sequencing.[60] This approach, used at the Mayo Clinic laboratory of Sommer and colleagues, has identified over 75 distinct mutations in over 100 different families.[13,14]

PCR with Chemical Cleavage of Mismatched Strands. In this procedure, amplified fragments from a normal (wild-type) gene are prepared, radiolabeled, and hybridized to the same amplified segments from a pa-

[55] C. M. Noyes, M. J. Griffith, H. R. Roberts, and R. L. Lundblad, *Proc. Natl. Acad. Sci. U.S.A.* **80,** 4200 (1983).
[56] D. L. Diuguid, M. J. Rabiet, B. C. Furie, H. A. Liebman, and B. Furie, *Proc. Natl. Acad. Sci. U.S.A.* **83,** 5803 (1986).
[57] M. Sugimoto, T. Miyata, S.-i. Kawabata, A. Yoshioka, H. Fukui, H. Takahashi, and S. Iwanaga, *J. Biochem. (Tokyo)* **104,** 878 (1988).
[58] K. Suihiro, S.-i. Kawabata, T. Miyata, H. Takeya, J. Takamatsu, K. Ogata, T. Kamiya, H. Saito, Y. Niho, and S. Iwanaga, *J. Biol. Chem.* **264,** 21257 (1989).
[59] D. L. Diuguid, M. J. Rabiet, B. C. Furie, H. A. Liebman, and B. Furie, *Blood* **74,** 193 (1989).
[60] E. S. Stoflet, D. D. Koeberl, G. Sakar, and S. S. Sommer, *Science* **239,** 419 (1988).

tient's gene. Where a mismatch occurs, either by a nucleotide substitution or a small deletion or insertion, these two strands will fail to pair completely, making them more susceptible to chemical cleavage. By using two chemical cleavage reactions, all possible nucleotide substitutions can be detected either in the coding (sense) or noncoding (antisense) strands. Hydroxylamine reacts with unpaired cytosine nucleotides and osmium tetroxide (in pyridine) reacts with unpaired thymidine nucleotides. After reaction with the alkaline agent, piperidine, all positions where there is a C or T mismatched base become cleaved.[61] Giannelli and colleagues in London[19] found that nearly all hemophilia B patients have detectable mutations.[30,31] An advantage of this approach is that from the size of the fragment, one can predict the number of bases from either end where the substitution occurs.

PCR with Single-Strand Conformation Mutations. The technique includes amplification and radiolabeling of DNA strands simultaneously by PCR. After heat denaturation, single-stranded DNA fragments are electrophoresed in a nondenaturing polyacrylamide gel. A DNA fragment with a single base difference usually has a sufficiently different conformation to alter its migration.[62] This screening procedure is usually not as sensitive as others but should detect the majority of single nucleotide substitutions and has been used to detect some factor IX mutations.[63-66] It is difficult to distinguish differences in fragments of more than about 200 nucleotides.

PCR with Denaturing Gradient Gel Electrophoresis. Gel electrophoresis in a concentration gradient of a denaturing agent is a method that separates DNA fragments according to their melting temperature.[67] Melting temperature is largely determined by a fragment's composition of bases. When a DNA fragment is electrophoresed through a gradient of a denaturing medium, the fragment remains double stranded until it reaches the region equivalent to a melting temperature. At that point, melting occurs and this decreases the mobility of the molecule in the gel. DNA fragments differing by as little as a single base can be distinguished. Attachment of a "GC clamp" to the 5' end of one the primers for the PCR reaction greatly

[61] R. G. H. Cotton, N. R. Rodrigues, and R. D. Campbell, *Proc. Natl. Acad. Sci. U.S.A.* **85,** 4397 (1988).
[62] M. Orita, H. Iwahana, H. Kanazawa, H. Hayashi, and T. Sekiya, *Proc. Natl. Acad. Sci. U.S.A.* **86,** 2766 (1989).
[63] D. B. Demers, S. J. Odelberg, and L. McA. Fisher, *Nucleic Acids Res.* **18,** 5575 (1990).
[64] S.-W. Lin and M.-C. Shen, *Thromb. Haemostasis* **65,** 965 (1991).
[65] B. M. Tam, R. T. A. MacGillivray, and B. C. Ritchie, *Thromb. Haemostasis* **65,** 968 (1991).
[66] B. M. Fraser, M.-C. Poon, and D. I. Hoar, *Hum. Genet.* **88,** 426 (1992).
[67] R. M. Myers, T. Maniatis, and L. S. Lerman, this series, Vol. 155, p. 501.

increases the sensitivity of this method.[68] This approach has distinguished a polymorphism in the Asian population in the first intron at base 192.[69]

Acknowledgments

Supported in part by the National Institutes of Health (HL-31193) and the March of Dimes Birth Defects Foundation (6-463).

[68] L. S. Lerman and K. Silverstein, this series, Vol. 155, p. 482.
[69] M. Tanimoto, M. Hamaguchi, T. Matsushita, M. Hamaguchi, T. Matsushita, K. Yamamoto, I. Sugiura, J. Takamatsu, and H. Saito, *Blood* **74**, 251a (1989).

[9] Characterization of Dysfunctional Factor VIII Molecules

By LEON W. HOYER

Introduction

Factor VIII (antihemophilic factor) participates in blood coagulation as an essential cofactor in the activation of factor X by factor IXa in the presence of phospholipid and calcium. It circulates in plasma as a large glycoprotein that is associated with von Willebrand factor in a noncovalent complex. The mature single-chain protein of 2332 amino acids is proteolytically processed before entering the circulation as a heterodimer consisting of amino-terminal heavy chain polypeptides of 92 to 200 kDa and an 80-kDa carboxyl-terminal light chain. Factor VIII procoagulant activity is generated by thrombin cleavage of the heavy chain at Arg-372 and Arg-740 to yield 54- and 44-kDa fragments, and thrombin cleavage of the 80-kDa chain at Arg-1689 to yield a 72-kDa fragment (Fig. 1).[1]

Hemophilia A (classic hemophilia) is an X chromosome-linked disorder of blood coagulation caused by deficient factor VIII activity. Although factor VIII procoagulant activity is consistently reduced or absent, hemophilia A is heterogeneous when characterized clinically or immunologically,[2] as well as when the specific molecular defects are identified.[3] Dysfunctional, immunoreactive factor VIII-like protein can be detected in

[1] W. H. Kane and E. W. Davie, *Blood* **71**, 539 (1988).
[2] L. W. Hoyer and R. T. Breckenridge, *Blood* **32**, 962 (1968).
[3] E. G. D. Tuddenham, D. N. Cooper, J. Gitschier, M. Higuchi, L. W. Hoyer, A. Yoshioka, I. R. Peake, R. Schwaab, K. Olek, H. H. Kazazian, Jr., J.-M. Lavergne, F. Giannelli, and S. E. Antonarakis, *Nucleic Acids Res.* **19**, 4821 (1991).

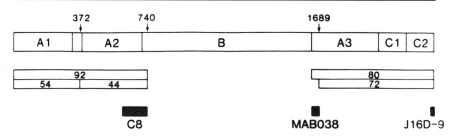

Fig. 1. Schematic representation of human factor VIII showing its domain structure (A1–A3, B, C1, and C2) and three thrombin cleavage sites (arrows). Below this are shown the thrombin cleavage fragments. The fragment mass values in kilodaltons correspond to those published by C. A. Fulcher, J. R. Roberts, and T. S. Zimmerman [*Blood* **61**, 807 (1983)]. The solid boxes represent the regions containing epitopes for the monoclonal antibodies used to identify factor VIII fragments. From Arai *et al.*[7] with permission.

10% of plasmas from patients with mild or moderate hemophilia A.[4] These plasmas are termed cross-reacting material (CRM) positive. The remaining hemophilic plasmas are either CRM reduced, in which case the reduction in immunologically detectable protein is comparable to reduction in functional activity, or CRM negative, i.e., there is no detectable factor VIII by sensitive immunoassays.

Although the factor VIII gene is large (186 kb), and unrelated individuals have different mutations, there has been considerable progress in identifying the causative point mutations or deletions.[3] Until recently, the specific point mutation or deletion could only be identified in half of the patients with severe hemophilia A using PCR amplification of genomic DNA followed by denaturing gradient gel electrophoresis.[5] Rapid screening of all essential regions of the factor VIII gene using an mRNA-based method has been more successful, however, as it also identified the presence of frequent intron 22 mutations that cause defective joining of exons 22 and 23 in the mRNA.[6] Because these studies do not necessarily establish the basis for reduced procoagulant activity, we have developed a simple and sensitive method for the immunoisolation of factor VIII from plasma.[7] This method has been used to identify the functional defect in the factor VIII-like protein recovered from plasmas of patients with CRM-positive hemophilia A.

[4] J. Lazarchick and L. W. Hoyer, *J. Clin. Invest.* **62**, 1048 (1978).
[5] M. Higuchi, H. H. Kazazian, Jr., L. Kasch, T. C. Warren, M. J. McGinniss, J. A. Phillips, III, C. K. Kasper, R. Janco, and S. E. Antonarakis, *Proc. Natl. Acad. Sci. U.S.A.* **88**, 7405 (1991).
[6] J. A. Naylor, P. M. Green, C. R. Rizza, and F. Giannelli, *Lancet* **340**, 1066 (1992).
[7] M. Arai, H. Inaba, M. Higuchi, S. E. Antonarakis, H. H. Kazazian, Jr., M. Fujimaki, and L. W. Hoyer, *Proc. Natl. Acad. Sci. U.S.A.* **86**, 4277 (1989).

Procedure

Preparation of Immunoadsorbent

Most studies have been carried out using IgG prepared by the caprylic acid method[8] from the plasma of a patient with a high-titer (3600 Bethesda units) factor VIII autoantibody. The IgG is coupled to cyanogen bromide-activated Sepharose CL-2B at a concentration of 4 mg/ml of settled gel volume.[9] Three other high-titer (570–3000 Bethesda units) human antifactor VIII antibodies (as well as autoantibodies) and a murine monoclonal antifactor VIII heavy chain antibody have also been successfully used as the immunoadsorbent when coupled to agarose beads.

Immunopurification of Factor VIII

After washing, the immunoadsorbent beads are suspended in Tris-buffered saline (TBS): 0.15 M NaCl, 4 mM CaCl$_2$, 20 mM Tris-HCl, pH 7.4. Prior to characterization, normal or hemophilia A patient plasmas are brought to 0.8 M NaCl and 0.5% Tween 80, and 0.6–2.5 ml is incubated with 20 μl of immunobeads overnight at room temperature with gentle aggitation. The amount of factor VIII protein immunoabsorbed on the beads can be estimated by measuring the difference between the factor VIII antigen (VIII : Ag) in the plasma and in the postadsorption supernatant fluid.[4] The beads are then washed into a 0.8 × 4-cm polypropylene column (Bio-Rad, Rockville Center, NY) using 20 ml of 50 mM imidazole, 40 mM CaCl$_2$, 5% (v/v) ethylene glycol, 0.5% (v/v) Tween 80, pH 6.4, and further washed with 5 ml TBS. The beads are then transferred into a 1.0-ml conical polystyrene tube, the wash buffer removed by centrifugation, and factor VIII eluted by adding 40 μl of 0.125 M Tris-HCl, pH 6.8, containing 20 mg/ml sodium dodecyl sulfate (SDS), 10% (v/v) glycerol, and 0.05 mg/ml bromphenol blue. After 1 hr incubation at 37° with gentle agitation, the tube is centrifuged and the eluted factor VIII is recovered in the supernatant.

Thrombin Treatment of Factor VIII

Some aliquots of immunopurified factor VIII were treated with thrombin before the SDS elution step. Thrombin, diluted to 10 units/ml with TBS, is added to these aliquots of washed factor VIII immunoadsorbent beads in sufficient quantity to achieve the desired concentration (0.2–2

[8] M. Steinbuch and R. Audran, *Arch. Biochem. Biophys.* **134**, 279 (1969).
[9] R. Axén, J. Poráth, and S. Ernbäck, *Nature (London)* **214**, 1302 (1967).

units of thrombin/unit VIII : Ag) and the mixture is incubated at room temperature for 0.5 to 60 min with gentle agitation before adding the SDS elution buffer.

Immunoblotting of Factor VIII

Eluted factor VIII (40 μl) is analyzed by discontinuous SDS-polyacrylamide gel electrophoresis (PAGE) using 1.5-mm-thick 5–12% (w/v) polyacrylamide slab gels[7] and is transferred to nitrocellulose sheets (Schleicher & Schuell, Inc. Keene, NH) using a Bio-Rad Trans-Blot cell.

After transfer, the nitrocellulose sheet is blocked by incubation at 37° for 1 hr with borate-buffered saline (BBS) pH 7.8,[4] containing 30 mg/ml bovine serum albumin (BSA) and 0.05% (v/v) Tween 20. The sheet is then incubated with antifactor VIII monoclonal antibodies (1 μg IgG/ml) in BBS containing 10 mg/ml bovine serum albumin and 0.05% (v/v) Tween 20 (dilution buffer). After overnight incubation at room temperature, the sheet is washed three times with BBS containing 0.05% (v/v) Tween 20 (washing buffer) and then incubated with ^{125}I-labeled affinity-purified sheep antimouse immunoglobulin in dilution buffer (80,000 cpm/ml). After a 4-hr incubation at room temperature, the sheet is again washed three times with washing buffer, dried, and visualized by autoradiography at −70° for 48 to 72 hr using Kodak XK-1 film (Eastman Kodak Co., Rochester, NY).

Comparable results have been obtained using chemiluminescence detection.[10] In this case, alkaline phosphatase–streptavidin and biotin-labeled goat antibody to mouse IgG are used to detect the monoclonal antibodies, and the enzyme complex is identified with Lumi-Phos 530 (Lumigen, Inc., Detroit, MI).

Amplification of Genomic DNA and Sequencing

Sequence analysis of apparent thrombin cleavage site mutations is carried out with leukocyte DNA amplified using Taq DNA polymerase (Perkin-Elmer Cetus, Norwalk, CT).[11] The primers are chosen so that the specific factor VIII cleavage site region will be amplified.[7,12]

Polymerase chain reaction (PCR) is performed in a 100-μl volume containing 200–400 ng of genomic DNA, 400 nM of each PCR primer, 200 μM of each dNTP, and 2 units of Taq DNA polymerase in a buffer

[10] A. P. Schaap, H. Akhavan, and L. J. Romano, Clin. Chem. (Winston-Salem, N.C.) 35, 1863 (1989).
[11] R. K. Saiki, D. H. Gelfand, S. Stoffel, et al., Science 239, 487 (1988).
[12] M. Arai, M. Higuchi, S. E. Antonarakis, et al., Blood 75, 384 (1990).

composed of 10 mM Tris-HCl (pH 8.3), 50 mM KCl, 1.5 mM MgCl$_2$, and 0.02% gelatin. Each of the 35 cycles consists of a 30-sec denaturation at 94°, thermal transition from 94° to the annealing temperature over 2 min, 45 sec of reannealing at 52–55°, and a 90-sec extension at 72° using a DNA Thermal Cycler (Perkin-Elmer Cetus, Norwalk, CT). Amplified DNA is desalted and excess dNTPs are removed by spin dialysis on a Centricon 30 (Amicon, Danvers, MA). The purified DNA is sequenced directly: 10 ng of SP7 sequencing primer (5'-GAT TTT GAC ATT TAT GAT-3'), end labeled with [γ-^{32}P]ATP using T4 polynucleotide kinase, is annealed with 80 ng of PCR product on ice after heat denaturation at 95° for 5 min. The reaction mixture is divided into four tubes containing 62 μM unlabeled dNTPs, 6.2 μM dideoxy (dd)NTPs, and 2 units of T7 DNA polymerase (Sequenase USB, Cleveland, OH) in the sequencing reaction buffer [25 mM Tris-HCl, pH 7.5, 10 mM MgCl$_2$, 70 mM NaCl, 7 mM dithiothreitol (DTT)]. After incubation at 37°C for 15 min, the reaction is stopped with 3 μl of a solution containing 95% formamide, 20 mM EDTA, 0.5 mg/ml bromphenol blue, 0.05% xylene cyanol FF. Samples are boiled for 3 min and electrophoresed in a 6% polyacrylamide/8 M urea gel at 58 W for 2 hr. Gels are then dried and exposed to Kodak XAR-5 film (Eastman Kodak Co., Rochester, NY) for 16 hr.

Denaturing Gradient Gel Electrophoresis Analysis of Amplified Genomic DNA

For evaluation of factor VIII mutations that do not prevent thrombin cleavage, the initial evaluation is carried out using denaturing gradient gel electrophoresis (DGGE) of DNA fragments to which are attached a "GC-clamp".[5] PCR products (~ 80 ng of each) from two patients are combined to form heteroduplexes. After heat denaturation at 95° for 5 min, the DNA solution is slowly cooled to room temperature (> 30 min) and subjected to DGGE under conditions determined empirically for each PCR product.[5] DNA is loaded onto a 6.5% polyacrylamide gel (14 × 19 cm, 0.75-mm thick) containing a linear gradient of denaturants and electrophoresed at 2–4 V/cm for 16–23 hr. The gradient difference in denaturants is 20% [100% denaturants = 7 M urea/40% (v/v) formamide]. Gels are then stained in ethidium bromide and photographed with a UV transilluminator. Purified GC-clamped PCR products that show abnormal migrating patterns on DGGE are then directly sequenced.[6,11,13]

[13] C. Wong, C. D. Dowling, R. K. Saiki, R. G. Higuchi, H. A. Erlich, and H. H. Kazazian, Jr., *Nature (London)* **330**, 384 (1987).

Results

Immunoisolation of Factor VIII Protein

Factor VIII is immunoadsorbed from plasma using a human antibody to factor VIII that reacts with both heavy and light chain determinants.[14] The adsorbent removes 65–95% of VIII:Ag in normal or CRM-positive plasmas. Because the interaction of human factor VIII with these antibodies is very avid, usual elution techniques such as low pH and chaotropic agents are rarely successful in separating the antifactor VIII from an immunoadsorbent.[15] For this reason, and because pH extremes destroy factor VIII immunogenicity, SDS is used to separate the bound factor VIII from the immunoadsorbent. Protein characterization is then accomplished by SDS-PAGE, followed by immunoblotting using monoclonal antifactor VIII. In general, human antifactor VIII antibodies are not satisfactory for this purpose because these plasmas have additional immunologic reactivities not related to factor VIII. Three monoclonal antibodies have been used in most studies: MAB038 (Chemicon, El Segundo, CA) binds to the 80-kDa light chain fragment but does not react with the thrombin-cleaved 72-kDa light chain fragment; J16D-9 [an antibody prepared by Dr. C. Fulcher (Scripps Research Institute, La Jolla, CA) by immunization with a synthetic peptide containing factor VIII residues 2318–2332] detects both 80- and 72-kDa light chain fragments; and C8 (JR Scientific, Woodland, CA) detects factor VIII heavy chain fragments as large as 200 kDa and reacts strongly with the 92- and 44-kDa thrombin-cleaved heavy chain fragments (Fig. 1).[7]

The sensitivity of the method is such that both heavy and light chain factor VIII determinants can be detected when at least 0.5 units of VIII:Ag are present in a 20-ml plasma sample.

Characterization of CRM-Positive Hemophilic Plasmas

In most cases, the monoclonal antibodies are used together to characterize the immunoadsorbent eluates for CRM-positive plasmas. Figure 2 illustrates a typical study. The factor VIII fragments (before and after thrombin treatment) of plasmas ARC-2 and ARC-3 cannot be distinguished from those of normal plasma. In contrast, the ARC-1 pattern after thrombin treatment is distinctly different, even though the factor VIII protein directly immunoisolated from ARC-1 plasma has apparently normal fragments (lane 3, Fig. 2). The 92-kDa band is more prominent than that for normal factor VIII after incubation with thrombin, and the 44-kDa

---

[14] C. A. Fulcher, S. D. G. Mahoney, J. R. Roberts, C. K. Kasper, and T. S. Zimmerman, *Proc. Natl. Acad. Sci. U.S.A.* **82**, 7728 (1985).

[15] J. P. Allain and D. Frommel, *Blood* **42**, 437 (1973).

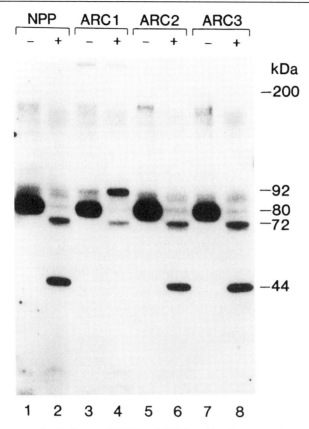

FIG. 2. Immunoadsorbed normal (NPP) and CRM-positive factor VIII fragments detected using the three monoclonal antibodies shown in Fig. 1. Thrombin-treated immunoisolated factor VIII is indicated as +; the − lanes are factor VIII not incubated with thrombin. The patterns for ARC-2 and ARC-3 are the same as those for a normal plasma. The ARC-1 92-kDa factor VIII heavy chain fragment is not cleaved by thrombin at Arg-372 so that the 44-kDa fragment is not generated. From Arai *et al.*[7] with permission.

fragment band is missing (lane 4, Fig. 2). ARC-1 light chain cleavage is normal, with loss of the 80-kDa band and the appearance of the 72-kDa fragment.

This pattern identifies the failure of thrombin to cleave the ARC-1 factor VIII heavy chain, and it suggests that the absence of procoagulant activity is a result of this defect. PCR amplification of exon 8 of the patient's factor VIII gene identified a missense mutation that causes a substitution of histidine for Arg-372, abolishing thrombin cleavage at this site.[7]

We have also identified two patients with a factor VIII light chain thrombin cleavage site mutation. In these instances, immunopurified factor VIII has normal heavy chain cleavage, but no conversion of the 80-kDa light chain to its 72-kDa fragment. In both patients, PCR amplification of exon 14 of the factor VIII gene detected a missense mutation that causes a cysteine substitution for Arg-1689. This mutation prevents factor VIII light chain thrombin cleavage.[12]

Mutations that change the factor VIII chain mass are also susceptible to characterization by this technique. To date, we have identified two non-functional factor VIII-like proteins with abnormal, slower moving heavy or light chains on SDS-PAGE.[16] Subsequently, the molecular defects have been identified by denaturing gradient gel electrophoresis screening of PCR-amplified factor VIII cDNA and sequencing the abnormal PCR products. In both cases, new N-glycosylation sites were identified. In one, substitution of threonine for Met-1772 in the factor VIII light chain creates a potential new N-glycosylation site at Asp-1770. A threonine substitution for Ile-566 generates a potential new N-glycosylation site in the second patient's factor VIII. In this case there is carbohydrate addition at Asp-564 in the A2 domain of the factor VIII heavy chain. The abnormal glycosylation was shown to be responsible for the reduced mobilities of the factor VIII chains on SDS-PAGE, as well as the loss of procoagulant activity.[16]

Summary

Immunopurification and characterization of dysfunctional factor VIII-like molecules in CRM-positive and CRM-reduced hemophilia A permit correlation of structural changes with molecular defects. The technique described here is sufficiently sensitive to characterize the molecular mass and enzymatic fragments of the factor VIII chains in patients with as little VIII : Ag as 0.05 units/ml. Specific abnormalities have been identified in 5 of the first 24 samples tested. In each case, the mutation responsible for factor VIII dysfunction has been determined by sequencing a part of the abnormal gene. Mutations have been identified that abolish critical thrombin cleavage sites or which generate new N-glycosylation sites. The technique provides a useful approach to the study of factor VIII structure–function relationships, and it has the potential to clarify further the molecular basis of factor VIII procoagulant activity.

Acknowledgments

This work was supported in part by U.S. Public Health Service Grants HL36099 and HL 44336.

[16] A. M. Aly, M. Higuchi, C. K. Kasper, H. H. Kazazian, Jr., S. E. Antonarakis, and L. W. Hoyer, *Proc. Natl. Acad. Sci. U.S.A.* **89,** 4933 (1992).

[10] Extrinsic Pathway Proteolytic Activity

By Jeffrey H. Lawson, Sriram Krishnaswamy, Saulius Butenas, and Kenneth G. Mann

Introduction

The activation of factors IX and X by the extrinsic pathway of coagulation is catalyzed by an enzyme complex composed of the integral membrane protein tissue factor (TF), activated factor VII (VIIa), calcium ions, and a phospholipid or cell surface. The serine protease, factor VIIa, possesses minimal activity toward its biological substrates in the absence of the other components of the complex.[1] The Ca^{2+}-dependent interaction between VIIa and TF anchored on the membrane surface results in a dramatic increase in the catalytic efficiency of the activation of factors IX and X.

Significant advances have been made in the isolation and purification of factor VII/VIIa and TF from bovine and human sources.[2,3] Well-characterized recombinant forms of TF and VIIa have become widely available.[4,5] In this chapter, we outline the methodology for studies of the catalytic properties of the human extrinsic "Xase" complex with respect to its biological substrates, human factors IX and X, and toward a synthetic fluorescent substrate for factor VIIa.

Reconstitution of Tissue Factor Apoprotein into Membranes

Incorporation into Mixed Detergent – Phospholipid Micelles

Functional TF can be reconstituted into mixed detergent–phospholipid micelles by incubating TF with preformed small unilamellar vesicles composed of phosphatidylcholine (75%) and phosphatidylserine (25%) (PCPS) in the presence of calcium ions and Tween 80. Analogous

[1] Y. Nemerson and R. Bach, *Prog. Hemostasis Thromb.* **6**, 237 (1982).
[2] A. Guha, R. Bach, W. Konigsberg, and Y. Nemerson, *Proc. Natl. Acad. Sci. U.S.A.* **83**, 299 (1986).
[3] R. Bach, Y. Nemerson, and W. Konigsberg, *J. Biol. Chem.* **256**, 8324 (1981).
[4] L. R. Paborsky, K. M. Tate, R. J. Harris, D. G. Yansura, L. Band, G. McCray, C. M. Gorman, D. P. O'Brien, J. Y. Chang, J. R. Swartz, V. P. Fung, J. N. Thomas, and G. A. Vehar, *Biochemistry* **28**, 8072 (1989).
[5] W. Ruf, A. Rehemtulla, and T. S. Edgington, *J. Biol. Chem.* **266**, 2158 (1991).

METHODS IN ENZYMOLOGY, VOL. 222

results have been reported by incubating TF and vesicles in the presence of cadmium ion.[6]

Small unilamellar PCPS vesicles prepared by sonication and centrifugation as previously described[7] are mixed with TF in 20 mM HEPES, 0.15 M NaCl, 5.0 mM CaCl$_2$, pH 7.4, in the presence of 0.03% (v/v) Tween 80. Maximal TF cofactor activity, as judged by initial rates of factor X activation (see below), is achieved following a 30-min incubation at 37° at concentrations of TF ranging from 50 pM to 5 nM and concentrations of PCPS ranging from 10 to 800 μM. Omission of Tween 80 or CaCl$_2$ from the reconstitution mixture results in minimal TF functional activity. Similar results have also been obtained with equivalent concentrations of Triton X-100.

This method yields reproducible results with minimal sample manipulation without requiring corrections for protein loss and accessibility following membrane reconstitution by more conventional procedures.[8] The results of kinetic studies performed with this type of TF/PCPS preparation are consistent with the accessibility of all added TF to factor VIIa and with the conclusion that this method of phospholipid reconstitution does not significantly alter the kinetic properties of the subsequently assembled enzyme. However, the physical state of TF in these detergent–phospholipid solutions is unclear.

Membrane Reconstitution by Detergent Dialysis

We have employed modifications to previously described detergent dialysis techniques to incorporate TF into PCPS membranes.[8,9] These procedures rely on the removal of octylglucoside at a controlled rate from detergent-solubilized phospholipid solutions in the presence of TF.

A thin film of phospholipids (12 mg PC and 4 mg PS in organic solvent) is deposited on the walls of a 13 × 100-mm borosilicate glass tube under a stream of dry N$_2$ and evacuated for 30 min to remove all traces of solvent. The lipid film is resuspended with 3 ml of 20 mM HEPES, 0.15 M NaCl, 80 mM octylglucoside, pH 7.0, with gentle mixing at 37°. The buffers for these procedures are prepared from a 200 mM HEPES stock solution pretreated with Chelex 100 resin (Bio-Rad, Richmond, CA) to remove adventitious divalent metal ions that can interfere with vesicle integrity. TF (30 μg) is separated from detergents by acetone precipitation.

[6] S. D. Carson and W. H. Konigsberg, *Science* **208**, 307 (1980).

[7] D. L. Higgins and K. G. Mann, *J. Biol. Chem.* **258**, 6503 (1983).

[8] R. Bach, R. Gentry, and Y. Nemerson, *Biochemistry* **25**, 4007 (1986).

[9] J. B. Galvin, S. Kurosawa, K. Moore, C. T. Esmon, and N. L. Esmon, *J. Biol. Chem.* **262**, 2199 (1987).

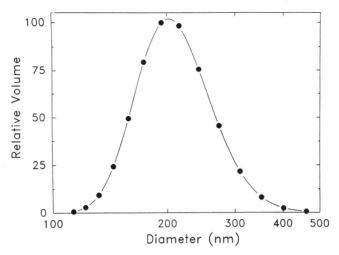

FIG. 1. Characteristics of TF-containing vesicles prepared by controlled detergent dialysis. The size distribution of PCPS vesicles containing TF (6 mM phospholipid, 250 nM TF) was characterized by quasi-elastic light scattering. The frequency distribution represents a plot of relative scattering volume versus vesicle diameter. The data are reasonably described by a gaussian distribution (mean diameter = 209.5 ± 47.8 nm).

For this step, TF is treated with 80% ice-cold acetone (1 volume TF, 4 volumes HPLC-grade acetone) and incubated for 30 min at −20°. Following centrifugation (10,000 g, 10 min), the supernatant is carefully decanted and residual solvent is removed from the pellet by evacuation for 30 min. The pellet is resuspended in 100 μl HEPES buffer without octylglucoside, added to the detergent–phospholipid mixture and dialyzed at a controlled rate against 20 mM HEPES, 0.15 M NaCl, pH 7.0, for 60 hr at room temperature. Controlled dialysis is achieved by placing the sample in wells (1.5 ml) of a microdialysis chamber (BRL Laboratories, Gaithersburg, MD) and by pumping buffer through the opposite chamber at a flow rate of 1.5 ml/min. Sample mixing during dialysis is aided by gentle agitation of the dialysis chamber with a rocker platform. Following dialysis, the cloudy samples are centrifuged (56,000 g, 15 min) to remove particulates and multilamellar vesicles. Phospholipid concentration is determined by phosphate analysis,[7] and TF recovery is quantitated using [^{125}I]TF added as a tracer. The size distribution of these vesicle preparations determined by quasi-elastic light scattering is illustrated in Fig. 1. The vesicles prepared by this procedure are characterized by a unimodal gaussian distribution centered at 210 ± 48 nm. Calculations based on recovery of phospholipid and TF, vesicle size, and functional measurements indicate approximately 10 TF molecules per vesicle with 50% of the TF molecules accessible to externally added components.[8] We have observed that these preparations are stable for at least 2 weeks when stored in the dark at room temperature.

Vesicle size, homogeneity, and phospholipid recovery are dependent on pH and ionic strength of the dialysis medium, dialysis rate, and the initial concentrations of phospholipid and detergent. One distinct disadvantage of this type of membrane reconstitution is that phospholipid and TF concentrations cannot be easily decoupled and varied independently. Previous studies have diluted TF-containing vesicles with appropriate concentrations of vesicles lacking TF to achieve the desired final phospholipid concentration.[10] However, this manipulation can create difficulties arising from the partitioning of membrane-binding reactants of the extrinsic Xase complex between vesicles that do not contain TF and therefore cannot support Xase assembly.

Factor X Activation

The conversion of factor X to factor Xa results from the cleavage of a single peptide bond[10a] (Arg^{51}-Ile^{52}) in the heavy chain of the molecule that releases an activation peptide from the disulfide-linked heavy and light chains of Xaα (Fig. 2).[11] An added complexity of this reaction results from the feedback action of product to cleave another peptide bond (Arg^{290}-Gly^{291}) at the COOH terminus of the heavy chain to yield Xβ and Xaβ. Additional feedback cleavage(s) by factor Xa in the heavy chain yielding Xaγ have also been reported.[12]

Discontinuous radiometric assays based on activation peptide release by reductive tritiation of carbohydrate residues have been described in detail and used extensively in measurements of factor X activation.[13,14] However, TCA-soluble tritiated peptides are released both on the formation of Xaα and by the feedback reactions of the product. Although the use of inhibitors and the appropriate corrections to avoid complexities from the second step have been described,[14] we have focused on methods that directly evaluate the expression of the active site of factor Xa and hence are confined to the measurements of Xase activity.

---

[10] S. D. Forman and Y. Nemerson, *Proc. Natl. Acad. Sci. U.S.A.* **83**, 4675 (1986).

[10a] Residue numbers for cleavage and glycosylation sites unambiguously identified by chemical means for bovine factor X/Xa are illustrated in Fig. 2. In human factor X, activation results from cleavage at Arg^{52}-Ile^{53}, two potential N-linked glycosylation sites have been proposed (Asn^{39} and Asn^{49}) and the cleavage site yielding the β species has not been identified.

[11] C. M. Jackson, *Prog. Hemostasis Thromb.* **7**, 55 (1984).

[12] K. Mertens and R. M. Bertina, *Biochem. J.* **185**, 647 (1980).

[13] M. Zur and Y. Nemerson, this series, Vol. 80, p. 237.

[14] Y. Nemerson and R. Gentry, *Biochemistry* **25**, 4020 (1986).

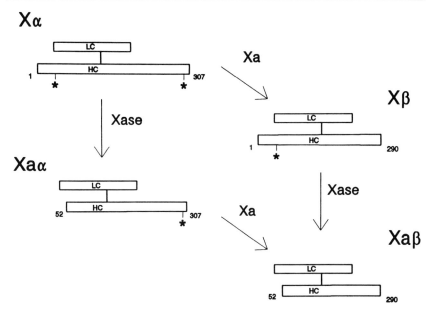

FIG. 2. Schematic diagram of bovine factor X activation. The disulfide-bonded heavy chain (HC) and light chain (LC) of factor X are illustrated. The extrinsic Xase complex catalyzes the conversion of factor Xα to factor Xaα by peptide bond cleavage at Arg-51. Feedback action of factor Xa on both zymogen and protease results in cleavage at Arg-290 on the heavy chain to yield factors Xβ and Xaβ. The positions of N-linked glycosylation at Asn-36 and O-linked glycosylation at Thr-300 are illustrated with an asterisk.

Discontinuous Measurements of Factor X Activation

Factor X activation is assessed by the increase in factor Xa-specific amidolytic activity following quenching of extrinsic Xase activity with EDTA. Factor X activation is initiated by mixing equal volumes of enzyme and substrate solution at 37°, with the enzyme defined as a stoichiometric complex of TF and factor VIIa on a membrane surface. Typical final concentrations of reactants are 10^{-7}–$10^{-6} M$ factor X, 10^{-10}–$10^{-9} M$ VIIa and TF, 10^{-5}–$10^{-4} M$ phospholipid in 20 mM HEPES, 0.15 M NaCl, 5 mM CaCl$_2$, pH 7.4. The enzyme solution is prepared by adding microliter quantities of factor VIIa to the reconstituted TF/membrane mixture in 20 mM HEPES, 0.15 M NaCl, 5 mM CaCl$_2$, pH 7.4, followed by a 10-min incubation at 37°. The substrate solution contains factor X in the same buffer. All procedures are conducted using Minisorb tubes (Nunc, Roskilde, Denmark) with low protein-binding properties.

The concentration of Xase in the reaction mixture is determined by the concentrations of TF, VIIa, and the K_d for the VIIa–TF interaction ($\approx 10^{-10} M$)[8,14]. Thus, at $10^{-9} M$ concentrations of enzyme constituents, the final concentration of enzyme will approximate the concentration of the limiting species. For any given concentrations of TF and VIIa, the concentration of the enzyme complex can be calculated with the appropriate quadratic expression.

Following initiation, samples are withdrawn at various time intervals and quenched by mixing with 3 volumes of buffer containing 25 mM EDTA and 0.1% (w/v) polyethylene glycol (PEG) 8000. For initial velocity measurements, withdrawn aliquots (50 μl) are manually mixed with 150 μl of quenching solution at 30- to 90-sec intervals following initiation. Typically, six to seven timed aliquots are used to obtain initial velocities of factor Xa formation. For the generation of complete progress curves, particularly at higher enzyme concentrations, timed aliquots are quenched automatically at intervals as short as 10 sec using a stepper motor-driven, five-syringe quenched-flow module (QFM-5, BioLogic, France) maintained at 37°. The reaction is initiated ($t = 0$) by loading the reaction syringe with 2.2 ml substrate solution and 2.2 ml enzyme solution under turbulent flow to ensure thorough mixing. At various times, 100 μl of the reaction mixture is expelled into an aging loop and quenched with 300 μl of quenching solution within 100 msec. The 400-μl quenched sample is expelled from the mixing lines and collected in Minisorb tubes. Typically, between 30 and 40 quenched samples are used to describe a progress curve.

Quenched samples from either of the two methods are subsequently analyzed by determining initial rates of Spectrozyme Xa hydrolysis. Appropriate aliquots of the quenched samples (2–20 μl) are applied in duplicate to wells of a 96-well plate (assay plate, Corning, Corning, NY). The entire plate is initiated with 200 μl/well chromogenic substrate solution (200 μM Spectrozyme Xa in 20 mM HEPES, 0.15 M NaCl, 25 mM EDTA, 0.1% PEG, pH 7.4). Hydrolysis of the synthetic substrate is continuously monitored at ambient temperature by following the increase in absorbance at 405 nm using a V_{max} plate reader (Molecular Devices, Menlo Park, CA). Initial rates are determined from the initial, steady-state portion of plots of A_{405} versus time and replicates are averaged. Initial rates observed with known concentrations of factor Xa (0–20 nM) applied to wells of the same plate are used to construct linear plots of initial velocity versus concentration of enzyme and permit direct interpolation of the concentration of factor Xa in the original reaction mixture at the time of quenching.

The discontinuous measurements permit the analysis of factor X activation at high sensitivity with a large dynamic range of detection. How-

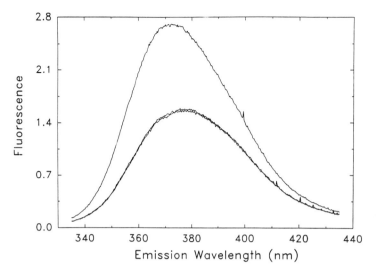

FIG. 3. Fluorescence emission spectra of 4-aminobenzamidine in the presence or absence of factors X and Xa. Reaction mixtures contained 25 μM pAB in the presence or absence of 1.6 μM factor X or factor Xa in 20 mM HEPES, 0.15 M NaCl, 5 mM CaCl$_2$, pH 7.4, at 37°. Emission spectra (band pass 2 nm) were collected using λ_{ex} = 320 nm. A long-pass filter (KV-330) was used to isolate scattered light. The lower spectra were obtained using pAB in the presence or absence of factor X. The upper spectrum illustrates the results obtained in the presence of factor Xa.

ever, for detailed kinetic studies, discontinuous measurements of product accumulation are labor intensive with obvious disadvantages.

Continuous Measurements of Factor X Activation

Continuous measurements of factor X activation have been conducted by employing the fluorescent serine protease inhibitor 4-aminobenzami-dine (pAB).[15] A similar approach has been used to monitor factor X activation by other enzymes.[16]

The methodology is based on the enhanced fluorescence of pAB on binding to the active site of serine proteases and the differential sensitivity of factors Xa and VIIa to inhibition by benzamidines.[17] Fluorescence emission spectra for pAB in the presence of factor X or factor Xa are illustrated in Fig. 3. The enhancement in fluorescence results from a

[15] S. A. Evans, S. T. Olson, and J. D. Shore, *J. Biol. Chem.* **257**, 3014 (1982).

[16] D. M. Monroe, G. B. Sherrill, and H. R. Roberts, *Anal. Biochem.* **172**, 427 (1988).

[17] A. H. Pedersen, T. Lund-Hansen, H. Bisgaard-Frantzen, F. Olsen, and L. C. Petersen, *Biochemistry* **28**, 9331 (1989).

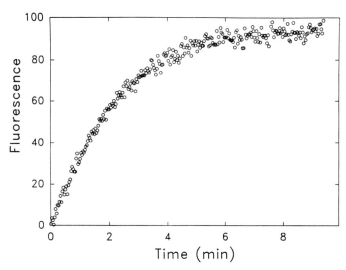

FIG. 4. Progress curve for factor X activation monitored using pAB. A reaction mixture containing 1.3 μM factor X, 100 μM PCPS/5 nM TF in 20 mM HEPES, 0.15 M NaCl, 5 mM CaCl$_2$, pH 7.4, at 37°, was initiated with 0.5 nM VIIa. Reaction progress was monitored using λ_{ex} = 320 nm and λ_{em} = 370 nm.

reversible interaction of pAB with the active site of factor Xa ($K_d \simeq 35$ μM) resulting in the inhibition of its proteolytic activity.[18] In contrast, the activity of factor VIIa is not significantly altered by this fluorophore at concentrations as high as 50 μM, which is consistent with a reported K_i of approximately 50 mM for the inhibition of VIIa by other benzamidines.[17]

Factor X activation is monitored in the presence of 25 μM pAB using an excitation wavelength of 320 nm (band pass 2 nm) and an emission wavelength of 370 nm (band pass 16 nm). Stray and scattered light is minimized with a long-pass filter KV-324 (Schott, Duryea, PA) in the emission beam. The reaction mixture (2.0 ml) in a stirred quartz cuvette (1 × 1 cm) contains 25 μM pAB, 0.4–2 μM factor X, and 10^{-9} M TF/ 10^{-5}–10^{-4} M PCPS in 20 mM HEPES, 0.15 M NaCl, 5.0 mM CaCl$_2$, pH 7.4, with the temperature regulated at 37°. The fluorescence signal is offset with this sample and the dynamic photomultiplier range is adjusted to give an approximately 80% full-scale signal with an identical reaction mixture containing factor Xa instead of factor X. Factor X activation is initiated by the addition of microliter amounts of a VIIa stock solution. A typical progress curve is illustrated in Fig. 4. The limiting amplitude reached in all cases is identical to the fluorescence signal of reaction mixtures containing

[18] P. A. Craig, S. T. Olson, and J. D. Shore, *J. Biol. Chem.* **264,** 5452 (1989).

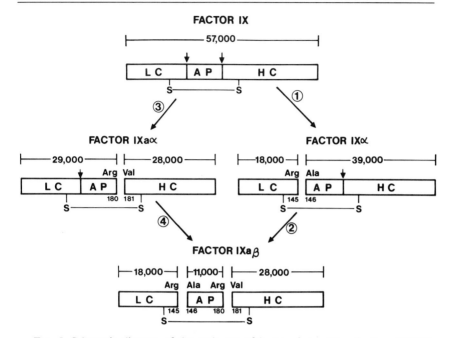

FIG. 5. Schematic diagram of the pathways of human factor IX activation. TF/VIIa catalyzes the cleavage of two peptide bonds in factor IX to yield factor IXaβ. The reaction proceeds via steps 1 and 2, where factor IX is first cleaved at the Arg[145]-Ala[146] peptide bond followed by cleavage at Arg[180]-Val[181], generating factor IXα as the reaction intermediate. LC, Light chain; AP, activation peptide; HC, heavy chain. Reproduced with permission from Lawson and Mann.[22]

the same concentration of factor Xa, indicating the quantitative conversion of substrate to product. Fluorescence is related to product concentration by using the initial and limiting signal to signify 0 and 100% product accumulation to permit extraction of initial velocities or analysis of the entire progress curve according to the integrated Michaelis–Menten equation.[19] Kinetic parameters for factor X activation obtained by the continuous measurements are coincident with those obtained by the discontinuous amidolytic activity measurements.

Factor IX Activation

The activation of factor IX to IXaβ requires the cleavage of two peptide bonds (Fig. 5).[20] Factor IX activation by TF/VIIa proceeds in a sequential reaction via cleavage at Arg[145]-Ala[146] followed by proteolysis at Arg[180]-

[19] R. G. Duggleby, Biochem. J. 235, 613 (1986).
[20] R. G. Di Scipio, K. Kurachi, and E. W. Davie, J. Clin. Invest. 61, 1528 (1978).

Val[181].[21,22] The second cleavage results in the release of a 35-amino-acid activation peptide (AP) from the factor IX molecule, which contains amino acid residues 146–180.

Measurements of factor IXaβ formation by TF/VIIa have relied on one of two experimental systems: clotting assays using factor IX-deficient plasma, or monitoring the proteolytic fragments of factor IX. Both systems are based on discontinuous measurements of factor IX activation. The methods for generating the quenched samples and performing the different assays are described below.

Discontinuous Measurements of Factor IX Activation

Tissue factor is relipidated into the appropriate lipid source as described above. After relipidation of the TF, factor VIIa is added to the reaction and incubated for 10 min at 37° in 20 mM HEPES, 0.15 M NaCl, 5 mM CaCl$_2$, pH 7.4. Typically the final concentrations of reagents are TF 10^{-9} M, factor VIIa 10^{-9} M, and PCPS 10^{-5}–10^{-4} M. The reaction is initiated by the addition of zymogen factor IX ranging in concentration from 10^{-7}–10^{-6} M. Following initiation, aliquots of the reaction mixture are removed at various times and quenched in an equal volume of 50 mM EDTA. Progress curves for factor IX activation are constructed from measurements of product concentration in the quenched samples by the methods described below.

Coagulation Assays

The clotting activities of factor IX activation reaction samples (described above) are evaluated by mixing 75 μl of a test sample, 75 μl of human factor IX-deficient plasma (George King Biomedical), and 75 μl of thromboplastin (Baxter). The mixture is warmed to 37° for 1 min followed by the addition of 75 μl of 25 mM CaCl$_2$. The clotting time is recorded in seconds by the formation of a fibrin clot in the test plasma (after the addition of CaCl$_2$). The clotting time of each test sample is compared to a factor IXaβ standard curve prepared with known concentrations of factor IXaβ (1 to 200 nM). This assay is based on the ability of activated factor IX to shorten the clotting time of factor IX-deficient plasma. In spite of the reported sensitivity of the method (1–100 ng/ml IXaβ added to plasma),[23] the use of this approach to derive mechanistic conclusions is limited by the complexities of plasma-based coagulation assays.

[21] S. P. Bajaj, S. I. Rapaport, and W. A. Russell, *Biochemistry* **22**, 4047 (1983).

[22] J. H. Lawson and K. G. Mann, *J. Biol. Chem* **266**, 11317 (1991).

[23] Y. Komiyama, A. H. Pedersen, and W. Kisiel, *Biochemistry* **29**, 9418 (1990).

Radiometric Assay

A radiometric assay based on the release of ^{3}H-labeled factor IX activation peptide has been described and used extensively in studies of factor IX activation.[24] This method labels the sialic acid residues of factor IX using NaB^{3}H$_4$ and exploits two additional facts: (1) the activation peptide released during factor IX activation is soluble in 5% trichloroacetic acid (TCA) while intact factor IX is not[13]; and (2) approximately 40% of the sialic acid residues located on the factor IX molecule are contained within the activation peptide region.[25] Thus, ^{3}H-labeled factor IX activation is accompanied by an increase in TCA-soluble radioactivity.

This assay has the advantage of being both sensitive and specific for evaluating factor IX activation in complex biological systems as exemplified by kinetic measurements of ^{3}H-labeled factor IX activation in plasma described by Jesty and Silverberg.[26] However, the process of oxidation followed by reduction of the factor IX with boro[^{3}H]hydride can result in a significant loss in the bioactivity of factor IX when assayed in a one-step clotting assay. Furthermore, kinetic interpretations of factor IX activation are complicated by the fact that this assay method only evaluates the second proteolytic step (AP release) in a two-step reaction process.

SDS-PAGE

SDS-PAGE analysis of factor IX activation is performed by taking quenched time points of a factor IX activation reaction, treating each sample with 2% (w/v) SDS, 0.01 M Tris-HCl, pH 6.8, 10% (v/v) glycerol, and 2% (v/v) 2-mercaptoethanol. Reaction subsamples are analyzed by SDS-PAGE (10% polyacrylamide gels) using the general methods of Laemmli.[27] Protein bands are identified by staining the gels with Coomassie brilliant blue and destaining by diffusion. The level of factor IX activation is then quantitated by scanning densitometry of the Coomassie brilliant blue-stained gel (Fig. 6). Densitometric analysis of the stained gel is performed using a Microscan 1000 scanning densitometer (TRI Inc.) equipped with a solid-state linear diode array camera to digitize images through a photographic lens. Digitized images contain 2.5 × 10^5 pixels, with each pixel measuring 0.16 × 0.16 mm. Each pixel in the image is an absorbance value which is linear over the range of 0–2.5 absorbance units. Data are expressed as integrated volumes for each protein band using the

[24] M. Zur and Y. Nemerson, *J. Biol. Chem.* **255,** 5703 (1980).
[25] K. Fujikawa, M. E. Legaz, H. Kata, and E. W. Davie, *Biochemistry* **13,** 4508 (1974).
[26] J. Jesty and S. A. Silverberg, *J. Biol. Chem.* **254,** 12337 (1979).
[27] U. K. Laemmli, *Nature (London)* **227,** 680 (1970).

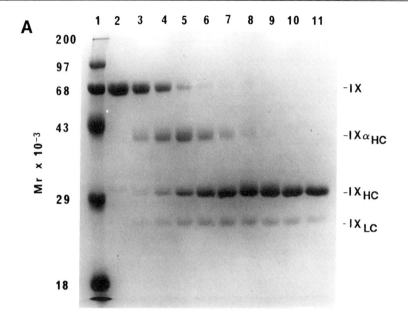

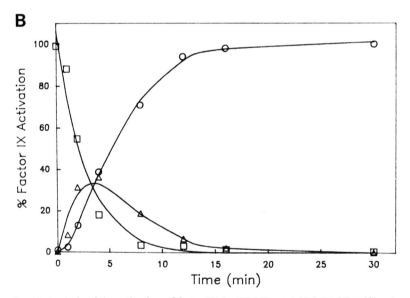

Fig. 6. Analysis of the activation of factor IX by TF/VIIa. (A) SDS-PAGE (10% polyacrylamide gel) analysis of various quenched time points from an ongoing factor IX activation reaction. Lane 1, molecular weight markers with designated molecular weights; lanes 2–11, samples quenched at time 0, 0.5, 1, 2, 4, 8, 12, 16, 30, and 60 min after the addition of factor

arbitrary density units of the scanning system. This method has the advantage of being able to visualize all of the proteolytic products of the factor IX activation reaction. This method is most useful at high substrate concentrations ($> 1 \mu M$, 4 μg of protein per lane of the gel) and therefore has limited application for rigorous kinetic analysis of factor IX activation.

Active Site Labeling of Factor IXaβ

Progress curves of factor IX activation can be analyzed using lissamine–rhodamine-labeled D-phenylalanyl-L-prolyl-L-arginine chloromethyl ketone (lr-FPR-ck). The synthesis of fluorescently labeled chloromethyl ketones is described in an accompanying chapter of this series.[28] For treatment of factor IX activation samples with lr-FPR-ck, 100-μl reaction samples are incubated with 50 mM Tris-HCl, pH 7.4, and 30 μM lr-FPR-ck for 30 min at 37°. Each reaction sample is then subjected to SDS-PAGE as described above. Protein bands which specifically incorporate lr-FPR-ck are visualized by excitation with a long-wavelength UV light box (Transilluminator, UVP Inc.) and photographed through a 540-nm cutoff filter. Total protein from each reaction point is then visualized on the gel by staining with Coomassie brilliant blue. This method is useful for specifically identifying proteolytic steps in the factor IX activation process which form a serine protease active site in the heavy chain of factor IX.[22]

Synthetic Substrates for Factor VIIa

Various investigators have identified potential substrates for evaluating the activity of factor VIIa using synthetic substrates.[5,29-31] However, the *p*-nitrobenzyl ester,[29] thioester,[30] and *p*-nitroanilide substrates[5,31] exhibit limited sensitivity for evaluating factor VIIa activity in the absence of tissue factor. A fluorescent substrate for directly evaluating the enzymatic

[28] E. B. Williams and K. G. Mann, this volume [28].
[29] M. Zur and Y. Nemerson, *J. Biol. Chem.* **253,** 2203 (1978).
[30] C. Kam, G. P. Valsuk, D. E. Smith, K. E. Arcuri, and J. C. Powers, *Thromb. Haemostasis* **64,** 133 (1980).
[31] A. H. Pedersen, O. Nordfang, F. Norris, F. C. Wiberg, P. M. Christensen, K. B. Moeller, J. Meidahl-Pedersen, T. C. Bech, K. Norris, U. Hedner, and W. Kisiel, *J. Biol. Chem.* **265,** 16786 (1990).

IX to the TF/VIIa complex. Protein bands are identified as factor IX, factor IXα$_{HC}$, factor IXaβ$_{HC}$, and factor IX$_{LC}$. (B) Relative concentrations of factor IX, factor IXα, and factor IXaβ as determined by densitometric analysis of the Coomassie brilliant blue-stained gels. Factor IX (□), factor IXα (△), and factor IXaβ (○) are shown. Reproduced with permission from Lawson and Mann.[22]

activity of factor VIIa has been described.[32] The synthesis of this compound and its utility for evaluating the enzymatic activity of factor VIIa in the presence and absence of tissue factor and phospholipid are described below.

Synthesis of 6-(Mes-D-Leu-Gly-Arg)amino-1-naphthalenediethylsulfonamide

The synthesis of 6-(Mes-D-Leu-Gly-Arg)amino-1-naphthalenediethylsulfonamide (M-LGR-nds) is begun by boiling 1 mol (223 g) of 6-amino-1-naphthalenesulfonic acid with 148 g (1 mol) of phthalic anhydride for 0.5 hr in 1 liter of pyridine. The reaction mixture is left at room temperature for 16 hr, and the precipitated product is filtered, washed with pyridine and water, and recrystallized from boiling water [yield 326 g (75%); mp 234–237°]. The formed 6-phthalimido-1-naphthalenesulfonic acid pyridinium salt (43.2 g, 0.1 mol) and 62.5 g (0.3 mol) of phosphorus pentachloride are boiled for 4 hr in 600 ml of chloroform. After evaporation of the solvent, the product is added to 2 liters of ice water. The precipitate is filtered, washed with water, dried, and recrystallized from toluene [yield 36.8 g (99%); mp 248–251°]. The formed 6-phthalimido-1-naphthalenesulfonyl chloride (37.2 g, 0.1 mol) is added to a solution of 10.3 ml (0.1 mol) diethylamine and 13.9 ml (0.1 mol) of triethylamine in 500 ml of acetone. After 4 hr at 20° the acetone is evaporated and the residual is added to 1 liter of water. The precipitate is filtered, washed with water, dried, and recrystallized from methanol [yield 35.5 g (87%); mp 199–203°]. The formed 6-phthalimido-1-naphthalenediethylsulfonamide (40.8 g, 0.1 mol) is boiled for 4.5 hr in 500 ml of methanol and 4.9 ml (0.1 mol) of hydrazine monohydrate. The methanol is evaporated and the residue is extracted with boiling chloroform. The chloroform is evaporated and the product is recrystallized from methanol [yield 21.7 g (78%); mp 106–108°]. The formed 6-amino-1-naphthalenediethylsulfonamide (2.78 g, 0.01 mol) and N-carbobenzyloxy-L-arginine hydrobromide (3.89 g, 0.01 mol) are dissolved in 15 ml of dry pyridine followed by the addition of 30 ml of dry toluene to this solution. The toluene is evaporated and the reaction mixture is cooled to −20°. 1,3-Dicyclohexylcarbodiimide solution (2.68 g, 0.013 mol) in 6 ml of dry pyridine is added to the reaction mixture. The mixture is held at −20° for 30 min, warmed to 4° for 1 hr, and then mixed by stirring at 20° for 20 hr. The precipitated dicyclohexylurea is filtered, the pyridine evaporated, and the remaining oil dissolved in 40 ml of 1-propanol–chloroform (1 : 3). This solution is washed with water and saturated sodium chloride in 2% HCl, 2% ammonia in water, and

[32] J. H. Lawson, S. Butenas, and K. G. Mann, *J. Biol. Chem.* **267,** 4834 (1992).

water. The organic layer is dried over anhydrous Na_2SO_4, the solvents are evaporated, and residual oil is hardened by grinding with dry toluene. The reaction product is filtered, washed with dry toluene and dry diethyl ether, and recrystallized from 1-propanol [yield 5.64 g (84%); mp 104–112°]. The formed 6-(N^α-carbobenzyloxy-L-arginyl)amino-1-naphthalenediethyl-sulfonamide hydrobromide (6.50 g, 0.01 mol) is dissolved in 10 ml of 3 N HBr in glacial acetic acid and kept at room temperature for 0.5 hr. The reaction mixture is added to 100 ml of dry diethyl ether. The precipitate is filtered, washed with diethyl ether, and dried. The dry 6-(L-arginyl)amino-1-naphthalenediethylsulfonamide dihydrobromide is dissolved in 10 ml of water. The solution is poured into a separation funnel, and 50 ml of 1-butanol is added followed by the addition of 5% $NaHCO_3$ (by shaking) until the water layer is pH 7.5. The organic layer is washed with water and concentrated to a final volume of 3–5 ml. The product is precipitated with dry diethyl ether, filtered, washed with diethyl ether, dried, and the 6-(L-arginyl)amino-1-naphthalenediethylsulfonamide hydrobromide is isolated [yield 4.73 g (92%); mp 133–137°].

For synthesis of the Mes-D-Leu-GlyOH, 9.1 g (0.05 mol) of D-Leu-$OCH_3 \cdot HCl$ is added with stirring to 13.9 ml (0.1 mol) of triethylamine in 100 ml of dry chloroform at 0°. This is followed by the addition of 4.5 ml (0.055 mol) of methane sulfonyl chloride to the solution at 0°, keeping the reaction mixture above pH 8. After stirring at 20° for 3 hr, the precipitated triethylamine hydrochloride is filtered and washed with chloroform. The filtrate is washed with 5% $NaHCO_3$, 10% $KHSO_4$, and H_2O. The organic layer is dried over anhydrous $MgSO_4$, the solvent is evaporated, and the residual oil (Mes-D-Leu-OCH_3, 10 g, 0.0446 mol) is dissolved at 10° in 50 ml of dry methanol containing 1.5 g of sodium. This is followed by the addition of 3 ml of H_2O to the solution and the reaction mixture is left at 20° for 72 hr. The precipitated Mes-D-Leu-ONa is filtered, dried, dissolved in 50 ml of H_2O, and then acidified to pH 2 by the addition of concentrated HCl. The solution is extracted with chloroform and the organic layer is dried over anhydrous $MgSO_4$ and evaporated. The residual oil (Mes-D-Leu-OH, 7.5 g, 0.0375 mol) is dissolved in 75 ml of chloroform containing 5.0 g (0.0375 mol) of Gly-$OC_2H_5 \cdot HCl$ at 0°, maintaining the solution at pH 8 with triethylamine. This is followed by the addition of 8.2 g (0.04 mol) of 1,3-dicyclohexylcarbodiimide (DCC) in 15 ml of dry chloroform. The reaction mixture is held at 0° for 4 hr and then raised to 20° for 20 hr. The precipitated dicyclohexylurea is filtered and washed with $CHCl_3$. The filtrate is washed with 5% $NaHCO_3$, 10% $KHSO_4$, and H_2O. The organic layer is dried over anhydrous $MgSO_4$ and evaporated. The residual oil (Mes-D-Leu-Gly-OC_2H_5, 8.4 g, 0.03 mol) is dissolved in 45 ml of methanol and 20 ml of 2 N NaOH in water. The solution is held at 20° for 24 hr and then diluted with water to a volume of 100 ml. The solution is concentrated to a volume of 35 ml, washed with 15 ml of chloroform, and acidified with concentrated HCl to pH 3. Following evaporation of the water solution, the residual is dissolved in 50 ml of acetone. Insoluble NaCl is removed by filtration and the acetone is evaporated under vacuum (yield of the product 47.4%; mp 122–126°). Finally, 2.66 g (0.01 mol) of Mes-D-Leu-GlyOH, 2.68 g (0.013 mol) of DCC, and 1.35 g (0.01 mol) 1-hydroxybenzotriazole hydrate are added to 25 ml of dry N,N-dimethyl-formamide (DMFA). The reaction mixture is kept at 0° for 1 hr followed by the addition of 5.15 g (0.01 mol) of 6-(L-arginyl)amino-1-naphthalene-diethylsulfonamide hydrobromide in 15 ml of DMFA. The reaction mix-ture is kept at 0° for 1 hr and then warmed to 20° for 20 hr with stirring. The precipitated dicyclohexylurea is filtered and the reaction solution is added to 150 ml of water. This solution is extracted with a mixture of 1-butanol–ethyl acetate (1:1). The organic layer is washed with 5% $NaHCO_3$, 10% $KHSO_4$, and water. The organic layer is concentrated, the product precipitated with dry diethyl ether, filtered, washed with diethyl ether, and dried [yield 5.94 g (87%); mp 190–205°; $R_f = 0.62$ (BAW 412); $[\alpha]_D^{20} = -12.9°$ (c 1; dimethyl sulfoxide (DMSO))]. We have observed that the dried compound is stable at 22° for at least 2 years.

Measurements of Factor VIIa Activity Using M-LGR-nds

The synthetic substrate, M-LGR-nds, is hydrolyzed by serine proteases to yield the tripeptide and 6-amino-1-naphthalenediethylsulfonamide. Measurements of the hydrolysis of M-LGR-nds by factor VIIa are based on the increase in fluorescence of the naphthalene moiety following substrate hydrolysis and the release of the 6-amino-1-naphthalenediethylsulfona-mide.

Stock solutions of M-LGR-nds (10 mM) are prepared in DMSO and diluted to final working solutions in aqueous buffers prior to use. Typical reaction mixtures contain 10^{-9}–10^{-8} M factor VIIa in 20 mM HEPES, 0.15 M NaCl, 5 mM $CaCl_2$, pH 7.4, and are initiated by the addition of M-LGR-nds to a final concentration of 40 μM. Reaction progress is fol-lowed in a fluorescence spectrophotometer thermostatted at 37° using λ_{ex} = 352 nm and measuring the emission signal at 470 nm. Scattered-light artifacts are minimized with a long-pass filter (Schott KV-399) in the emission beam. Fluorimeter settings should be adjusted such that the fluorescence signal of a 40 μM solution of M-LGR-nds (diluted in the HBS) is offset to zero. The fluorescent signal of a 1 μM solution of 6-amino-1-naphthalenediethylsulfonamide (in the same buffer) should rep-resent the full-scale fluorescence signal. Initial rates of substrate hydrolysis are obtained from fluorescence tracings, where the change in fluorescence

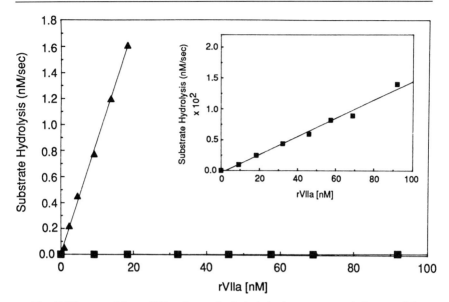

FIG. 7. The rate of factor VIIa substrate hydrolysis in the presence and absence of tissue factor. The rate of substrate hydrolysis is shown (▲) as a function of increasing factor VIIa concentrations with an excess of tissue factor (50 nM) in the experimental system. The rate of factor VIIa substrate hydrolysis is shown (■) in the absence of tissue factor. The substrate concentration used for each experiment was 40 μM. These data illustrate that tissue factor dramatically enhances (> 100-fold) the rate of factor VIIa substrate hydrolysis. Inset: Rate of factor VIIa substrate hydrolysis as a function of increasing concentrations of factor VIIa in HBS, 5 mM Ca^{2+}, pH 7.4, at 37°. Reproduced with permission from Lawson *et al.*[32]

per unit time is converted to concentration terms using the known fluorescence intensity of the 1 μM solution of the 6-amino-1-naphthalenediethylsulfonamide. In experiments conducted in the presence of TF and/or phospholipids, the final concentrations of tissue factor and factor VIIa are typically 10^{-10}–10^{-9} M. Under these conditions, both detergent-soluble and relipidated TF can function as effective cofactors. To use detergent-soluble TF, typically, 30 μM TF in the presence of 30 mM octylglucoside or 1 mM *n*-dodecyloctaethylene glycol is diluted to a final TF concentration of 50 nM with no detectable change in TF solubility. The enzyme components are incubated to permit assembly of the Xase complex, and substrate hydrolysis is initiated by the addition of 40 μM M-LGR-nds.

The sensitivity of this substrate for measuring factor VIIa activity is illustrated on Fig. 7. The inset in Fig. 7 illustrates the linear dependence of initial velocity on VIIa concentration and demonstrates that reliable initial rate measurements can be obtained at 10^{-9} M concentrations of enzyme. The rate of M-LGR-nds hydrolysis by factor VIIa is enhanced by the

TABLE I
KINETIC CONSTANTS OF M-LGR-nds SUBSTRATE HYDROLYSIS AS FUNCTION OF FACTOR
VIIa/TISSUE FACTOR COMPLEX ASSEMBLY

Enzyme	K_m ($\mu M \pm$ S.E.)	k_{cat} ($sec^{-1} \pm$ S.E.)	k_{cat}/K_m ($M^{-1} sec^{-1}$)
PCPS/Ca^{2+}	N.H.[a]	N.H.	—
TF/Ca^{2+}	N.H.	N.H.	—
TF/PCPS/Ca^{2+}	N.H.	N.H.	—
VIIa[b]	N.H.	N.H.	—
VIIa/Ca^{2+}	249 ± 88.5	0.010 ± 0.002	41.4
VIIa/PCPS/Ca^{2+}	580 ± 101	0.019 ± 0.003	32.6
VIIa/TF/Ca^{2+c}	180 ± 66	0.784 ± 0.143	4356
VIIa/TF/PCPS/Ca^{2+}	162 ± 16	0.675 ± 0.046	4166
VIIa/TF/PC/Ca^{2+}	208 ± 12	0.875 ± 0.032	4206

[a] N.H., No substrate hydrolysis observed.
[b] In the presence of 1 mM EDTA.
[c] Tissue factor solubilized in n-dodecyloctaethylene glycol micelles.

presence of saturating concentrations of TF even though the cofactor possesses no activity toward this substrate (Fig. 7). One consequence of the cofactor-dependent enhancement in rate is that this substrate is sensitive enough to reliably quantitate factor VIIa activity in the presence of TF at factor VIIa concentrations below 10^{-10} M.

Kinetic constants of factor VIIa-dependent M-LGR-nds substrate hydrolysis as a function of complex assembly can also be determined (Table I). Factor VIIa is added to relipidated or detergent-soluble TF as described above. After formation of the catalytic complex, the reaction is started by the addition of substrate. Typically, substrate concentrations are varied from 10^{-8} to 10^{-4} M M-LGR-nds. Initial rates of substrate hydrolysis are recorded at each M-LGR-nds concentration, and the kinetic constants of factor VIIa-dependent substrate hydrolysis are determined. Data presented in Table I illustrate that the K_m for M-LGR-nds hydrolysis is not dramatically influenced by the presence of TF or phospholipid. However, the k_{cat} for the reaction is increased by 70-fold in the presence of TF. Thus, the overall second-order rate constant for substrate hydrolysis (k_{cat}/K_m) is increased by 100-fold when factor VIIa is incorporated into the complete Xase complex.

Tissue factor solubilized in detergent micelles and relipidated in phospholipid vesicles composed of 100% phosphatidylcholine (PC) or vesicles containing 25% acidic phospholipid (PCPS) has the same effect on the enzymatic activity of factor VIIa. These data indicate that the TF-dependent increase in factor VIIa substrate hydrolysis is independent of the nature of the supporting surface. These data also suggest that direct binding of factor VIIa to tissue factor induces an alteration in the active site of factor VIIa which allows for more efficient hydrolysis of the substrate arginylamide bond.

This method of evaluating factor VIIa activity has the advantage of being both rapid and sensitive. Because this method directly evaluates factor VIIa activity it eliminates potential problems which can arise from the use of natural substrates, such as factor X activation, which functions as a reporter of factor VIIa activity. Furthermore, because this method can detect changes in factor VIIa activity in the presence of TF, it appears to be a useful method for evaluating the catalytic activity of mutant forms of factor VIIa and TF as these proteins are expressed by recombinant technology.

Acknowledgment

Supported by NIH Grants HL-34575, HL-35058 (K. G. M.), and HL-38337 (S. K.).

[11] Tissue Factor Pathway Inhibitor

By Thomas J. Girard and George J. Broze, Jr.

Plasma contains a tissue factor pathway inhibitor (TFPI), previously referred to as lipoprotein-associated coagulation inhibitor (LACI)[1] and extrinsic pathway inhibitor (EPI).[2] TFPI functions as a multivalent protease inhibitor; it inhibits activated factor X (Xa) directly and inhibits factor VIIa–tissue factor activity in a factor Xa-dependent fashion by forming an inhibitory complex consisting of factor Xa, TFPI, factor VIIa, and tissue factor.[3] These activities suggest that TFPI serves a regulatory role in hemostasis.

The first section in this chapter describes several assays useful for studying the various facets of TFPI function; the second section contains procedures for the purification of TFPI from mammalian cell culture, and the third section summarizes the properties of the TFPI molecule.

[1] G. J. Broze, Jr. and J. P. Miletich, *Blood* **69**, 150 (1987).
[2] L. V. M. Rao and S. I. Rapaport, *Blood* **69**, 645 (1987).
[3] G. J. Broze, Jr., L. A. Warren, W. F., Novotny, D. A. Higuchi, J. J. Girard, and J. P. Miletich, *Blood* **71**, 335 (1988).

Assays

A unique feature of TFPI is its factor Xa-dependent inhibition of factor VIIa–tissue factor activity, and the most reliable assays for identifying TFPI are assays which measure this inhibition. Two such assays are described below. They are sensitive, specific, and allow for the identification of TFPI in plasma, serum, or culture media. TFPI also directly inhibits factor Xa activity, and assays which measure inhibition of factor Xa activity are useful for the characterization of TFPI. These assays are less complex than the factor VIIa–tissue factor inhibition assays; however, they do not discriminate TFPI from other known factor Xa inhibitors and therefore are not used to determine TFPI activity in complex mixtures. Ligand blotting with radiolabeled factor Xa is useful in identifying TFPI in complex mixtures. This is because other inhibitors that bind factor Xa appear to be irreversibly denatured by the sodium dodecyl sulfate (SDS) used in the procedure.

The generation of polyclonal and monoclonal antibodies to TFPI has allowed for the identification of TFPI using immunoassays and Western blot analysis.[4]

TFPI Inhibition of Factor VIIa–Tissue Factor

The two assays for measuring TFPI inhibition of factor VIIa–tissue factor activity are a radiometric assay and a three-stage clotting-time assay. The first stage of each assay allows for the inhibition of factor VIIa–tissue factor activity by TFPI and the remaining stages measure residual factor VIIa–tissue factor activity. In the radiometric assay, factor VIIa–tissue factor activity remaining after the first stage is measured by its ability to cleave ³H-labeled factor X to factor Xa or ³H-labeled factor IX to factor IXa, releasing ³H-labeled activation peptide.[5] In the three-stage clotting-time assay, factor VIIa-tissue factor activity remaining from the first stage activates factor X in the second stage, and the amount of factor Xa produced is determined by a bioassay in the third stage.

These assays can be used to detect from 1 to 150 ng/ml equivalent TFPI activity in complex mixtures and can be used to monitor TFPI activity throughout its purification. One unit of TFPI activity is defined as that activity present in 1 ml of pooled normal plasma (George King Biomedical, Overland Park, KS).

Method I: Radiometric Assay. A mixture containing factor VII or factor VIIa (0.4 μg/ml), factor X (2 μg/ml), EDTA-washed tissue factor (10 μl/ml),[1] and 8 mM $CaCl_2$ in TBSA (100 mM NaCl, 50 mM Tris-HCl,

[4] W. F. Novotny, S. G. Brown, J. P. Miletich, D. J. Rader, and G. J. Broze, Jr., *Blood* **78**, 387 (1991).
[5] M. Zur and Y. Nemerson, this series, Vol. 80, p. 237.

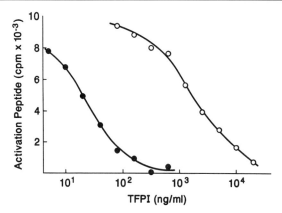

FIG. 1. Dependence of factor Xa for the inhibition of factor VIIa–tissue factor activity by TFPI. TFPI inhibition of factor VIIa–tissue factor activity was measured using the radiometric assay and ^{3}H-labeled factor IX in the absence (O) and presence of factor Xa (●). The x axis is the concentration of TFPI in the test sample.

pH 7.5, 0.1% (w/v) bovine serum albumin) is incubated for 10 min to allow for the activation of factors VII and X. Of this mixture, 50 μl is then added to 25 μl of TFPI test sample (diluted in TBSA) followed by incubation for 30 min at room temperature. Then 25 μl of ^{3}H-labeled factor X or ^{3}H-labeled factor IX[5] [20 μg/ml; 50–100 disintegrations per minute (dpm)/ng] is added and after exactly 10 min of incubation at room temperature, 200 μl of ice-cold 7.5% (w/v) trichloroacetic acid (TCA) is added. The tubes are vortexed, placed in ice for 10 min, and then centrifuged for 5 min (12,000 g) to pellet the TCA-precipitable material (which includes labeled factors X and Xa or IX and IXa). Supernatant fluid (200 μl), which contains the released labeled activation peptide, is placed in scintillation fluid and counted. A standard curve constructed using purified TFPI is linear from 10 to 80 ng/ml TFPI in the test sample when plotted as the log TFPI concentration versus soluble ^{3}H-labeled activation peptide (Fig. 1).

Method II: Three-Stage Clotting-Time Assay. In the first stage, 10 μl of factor VII (1 μg/ml), 10 μl of factor X (10 μg/ml), 10 μl of $CaCl_2$ (40 mM), 20 μl of EDTA-washed tissue factor[1] (5%, v/v), and 50 μl of TFPI test sample in TBSA are mixed and incubated at room temperature. After 30 min, a 10-μl sample is diluted 100-fold into TBSA containing 5 mM $CaCl_2$. In the second stage, 50 μl of this diluted sample, 50 μl factor VIIa (1 μg/ml),[3] 50 μl of factor X (10 μg/ml), and 50 μl of $CaCl_2$ (25 mM) are added to a fibrometer cup and incubated at 37° for exactly 1 min. In the third stage, 50 μl of a mixture containing 10 parts factor X-deficient plasma (George King) and 1 part rabbit brain cephalin (prepared as described by Sigma Chemical Co., St. Louis, MO) is added, and the time to clot formation is determined with a fibrometer (Baltimore Biological Lab-

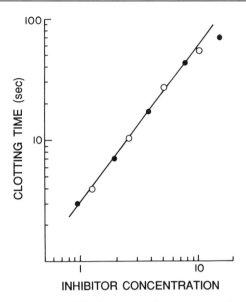

Fig. 2. Comparison of purified TFPI and normal human serum in the three-stage clotting-time assay. Dilutions of normal human serum were assayed in the three-stage clotting-time assay and compared to a standard curve constructed using purified TFPI. The y axis is prolongation of the clotting beyond the control value (39 sec), and the x axis is the final concentration of TFPI in ng/ml (●), or normal human serum in percent (v/v) (O) in the first stage of the assay. Reproduced from Broze and Miletich.[15]

oratory, Cockeysville, MD). In the first stage a mixture containing TBSA rather than test sample serves as a control and the concentration of tissue factor is chosen to produce a control clotting time of 35–40 sec (Fig. 2). Plasma cannot be used directly in this assay as a clot will form in the first stage; however, TFPI in plasma samples can be determined using this assay by converting plasma samples into serum samples.[1]

A similar assay, in which factor Xa generation is detected by a chromogenic substrate assay, has also been described.[6]

TFPI Inhibition of Factor Xa

Two types of assays are routinely employed for measuring TFPI inhibition of factor Xa activity. Samples containing TFPI are incubated with factor Xa and remaining factor Xa activity is determined by its cleavage of a chromogenic substrate[7] or its induction of coagulation in plasma. The

[6] A. H. Pedersen, O. Nordfang, F. Norris, F. C. Wiberg, P. M. Christensen, K. B. Moller, J. Meidal-Pedersen, T. C. Beck, K. Norris, U. Hedner, and W. Kisiel, *J. Biol. Chem.* **265,** 16786 (1990).
[7] R. Lottenberg, U. Christensen, C. M. Jackson, and P. L. Coleman, this series, Vol. 80, p. 341.

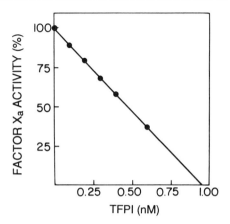

FIG. 3. Stoichiometry of factor Xa:TFPI binding. Bovine factor Xa (1 nM) and full-length rTFPI were incubated for 1 hr at room temperature and the remaining factor Xa activity was determined by the chromogenic assay. Reproduced from Wesselschmidt.[8]

chromogenic assay is used primarily for the generation of progress curves of factor Xa inhibition by TFPI; however, since the stoichiometry of the TFPI:factor Xa is 1:1, a modified chromogenic assay can be used to determine TFPI concentration. These assays are useful for the characterization of purified TFPI and can be used to monitor TFPI activity in the latter stages of purification. Neither assay is useful for the determination of TFPI activity in plasma, serum, or serum-containing culture media due to the presence of other factor Xa inhibitors.

Method I: Chromogenic Assay. For kinetic studies, 100 μl of factor Xa (10 nM) is added to a cuvette containing TFPI sample and 0.1 mM Spectrozyme Xa (methoxycarbonyl-D-cyclohexylglycylglycylarginine-p-nitroanilide acetate, American Diagnostica Inc., Greenwich, CT) in 900 μl of TBSA, and the change in absorbance at 405 nm is monitored spectrophotometrically.

To determine the concentration of TFPI, active site-titrated factor Xa[8] (1.0 nM) and different dilutions of the TFPI test sample are incubated in 900 μl of TBSA in a cuvette for 1 hr at room temperature (to allow for the factor Xa–TFPI complex to form), after which, 100 μl of 1 mM Spectrozyme Xa is added and the initial rate of change in absorbance at 405 nm per minute is determined spectrophotometrically. The level of TFPI in the test sample is determined by plotting the rate of spectrozyme Xa cleavage versus the concentration of test sample and extrapolating the line to the x axis, the point where 1 nM factor Xa is completely inhibited (Fig. 3). For

[8] R. L. Wesselschmidt, K. M. Likert, T. J. Girard, and G. J. Broze, Jr., *Blood* **79**, 2004 (1992).

this application the use of bovine factor Xa is superior to human factor Xa because human TFPI binds bovine factor Xa with 10-fold greater affinity than human factor Xa.[9] Altered recombinant TFPI molecules with very poor affinities for factor Xa cannot be quantitated in this fashion.

Method II: Coagulation Assay. A mixture of 50 μl of rabbit brain cephalin, 50 μl of CaCl$_2$ (25 mM), 50 μl of human factor Xa (1 nM), and 50 μl of TFPI test sample in TBSA is incubated at 37° for 1 min. Then 50 μl of factor X-deficient human plasma is added and the time to clot formation is determined in a fibrometer. The degree of apparent factor Xa inhibition is determined by comparison of the clotting time to a standard concentration curve constructed using purified TFPI (Fig. 4).

Ligand and Western Blots

TFPI binds to and inhibits factor Xa directly.[3] TFPI subjected to nonreducing SDS-PAGE followed by transfer to nitrocellulose can be visualized by probing with ^{125}I-labeled factor Xa or trypsin, whereas other factor Xa inhibitors are denatured by the conditions of SDS-PAGE (Fig. 4, inset). The binding of ^{125}I-labeled factor VIIa to TFPI in the presence of factor Xa, tissue factor, and calcium can also be demonstrated (Fig. 5), although this analysis is not recommended for routine identification of TFPI.

In addition to ligand blot analyses, anti-TFPI polyclonal and monoclonal antibodies can be used to identify TFPI by Western blot analysis. These blotting techniques can be used to follow TFPI purification. Detection of ~20 ng of TFPI antigen in complex mixtures can be accomplished using these techniques; the limitation is in the amount of total protein that can be applied to the gel.

Methods. TFPI samples in solution containing 0.83 M Tris-HCl, pH 6.8, 13.3% (v/v) glycerol, 2.67% (w/v) sodium dodecyl sulfate, and 0.00067% (w/v) bromphenol blue are subjected to SDS-polyacrylamide gel electrophoresis using a 4% stacking gel and a 15% separating gel in the buffer system of Laemmli.[10] Proteins in the gel are then electrophoretically transferred to nitrocellulose in blot buffer[11] (192 mM glycine, 25 mM Tris-HCl, pH 8.3, 20% (v/v) methanol) using a Bio-Rad Mini Trans-Blot cell at 100 V for 1 hr. The nitrocellulose is incubated at room temperature for 30 min in 3% nonfat dried milk containing 0.02% NaN$_3$ to reduce nonspecific binding.

[9] G. J. Broze, Jr., L. A. Warren, W. F. Novotny, K. M. Roesch, and J. P. Miletich, *Blood* **70**, 385a (1987).

[10] U. K. Laemmli, *Nature (London)* **227**, 680 (1970).

[11] H. Towbin, T. Staehelin, and J. Gordon, *Proc. Natl. Acad. Sci. U.S.A.* **76**, 4350 (1979).

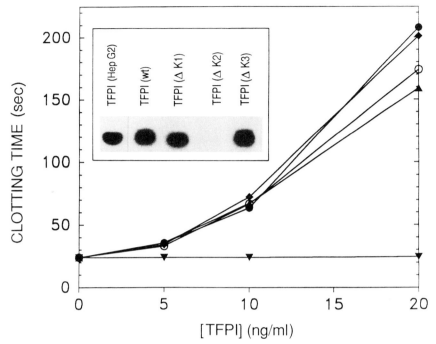

FIG. 4. Binding and inhibition of factor Xa by wild-type and mutant TFPIs. Inhibition of the factor Xa activity was determined using the coagulation assay. Mutagenesis was performed to modify the P_1 residue in the active site cleft of each Kunitz domain to produce the following mutants: TFPI(ΔK1) has Lys-36 changed to Ile, TFPI(ΔK2) has Arg-107 changed to Leu, and TFPI(ΔK3) has Arg-199 changed to leu. TFPI(wt) is recombinant wild-type TFPI. Samples are TFPI(Hep G2) (O), TFPI(wt) (●), TFPI(ΔK1) (▲), TFPI(ΔK2) (▼), and TFPI(ΔK3) (◆). The x axis is the concentration of TFPI in the test sample. Inset: ^{125}I-Labeled factor Xa blot analysis of TFPI mutants. Western analysis using rabbit polyclonal anti-TFPI antibodies was performed on a duplicate blot to confirm that equivalent amounts of TFPI were present in each sample (data not shown). Reprinted by permission from *Nature* **338,** 519, copyright 1989, Macmillan Magazines Ltd.

For factor Xa ligand blots, ^{125}I-labeled factor Xa (~10^3 dpm/ng; 10^5 dpm/ml) in TBSA is added, followed by overnight incubation at room temperature with gentle agitation. The next morning the nitrocellulose is washed three times with TBSA for 15 min per wash, air dried, and autoradiography is performed to detect bound ^{125}I-labeled factor Xa.

Factor VIIa blots can be performed similarly, except that the blot is incubated overnight in TBSA containing ^{125}I-labeled factor VIIa (~10^3 dpm/ng; 10^5 dpm/ml), factor Xa (100 ng/ml), EDTA-washed tissue factor (5% v/v),[1] and $CaCl_2$ (5 mM). Western blot techniques, using either

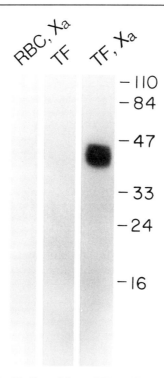

FIG. 5. Requirements for binding of factor VIIa to TFPI. Purified TFPI(Hep G2) was subjected to SDS-PAGE followed by transfer to nitrocellulose. Ligand blot analysis was performed in 0.1 M NaCl, 0.05 M Tris-HCl, pH 7.5, containing 0.1% bovine serum albumin, 5 mM CaCl$_2$, ^{125}I-labeled factor VIIa, and the indicated additions. RBC, Rabbit brain cephalin; Xa, factor Xa; TF, tissue factor. TF contains phospholipids; RBCs, which contain phopholipids but no tissue factor, serve as a control.

mouse monoclonal antibody to TFPI or rabbit polyclonal antibodies to TFPI as probe, followed by the appropriate secondary antibody–alkaline phosphatase conjugate and colorimetric development,[12] may also be used to detect TFPI in complex mixtures.

Purification

Due to its low concentration and its association with lipoproteins, the purification of TFPI from plasma is difficult.[13] The isolation of the TFPI

[12] M. L. Billingsley, K. R. Pennypacker, C. G. Hoover, and R. L. Kincaid, *BioTechniques* **5**, 22 (1987).
[13] W. F. Novotny, T. J. Girard, J. P. Miletich, and G. J. Broze, Jr., *J. Biol. Chem.* **264**, 18832 (1989).

released into plasma following heparin infusion is easier because this form of TFPI binds tightly to heparin and can be separated from the bulk of plasma proteins by heparin–agarose chromatography.[14] The use of this procedure, however, is limited by the need to infuse several volunteers with heparin to obtain sufficient starting material.

The human hepatoma cell line Hep G2 grows in serum-free media and secretes TFPI into the culture media; TFPI was initially purified from Hep G2 culture media.[15] The procedure involves precipitating TFPI from the conditioned media with $CdCl_2$, followed by bovine factor Xa affinity chromatography and gel filtration.[16] With the generation of monoclonal antibodies to TFPI, immunoaffinity chromatography can replace factor Xa affinity chromatography.

Cloning of TFPI cDNAs has allowed the expression of recombinant TFPI (rTFPI) in a number of different types of transfected cells[6,17] and rTFPI can also be purified using this purification scheme.

TFPI Purification

Growth of Hep G2 Cells. Hep G2 cells are grown in roller bottles at 37° in Hams F12 media supplemented with 0.5% (w/v) lactalbumin, 25 mM HEPES, pH 7.0, 2 mM L-glutamine, 1 mM sodium pyruvate, 1× nonessential amino acids solution (#320-1140AG, GIBCO, Grand Island, NY), 1× insulin–transferrin–sodium selenite media supplement (#I1884, Sigma), 1× antibiotic–antimycotic (#600-5240AG, GIBCO), and 20 μg/liter liver cell growth factor (#7387, Sigma). Every third or fourth day conditioned media are harvested and cells are refed with fresh media. Cells cultured in this fashion have been maintained for greater than 2 years. Harvested conditioned media are adjusted to 0.02% (w/v) sodium azide, 200 nM phenylmethylsulfonyl fluoride (PMSF), and 22 trypsin inhibitor units/liter aprotinin (#A-6012, Sigma) and stored at 4°.

Cadmium Chloride Precipitation. From 4 to 15 liters of conditioned media are clarified by centrifugation at 2500 g for 30 min. The supernatant fluid is adjusted to 10 mM $CdCl_2$ by the dropwise addition of 1 M $CdCl_2$ over a 5-min period, with stirring. Stirring is continued for 30 min, after which the precipitated material is collected by centrifugation at 10,000 g for 20 min. The supernatant fluid is decanted and the pelleted material is

[14] W. F. Novotny, M. Palmier, T.-C. Wun, G. J. Broze, Jr., and J. P. Miletich, *Blood* **78**, 394 (1991).
[15] G. J. Broze, Jr. and J. P. Miletich, *Proc. Natl. Acad. Sci. U.S.A.* **84**, 1886 (1987).
[16] G. J. Broze, Jr., L. A. Warren, J. J. Girard, and J. P. Miletich, *Thromb. Res.* **48**, 253 (1987).
[17] T. J. Girard, L. A. Warren, W. F. Novotny, J. P. Miletich, and G. J. Broze, Jr., *Thromb. Res.* **55**, 37 (1989).

TABLE I
PURIFICATION OF TFPI (Hep G2)[a]

Step	Protein (mg)	Activity (units)[b]	Specific activity (units/mg)	Yield (%)	Purification (-fold)
Serum-free media	14,800	2310	0.156	100	1
CdCl$_2$ precipitation and elution	290	2125	7.33	92	47
Bovine factor Xa Affi-Gel	0.16	1410	8810	61	56,500
Superose 12	0.11	1160	10,500	50	67,300

[a] Reprinted with permission from Ref. 16, copyright 1986, Pergamon Press plc.

[b] One unit is defined as that amount present in 1 ml of pooled normal human plasma as determined in the three-stage clotting-time assay.

resuspended in 150 ml 0.5 M EDTA, pH 9.5, with stirring. The resulting suspension is clarified by centrifugation at 10,000 g for 20 min. TFPI is further purified using either factor Xa affinity or immunoaffinity chromatography.

Bovine Factor Xa Affi-Gel Chromatography. The CdCl$_2$ eluate is diluted with an equal volume of TS (100 mM NaCl, 50 mM Tris-HCl, pH 7.5) containing 1% (w/v) Lubrol and applied at a rate of 15 ml/hr to a 5-ml Sepharose CL-4B guard column connected in series to a 5-ml Affi-Gel 10 column containing bovine factor Xa linked at 1 mg/ml[16] and equilibrated in TS containing 0.5% (w/v) Lubrol. The guard column is removed and the factor Xa affinity column is washed at 3 ml/hr with 20 volumes of TS containing 2% (w/v) Lubrol, followed by 4 volumes of TS containing 1% (w/v) octylglucoside. TFPI is eluted with 0.5 M benzamidine in TS containing 1% (w/v) octylglucoside.

Superose 12 Chromatography. The fractions eluted from the factor Xa affinity chromatography column containing TFPI are pooled, concentrated to 1 ml (Amicon YM10, Beverly, MA), and applied to two 30 × 0.6-cm Superose 12 columns (Pharmacia, Piscataway, NJ) linked in series and equilibrated in 1 M NaSCN, 1% octylglucoside. TFPI-containing fractions are pooled, concentrated (Amicon YM10), and stored at −70°.

Table I summarizes the purification of TFPI from 4 liters of Hep G2 conditioned media. The final product contains two predominant forms of TFPI: one is full-length, whereas the other is proteolytically truncated at the carboxy terminus. These two forms of TFPI can be separated by heparin–agarose[8] or cation-exchange chromatography. The carboxy truncation of TFPI occurs in the conditioned media of many types of cultured cells, but the responsible enzyme(s) has not yet been identified; although

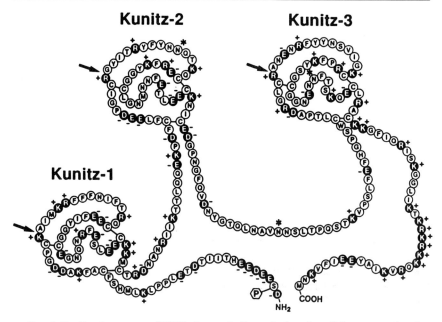

FIG. 6. Predicted structure of TFPI. Arrows indicate the location of the presumed active site inhibitor clefts ($P_1 - P_1'$) for each Kunitz-type domain. The charges of the amino acid side chains are shown (histidine side chains are considered as unchanged). Asterisks indicate the potential sites for N-linked glycosylation and P represents the site of partial phosphorylation. Reprinted by permission from *Nature* **338**, 518, copyright 1989, Macmillan Magazines Ltd.

the inclusion of serum in the culture media retards this process, it also complicates the purification procedure.

Characterization

Physical Properties

Based on cDNA sequences, mature TFPI contains 276 amino acids (32 kDa), whereas, by SDS-PAGE analysis, full-length TFPI has an apparent molecular mass of 42 kDa. The mature protein contains a negatively charged amino terminus, followed by three Kunitz-type protease inhibitory domains and a positively charged carboxy terminus. N-Linked glycosylation occurs at one or more of three potential sites (Fig. 6).[13] Other identified covalent modifications of TFPI isolated from cell cultures include the partial phosphorylation of serine residue 2, apparently by the action of casein kinase II,[18] and the sulfation of N-linked carbohydrate in TFPI expressed by endothelial cells and kidney cells (A293, ATCC CRL 1573; Caki, ATCC HTB46[19]). The role of these posttranslational modifications of the TFPI molecule in its physiologic function is not known.

Funtional Properties

TFPI Inhibition of Factor Xa. TFPI produces direct inhibition of factor Xa, by binding at or near the serine active site in factor Xa with 1:1 stoichiometry (Fig. 3). It is also a potent inhibitor of trypsin.

The interaction between TFPI and factor Xa demonstrates "slow," "tight-binding" kinetics, which appears to conform to the same mechanism described for other Kunitz-type protease inhibitors:

$$E + I \underset{k_2}{\overset{k_1}{\rightleftharpoons}} EI \underset{k_4}{\overset{k_3}{\rightleftharpoons}} EI^* \qquad (1)$$

where E is enzyme, I is inhibitor, EI is the initial collision complex with a K_i (initial) $= k_2/k_1$, and EI* is the final complex which develops "slowly" from EI and is of higher affinity.[20,21] The K_i(final) of the final EI* complex equals $K_i[k_4/(k_3 + k_4)]$. "Tight-binding" inhibitors produce significant inhibition at concentrations near that of the enzyme being inhibited, making the determination of kinetic constants technically difficult. However, specific techniques are available for appropriate kinetic analysis.[22,23]

The second Kunitz domain of TFPI is ultimately responsible for factor Xa inhibition[24] (Fig. 4), but other regions of the TFPI molecule participate as well. The basic, carboxy-terminal domain of TFPI is required for high-affinity binding of factor Xa in the immediate, encounter complex EI. Thus, the carboxy-terminal truncated forms of TFPI isolated from the conditioned media of cultured cells are considerably less potent inhibitors of factor Xa than full-length TFPI. In the absence of calcium, heparin enhances the inhibition of factor Xa by carboxy-truncated forms of TFPI, whereas it modestly diminishes factor Xa inhibition by full-length TFPI.[8] However, in the presence of calcium, heparin enhances the inhibition of

[18] T. J. Girard, D. McCourt, W. F. Novotny, L. A. MacPhail, K. M. Likert, and G. J. Broze, Jr., *Biochem. J.* **270**, 621 (1990).

[19] P. L. Smith, T. P. Skelton, D. Fiete, S. M. Dharmesh, M. C. Beranek, L. MacPhail, G. J. Broze, Jr., and J. U. Baenziger, *J. Biol. Chem.* **267**, 19140 (1992).

[20] J. F. Morrison, *Trends Biochem. Sci.* **7**, 102 (1982).

[21] E. Antonini, P. Ascenzi, E. Menegatti, and M. Guarneri, *Biopolymers* **22**, 363 (1983).

[22] J. F. Morrison and C. T. Walsh, *Adv. Enzymol.* **61**, 201 (1988).

[23] C. Longstaff and P. J. Gaffney, *Biochemistry* **30**, 979 (1991).

[24] T. J. Girard, L. A. Warren, W. F. Novotny, K. M. Likert, S. G. Brown, J. P. Miletich, and G. J. Broze, Jr., *Nature (London)* **338**, 518 (1989).

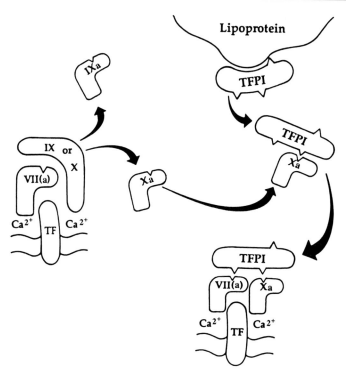

FIG. 7. Proposed mechanism for the inhibition of the factor VIIa–tissue factor complex by TFPI. VII(a) denotes either factor VII or factor VIIa. Tissue factor apoprotein is shown spanning a phospholipid membrane. The indentations represent the active sites for factor VIIa and factor Xa and the protrusions represent the three Kunitz-type domains of TFPI. In the factor Xa–TFPI complex, the active site of factor Xa is bound to the second Kunitz domain of TFPI. In the final quaternary factor Xa–TFPI–factor VIIa–tissue factor complex, factor Xa is bound at its active site to Kunitz-2 and factor VIIa is bound at its active site to Kunitz-1. Reprinted by permission from *Nature* **338**, 519, copyright 1989, Macmillan Magazines Ltd.

factor Xa by both full-length and carboxy-truncated TFPI (G. J. Broze, Jr., unpublished observations). The proteolytic cleavage of TFPI between Kunitz domains 1 and 2 induced by human leukocyte elastase reduces the inhibitory effect of TFPI against factor Xa, but not against trypsin.[25]

TFPI Inhibition of VIIa–Tissue Factor Activities. Factor Xa-dependent inhibition of factor VIIa–tissue factor by TFPI appears to involve the formation of a quaternary factor Xa–TFPI–factor VIIa–tissue factor complex (Fig. 7). This inhibitory complex could arise through the binding

[25] D. A. Higuchi, T.-C. Wun, K. M. Likert, and G. J. Broze, Jr., *Blood* **79**, 1712 (1992).

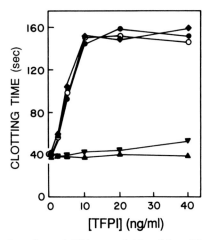

FIG. 8. Factor VIIa–tissue factor inhibitory activities of the wild-type and mutant TFPIs. Three-stage clotting-time assays were performed with TFPI(Hep G2) (O), TFPI(wt) (●), TFPI(ΔK1) (▲), TFPI(ΔK2) (▼), and TFPI(ΔK3) (◆) (see Fig. 4). The x axis is the final concentration of TFPI in the first stage of the assay. Reprinted by permission from *Nature* **338**, 519, copyright 1989, Macmillan Magazines Ltd.

of TFPI to a preformed factor Xa–factor VIIa–tissue factor complex, or by TFPI first binding factor Xa with subsequent binding of the factor Xa–TFPI complex to factor VIIa–tissue factor. The factor Xa requirement for factor VIIa–tissue factor inhibition by TFPI is not absolute, and high concentrations of TFPI have been shown to inhibit the activity of factor VIIa–tissue factor directly[6] (Fig. 1).

Studies using altered TFPI molecules in which the P_1 residue at the active site cleft of each Kunitz domain has been individually changed show that both Kunitz domains 1 and 2 are required for the inhibition of factor VIIa–tissue factor activity[24] (Fig. 8). Presumably in the quaternary inhibitory complex, the active site of factor Xa is bound by Kunitz-2 and the active site of factor VIIa is bound by Kunitz-1. A chimeric molecule containing the γ-carboxyglutamic acid domain of factor Xa and the Kunitz-1 domain of TFPI inhibits factor VIIa–tissue factor directly,[26] suggesting that the binding of factor Xa to TFPI serves to juxtapose these domains on separate molecules, thereby producing a complex with high affinity for factor VIIa–tissue factor. Alteration of the active site residue of the third Kunitz domain has no significant effect on either factor Xa or factor VIIa–tissue factor inhibition by TFPI, and a physiologic role for this domain has not yet been identified.

[26] T. J. Girard, L. A. MacPhail, K. M. Likert, W. F. Novotny, J. P. Miletich, and G. J. Broze, Jr., *Science* **248**, 1421 (1990).

Plasma TFPI

Plasma TFPI concentrations in normal individuals range from 60 to 180 ng/ml with a mean of 113 ng/ml.[4] When analyzed by SDS-PAGE, plasma TFPI consists of multiple forms, with the predominant proteins of M_r 34,000 and 40,000 and less abundant bands of higher apparent molecular weights. Western blot analysis indicates the 40,000 molecular weight and higher molecular weight forms contain apolipoprotein AII disulfide bonded to TFPI.[13] The discrepancy in apparent molecular weights between TFPI isolated from the conditioned media of cultured cells (42,000) and plasma TFPI (34,000) requires further evaluation.

Greater than 90% of the TFPI in the plasma of normal individuals circulates associated with lipoproteins (LDL > HDL ≫ VLDL).[1,27] Platelets also contain TFPI, which is released following stimulation with thrombin or the calcium ionophore A23187.[28] Administration of heparin *in vivo* increases the concentration of TFPI in plasma two- to fourfold[4,14,29] and appears to involve the release of TFPI from intra- or extracellular stores. Which cell type is responsible for the maintenance of the plasma levels of TFPI is not known, although the endothelium is an attractive candidate.

[27] N. L. Sanders, S. P. Bajaj, A. Zivelin, and S. I. Rapaport, *Blood* **66**, 204 (1985).
[28] W. F. Novotny, T. J. Girard, J. P. Miletich, and G. J. Broze, Jr., *Blood* **72**, 2020 (1988).
[29] P. M. Sandset, U. Abildgaard, and M. L. Larsen, *Thromb. Res.* **50**, 803 (1988).

[12] Mutational Analysis of Receptor and Cofactor Function of Tissue Factor

By WOLFRAM RUF, DAVID J. MILES, ALNAWAZ REHEMTULLA, and THOMAS S. EDGINGTON

Introduction

There are fundamentally important reasons to elucidate the structural basis of protein function. This includes an understanding of the rules of complex protein–protein assemblies. One successful strategy derives from the ability to modify efficiently the structure of proteins by site-directed exchange and deletional mutagenesis and to identify functional defects resulting from such modifications. This chapter outlines strategies and specific methods to analyze the structure–function relationships of the integral membrane glycoprotein tissue factor (TF). TF serves as a cellular receptor and enzymatic cofactor for the latent serine protease factor VIIa (VIIa) and its zymogen factor VII (VII). High-affinity binding of VIIa and VII and the support of efficient catalysis of substrates by the binary complex of TF and VIIa (TF · VIIa) are two discrete functional characteristics of TF that merit elucidation. Analytical methods will be described to detect and to analyze in detail the specific functional defects resulting from rational mutational change.

Expression of Recombinant Human Tissue Factor in Mammalian Cells

A number of considerations influence the choice of an expression system for mutational analysis. Lack of efficient disulfide bond formation and of typical posttranslational modification may result in a protein with reduced specific functional activity when a prokaryotic expression system is used. To minimize this potentially confounding factor, we have preferred eukaryotic expression systems for TF. Expression systems in yeast, in various mammalian cell lines, and in insect cells using baculovirus vectors are commonly adopted. Glycoproteins expressed in yeast and insect cells typically lack the same complex carbohydrate composition obtained from expression in mammalian cells; however, these expression systems have some advantages for protein structural studies. Because the emphasis is on the functional analysis of a membrane-anchored protein, mutational analysis mandates strategies that permit efficient screening. Both yeast and insect cell expression are incompatible with analysis of the protein on a viable cell because of cell lysis associated with high-level expression in insect cells, and because of the cell wall in yeast, which limits access to an integral membrane surface receptor such as TF. In contrast, mammalian cells can be transfected to express stably on their cell surface recombinant proteins, providing means to obtain a cell line with a quantified number of cell surface receptors. This allows efficient analysis of the receptor function in a physiological environment. Transient expression in mammalian cells further allows efficient and rapid screening for functional loss resulting from introduced mutations without time-consuming cloning and selection strategies, as for the insect cell expression system. Transient expression for initial screening followed by stable expression for detailed analysis has been adopted by us for structure–function analysis of TF.

Transient Expression of Tissue Factor

The DNA for the complete TF coding sequence, including the leader sequence, is cloned into the expression vector cDM8 to place transcription under the control of the strong cytomegalovirus promoter.[1] Plasmid DNA is isolated from transformed *Escherichia coli* MC1061 and purified by CsCl density gradient centrifugation. Mammalian cell lines that do not constitutively express TF are used for transfection. Chinese hamster ovary cells (CHO-Kl) are suitable in this respect. These cells are transfected using the calcium phosphate precipitation method,[2] resulting in a 10 to 20% transfection efficiency. After transfection, cells are grown for 60–72 hr in complete medium followed by detachment from the flasks with 10 mM EDTA in phosphate-buffered saline (PBS). After repeated washes in PBS, cells are counted, aliquoted, and pelleted, and the buffer is removed prior to storage at −70° until analysis.

Generation of Stable Cell Lines

Stable cell lines are generated when a dysfunctional mutant of interest is identified from analysis by transient expression. This initial screening also establishes that the mutant protein can be synthesized and expressed by the cells. Using the calcium phosphate transfection method,[2] the plasmid DNA and a neomycin resistance plasmid (e.g., pMAMneo) are cotransfected at a 20:1 ratio. The limiting concentration of the selection plasmid ensures that the majority of cells which incorporate the selection marker are also transfected with the plasmid DNA. One day after transfection of a 75-cm² flask, the confluent cells are detached using trypsin–EDTA and seeded into eight 100-mm petri dishes. Cells are grown in the presence of 600 μg/ml G418 (Geneticin, GIBCO-BRL, Grand Island, NY) in complete medium [Dulbecco's modified Eagle's medium (DMEM), 10% newborn calf serum, 0.1 mM proline] until individual colonies are detectable (approximately 2 to 3 weeks). During the selection period, the medium containing G418 is usually replaced once. Visible single colonies (~1–3 mm) are harvested from the petri dishes, transferred to 24-well plates, and passaged in G418 selection medium for 2 to 3 weeks. Homogeneity of the cell line with respect to the level of TF expression is then determined by flow cytometry. In case of heterogeneity with respect to TF expression, single-cell cloning or, alternatively, fluorescence-activated cell sorting (FACS) may be used to increase homogeneity and expression levels of TF.[3]

For functional analysis, cell lines stably expressing approximately 0.25 to 1 × 10⁶ molecules per cell are preferred. Higher expression levels result in reduced specific functional activity. This may result from limitations in

[1] B. Seed, *Nature (London)* **329,** 840 (1987).
[2] R. E. Kingston, *in* "Current Protocols in Molecular Biology" (F. M. Ausubel, ed.), p. 9.1.1. Wiley, New York, 1987.
[3] A. Rehemtulla, W. Ruf, and T. S. Edgington, *J. Biol. Chem.* **266,** 10294 (1991).

TABLE I
EXPRESSION OF TF AND TF(1-219) IN CHO AND Sf9 CELLS

Protein	Yield from CHO cells (mg/liter of culture)	Yield from Sf9 cells (mg/liter of culture)
Full-length TF	3-5	12-18
TF(1-219)	0.7-1	0.1-1.2

phospholipid binding sites for factor X,[4] although other possibilities have not been excluded. Production for the purpose of isolation of purified recombinant TF requires higher levels of cellular expression. Repeated sorting cycles using flow cytometry may be used to select for high levels of expression.[5] However, infection with mycoplasma is a frequent problem associated with repeated cell sorting, and both the levels of expression of the recombinant protein and the growth rate of cells infected with mycoplasma are significantly reduced. As an alternative strategy, expression in insect cells is effective. Because insect cell expression employs a transient viral infection, higher yield may be achieved compared to stable cell lines, which demonstrate significant growth inhibition and instability at very high levels of cell surface expression of recombinant TF. The expression of TF in insect cells surpasses the yield from mammalian cells (Table I). Compared to full-length wild-type TF, the soluble TF extracellular domain truncation mutant [TF(1-219)][4] was less efficiently expressed by both CHO cells and Sf9 insect cells (Table I), indicating that a high-yield expression system for a membrane glycoprotein may not be as suitable for a soluble truncated mutant of the same protein and that removal of domains which govern cellular traffic may result in inefficient expression of recombinant protein.

Mutagenesis of TF

Mutational strategies can be considered as random when performed as a search or as rational when guided by a hypothesized structural model or by experimental data which define an area of interest in the target protein. Mutational changes include domain deletion, domain exchange to or from a homologous protein, mutational elimination of disulfide bonds, conservative and nonconservative substitutions of specific residues, including ala-

[4] W. Ruf, A. Rehemtulla, J. H. Morrissey, and T. S. Edgington, *J. Biol. Chem.* **266**, 2158 (1991).
[5] A. Rehemtulla, M. Pepe, and T. S. Edgington, *Thromb. Haemostasis* **65**, 521 (1991).

TABLE II
SITE-DIRECTED MUTAGENESIS OF TF CODING SEQUENCE

1. Design and synthesize oligonucleotide
   Change codon to introduce desired mutation
   Incorporate new restriction enzyme recognition site by making mutations that do not change the translational sense of the gene
   Include at least eight nucleotides of wild-type sequence at the 5′ and 3′ ends of the oligonucleotide
2. Isolate uracil-incorporated single-stranded plasmid DNA
   Transform *E. coli* strain CJ236/P3 that has the *ung⁻ dut⁻* genotype with a plasmid containing the TF coding sequence
   Infect culture of transformed cells with M13 helper phage to produce single-stranded plasmid DNA with uracil incorporated
   Purify single-stranded DNA from phage precipitate
3. Anneal mutagenic oligonucleotide to single-stranded DNA
   Incubate (in 10 μl) 50 ng of phosphorylated oligonucleotide and 200–500 ng of single-stranded DNA in 20 mM Tris, 2 mM MgCl$_2$, 50 mM NaCl, pH 7.4, for 2 min in heating block at 72°
   Transfer heating block with samples to bench top and allow to cool slowly to room temperature (approximately 1 hr)
4. Synthesize second DNA strand using mutagenic oligonucleotide as primer
   Add 1 μl unmodified T7 DNA polymerase (0.5 U), 1 μl T4 DNA ligase (3–4 U), and 1 μl deoxynucleoside triphosphates (5 mM), ATP (10 mM) in 100 mM Tris, 50 mM MgCl$_2$, 20 mM dithiothreitol (DTT), pH 7.4
   Incubate 5 min on ice, 5 min at room temperature, and then 90 min at 37° followed by quenching with 0.3 μl of 0.5 M EDTA, pH 8.0
5. Identify mutants by restriction enzyme analysis
   Transform *E. coli* strain MC1061/P3 with product of second-strand synthesis reaction. This strand has an amber mutation in the P3 plasmid. The *supF* gene in CDM8 will complement the mutation and generate resistance for ampicillin and tetracycline encoded by the P3 plasmid
   Perform a small-scale plasmid DNA isolation procedure on individual MC1061/P3 transformants and screen plasmid isolations by restriction analysis to identify mutations in the coding sequence
   Purify plasmid DNA by CsCl density centrifugation for transfection experiments and for DNA sequencing to confirm that the mutation of interest was made

nine exchange, and random scanning nucleotide substitutions. For the mutational exchange of a single or several amino acid residues in TF, oligonucleotide-directed mutagenesis according to Kunkel[6] is used (protocol outline in Table II). Mutants are generated directly in the expression plasmid containing the DNA insert for the TF coding sequence. Oligonucleotides are designed to introduce both the desired mutation and one or more new restriction sites into the coding sequence. Restriction enzymes

[6] T. A. Kunkel, *Proc. Natl. Acad. Sci. U.S.A.* **82,** 488 (1985).

that are rare cutters in the plasmid, including the coding sequence, are preferred.

The quality of the single-stranded template is particularly important for efficient mutagenesis. The *E. coli* strain CJ236/P3 is used as the host strain for the generation of single-stranded DNA. Growth of this strain in chloramphenicol ensures maintenance of the F' encoding plasmid which is required for the infection with helper phage. For high yield of phage, bacteria are grown in rich medium (e.g., TYP broth) and infected with helper phage at a multiplicity of infection (MOI) of 10. Bacteria are separated from the phage by centrifugation (twice at 12000 *g*). Single-stranded template is prepared using standard protocols of polyethylene glycol (PEG) precipitation and organic extraction. Small oligonucleotide contaminants may act as random primers during the mutagenesis reactions, reducing the efficiency of mutant generation and isolation. Incubation with high salt buffers (300 m*M* NaCl) and treatment with RNase A significantly reduce these contaminants and improve the quality of the preparation. The yield of single-stranded plasmid template relative to single-stranded M13 DNA can be assessed by agarose gel electrophoresis and can serve as a control for the quality of the template preparation. Purified single-stranded DNA is stored at $-70°$ or is prepared in small batches freshly for the mutagenesis experiments.

The phosphorylation of the oligonucleotide, annealing, and second-strand synthesis must be executed with care using the procedure indicated in Table II. The buffers and enzymes for these reactions are commercially available (e.g., Bio-Rad, Richmond CA; United States Biochemical Corporation, Cleveland, OH). Half of the synthesis reaction is used for transformation of MC1061/P3 and several hundred colonies can be obtained. Out of the 10 to 30 colonies initially screened, 80% or more demonstrate the desired mutation when mutagenesis is optimally performed.

Analysis of TF Mutants

TF Immunoassay

Analysis of TF mutants involves an initial screening which aims to establish an estimate of specific functional activity. This requires reproducible and accurate analysis of the concentration of TF protein by immunoassay and the TF functional activity in the test samples. TF concentration is determined by an enzyme-linked immunosorbent assay (ELISA) using two monoclonal antibodies to nonoverlapping epitopes[7] in the pro-

---

[7] W. Ruf, A. Rehemtulla, and T. S. Edgington, *Biochem. J.* **278,** 729 (1991).

TABLE III
TWO-SITE ELISA FOR TISSUE FACTOR

Coat ELISA plate with 100 μl of capture MAb or affinity-purified polyclonal antibody (5 μg/ml) overnight at 4°

Lyse 3 × 10⁶ cells stored frozen as a pellet in 300 μl 6 mM CHAPS in Tris-buffered saline (TBS)

Block plate with 5% nonfat dry milk in TBS for 1 hr; wash 3× with TBS

Add to replicate wells either sample or TF standards (purified TF stored in detergent and diluted to 0.5 to 200 ng/ml), both diluted in 0.03% Tween 80 in TBS. Samples in 6 mM CHAPS should be diluted at least 1:6 in sample dilution buffer and several serial 1:2 dilutions of the sample are analyzed

Wash plate with 0.03% Tween 80 in TBS and add secondary antibody (biotinylated or peroxidase-conjugated MAb) at 0.5 to 2 μg/ml. With biotinylated MAb, an additional incubation step with streptavidin-conjugated enzyme needs to be performed

After extensive washes, an appropriate color reaction is used to detect bound secondary antibody, such as 4-nitrophenyl phosphate (1 mg/ml) in 0.1 M Tris, 0.1 M NaCl, 5 mM MgCl$_2$, pH 9.5, for the detection of bound alkaline phosphatase

---

tocol given in Table III. An accurate immunoassay of mutant proteins requires that both monoclonal antibody (MAb) epitopes are equally expressed on the mutant and the wild-type protein. Loss of MAb reactivity is readily recognized, because antigen is not detected by ELISA with the standard reagent MAbs. However, subtle local modification of an epitope may lead to an underestimate of TF concentration, which can be evaluated using different pairs of MAbs or by the following assay modification using polyclonal antibody. Affinity-purified, polyclonal antibody against TF is substituted as a capture antibody, and bound TF is detected by a series of different MAbs as secondary detection antibodies. This procedure not only provides an approach to accurate determination of TF concentration by ELISA, but also identifies which MAb epitopes are lost or modified in the mutant. Repeated undetectable expression of a mutant may be an indication for altered cellular processing and lack of proper protein folding. This is further supported by Western blot analysis of the cell lysate using polyclonal antibodies against TF, which frequently demonstrates traces of protein with an apparent lack of the complete glycosylation pattern typical for wild-type TF or structurally unaltered mutants.

Functional Clotting Assay

To determine functional activity of a mutant TF that has been transiently expressed, a clotting assay is used, based on the assumption that the rapid assembly of TF with VII/VIIa followed by the activation of substrate in the clotting assay provides a measure for both the affinity of the TF–VII interaction and the substrate recognition and efficiency of specific peptide bond hydrolysis by TF · VIIa. Cell pellets (1×10^6 cells), which are stored at $-70°$, are resuspended by the addition of 500 μl of 15 mM β-D-octylglucopyranoside in HEPES-buffered saline (HBS), pH 7.4, and vortexed for 1 min. Cells are further lysed for 12–18 min at $37°$ followed by a threefold dilution with HBS.[8] The cell lysate is transferred to ice until analysis, which should be carried out rapidly. A 30-min incubation on ice results in a 20% or more reduction in functional activity. TF function is analyzed in a one-stage clotting assay by mixing equal volumes (100 μl) of sample and citrated, pooled normal plasma in polystyrene tubes, equilibrating the mixture at $37°$ for 1 min and recording the clotting time after addition of 100 μl of 20 mM $CaCl_2$. Phospholipid-reconstituted purified natural or recombinant TF is used to establish a calibration curve from 0.5 pg/ml to 50 ng/ml. TF is reconstituted into phosphatidylserine/phosphatidylcholine vesicles (30/70, w/w) using detergent solubilization and dialysis. Rigorous acetone (80%, v/v, in water) washes (several times with 30-min incubations at $-20°$) of the lyophilized TF is used to remove residual detergent used during the purification. The protein is taken up in phospholipid solubilized with 0.25% deoxycholate in water followed by addition of $CdCl_2$ to 2.5 mM and a 10-fold dilution with HBS (140 mM NaCl, 10 mM HEPES, pH 6.0) to yield a TF concentration of approximately 100 nM with a 1:200 (w/w) protein:phospholipid ratio. After 1 hr at $37°$ the detergent is slowly removed by dialysis against HBS. Approximately 5 ng/ml[3] reconstituted TF in the sample yields a 50-sec clotting time and the TF functional activity corresponding to this clotting time is defined as 1 U/ml. A calibration curve of clotting time in seconds versus units per milliliter is established on a double-logarithmic scale. The functional activity of TF mutants is derived from this calibration curve and the specific activity is calculated based on the antigen determination by ELISA. The specific activity determined from 10 transient expression experiments was 465 mU/ng for wild-type TF, consistent with the activity (485 ± 50 mU/ng, $n = 7$) of a stable wild-type TF cell line which is subjected to the same lysis procedure as an internal standard.

Detailed Characterization of TF Mutants

Quantitation of TF Cell Surface Expression by Flow Cytometry

Cell lines which stably express TF mutants are analyzed by quantitative fluorescence-activated cell sorting using a set of standardized MAbs. This

[8] B. P. Tsao, D. S. Fair, L. K. Curtiss, and T. S. Edgington, *J. Exp. Med.* **159**, 1042 (1984).

TABLE IV

ANALYSIS OF TISSUE FACTOR EXPRESSION BY FLUORESCENCE-ACTIVATED CELL SORTING

Harvest 10^5 to 10^6 CHO cells stably expressing TF with ice-cold 10 mM EDTA in PBS. Wash 2 times in ice-cold PBS by repeated sedimentation and resuspension in fresh buffer. Resuspend in ice-cold PBS and divide into 4-ml tubes, one for each antibody reaction, including one nonspecific antibody

Centrifuge briefly to pellet the cells

For staining with primary antibody, resuspend the pelleted cells in 50 μl of DMEM with 10 to 30 μg/ml antibody specific for TF or a nonspecific antibody control of the same isotype. Hold on ice for 30 min, vortexing after 15 min

Add 200 μl of DMEM and then carefully add 200 μl of bovine calf serum as a cushion below the cell suspension. Centrifuge for 5 min at 500 g (4°) to pellet the cells through the serum cushion away from the antibody in the DMEM

Aspirate the supernatant and retain cell pellet for staining with fluorescein isothiocyanate (FITC)-labeled secondary antibody

Resuspend the cells in 50 μl of goat antimouse IgG labeled with FITC (20 to 50 μg/ml). Hold on ice for 30 min, vortexing once

Separate cells through a serum cushion as above

Aspirate and discard the supernatant. Resuspend cell pellet in 1 ml of ice-cold PBS and hold in the dark until analysis by flow cytometry

---

analysis provides quantitation of cell surface TF expression as well as reasonable evidence for the proper folding and conformation of MAb-defined discontinuous epitopes. A protocol for antibody staining is given in Table IV. Levels of cell surface expression determined by FACS and by radioligand binding analysis on the same cell lines gave concordant results.[3] This supports the utility of flow cytometry as a simple and rapid method to quantify cell surface expression of mutant molecules and to identify regions that may have undergone aberrant protein folding as a result of a mutation.

Receptor Function Evaluation by Radioligand Binding with VII/VIIa

Receptor function of TF mutants is evaluated by radioligand binding analysis with VII or VIIa as ligand. Binding constants for VII and VIIa are similar in various assays, with an approximately twofold lower estimated K_d for VIIa relative to VII.[9,10] High-affinity binding of VII/VIIa to TF is not significantly influenced by phospholipid, and affinities are similar for binding of VIIa to TF expressed on viable cell surfaces,[3] to TF reconsti-

[9] R. Bach, R. Gentry, and Y. Nemerson, *Biochemistry* **25**, 4007 (1986).
[10] W. Ruf, M. W. Kalnik, T. Lund-Hansen, and T. S. Edgington, *J. Biol. Chem.* **266**, 15719 (1991).

TABLE V
ANALYSIS OF FACTOR VII BINDING TO CELL SURFACE TISSUE FACTOR

CHO cells which stably express TF or TF mutants are seeded into 12-well plates 24–72 hr before analysis. There must be a confluent cell monolayer for the binding assay

The cells are gently washed with HBS (10 mM HEPES, 140 mM NaCl, pH 7.4) and the following addition is made after decanting the supernatant: 100 μl of 2% bovine serum albumin (BSA), 20 mM CaCl$_2$ in HBS; 100 μl of ^{125}I-labeled VII/VIIa (0.1 to 200 nM) diluted in HBS; and 200 μl of HBS or competitor in HBS for the determination of nonspecific binding (e.g., 50-fold molar excess of unlabeled VII or inhibitory MAb). Nonspecific binding may be alternatively determined from reactions which are incubated with nontransfected CHO cells seeded in the nonspecific control wells on the same plate with TF positive cells

The cells are incubated for 2 hr at 37° with rotary agitation at 50 rpm on a platform mixer

Aliquots of supernatant are collected to determine free ligand prior to 6 rapid washes of the cell monolayer with ice-cold HBS containing 0.5% BSA, 5 mM CaCl$_2$

Cells are lysed with 500 μl of 200 mM NaOH, 1% (w/v) SDS, 10 mM EDTA at 37°, and the lysate is transferred to a tube for gamma counting

The amount of bound and free ligand for each reaction is determined and Scatchard analysis is performed after correction for nonspecific binding

tuted into phospholipid vesicles,[9] or to phospholipid-free TF solubilized with detergent.[10] On cell surfaces and on negatively charged phospholipid, binding of VII and VIIa is characterized by positive cooperativity. This has been attributed to dimerization of TF on the cell surface,[11] as also indicated by chemical cross-linking.[12] The stoichiometry of the TF·VII complex has been shown by several studies to be 1:1.[3,9,13] In addition to equilibrium binding studies, dissociation analysis may be used to establish unaltered VII/VIIa binding to TF.[14] For this analysis, saturating concentrations of radiolabeled VII/VIIa are preincubated with cell surface TF and dissociation is initiated by dilution jump or by addition of a 100-fold molar excess of competitive inhibitor (e.g., unlabeled VII or inhibitory MAb), both of which prevent reassociation of VIIa that has been dissociated from the complex. The dissociation half-time in the presence of competitive inhibitor is approximately 30 min at 37°, and a slower dissociation rate is observed for dilution jump, consistent with the cooperative binding characteristics of the TF–VII interaction.

Radioligand binding analyses using cell surface-expressed TF (Table V) are more readily performed than the analysis with phospholipid vesicles,

[11] D. S. Fair and M. J. MacDonald, *J. Biol. Chem.* **262**, 11692 (1987).
[12] S. Roy, L. R. Paborsky, and G. A. Vehar, *J. Biol. Chem.* **266**, 4665 (1991).
[13] T. A. Drake, W. Ruf, J. H. Morrissey, and T. S. Edgington, *J. Cell Biol.* **109**, 389 (1989).
[14] W. Ruf, D. J. Miles, A. Rehemtulla, and T. S. Edgington, *J. Biol. Chem.* **267**, 6375 (1992).

TABLE VI

FACTOR VII BINDING ANALYSIS IN ABSENCE OF PHOSPHOLIPID

Coat Falcon microtest III flexible assay plate with 100 μl of purified anti-TF MAb TF9-10H10 at 10 μg/ml overnight at 4°. Only half of the plate is coated with antibody. Half of the wells remain empty for determination of nonspecific binding.

Block both MAb-coated and noncoated wells with 200 μl of 5% nonfat dry milk (Carnation) in TBS after decanting the plate

After three washes with TBS, add the following components to the wells (noncoated wells receive the complete reaction for the determination of nonspecific binding): 25 μl of 2% bovine serum albumin (BSA), 20 mM CaCl$_2$ in TBS; 25 μl of TF or TF mutant (8 nM) solubilized with CHAPS (0.1–5 mM) in TBS; 50 μl of ^{125}I-labeled VII or VIIa (0.05 to 200 nM) diluted in TBS

Incubate for 90 min at 37°

Collect aliquot of supernatant for determination of free ligand and wash plate rapidly 12$\times$ with ice-cold 5 mM CaCl$_2$, 0.5% BSA in TBS

Count ^{125}I in supernatant and in the washed wells and determine binding constants from free and bound ligand by Scatchard analysis. At 2 nM TF 50–100% of the receptor is bound to the capture antibody and the maximal number of sites should be approximately 100–200 fmol/well.

because of the relative ease of separation of bound and free ligand. Alternatively, a cell- and phospholipid-free binding assay provides a rapid alternative to analyze binding constants of purified or partially purified mutant TF (Table VI). Solubilization of TF with 10 mM CHAPS is recommended, because of low nonspecific binding in the presence of this detergent. Nonspecific binding in this assay compares favorably with cell-binding assays, and the cell-free assay may be preferably used when the affinity of VII binding to the mutant TF is considerably reduced. TF and VII are assembled in solution, followed by binding of TF and TF · ^{125}I-labeled VII complexes to the capture antibody, which binds free and complexed TF equally well. At the concentration of TF (2 nM) indicated in Table VI, most of the receptor is bound to the capture antibody and Scatchard analysis can be based on the amount of VII/VIIa specifically bound to the wells and on the free fraction determined from an aliquot of the supernatant.

Catalytic Activity of TF · VIIa via Chromogenic Substrate Hydrolysis

Chromogenic or fluorogenic peptidyl substrates are essential to assess the function of the catalytic site of VIIa independent of the effects of extended substrate recognition necessary for specific proteolysis. In a purified protein system, several substrates more specific for other serine proteases are also found to be suitable to analyze catalytic function of VIIa

TABLE VII
CHROMOGENIC SUBSTRATES FOR ANALYSIS OF FACTOR VIIa CATALYTIC FUNCTION

Substrate	Concentration[c]	Hydrolysis rate by		
		VIIa[a] (mOD/min[d])	TF·VIIa[b] (mOD/min)	No enzyme (mOD/min)
Spectrozyme FXa	2 mM	1.1	45.0	0.0
	1 mM	0.6	23.8	0.0
Pefachrome VIIa	2 mM	11.0	108.0	0.7
	1 mM	6.6	61.3	0.0
S-2288	2 mM	4.9	73.9	0.0
	1 mM	2.2	41.8	0.0

[a] The final concentration of VIIa in a 100-μl reaction was 200 nM.
[b] VIIa at 200 nM and TF(1–219) at 1 μM were present in a typical reaction, as described in the text.
[c] Final substrate concentration in the assay.
[d] Rate increase in absorbance (100 μl reaction).

(commercially available from Novo Nordisk). A kinetic microtiter-plate reader (e.g., Thermomax, Molecular Devices, Menlo Park, CA) is used for monitoring the rate of hydrolysis of the chromogenic substrate. In a typical assay, 25 μl of 2% bovine serum albumin and 20 mM CaCl$_2$ in TBS (20 mM Tris, 140 mM NaCl, pH 7.4) are added to the microtiter wells (e.g., tissue culture 96-well plate) to prevent nonspecific adsorption and to provide the Ca^{2+} necessary for the TF–VIIa interaction. VIIa (25 μl at 400–800 nM) and 25 μl of TF are then added, followed by 25 μl of chromogenic substrate (2–8 mM). The rate of increase of absorbance (ΔOD) can immediately be monitored in the kinetic plate reader and is usually stable for 10 min at saturating concentrations of chromogenic substrate. Certain detergents (e.g., Triton X-100) at higher concentrations interfere with the quantitative analysis; CHAPS is suitable and can be included at up to 4 mM final concentration without significant error.[14] All chromogenic substrates cleaved by free VIIa demonstrated a faster hydrolysis by VIIa in the presence of TF. Spectrozyme FXa appears best suited to discriminate between free VIIa and TF·VIIa complex, with a 40-fold increase in the rate of hydrolysis in the presence of TF compared to the 10- to 20-fold increase observed with Pefachrome VIIa and S-2288. These latter substrates are more suitable to analyze free VIIa. Typical rates of chromogenic substrate hydrolysis (expressed in mOD/min) in the presence and absence of TF are given in Table VII.

The analysis of chromogenic substrate hydrolysis by mutant TF·VIIa complexes provides valuable information on the TF–VIIa interactions which result in enhanced catalytic activity of the VIIa catalytic site. This analysis is also critical to identify selective defects in extended protein substrate recognition by TF·VIIa, which is characterized by normal small substrate hydrolysis, but deficient proteolytic activation of factor X.[14] Conversion of VII to VIIa is required for the cleavage of chromogenic substrates, and the activation of VII is markedly accelerated by TF.[15,16] It is possible to analyze VII to VIIa activation using modifications of the chromogenic assay.[17]

Analysis of Proteolytic Activity of TF·VIIa

Efficient extended protein substrate recognition by TF·VIIa is mediated by multiple contacts of VIIa with TF,[10] TF·VIIa with substrate, and interactions of factor X with charged phospholipid.[4] Simultaneous reaction of factors X and IX with the TF·VIIa complex has been observed to result in a more efficient activation of factor IX, suggesting a number of complex macromolecular assembly events during the proteolytic activation of these coagulation zymogens.[18] It must be considered whether TF contributes directly to extended substrate recognition by providing contacts with substrate residues or only indirectly by interactions with VIIa, which in turn induce specific protein substrate recognition by the protease. On formation of the TF·VIIa complex, a new thermodynamic force field is created by residues of both proteins, thus a new interactive surface may be created which deviates slightly from the surfaces of the uncomplexed proteins.[19] Analysis of factor X activation provides a simple method for the study of specific proteolytic activity of the TF·VIIa complex, because of the availability of chromogenic substrates for quantitation of activated factor X. Calculation of Michaelis–Menten parameters for the activation of factor X requires that the concentration of the TF·VIIa complex is eliminated as a variable, thus an excess of either enzyme or cofactor is added to the reaction.[20] Suitable concentrations for the TF·VIIa complex are 0.02– 0.5 nM in the presence of phospholipid and 2–50 nM in the absence of phospholipid. Detergent-solubilized TF, phospholipid-reconstituted TF, or TF expressed on cell surfaces is assembled with VIIa in the presence of 5 mM CaCl$_2$ during a short preincubation (5 to 10 min) followed by the

[15] Y. Nemerson and D. Repke, *Thromb. Res.* **40**, 351 (1985).
[16] L. V. M. Rao and S. I. Rapaport, *Proc. Natl. Acad. Sci. U.S.A.* **85**, 6687 (1988).
[17] A. H. Pedersen, T. Lund-Hansen, H. Bisgaard-Frantzen, F. Olsen, and L. C. Petersen, *Biochemistry* **28**, 9331, (1989).
[18] J. H. Lawson and K. G. Mann, *J. Biol. Chem.* **266**, 11317 (1991).
[19] J. Janin and C. Chlothia, *J. Biol. Chem.* **265**, 16027 (1990).
[20] Y. Nemerson and R. Gentry, *Biochemistry* **25**, 4020 (1986).

addition of factor X. Samples are removed from the reaction and rapidly quenched with EDTA (100 mM). The factor Xa formed is quantified with chromogenic substrate (e.g., Spectrozyme FXa or S-2222). S-2222 is specific for factor Xa and is not cleaved by TF·VIIa, thus providing an opportunity for accurate determination of factor Xa in the presence of functional TF·VIIa without quenching of the reaction. Typically samples taken and quenched at 1-min intervals are used to calculate an initial rate of factor Xa formation. Kinetic parameters are calculated using linear transforms of the Michaelis–Menten equation. The $K_{m(app)}$ determined is influenced by the phospholipid composition and for comparison of wild-type and mutant TF, this variable must be considered. In the absence of phospholipid, detergent (e.g., Triton X-100) may significantly influence the assembly of factor X with TF·VIIa; thus the detergent concentration must be controlled. Low concentrations of CHAPS appear to be less perturbing than Triton X-100 and this detergent may be more suitable for such analyses.

Considerations for Interpretation of Results

Alterations of Global Fold by Mutations

Global structural alterations can be inadvertently introduced by any of the mutational strategies and may not only result in lack of expression of the mutant, but may account for loss of function. Cellular expression provides a diagnostic procedure to detect global conformational alterations resulting from mutagenesis. The level of expression of the recombinant protein gives an initial indication for proper global folding of the protein. Structural alterations commonly lead to intracellular degradation of the translated protein[21,22] and result in low-level cell surface expression. This is illustrated by selected mutations of Trp-14 in TF[23]: Charged substitutions at this position were incompatible with expression of the mutant TF, whereas conservative hydrophobic substitutions did not diminish expression levels or function, consistent with participation of Trp-14 in hydrophobic core structures necessary for native folding of TF. Analysis of posttranslational modifications during expression in mammalian cells provides additional markers indicative of proper global folding. As an example, glycosylation of the TF extracellular domain is reduced in the TF(S186/S209) mutant, which has the carboxyl disulfide bond eliminated

[21] A. A. Pakula, V. B. Young, and R. T. Sauer, *Proc. Natl. Acad. Sci. U.S.A.* **83,** 8829 (1986).
[22] S. H. Bass, M. G. Mulkerrin, and J. A. Wells, *Proc. Natl. Acad. Sci. U.S.A.* **88,** 4498 (1991).
[23] A. Rehemtulla, W. Ruf, D. J. Miles, and T. S. Edgington, *Biochem. J.* **282,** 737 (1992).

and which also demonstrates a selective epitope loss for a monoclonal antibody directed to the carboxyl aspect of the TF extracellular domain.[3]

Identification of Functional Sites in TF

Various mutational strategies have been employed to analyze TF structure–function relationships. Whereas the deletion of the transmembrane and cytoplasmic domain is compatible with secretion and high-level expression of the protein,[4] extracellular domain deletion mutants which were designed based on the exon–intron organization have proved in our hands to be unsuccessful. Only traces of protein were detected by Western blot analysis of cell lysate, with mutants corresponding to domains encoded by exon 1–3 and exon 4–6, and the proteins were neither cell surface expressed nor secreted. In contrast to the limited usefulness of domain deletion mutants in TF, selective mutational disruption of the two disulfide bonds in TF by Cys to Ser exchange was a successful strategy to establish a disulfide bond requirement in the carboxyl aspect of the TF extracellular domain.[3]

Single amino acid substitutions are the most informative approach to relate structure to specific function. Areas of interest for mutational exchange are defined by antibody epitope mapping, cross-linking analysis using the natural ligand,[24] or by selective chemical modification to search for functionally important side-chain groups. As a search strategy, changes in the charge properties of a side chain by Arg for Trp successfully identified the functional importance of residues in the third Trp-Lys-Ser motif. However, this substitution was not compatible with expression of protein at the Trp-14 position, as discussed above. Ala substitutions carry a lower risk of conformational effects, because they result only in the removal of the side chain beyond the C^β of the residue which has been replaced. No side-chain groups or charges are introduced at the position, nor is there relaxation of the peptide backbone to the extent associated with Gly substitutions, which lack the C^β atom. Multiple Ala substitution may be used for the initial search. Single Ala substitutions, which in general do not perturb the global folding of mutant proteins, then allow the identification of single functional residues (e.g., Tyr-157, Lys-159 in TF), and conservative substitutions (e.g., Tyr to Phe) may be useful to define the functional group in the amino acid side chain. The detailed analysis of these single amino acid substitutions will enable dissection of the contribution of cofactor residues to protease binding, catalytic function, and extended substrate recognition.

---

[24] W. Ruf and T. S. Edgington, *Proc. Natl. Acad. Sci. U.S.A.* **88,** 8430 (1991).

Acknowledgments

This is publication 7113-IMM from the Department of Immunology, The Scripps Research Institute. This work was supported in part by NIH Grants P01 HL-16411 and P50 MH-47680, and by fellowships from the American Heart Association, California Chapter (to W.R. and A.R.).

[13] Factor V

By Michael Kalafatis, Sriram Krishnaswamy, Matthew D. Rand, and Kenneth G. Mann

Introduction

The proteolytic activation of prothrombin is catalyzed by an enzyme complex composed of the serine protease factor Xa reversibly associated with the cofactor factor Va on membranes containing negatively charged phospholipid in the presence of calcium ions. The cofactor is derived from the inactive procofactor factor V, which circulates in plasma as a single-chain protein (M_r 330,000). The conversion of factor V to Va results from proteolytic activation by the action of thrombin or factor Xa to yield a heterodimeric cofactor species.[1]

The sequence for human factor V indicates the structural organization schematized in Fig. 1.[2-5] The procofactor possesses three A domains, two C domains, and a connecting B region. Proteolysis by thrombin at three sites results in the release of the B domain in the form of an activation peptide with the concomitant generation of cofactor activity. The heavily glycosylated activation peptide is released in two fragments (M_r 71,000 and 150,000). Factor Va is composed of a heavy chain (Va_{HC}) derived from the NH$_2$ terminus of the procofactor (A1–A2 domains) and a light chain (Va_{LC}) derived from the COOH terminus (A3–C1–C2 domains, Fig. 1). Additional cleavages occur, and several studies indicate that both subunits

[1] K. G. Mann, R. J. Jenny, and S. Krishnaswamy, *Annu. Rev. Biochem.* **57,** 915 (1988).

[2] R. J. Jenny, D. D. Pittman, J. J. Toole, R. W. Kriz, R. A. Aldape, R. M. Hewick, R. J. Kaufman, and K. G. Mann, *Proc. Natl. Acad. Sci. U.S.A.* **84,** 4846 (1987).

[3] E. R. Guinto, C. T. Esmon, K. G. Mann, and R. T. A. MacGillivray, *J. Biol. Chem.* **267,** 2971 (1992).

[4] B. H. Odegaard and K. G. Mann, *J. Biol. Chem.* **262,** 11233 (1987).

[5] M. Kalafatis, M. D. Rand, R. J. Jenny, Y. H. Ehrlich, and K. G. Mann, *Blood* **81,** 704 (1993).

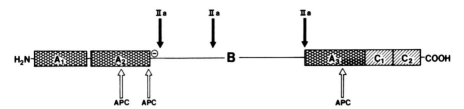

FIG. 1. Structural features of factor V. Thrombin (IIa) catalyzes the activation of human factor V at positions Arg^{709}-Ser^{710}, Arg^{1545}-Ser^{1546}.[2] Thrombin also activates bovine factor V by cleavage at Arg^{713}-Ser^{714}, Arg^{1536}-Ser^{1537}.[3] The cleavage of the B region by thrombin in human factor V (Arg-1018) or bovine factor V (Arg-1006) is not necessary for cofactor activation. APC cleavage sites are depicted by the open arrows. APC cleaves bovine factor Va at two known positions[4]: Arg^{505}-Gly^{506} and Arg^{1753}-Ala^{1754}. These cleavages may occur within the human factor V molecule because the cleavage sites exist (Arg^{506}-Gly^{507} and Arg^{1765}-Leu^{1766}). A third APC cleavage occurs at the COOH-terminal end of the factor Va heavy chain[4] and the specific site within the bovine molecule is most likely Arg^{662}-Asn^{663},[5] whereas the specific site remains to be identified within the human molecule. The acidic region, contained within the human or bovine factor V molecule (653–709 and 664–704, respectively) and located between a thrombin and an APC cleavage site, is indicated (⊖).

migrate as doublets on SDS-PAGE with apparent molecular weights of M_r 94,000/90,000 for Va_{HC} and M_r 74,000/72,000 for Va_{LC}. The difference between the two forms of each subunit has been found at their carboxyl-terminal ends.[4]

The inactivation of factor Va by activated protein C (APC) is achieved by proteolysis of both chains of the cofactor. The cleavage of Va_{LC} by APC and other proteases, including factor Xa and plasmin, yields two fragments (M_r 30,000 and 48,000) and does not inactivate the cofactor.[4,6] Inactivation of factor Va by APC correlates best with the cleavage of Va_{HC}.[7,8] However, it is not clear whether both sites of proteolysis in Va_{HC} are responsible for the loss in activity.

Studies of the structure–function relationships in factor Va have been aided by the availability of the sequence of the cDNA coding for factor V,[2,3,9] by the development of binding and kinetic measurements to examine individually the discrete interactions of the cofactor within prothrombinase, and by the isolation and characterization of proteolytic fragments through protein chemistry techniques. It is well established that the bind-

[6] M. N. Omar and K. G. Mann, J. Biol. Chem. 262, 9750 (1987).
[7] P. B. Tracy, M. E. Nesheim, and K. G. Mann, J. Biol. Chem. 258, 662 (1983).
[8] K. Suzuki, B. Dahlbäck, and J. Stenflo, J. Biol. Chem. 257, 6556 (1982).
[9] W. H. Kane and E. W. Davie, Proc. Natl. Acad. Sci. U.S.A. 83, 6800 (1986).

ing of factor Va to phospholipid only involves the Va_{LC} of the cofactor and is independent of the presence of calcium ions.[10-13] Conversely, the binding of factor Va to factor Xa involves both the Va_{HC} and the Va_{LC} of the cofactor and is dependent on the presence of calcium ions.[14,15] Finally, it has been demonstrated that the Va_{HC} of the cofactor is involved in the interaction with prothrombin.[14] In this article, we summarize the methodology for the isolation of factor Va, its subunits, and proteolytic derivatives of the cofactor, as well as the various steps of the purification and partial characterization of a small region of the Va_{LC} of the cofactor which possess a phospholipid binding domain.

Isolation and Characterization of Factor V, Factor Va, and Factor Va Components

Purification of Human Factor V

Human plasma factor V is purified as previously reported by our laboratory using an antifactor Va light chain immunoaffinity column (αHFV-1).[16] This antibody recognizes human factor V, human factor Va, and bovine factor Va with different affinities. Purified human factor V is stored at $-20°$ in 10 mM boric acid, 10 mM Trizima Base, 2 mM CaCl$_2$, pH 8.3, and 50% glycerol at a concentration varying from 2 to 4 mg/ml.

Isolation of Human Factor Va Light and Heavy Chains

Purified single-chain human factor V (Fig. 2, lane 1) is diluted 10-fold in 20 mM HEPES, 0.15 M NaCl, 2 mM CaCl$_2$, pH 7.4, to 1 mg/ml ($\sim 3 \mu M$). The activation of factor V is initiated by the addition of 18 nM human α-thrombin and allowed to proceed at 37° for 10 min followed by the addition of 20 μM D-Phe-Pro-Arg-chloromethyl ketone (FPR-ck) and further incubation at room temperature for 5 min to inactivate thrombin. The partially activated sample (Fig. 2, lane 2) is applied to 3 ml of αHFV-1 coupled to Sepharose CL-4B (~ 3 mg of IgG/ml of resin) followed by washing with the same buffer. The flow-through of the immunoaffinity

[10] S. Krishnaswamy and K. G. Mann, *J. Biol. Chem.* **263**, 5714 (1988).
[11] D. L. Higgins and K. G. Mann, *J. Biol. Chem.* **258**, 6503 (1983).
[12] P. B. Tracy and K. G. Mann, *Proc. Natl. Acad. Sci. U.S.A.* **80**, 2380 (1983).
[13] M. L. Pusey and G. L. Nelsestuen, *Biochemistry* **23**, 6202 (1984).
[14] E. R. Guinto and C. T. Esmon, *J. Biol. Chem.* **259**, 13986 (1984).
[15] A. E. Annamalai, A. K. Rao, H. C. C. Chiu, D. Wang, A. K. Dutta-Roy, P. N. Walsh, and R. W. Colman, *Blood* **70**, 139 (1987).
[16] J. A. Katzmann, M. E. Nesheim, L. S. Hibbard, and K. G. Mann, *Proc. Natl. Acad. Sci. U.S.A.* **78**, 162 (1981).

$M_r \times 10^{-3}$

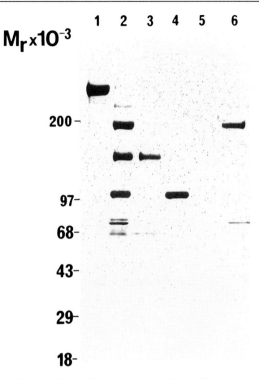

FIG. 2. Isolation of human factor Va components. Factor Va components are separated using the anti-light-chain immunoaffinity column. The products of each elution step are analyzed on a 4–12% linear gradient, silver-stained SDS-PAGE: lane 1, starting material, single-chain human factor V; lane 2, activation intermediates of factor V; lane 3, heavily glycosylated activation peptides (M_r 150,000 and 71,000); lane 4, fragment eluted in the presence of 10 mM EDTA (human Va$_{HC}$, M_r 105,000); lane 5, elution with 1.7 M NaCl; lane 6, fragments eluted with 3 M NaSCN (human Va$_{LC}$, doublet of M_r 72,000/74,000) and activation intermediate of M_r 220,000, which is the precursor of the Va$_{LC}$. Position of the molecular weight markers is indicated at left.

column contains the M_r 150,000 and M_r 71,000 heavily glycosylated activation peptides (Fig. 2, lane 3), which can only be visualized following SDS-PAGE by staining with silver. The elution of the Va$_{HC}$ of the cofactor is performed using 10 mM EDTA in 20 mM HEPES, 0.15 M NaCl, pH 7.4. Under these conditions only one fragment of M_r 105,000 is eluted from the column (Fig. 2, lane 4). Nothing is eluted from the column in the presence of 1.7 M NaCl (Fig. 2, lane 5). Va$_{LC}$ of the cofactor (the M_r 72,000/74,000 doublet), as well as its precursor (M_r 220,000), are eluted from the column in the presence of 3 M sodium thiocyanate (NaSCN) (Fig. 2, lane 6).

Human factor Va is prepared as described above with the omission of the EDTA step and is eluted from the column with 1.7 M NaCl. It is important to note that factor V loses more than 90% of its coagulant cofactor activity when it is eluted from the column in the presence of 3 M NaSCN. Control experiments showed that bovine factor Va can be eluted from the immunoaffinity column and preserves its coagulant activity with 50% ethylene glycol.

The purified components of the human factor Va (Fig. 2) are visualized on a 0.75 × 16-cm 4–12% gradient SDS-PAGE. The SDS-PAGE is silver stained according to a previously described method.[17] The gels are stored in the dark in a solution (12 g/250 ml water) of Hypo Clearing Agent (Kodak, Rochester, NY).

Isolation of Bovine Factor Va Light and Heavy Chains

Bovine factor Va is purified according to the method earlier described by our laboratory, with the omission of the Cibacron blue step.[18] Approximately 50 mg (~ 1 mg/ml) of partially purified factor V is activated with thrombin (110 nM at 37° for 15 min) and the reaction is stopped by adding 3 μM FPR-ck. CaCl$_2$ is added (2 mM final concentration) and the solution is diluted 3- to 5-fold in 20 mM HEPES, 0.15 M NaCl pH 7.4, 2 mM CaCl$_2$, and applied to a 300-ml anti-light-chain immunoaffinity column (αHFV-1, 3 mg of antibody/mg of resin) equilibrated in the same buffer. The column is thoroughly washed with the above buffer and the fractions are tested for biological activity using factor V-deficient plasma.[18] Bovine factor Va is eluted from the column using 20 mM HEPES, 2 mM CaCl$_2$, 1.7 M NaCl, pH 7.4. The fractions containing factor Va activity are pooled following analysis by SDS-PAGE, and precipitated by the addition of solid ammonium sulfate (80% saturation). Precipitated protein is collected by centrifugation (10,000 g, 4°, 30 min), resuspended in 10 mM boric acid, 10 mM Trizima Base, 2 mM CaCl$_2$, 50% (v/v) glycerol, pH 8.3, and stored at −20°.

Purified bovine factor Va consists of a heavy and light chain, each of which migrate as doublets on SDS-PAGE (Fig. 3A, inset lane 1). The relative heterogeneity within each doublet varies with different preparations. The subunits of factor Va are isolated by anion-exchange chromatography following subunit dissociation by treatment of the cofactor with EDTA. Subunit dissociation is achieved by treating 30 mg of factor Va with 10 mM EDTA, followed by extensive dialysis (16 hr, 4°) versus 20 mM HEPES, 2 mM EDTA, pH 7.4. The dialyzed protein is applied at room temperature to a Fast-Flow Q-Sepharose column (Pharmacia, Pisca-

[17] C. R. Merril, M. L. Dunau, and D. Goldman, *Anal. Biochem.* **110**, 201 (1981).
[18] M. E. Nesheim, J. A. Katzmann, P. B. Tracy, and K. G. Mann, this series, Vol. 80, p. 249.

taway, NJ) (2.5 × 10 cm, 50 ml) equilibrated in the same buffer. The column is washed with 250 ml of equilibration buffer and bound proteins are eluted with a linear gradient of increasing NaCl (0–0.6 M) prepared in the same buffer. Fractions (3 ml) containing protein are further characterized by SDS-PAGE prior to pooling. A typical chromatogram is illustrated in Fig. 3A. Peak 2 (Fig. 3A) contains the Va$_{LC}$ doublet (M_r 74,000/72,000), which is eluted at low NaCl concentration. The appearance of the Va$_{LC}$ as a doublet is the result of the cleavage at the carboxyl-terminal end of the intact factor V molecule by an as-yet unidentified protease. The doublet of M_r 48,000/46,000 and the fragment of M_r 30,000 (Fig. 3A, inset lane 2) are the result of the factor Va$_{LC}$ degradation during the purification procedures. A minor fragment of M_r 39,000 is coeluted with the Va$_{LC}$ and is most probably the result of plasmin cleavage of the factor Va$_{HC}$ during the purification procedures. Peak 3 (Fig. 3A) contains the Va$_{HC}$ of the cofactor (M_r 94,000) and trace amounts of several high-molecular-weight products which are the consequence of the incomplete activation of factor V by thrombin (Fig. 3A, inset lane 3).

It is common for factor Va preparations to contain a Va$_{HC}$ with substantial heterogeneity (Fig. 3B, inset lane 1). This heterogeneity of the Va$_{HC}$ (doublet of M_r 94,000/90,000) is the result of an unidentified contaminating proteolytic activity during the last steps of the factor Va purification. When such factor Va preparations are subjected to anion-exchange chromatography using Q-Sepharose in the presence of EDTA, three different peaks are obtained (Fig. 3B) instead of two, as compared with Fig. 3A. Peak 2 (Fig. 3B) contains the Va$_{LC}$ and the above-mentioned proteolytic fragments (Fig. 3B, inset lane 2). Peak 3a contains the M_r 90,000 "light" heavy chain (Va$_{H90}$) and some contaminants of peak 2 (Fig. 3B, inset lane 3a). Finally, peak 3b was composed exclusively of the intact Va$_{HC}$ of the cofactor (Fig. 3B, inset lane 3b). These data together suggest that Va$_{HC}$ contains a small peptide at one of its extremities, which is negatively charged at pH 7.4. On removal of this peptide the behavior of the remaining factor Va heavy chain changes considerably when applied on an anion-exchange column. The only region of the Va$_{HC}$ consistent with this observation is the COOH terminus of the molecule. The region 653–709 of the human Va$_{HC}$ (664–704 of the bovine Va$_{HC}$), which precedes an important thrombin activation cleavage site, contains acidic clusters, which can account for the capability of the Va$_{HC}$ of the cofactor to bind to an anion-exchange matrix. The two components of factor Va which are eluted close to each other (i.e., Va$_{LC}$ and Va$_{H90}$) are further purified using the anti-light-chain immunoaffinity column in the presence of 2 mM EDTA. Using these conditions, Va$_{H90}$ does not bind to the column. The bound Va$_{LC}$ and its degradation products are subsequently eluted in the presence of 1.7 M NaCl.

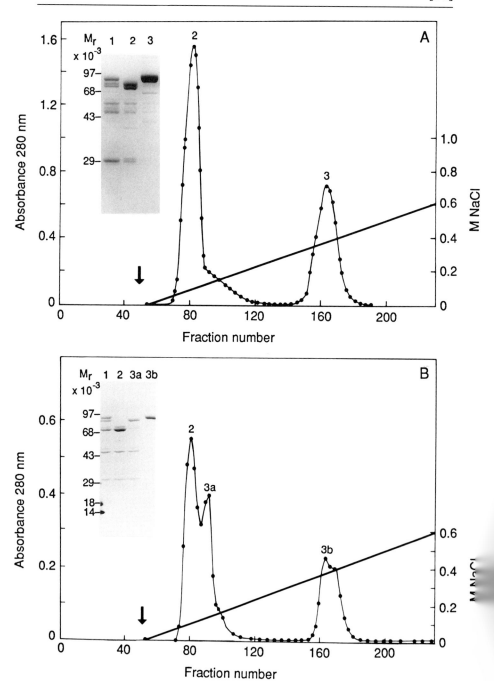

The separation of the Va_{LC} doublet is achieved by ion-exchange chromatography. The doublet representing the Va_{LC} of the cofactor (0.5–1 mg) is extensively dialyzed against 20 mM 2-(N-morpholino)ethanesulfonic acid (MES), 2 mM EDTA, pH 6.0, and applied to a 1-ml prepacked Mono S column (Pharmacia). Elution is performed using a linear gradient of NaCl (0–0.8 M NaCl) in the above buffer. Two peaks are obtained under these conditions containing the two separate elements of the Va_{LC}. All the isolated factor Va components are precipitated with ammonium sulfate (70%) centrifuged at 100,000 g (4°, 30 min) and stored at $-20°$ in 10 mM borate, 10 mM Tris, 2 mM $CaCl_2$, 50% (v/v) glycerol, pH 8.3.

Enzymatic Degradation of Light Chain

Because of the availability of bovine plasma, most studies, including the description of the following experiments, were conducted using bovine factor Va components. The results can be compared to the human molecule because there is 88% identity between the two molecules (only between the Va_{HC} and the Va_{LC} of the two cofactors).[3]

The cleavage of the isolated Va_{HC} of the cofactor with bovine APC gives rise to products of M_r 70,000 (NH_2 terminal), M_r 24,000, and M_r 20,000 (COOH terminal), whereas factor Xa cleavage results in a M_r 56,000 NH_2-terminal fragment and a M_r 46,000 COOH-terminal fragment.[4] The functions of these proteolytic fragments have not yet been determined.

Separation of the Va_{LC} cleavage products is achieved by reversed-phase HPLC. Purified factor Va_{LC} (doublet) is diluted in HBS (2.23 mg in 2.5 ml of HBS) and incubated with APC (44.5 μg) at 37° for 5 hr. Both APC and factor Xa cleave the Va_{LC} at Arg-1765 and give rise to an NH_2-terminal M_r 30,000 fragment and to a COOH-terminal doublet (M_r 46,000/48,000). The reaction is stopped by freezing the sample in dry ice. The sample is lyophilized overnight and dissolved in 1 ml of 0.05% trifluoroacetic acid (v/v) (TFA)/H_2O. The sample is centrifuged (10,000 rpm, 5 min), filtered through 0.45-μm filters (Gelman Sciences, Ann Arbor, MI), and loaded onto a 0.46 × 7.5-cm Beckman Ultrapore RPSC-C3 column operated at 1 ml/min with a gradient of 0% buffer B to 30% buffer B in 10 min,

---

FIG. 3. Isolation of bovine factor Va constitutents. Factor Va components are separated using an anion-exchange column (Q-Sepharose). (A) Routine separation of the factor Va components; (inset) Coomassie blue-stained 10% SDS-PAGE: lane 1, starting material, factor Va; lane 2, bovine Va_{LC} (peak 2); lane 3, bovine Va_{HC} (peak 3). (B) Separation of the heterogeneous factor Va components; (inset) Coomassie blue-stained 5–15% SDS-PAGE: lane 1, starting material, heterogeneous factor Va; lane 2, Va_{LC} (peak 2); lane 3a, factor Va "light" heavy chain (Va_{H90}, peak 3a); lane 3b, factor Va_{HC} (peak 3b). Position of the molecular weight markers is indicated at left.

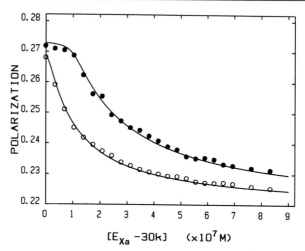

FIG. 4. Competitive displacement of PCPS-bound Va by the isolated M_r 30,000 peptide. Reaction mixtures containing 150 nM pyrene malemide-labeled Va (Pyr-Va) in 20 mM HEPES, 0.15 M NaCl, 2 mM CaCl$_2$, pH 7.4, and either 16 μM (●) or 8 μM (O) PCPS, were titrated with increasing concentrations of the isolated 30K peptide. Fluorescence anisotropy was measured using λ_{ex} 330 nm, λ_{em} 396 nm. The lines are drawn according to the constants $K_{d(Pyr-Va-PCPS)} = 3.18 \pm 1.15$ nM, $K_{d(30K-PCPS)} = 1.55 \pm 0.55$ nM, and stoichiometry $= 42 \pm 0.5$ mol PCPS/mol Va at saturation. Adapted from Krishnaswamy and Mann.[10]

followed by a gradient elution of 30% buffer B to 60% buffer B over 25 min (buffer A is 0.05% TFA in H$_2$O and buffer B is 0.05% TFA in CH$_3$CN). The column effluent is monitored at 214 nm and the data are collected on a strip chart recorder operated at 5 mm/min. Fractions are collected manually and contain the totality of the desired peak. The M_r 30,000 fragment is eluted from the column at 40% buffer B, while the M_r 46,000/48,000 doublet is eluted at 45% buffer B. The absorbance of the proteins contained within each peak is measured at 280/320 nm. An aliquot of each peak representing 1–3 μg of protein is dried under vacuum, dissolved in 30 μl of 62.5 mM Tris, 480 mM glycine, 0.25% SDS, and the purity of the fractions is verified by a 5–15% linear gradient silver-stained SDS-PAGE. On cleavage at Arg-1765 of the human factor V the M_r 46,000/48,000 doublet is found to be insoluble above pH 4, consequently the study of its function is not yet possible. Conversely, the M_r 30,000 fragment is soluble at pH 7.4 and its function can be analyzed. Competition studies using proteolytic fragments of Va$_{LC}$ and Va labeled with pyrene maleimide (Fig. 4) indicate that the NH$_2$-terminal portion of Va$_{LC}$ (i.e., the M_r 30,000 fragment, residues 1546–1765) of the human factor V molecule is responsible for the protein–membrane interaction in the intact cofactor.[10]

Enzymatic Degradation of Va_{LC} in Presence of Phospholipid Vesicles:
Va_{LC} Footprint

Control experiments using a mixture of iodinated and unlabeled Va_{LC} demonstrated that only the Va_{LC} incubated with PCPS is protected from enzymatic digestion (using trypsin or chymotrypsin). We further found that in both cases two small peptides of M_r 14,000 and 18,000 were still present after a 45-min digestion at 37°.[19] These peptides are purified and partially characterized as described below.

Va_{LC} (3 μM) in HBS is incubated for 5 min at 37° with phospholipid vesicles (750 μM) composed of 25% phosphatidylserine and 75% phosphatidylcholine (PCPS). TPCK-treated trypsin or chymotrypsin (Worthington, Freehold, NJ) are then added to separate mixtures at a final concentration of 6.4 and 8 μM, respectively (45 min, 37°), in a final volume of 2 ml. The reaction is stopped by injecting the mixture onto a 7.5-mm × 30-cm TSK-3000 HPLC gel filtration column (Beckman) equilibrated in 20 mM HEPES, 0.15 M NaCl, pH 7.4, operated at 1 ml/min. Two major peaks are obtained: the peak corresponding to the void volume of the column contains the peptides which are bound to phospholipid vesicles and subsequently protected from enzymatic digestion; a second peak is observed and contains trypsin and residual peptides corresponding to the regions of the light chain that are not protected by the PCPS vesicles and are digested by trypsin. The first peak is collected during 3 min (4–7 min after injection). Using dry ice, the totality of the peak is immediately frozen and lyophilized. The white powder is redissolved in 2 ml of 0.05% TFA/ water centrifuged in 1.5-ml Eppendorf tubes at 10,000 g (22°, 10 min), filtered through 0.45-μm sterile Acrodisc filters (Gelman Sciences), and subjected to reversed-phase HPLC chromatography using a 4.6-mm × 7.5-cm Beckman Ultrapore RPSC-C3 column (Beckman). Fractions are collected manually and in all cases contain the totality of the peak. It should be noted that the peak height is collected in a separate acid-washed Eppendorf tube (0.2–0.5 ml) and the corresponding protein is used for amino acid sequence and amino acid composition as follows: the concentration of the fragment is determined by measuring the absorbance at 280/320 nm assuming an extinction coefficient of $1 ml \cdot mg^{-1} \cdot cm^{-1}$. An aliquot of the fraction (usually 1–3 μg) is dried completely using a Savant SpeedVac, dissolved in 30 μl of 62.5 mM Tris, 480 mM glycine, 0.25% SDS, and analyzed on 10–20% SDS-PAGE stained with silver. Once the purity of the fragments is established, 100–150 pmol of peptide (usually 100–200 μl of peptide solution) are subjected to NH$_2$-terminal sequencing using an Applied Biosystem (Foster City, CA) 475A protein sequencing

[19] M. Kalafatis, R. J. Jenny, and K. G. Mann, *J. Biol. Chem.* **265**, 21580 (1990).

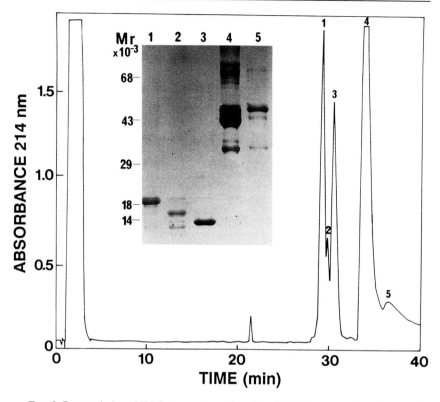

Fig. 5. Reversed-phase HPLC chromatography of the PCPS/Va$_{LC}$ complex after trypsin treatment and gel filtration. The lyophilized sample is dissolved in 1 ml of 0.05% TFA/water and injected on an RPSC-C3 reversed-phase HPLC column. (Inset) 8–20% SDS-PAGE stained with silver. Lanes 1–5 correspond to the peaks 1–5 of the chromatogram. The position of the molecular weight standards is indicated at left. Adapted from Kalafatis *et al.*[19]

system. The peptide solution is concentrated to 30–40 μl using a Savant SpeedVac concentrator in an acid-washed 1.5-ml Eppendorf tube. The solution is applied to a glass fiber filter precoated with Polybrene (Applied Biosystems) and dried in the oven of the sequencer at 55°. The tube is washed with 30 μl of 10% acetic acid (Baker, Phillipsburg, NJ) in HPLC-grade water (Baker) and the solution is applied to the glass fiber filter and treated as above. The phenylthiohydantoin amino acids are identified after reversed-phase HPLC chromatography and compared to a chromatogram containing all the amino acid derivatives (except cysteine). For amino acid composition only 30–50 pmol is necessary. About 50 μl of the purified peptide solution is mixed with 1 nmol of norleucine in an acid-washed borosilicate glass culture tube (6 × 50 mm) and the mixture is completely dried using a Savant SpeedVac concentrator. The tube is then transferred into an acid-washed glass culture tube containing 600 μl of 6 N hydrochloric acid (Pierce Chemical Co., Rockford, IL) and 2% thiodiglycolic acid. The latter is added in order to avoid oxidation of methionines and tyrosines. The tubes are purged with nitrogen, sealed under vacuum, heated at 110°, and hydrolysis is allowed to proceed in a vapor state for 24 hr. The samples are dried completely, resuspended in 200 μl of 0.2 N sodium citrate buffer, pH 2.2, filtered through 0.45-μm filters (Gelman Sciences), and injected onto a cation-exchange column (Interaction Chemicals AA 911). The amino acid composition is determined using a pH gradient from 3.17 to 9.94, with postcolumn derivatization/detection using o-phthalaldehyde.

Figure 5 shows a characteristic profile and the corresponding gradient gel (Fig. 5, inset), obtained after chromatography of the phospholipid/light chain mixture after trypsin treatment and gel filtration. The two peptides of M_r 18,000 and M_r 14,000 contained in the first and third peak, respectively (Fig. 5), are collected, concentrated using a Savant SpeedVac, and extensively dialyzed against 20 mM HEPES, 0.15 M NaCl, pH 7.4. When assayed for binding to PCPS only the M_r 18,000 peptide shows to interact with PCPS vesicles. Direct NH$_2$-terminal amino acid sequence of the M_r 18,000 fragment revealed the sequence -YEDD-PE-F-EDNAIQP, which corresponds to a region of the human factor V molecule starting at residue

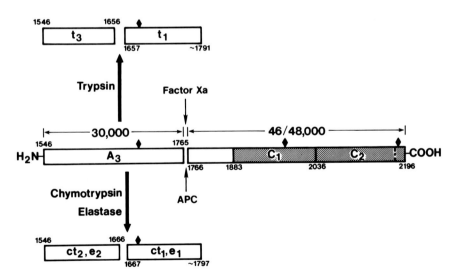

FIG. 6. Schematic representation of the Va$_{LC}$ processing by various proteolytic enzymes. The trypsin, chymotrypsin, and elastase cleavage sites in the human factor V molecule are indicated based on the NH$_2$-terminal sequence analysis and the amino acid composition of each peptide. The cleavage site for factor Xa and APC is also indicated. The filled diamonds show the three potential glycosylation sites within the Va$_{LC}$. Adapted from Kalafatis et al.[19]

1657 (Fig. 6), whereas the smaller peptide has the NH_2-terminal sequence of the intact light chain (Fig. 6). Similar experiments conducted with the peptides derived after treatment of the phospholipid/Va_{LC} with chymotrypsin resulted in a lipid-binding peptide possessing an NH_2-terminal sequence, which agrees with a portion of human factor V starting at residue 1667.

The results obtained after digestion of the factor Va_{LC} with Xa and/or APC and those obtained using other proteolytic enzymes (trypsin, chymotrypsin) in the presence of PCPS vesicles lead us to the conclusion that a peptide of 99 amino acids located between residues 1667–1765 contains a phospholipid binding domain of factor Va (Fig. 6). However, it should be noted that the presence of a phospholipid binding domain within the other half of the factor Va light chain (fragment of M_r 46,000/48,000) should not be excluded, because the binding capacities of this fragment cannot be studied under the conditions described above, owing to its insolubility when separated from the rest of the Va_{LC} (i.e., the M_r 30,000 fragment). However, recent data using recombinant DNA technology[20] have suggested a second phospholipid binding site on the COOH-terminal portion of the light chain (C_2 domain, Fig. 6).

Acknowledgments

This work was supported by Grants HL34575 (K.G.M.) and HL-38337 (S.K.) from the National Institutes of Health.

[20] T. L. Ortel, D. Devore-Carter, M. A. Quinn-Allen, and W. H. Kane, *J. Biol. Chem.* **267**, 4189 (1992).

[14] Site-Directed Mutagenesis and Expression of Coagulation Factors VIII and V in Mammalian Cells

By DEBRA D. PITTMAN and RANDAL J. KAUFMAN

Introduction

Coagulation factors VIII and V are very large glycoproteins which function as cofactors for the proteolytic activation of factor X and prothrombin, respectively. These proteins circulate in plasma as inactive precursors that require proteolysis at specific sites mediated by factor Xa or thrombin to elicit their full functional activity.[1,2] The activated forms of these glycoproteins increase the V_{max} of zymogen activation by interaction with calcium, negatively charged phospholipid, substrate (factor X or prothrombin, respectively), and enzyme (factor IXa or factor Xa, respectively).[3-6] In addition, both factors are inactivated by proteolysis mediated by activated protein C.[7]

Isolation of the cDNA clones for these proteins indicated that both proteins have a domain structure of A1–A2–B–A3–C1–C2.[8-11] The triplicated A domain has homology to the copper–binding protein in plasma, ceruloplasmin. The duplicated C domain in the C terminus of the protein has homology to phospholipid-binding proteins characterized from *Dictyostelium discoidium*[12] and to a mammary epithelial cell surface protein.[13] Separating domains A2 and A3, both proteins have a large B domain that shares no homology with known proteins.[6,12] The primary amino acid sequences of factor VIII and factor V share 40% homology in their A and C domains, whereas the B domains have significantly diverged.[6,12] The B domains for both cofactors reside on single exons, suggesting that the two factors evolved by gene duplication after insertion of a DNA segment encoding the B domain.[14,15]

[1] K. G. Mann, M. E. Nesheim, and P. B. Tracy, "Blood Coagulation" (R. F. A. Zwaal and H. C. Hemker, eds.), p 15. Elsevier, Amsterdam, 1986.

[2] P. F. Fay, "Blood Coagulation" (R. F. A. Zwaal and H. C. Hemker, eds.), p. 35. Elsevier, Amsterdam, 1986.

[3] M. E. Nesheim, J. B. Taswell, and K. G. Mann, *J. Biol. Chem.* **254**, 10952 (1979).

[4] J. Rosing, G. Tans, J. W. P. Grovers-Riemslag, R. F. A. Zwaal, and H. C. Hemker, *J. Biol. Chem.* **255**, 274 (1980).

[5] G. van Dieijen, G. Tans, J. Rosing, and H. C. Hemker, *J. Biol. Chem.* **256**, 3433 (1981).

[6] K. G. Mann, R. J. Jenny, and S. Krishnaswamy, *Annu. Rev. Biochem.* **57**, 915 (1988).

[7] C. T. Esmon, *Science* **235**, 1348 (1987).

[8] G. A. Vehar, B. Keyt, D. Eaton, H. Rodriguez, D. P. O'Brien, F. Rotblat, H. Oppermann, R. Keck, W. I. Wood, R. N. Harkins, E. G. D. Tuddenham, R. M. Lawn, and D. J. Capon, *Nature (London)* **312**, 337 (1984).

[9] J. J. Toole, J. L. Knopf, J. M. Wozney, L. A. Sultzman, J. L. Buecker, D. D. Pittman, R. J. Kaufman, E. Brown, C. Shoemaker, E. C. Orr, G. W. Amphlett, W. B. Foster, M. L. Coe, G. J. Knutson, D. N. Fass, and R. M. Hewick, *Nature (London)* **312**, 342 (1984).

[10] R. J. Jenny, D. D. Pittman, J. J. Toole, R. W. Kritz, R. A. Aldape, R. M. Hewick, R. J. Kaufman, and K. G. Mann, *Proc. Natl. Acad. Sci. U.S.A.* **84**, 4846 (1987).

[11] W. H. Kane, A. Ichinose, F. S. Hagen, and E. W. Davie, *Biochemistry* **26**, 6508 (1987).

[12] W. H. Kane and E. W. Davie, *Blood* **71**, 539 (1988).

[13] J. D. Stubbs, C. Lekutis, K. L. Singer, A. Bui, D. Yuzuki, U. Srinivasan, and G. Parry, *Proc. Natl. Acad. Sci. U.S.A.* **87**, 8417 (1990).

[14] J. Gitscher, W. I. Wood, J. M. Goralka, K. L. Wion, E. Y. Chen, D. H. Eaton, G. A. Vehar, D. J. Capon, and R. M. Lawn, *Nature (London)* **312**, 326 (1984).

[15] L. D. Cripe, K. D. Moore, and W. H. Kane, *Blood* **78**, 181a (1991).

Factor V synthesis has been demonstrated in human[16] and guinea pig megakaryocytes[17] and in a human hepatoma cell line, Hep G2.[18] The probable site of factor V synthesis *in vivo* is the liver.[18] The most likely site for factor VIII synthesis is the hepatocyte.[19-22] Factor VIII circulates as a heterodimer composed of the amino-terminal-derived 200-kDa fragment in a metal ion complex with the carboxyl-terminal–derived 80–kDa light chain fragment.[23,24] This heterodimer is bound to von Willebrand factor by hydrophobic and hydrophilic interactions at the amino terminus of the factor VIII light chain.[25-30] In contrast, factor V circulates in plasma as a single-chain polypeptide, at a 30-fold greater level than factor VIII.[31,32] The activation of both proteins requires specific cleavages which release the B domains as large activation peptides.[33-35]

Our understanding of the requirements for factor VIII and factor V function was greatly facilitated by the ability to express these large proteins in heterologous cells and to design specific mutations to assist in

[16] A. M. Gerwrtz, M. Keefer, K. Doshi, A. E. Annamalai, H. C. Chiu, and R. W. Colman, *Blood* **67**, 1639 (1986).
[17] H. C. Chui, P. K. Schick, and R. W. Colman, *J. Clin. Invest.* **75**, 339 (1985).
[18] D. B. Wilson, H. H. Salem, J. S. Mruk, I. Maruyama, and P. W. Majerus, *J. Clin. Invest.* **73**, 654 (1983).
[19] D. A. Kelly, J. A. Summerfild, and E. G. D. Tuddenham, *Br. J. Haematol.* **56**, 535 (1984).
[20] J. H. Lewis, F. A. Bontempo, and J. A. Sperio, *N. Engl. J. Med.* **321**, 1189 (1985).
[21] Zelechowska, J. A. van Mourik, and T. Brodniewic-Proba, *Nature (London)* **317**, 729 (1985).
[22] E. A. Bontempo, J. H. Lewis, and T. J. Gorenc, *Blood* **69**, 1721 (1987).
[23] C. A. Fulcher, J. R. Roberts, and T. S. Zimmerman, *Proc. Natl. Acad. Sci. U.S.A.* **79**, 1648 (1982).
[24] F. Roblat, D. P. O'Brien, F. J. O'Brien, A. H. Goodall, and E. G. D. Tuddenham, *Biochemistry* **24**, 4294 (1985).
[25] H. J. Weiss, L. L. Sussman, and L. W. Hoyer, *J. Clin. Invest.* **60**, 390 (1977).
[26] R. J. Hamer, J. A. Koedam, N. H. Besser-Visser, R. M. Bertina, J. A. van Mourik, and J. J. Sixma, *Eur. J. Biochem.* **166**, 37 (1987).
[27] P. A. Foster, C. A. Fulcher, R. A. Houghten, and T. S. Zimmerman, *J. Biol. Chem.* **263**, 5230 (1988).
[28] P. Lollar, D. C. Hill-Eubanks, and C. G. Parker, *J. Biol. Chem.* **263**, 10451 (1988).
[29] R. J. Kaufman, L. C. Wasley, M. V. Davies, R. J. Wise, and A. J. Dorner, *Mol. Cell. Biol.* **9**, 1233 (1989).
[30] A. Leyte, M. P. Verbeet, and T. Brodniewic-Proba, *Biochem. J.* **257**, 679 (1989).
[31] B. Dahlbäck, *J. Clin. Invest.* **66**, 583 (1980).
[32] J. Katzmann, M. E. Nesheim, L. S. Hibbard, and K. G. Mann, *Proc. Natl. Acad. Sci. U.S.A.* **78**, 162 (1981).
[33] K. Suzuki, B. Dahlbäck, and J. Stenflo, *J. Biol. Chem.* **257**, 5230 (1982).
[34] D. D. Monovic and P. B. Tracy, *Biochemistry* **29**, 1118 (1990).
[35] C. A. Fulcher, J. R. Roberts, and T. S. Zimmerman, *Blood* **61**, 807 (1983).

structure–function analysis.[36-47] However, due to their extremely large size and the complexity of the post-translational modifications required for biological activity, expression of functional protein has only been accomplished in mammalian cells. This chapter will review strategies and provide protocols that have proved successful for the expression and characterization of these proteins in mammalian cells.

Expression of foreign genes in mammalian cells has become an increasingly important technology in biochemistry and molecular and cellular biology. The choice of a particular expression strategy is dependent on the particular objectives of the study. The criteria in evaluating what expression scheme to employ include (1) the method desired for introducing the foreign gene into the cell, (2) the particular requirement for a specific cell type in which to obtain expression, (3) the amount of protein expression required to achieve the goals of the study, and (4) the particular need for an inducible vector to obtain expression of proteins that are potentially toxic. This review will discuss two host–vector systems which have proved most successful for obtaining high-level expression of factor V and factor VIII in mammalian cells. In addition, methods for oligonucleotide site-directed mutagenesis and analysis of factor V and factor VIII transfected cells are described.

Transient Expression of Transfected DNA in Mammalian Cells

Transient expression has become a convenient, rapid method to study expression of foreign genes in mammalian cells. When cells take up DNA

[36] R. J. Kaufman, L. C. Wasley, and A. J. Dorner, *J. Biol. Chem.* **263,** 6352 (1988).
[37] D. D. Pittman, K. A. Marquette, R. J. Jenny, K. G. Mann, and R. J. Kaufman, *Thromb. Haemostasis* **62,** 1076 (1989).
[38] W. H. Kane, D. Devore-Carter, and T. L. Ortel, *Biochemistry* **29,** 6762 (1990).
[39] J. J. Toole, D. D. Pittman, E. C. Orr, P. Murtha, L. C. Wasley, and R. J. Kaufman, *Proc. Natl. Acad. Sci. U.S.A.* **83,** 5939 (1986).
[40] D. L. Eaton, W. I. Wood, D. Eaton, P. E. Hass, P. Hollinghead, K. Wion, J. Mather, R. M. Lawn, G. A. Vehar, and C. Gorman, *Biochemistry* **235,** 8343 (1986).
[41] D. D. Pittman and R. J. Kaufman, *Proc. Natl. Acad. Sci. U.S.A.* **85,** 2429 (1988).
[42] D. D. Pittman and R. J. Kaufman, *Thromb. Haemostasis* **61,** 162 (1989).
[43] K. A. Marquette, D. D. Pittman, and R. J. Kaufman, *Blood* **76,** 429a (1190).
[44] A. Leyte, H. B. van Schijndel, C. Niehrs, W. B. Huttner, M. P. Verbeet, K. Mertens, and J. A. van Mourik, *J. Biol. Chem.* **266,** 740 (1991).
[45] D. D. Pittman, M. Millenson, K. Marquette, K. Bauer, and R. J. Kaufman, *Blood* **79,** 389 (1992).
[46] G. Hortin, *Blood* **76,** 946 (1990).
[47] D. D. Pittman, J. Wang, and R. J. Kaufman, *Biochemistry* **31,** 3315 (1992).

by a transfection process, they express it transiently over a period of several days to several weeks. Eventually the DNA is lost from the population. The efficiency of expression after transfection of plasmid DNA is dependent on the number of cells which acquire the DNA, the gene copy number, and the expression level per gene. For a number of cell lines it is possible to directly introduce plasmid DNA into 5–50% of the cells in a population. A variety of methods for introducing DNA was recently reviewed.[48] The most convenient and reproducible methods are DNA transfection mediated by DEAE-dextran,[49] electroporation,[50,51] and lipofection with cationic phospholipids.[52] Certain cell types may be more amenable to transfection by one procedure compared to another. Expression of a gene transiently offers a convenient means to compare different vectors. Expression vectors should be tested by transient transfection prior to the more laborious procedure of isolating and characterizing stably transfected cell lines. Transient expression experiments obviate the effects of integration sites on expression and the possibility of selecting cells which harbor mutations in the transfected DNA. For mutagenesis studies, transient transfection is particularly advantageous because preliminary studies may be rapidly conducted to determine whether a particular mutation can yield a secreted and/or functional molecule.

A large variety of host–vector systems for transient expression have been described. Most useful vectors contain multiple elements, including (1) an SV40 origin of replication for amplification to high copy number in COS-1 monkey cells, (2) an efficient promoter element for transcription initiation, (3) mRNA processing signals, which include mRNA cleavage and polyadenylation sequences and frequently intervening sequences, (4) polylinkers that contain multiple endonuclease restriction sites for insertion of foreign DNA, and (5) selectable markers that can be used to select cells that have stably integrated the plasmid DNA. If a relatively efficient expression vector is used, then the expression level obtained from a particular insert is most dependent on the particular gene insert and to a lesser extent on the particular vector. Thus, it is not possible to generalize results obtained from one insert to another.

[48] W. A. Keown, C. R. Campbell, and R. S. Kucherlapati, this series, Vol. 185, p. 527.
[49] J. H. McCutchan and J. S. Pagano, J. Natl. Cancer Inst. 41, 351 (1968).
[50] H. Potter, L. Weir, and P. Leder, Proc. Natl. Acad. Sci. U.S.A. 81, 7161 (1984).
[51] G. Chu, H. Hayakana, and P. Berg, Nucleic Acids Res. 15, 1311 (1987).
[52] P. L. Felgner, T. R. Gadek, M. Holm, R. Roman, H. W. Chan, M. Wenz, J. P. Northrop, G. M. Ringold, and M. Danielsen, Proc. Natl. Acad. Sci. U.S.A. 84, 7413 (1987).

COS Transfection Protocol for 100-mm Plates

Reagents

Stock (10X) DEAE-dextran (Pharmacia, Piscataway, NJ MW 500,000) 2.5mg/ml in Dulbecco's modified Eagle's medium (DMEM). Store at 4°

(10X) Tris, 1 M (pH 7.3), store at 4°

(1000x) Chloroquine, 0.1 M (Sigma, St. Louis, MO), store at $-20°$ in dark

(1X) 10% DMSO reagent (1 liter):

137 mM NaCl (8 g)
5 mM KCl (0.37 g)
0.7 mM NH$_2$HPO$_4$ (0.1 g)
21 mM HEPES (5 g)
6 mM D-glucose (1.08 g)

Take the pH to 7.1, add 100 ml DMSO, and sterile filter through 0.2-μm filter

DMEM

Cells. COS-1 cells are grown in DMEM, supplemented with 10% heat-inactivated fetal calf serum (FCS), 1 mM glutamine, 100 U/ml streptomycin, 100 μg/ml penicillin. Cells are subcultured twice a week at a split ratio of 1:8.

Transfection

1. A confluent plate of cells is trypsinized and $\frac{1}{6}$ of the cells are seeded into each 100 mM tissue culture plate   hr prior to transfection. Cells should be 80–90% confluent at time of transfection.

2. Aspirate plates and wash 2 times each with 7 ml serum-free DMEM. To maintain cell viability during the transfection process it is essential that all the serum is removed before adding DEAE-dextran.

3. Feed the cells with 4 ml of the DNA–medium mixture (2 μg/ml), which is prepared as follows: (A) In a sterile tube add 0.4 ml 1 M Tris-HCl (pH 7.3). (B) Add 8 μg of DNA to the tube and mix well. (C) To the tube add 3.2 ml of DMEM containing 2 mM glutamine, 100 U/ml streptomycin, and 100 μg/ml penicillin. (D) Add 0.4 ml of 2.5 mg/ml DEAE-dextran and mix well. All reagents should be at room temperature or at 37°.

4. Incubate cells at 37° for 5–8 hr.

5. Aspirate the DNA mixture and rinse once with 7 ml of serum-free DMEM. At this point approximately 50% of the cells may have rounded up and/or detached from the plate.

6. Treat the cells with 2 ml of 10% DMSO reagent for 2 min at room temperature. Aspirate.

7. Add 4 ml/plate of 0.1 mM chloroquine in DMEM, 10% FCS and incubate at 37°.

8. After 1.5-2 hr, rinse once with serum-free DMEM or phosphate-buffered saline (PBS). Add 10 ml of DMEM, 10% FCS per plate and incubate at 37°.

9. After 24–30 hr remove medium and feed with 10 ml of fresh DMEM, 10% FCS.

10. Harvest at 54–60 hr, and analyze.

Generation of Stably Transfected High-Level Stable Cell Lines

The ability to select for stable integration of plasmid DNA into the host chromosome has permitted the generation of stably transfected cell lines which indefinitely express a desired gene product. By cotransfection of a selectable marker gene with a nonselectable gene of interest, it is possible to obtain cells that express the desired gene through selection of those cells which express the selectable marker. This ability to select for incorporation of one gene by selection for a second unlinked gene has been termed cotransformation.[53] In cotransformation, separate DNA molecules become ligated together inside the cell and subsequently cointegrate as a unit via nonhomologous recombination into the host chromosome. Different cell lines and different DNA transfection methods yield dramatically different frequencies of cotransformation. As a consequence, vectors have been constructed which contain both the selectable gene and a transcription unit for the desired gene within the same plasmid to obviate the requirement for cotransformation.

High-level expression of the transfected DNA can be obtained by amplification of the selectable marker. Although many schemes have been devised to amplify specific genes, direct selection methods are frequently not available for most genes of interest. For these cases, it is possible to introduce the gene of interest with a selectable and amplifiable marker gene to generate cells that coamplify the desired gene. A number of different selectable and amplifiable marker genes have been reviewed.[54]

Of all the selectable markers, dihydrofolate reductase (DHFR) is the most widely used, and thus further discussion will be limited to its use. Methotrexate (MTX) is a folic acid analog which binds and inhibits DHFR, leading to cell death. When cells are selected for growth in sequentially increasing concentrations of MTX, the surviving population contains increased levels of DHFR, which results from amplification of the DHFR

[53] M. Wigler, R. Sweet, G. K. Sim, B. Wold, A. Pellicer, E. Lacy, T. Maniatis, S. Silverstein, and R. Axel, *Cell (Cambridge, Mass.)* **16,** 777 (1979).
[54] R. J. Kaufman, this series, Vol. 185, p. 487.

gene. The wide utility of the DHFR selection system relies on the availability of Chinese hamster ovary (CHO) cells that are deficient in DHFR.[55] The DHFR-deficient cells which survive require thymidine, glycine, and hypoxanthine for growth, and do not grow without added nucleosides unless they acquire a functional DHFR gene. Most commonly used clones are the DUKX-B11 (DXB-11) isolated from the proline auxotroph CHO-K1 and the DG44 isolated from the CHO Toronto cell line. (These lines can be obtained from Dr. Lawrence Chasin, Columbia University, New York, NY.) Using CHO cells for the expression of heterologous genes affords the following advantages: (1) the amplified genes are integrated into the host chromosome and are stably maintained even in the absence of continued drug selection; (2) a variety of proteins have been properly expressed at high levels in CHO cells[56-63]; (3) CHO cells adapt well to growth in the absence of serum and can grow either attached or in suspension; and (4) CHO cells have been scaled to volumes greater than 5000 liters.

A variety of coamplification vectors have been constructed in which the product gene transcription unit and the selection gene transcription unit are contained within the same vector. Dicistronic expression vectors have been constructed to express the selection gene from the same transcription unit as the product gene.[64-68]

An improved dicistronic mRNA expression vector has been derived by the use of an internal ribosome binding site to promote internal translation

[55] G. Urlaub, and L. A. Chasin, *Proc. Natl. Acad. Sci. U.S.A.* **77**, 4216 (1980).
[56] S. J. Schahill, R. Devos, J. V. Heyden, and W. Fiers, *Proc. Natl. Acad. Sci. U.S.A.* **80**, 4654 (1983).
[57] P. W. Berman, D. Dowbenko, C. C. Simonsen, and L. A. Laskey, *Science* **222**, 524 (1983).
[58] F. McCormack, M. Trahey, M. Innis, B. Dieckmann, and G. Ringold, *Mol. Cell. Biol.* **4**, 166 (1984).
[59] R. J. Kaufman, L. C. Wasley, A. T. Spiliotes, S. D. Gossels, S. A. Latt, G. R. Larsen, and R. M. Kay, *Mol. Cell. Biol.* **5**, 1730 (1985).
[60] J. Haynes and C. Weissman, *Nucleic Acids Res.* **11**, 687 (1983).
[61] D. M. Kaetzel, J. K. Browne, F. Wondisford, T. M. Nett, A. R. Thomason, and J. H. Nilson, *Proc. Natl. Acad. Sci. U.S.A.* **82**, 7280 (1985).
[62] R. J. Kaufman, L. C. Wasley, B. C. Furie, B. Furie, and C. J. Shoemaker, *J. Biol. Chem.* **261**, 9622 (1986).
[63] R. J. Kaufman, L. C. Wasley, and A. J. Dorner, *J. Biol. Chem.* **263**, 6352 (1988).
[64] R. J. Kaufman, P. Murtha, and M. Davies, *EMBO J.* **6**, 187 (1987).
[65] A. Balland, T. Faure, D. Carvallo, P. Cordier, P. Ulrich, B. Fournet, H. De La Salle, and J.-P. Lecocq, *Eur. J. Biochem.* **172**, 565 (1988).
[66] M. Kozak, *Mol. Cell. Biol.* **7**, 3438 (1987).
[67] R. J. Kaufman, M. V. Davies, V. Pathak, and J. W. B. Hershey, *Mol. Cell. Biol.* **9**, 946 (1989).
[68] R. J. Kaufman, M. V. Davies, L. C. Wasley, and D. Michnick, *Nucleic Acids Res.* **19**, 4484 (1991).

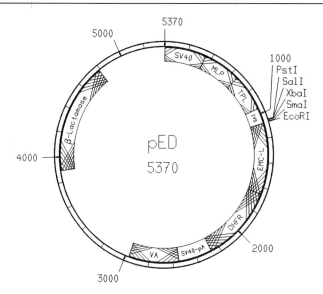

FIG. 1. An efficient vector for cofactor expression in transient and stably transfected cells, pED,[68] is an improved version of pMT2[67]; it contains plasmid sequences from puc18 that allow for propagation and selection for ampicillin resistance in *E. coli.* It contains the SV40 origin of replication and enhancer element and utilizes the adenovirus major late promoter for transcription initiation. Contained within the 5' end of the mRNA is the tripartite leader from adenovirus late mRNA and a small intervening sequence. There are cloning sites included for insertion of foreign DNA. In the 3' end of the transcript there is a cleavage and polyadenylation signal from the SV40 early region. The EMC virus internal ribosome binding site was introduced immediately upstream of the initiation site for DHFR translation. pED also contains the adenovirus virus-associated (VAI) gene, which is included to improve translation of plasmid-derived mRNA.[67,74,75]

initiation within the mRNA. The 5' untranslated region from poliovirus[69] and encephalomyocarditis virus (EMC)[70] can promote ribosome binding at internal sites within an mRNA. The improved set of mammalian cell dicistronic mRNA expression vectors utilizes the 5' untranslated region from EMC virus to promote efficient internal translation initiation of selectable markers encoding DHFR (pED) (Fig. 1),[68] a methotrexate-resistant DHFR (pED-MTX[r]), neomycin phosphotransferase (pED4-Neo),[71,72]

[69] J. Pelletier and N. Sonenberg, *Nature (London)* **334,** 320 (1988).
[70] S. K. Jang, M. V. Davies, R. J. Kaufman, and E. Wimmer, *J. Virol.* **63,** 1651 (1989).
[71] C. R. Wood, G. E. Morris, E. M. Alderman, L. Fouser, and R. J. Kaufman, *Proc. Natl. Acad. Sci. U.S.A.* **88,** 7280 (1991).
[72] H.-M. Koo, A. M. C. Brown, R. J. Kaufman, C. M. Prorock, Y. Ron, and J. P. Dougherty, *Virology* **186,** 669 (1992).

and adenosine deaminase (pEA).[73] The use of pED is limited to DHFR-deficient cells. pED4-Neo, pEA, and pED-MTX^r may be used as dominant markers in different cell types. These vectors permit the rapid derivation of stable cell lines which express high levels of the desired product.[74,75]

CHO Transfection Protocol

Numerous procedures are available to introduce DNA into mammalian cells for the generation of stable cell lines. The most commonly used method is the introduction of the DNA as a coprecipitate with calcium phosphate.[54] Higher transfection efficiency can be obtained with electroporation[50,51,76] or lipofection.[52] Linearization of the DNA prior to electroporation or lipofection results in even higher efficiencies and is essential for efficient cotransfection of two plasmid DNAs.

Electroporation

Electroporation involves pulsing the cells with a high-voltage electrical field. Electroporation causes a transient formation of large pores within the cell membrane through which DNA molecules may pass.

1. Exponentially growing CHO DHFR-deficient cells are harvested by trypsinization. They are resuspended in 10 ml of prewarmed sterile phosphate-buffered saline, transferred to a sterile centrifuge tube, and centrifuged for 5 min at a low speed (700 rpm). Resuspend the cells in prewarmed PBS and count. Centrifuge again at a low speed. All procedures are performed at room temperature.

2. Cells are resuspended in PBS to a final concentration of 2×10^6 cells/ml.

3. Add 1.0 ml of cells to Bio-Rad (Richmond, CA) sterile cuvettes.

4. Linearize 100 μg of plasmid DNA, extract once with an equal volume phenol–chloroform, once with chloroform, and precipitate at $-20°$ with 2 volumes of ethanol. Resuspend the dried pellet in 90 μl of sterile water and add 10 ml of $10\times$ sterile PBS. Add 100 μg of linearized DNA to the cells.

5. Turn on the pulse generator (Gene Pulser, Bio-Rad, Richmond, CA): 1250 μF on the capacitor extender, 200 V on the pulse control. The voltage and capacitance should be optimized such that approximately 50% of the cells survive the electroporation.

[73] R. J. Kaufman, P. Murtha, D. E. Ingolia, C.-Y. Yeung, and R. E. Kellems, *Proc. Natl. Acad. Sci. U.S.A.* **83,** 3136 (1986).

[74] J. B. W. Hershey, *Annu. Rev. Biochem.* **60,** 717 (1991).

[75] R. J. Kaufman and P. Murtha, *Mol. Cell. Biol.* **7,** 1568 (1987).

[76] K. Shigekawa and W. J. Dower, *BioTechniques* **6,** 742 (1988).

6. Place the cuvette containing cells and linearized DNA into the electrode assembly. Make sure that the foil on the cuvette makes contact with the foil of the cuvette holder.

8. Discharge; allow cells to remain in the cuvette for at least 10 min before plating into alpha growth medium containing ADT (Sigma) (10 μs/ml each of adenosine, deoxyadenosine, thymidine), and replace at 37°.

9. After 48 hr remove the medium and add alpha growth medium without ADT.

10. Isolate colonies 14 days later (see lipofection section).

Lipofection

This efficient method of transfection utilizes liposomes to introduce DNA into cells. The reagent consists of a cationic lipid, N-[1-(2,3-dio-leyloxy)propyl]-N,N,N-trimethylammonium chloride (DOTMA), and dioleoylphosphatidylethanolamine. The lipofection reagent is available from GIBCO-BRL (Gaithersburg, MD). The positively charged lipid forms a complex with the DNA and during transfection fuses with the lipid bilayer of the cell membrane.

1. CHO DHFR-deficient cells are subcultured 24 hr prior to transfection into a 100-mm tissue culture plate. The cells should be at a density of $1-2 \times 10^6$/100-mm plate at the time of transfection. The growth medium for CHO DHFR-deficient cells is alpha minimum essential medium, 10% (v/v) dialyzed fetal calf serum,[77] 100 U/ml penicillin, 100 μg/ml strepto-mycin, 1 mM glutamine, and 10 μg/ml each of adenosine, deoxyadeno-sine, and thymidine.

2. The medium used for the transfection is Opti-MEM (available from GIBCO-BRL, Gaithersburg, MD). Prewarm the Opti-MEM containing, 1 mM glutamine, 100 U/ml penicillin, and 100 μg/ml streptomycin at 37°. In a sterile polystyrene tube add 20 μg of linearized DNA to 1.5 ml of Opti-MEM. In a separate sterile polystyrene tube mix 80 μg of lipid with 1.5 ml of Opti-MEM.

3. Mix the lipid and DNA and incubate for 30 min at room temperature.

4. Aspirate the medium from the cells and rinse the monolayer twice with 7 ml of prewarmed Opti-MEM.

5. Add the 3 ml of DNA–lipid mixture to the cells and incubate at 37° for 3 hr. Check the cells frequently. If the cells appear to be dying, the transfection time may need to be shortened.

6. Add 10 ml of alpha growth medium with ADT. Incubate the cells overnight at 37°.

[77] A. J. Dorner and R. J. Kaufman, this series, Vol. 185, p. 577.

7. Remove the medium and wash the cells twice with prewarmed phosphate-buffered saline. Add 10 ml of alpha growth medium with ADT.

8. Two days later, subculture the cells at a split ratio of 1 : 10 or 1 : 15 into alpha medium containing 10% dialyzed fetal calf serum with no ADT.

10. After 14 days, colonies are ready to be picked. With factor V and Factor VIII expression vectors, the best results have been obtained by isolation of individual clones and then subsequent selection of those clones for resistance to increasing concentrations of methotrexate.

Strategical Considerations

The most convenient expression system which can yield rapid results is transient expression in COS-1 monkey kidney cells. This system should be used to verify that the cDNA obtained can direct synthesis of the desired product or mutated protein.

Several steps should be taken if expression of the heterologous gene cannot be detected compared to the positive control. First, one must ensure that the vector was properly assembled and the mRNA was properly expressed. If the mRNA is of the expected size and of the appropriate amount, then it is likely that transcription and mRNA processing are occurring correctly. Generally, as the size of the cDNA increases, the mRNA expression level is reduced. For example, insertion of the 7-kilobase (kb) factor VIII cDNA may reduce mRNA level 10-fold compared to DHFR mRNA from the original pED vector without insert. If the mRNA is present but the protein is not detected, then the intactness of the coding region should be evaluated by translation of the transfected COS-1 cell RNA in an *in vitro* translation system, such as the reticulocyte lysate.

Expression of factor VIII by transient transfection using the pED vector yields approximately 0.3 U/ml factor VIII in the conditioned medium. In contrast, expression of factor V by the same methodology produces approximately 1 U/ml.[37] Derivation of stably transfected and DHFR-coamplified CHO cell lines has yielded expression of factor VIII at 1 U/ml (0.2 μg/ml) and factor V at 1 U/ml (7 μg/ml). The reasons for the relatively low level of expression for factor VIII involve inefficient secretion[78] and requirement for high levels of von Willebrand factor for stable accumulation in the conditioned medium.[29,36]

Insight into structure–function properties and mechanisms limiting its secretion were elucidated by construction of deletion mutants (Fig. 2). Deletion of the B domain (residues 741–1648) resulted in a fully functional protein with expression improved approximately 5- to 10-fold.[42] The increased expression resulted primarily from more efficient secretion due

---

[78] A. J. Dorner, D. G. Bole, and R. J. Kaufman, *J. Cell Biol.* **105,** 2665 (1987).

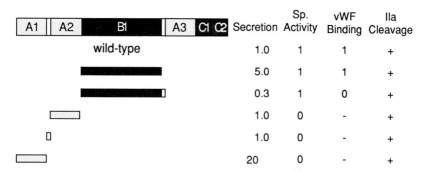

	Sp. Secretion	Activity	vWF Binding	IIa Cleavage
wild-type	1.0	1	1	+
	5.0	1	1	+
	0.3	1	0	+
	1.0	0	-	+
	1.0	0	-	+
	20	0	-	+

FIG. 2. Deletion analysis of factor VIII. The domain structure of human factor VIII is depicted. Site-directed DNA-mediated mutagenesis was performed to create deletions shown. The B domain was deleted from residues 741 to 1648. This deletion was extended to residue 1689 to remove the von Willebrand factor (vWf) binding site. The A2 domain was removed from residues 373 to 740. The A1 domain was deleted from 1 to 372. All amino acid numbers were with respect to the amino terminus of mature factor VIII. Relative values for secretion, specific activity, and vWF binding are given with reference to wild-type factor VIII. (−), Not performed; (+), correct cleavage products.

to a decrease in the interaction of the mutant protein with a lumenal protein of the endoplasmic reticulum, the glucose-regulated protein of 78 kDa (GRP78), also known as immunoglobulin binding protein (BiP).[78] Extending this deletion to residue 1689 removed the von Willebrand factor binding site and yielded a protein which had functional activity, but could not bind vWF. In addition, vWF could not promote stable accumulation of this mutant factor VIII in the conditioned medium. Deletion of domain A2 (residues 373–740) did not affect secretion, although the protein had lost functional activity. In contrast, deletion of domain A1 (residues 1–372) resulted in a 20-fold increase in secretion.[43] The increased secretion correlated with a reduced interaction with BiP. It is thought that a primary BiP binding site resides within the A1 domain of factor VIII.

Methods for Analysis and Mutagenesis

Radiolabeling Cells

The method most frequently used to monitor protein synthesis is to label the cells metabolically with [35S]methionine or [35S]cysteine. Factor VIII and factor V expression are easily analyzed with [35S]methionine. Transiently transfected COS cells are labeled at a density of $2-5 \times 10^6$ cells/100-mm dish at 48–60 hr posttransfection. CHO cells are subcultured at a density $0.2-2.0 \times 10^7$/100-mm dish at 24–48 hr prior to labeling. The cells should be growing logarithmically to be metabolically active. The growth medium for the cells is supplemented with 10% dialyzed fetal calf serum, 1 mM glutamine, 100 μg/ml streptomycin, 100 U/ml penicillin, and 0.1% aprotinin (Sigma, St. Louis, MO).

1. The medium is prewarmed to 37° before labeling. The conditioned medium is removed by aspiration, and cells are washed once with methionine and/or cysteine-free medium containing growth supplements. Intracellular pools of cold amino acids are depleted by incubating the cells in 1 ml of methionine-free medium containing 10% dialyzed fetal calf serum (v/v), 1 mM glutamine, 0.1% aprotinin, 100 μg/ml penicillin, and 100 U/ml streptomycin at 37° for 15 min. Methionine-free minimal essential medium (MEM) can be obtained from Flow Laboratories (McLean, VA), or Specialty Media (Lavallette, NJ). Cysteine-free medium can be obtained from Specialty Media or prepared with the MEM Select-amine Kit available from GIBCO (Grand Island, NY).

2. The medium is removed and the cells are fed 1.5 ml of fresh methionine-free medium containing 200–300 μCi [^{35}S]methionine or [^{35}S]cysteine containing growth supplements and 0.1% aprotinin.

3. To obtain protein with a high specific radioactivity, a labeling period of 2 hr is optimal for factor VIII and factor V. Labeling times greater than 2 hr deplete the radiolabeled methionine or cysteine from the medium. After 2 hr, 1.5 ml of fresh growth medium is added to the plate. Incubate the cells for an additional 4 hr.

4. To analyze the primary translation product and to follow the intracellular processing of the protein, the cells are pulse-labeled with the desired radiolabeled amino acid and chased for various times in medium containing excess unlabeled amino acid. The length of the pulse and chase time may varying depending on the experiment. After a short pulse of 15 min, remove the radiolabel. Rinse the cells once with 5 ml of prewarmed growth medium. The chase is performed by immediately adding 3 ml of prewarmed complete medium containing 100-fold excess of cold methionine or cysteine, depending on the label. The length of the chase time may vary from 30 min to 24 hr.

5. The conditioned medium is harvested by centrifugation at 4° at a low speed (500 rpm) for 5 min to remove cellular debris. The conditioned medium is transferred to a fresh tube and soybean trypsin inhibitor and phenylmethylsulfonyl fluoride (PMSF) are added to 1 mg/ml and 0.1 mM, respectively.

6. The cell extract can be prepared by lysis in nonionic or ionic detergent buffers. Proteins become susceptible to proteases after lysis. It is important to add protease inhibitors to the buffer and keep the cell extracts

cold. Lysis in a Nonidet-40 (NP-40) lysis buffer allows for good recovery of protein, does not lyse the nuclei, and does not interfere with the binding of low-affinity binding antibodies. The buffer consists of 0.15 M NaCl, 50 mM Tris-HC1, pH 7.5, 0.05% SDS, 1% NP-40 (v/v). Add PMSF (0.1 mM) and soybean trypsin inhibitor (1 mg/ml) immediately before lysing the cells.

For a 100-mm plate the cells are rinsed twice with cold PBS and kept on ice. The PBS is immediately removed and 1 ml of lysis buffer with protease inhibitors is added. The cells are scraped with a rubber policeman and transferred to a 1.5-ml microcentrifuge tube. Cellular debris is removed by centrifugation for 15 min in a microcentrifuge. The cell extract is transferred to a fresh tube and stored at $-20°$ or $-80°$.

Immunoprecipitation

Immunoprecipitation of radiolabeled protein is an essential tool for analysis of protein expression. Coupling the antibody to Sepharose CL-4B (Pharmacia, Piscataway, NJ) or Affi-Gel (Bio-Rad, Richmond, CA) (according to the manufacturer's recommendation) eliminates the need for a secondary antibody or protein A-Sepharose to precipitate the immune complexes, and in many cases reduces background.

1. The conditioned medium and cell extracts are prepared as described above.
2. The amount of antibody to be used will depend on its titer. The optimal dilution for immunoprecipitation should be determined by performing a series of immunoprecipitations in which the antibody concentration is varied with a constant amount of sample. Antibody is added to the sample in a 1.5-ml microcentrifuge tube and the reaction is incubated 2–18 hr at 4° on a rotating platform, such as hematology–chemistry mixer model 346 (Fisher, Fairlawn, NJ). To reduce nonspecific binding the conditioned medium or cell extract may be precleared by pretreatment with 100 μl of protein A-Sepharose (Pharmacia, Piscataway, NJ) for 1 hr at 4° on a rotating platform. Protein A-Sepharose is prepared as follows: 0.1 g of protein A-Sepharose beads is suspended in 20 ml of 50 mM Tris-HC1 (pH 7.5), 5 mM EDTA (pH 8.0), 0.15 M NaCl, and 0.5% NP-40, and the beads are allowed to settle. This is repeated twice and the final pellet is resuspended in 2.5 ml of buffer at a final concentration of 40 mg/ml. The protein A-Sepharose is removed by centrifugation in a microcentrifuge at 12,000–14,000 g for 3 min at 4° and the supernatant is transferred to a fresh 1.5-ml centrifuge tube.
3. For monoclonal antibodies not coupled to a solid support, a second-

ary antibody is added as a precipitating agent. An excess of 1 mg/ml rabbit antimouse IgG, IgM, IgA (H + L) (Zymed Laboratories, San Francisco, CA) is added and the sample is incubated for 1 hr on a rotating platform at 4°. The secondary antibody should be titrated in experiments where increasing concentrations are added to a fixed volume of sample.

4. To precipitate a rabbit polyclonal antibody or a monoclonal antibody–rabbit antimouse complex, an excess of protein A-Sepharose is added (the amount necessary to bind the complex should be determined in a pilot experiment). The sample is incubated on a rotating platform for 2–18 hr at 4°. Fresh protease inhibitors are added when the protein A-Sepharose is added. The relative binding of protein A to immunoglobulins varies for different species.[77]

5. Immunocomplexes are collected by centrifugation at 12,000–14,000 g in a microcentrifuge for 3 min at 4°. The supernatant is aspirated. The pellets are washed once each in PBS containing 1% Triton-X 100, PBS containing 0.1% Triton X-100, and PBS containing 0.05% Triton-X 100.

6. Laemmli sample buffer[79] is added to the pellet before loading on a low % bis-acrylamide containing SDS-8% polyacrylamide gel.[80]

Protease Sensitivity

To examine if a mutation affects susceptibility to thrombin or factor Xa digestion, experiments may be performed using the conditioned medium as follows:

1. Immunoprecipitates of the radiolabeled protein are resuspended in 50 mM Tris-HCl (pH 7.5), 0.15 M NaCl, 2.5 mM CaCl$_2$, and 5% glycerol and divided into 2 aliquots.

2. To one aliquot, human thrombin is added to a final concentration of 5–10 U/ml or factor Xa is added to a final concentration of 0.5–2 μg/ml. Human thrombin and factor Xa can be obtained from Haematologic Technologies, Inc. (Essex Junction, VT). Both aliquots are incubated at 37° for 1 hr. For some experiments it may be desirable to perform a time course of activation. In this case human thrombin is added at 1–2 U/ml and the reaction is incubated at 37°. At various times an aliquot is taken and terminated by the addition of 2.5% 2-mercaptoethanol and 0.5% SDS. The samples are heated to 85° for 5 min prior to electrophoresis on a 7–8% low-bis-SDS-polyacrylamide gel.[80] The proteins are visualized by autoradiography after treatment with En³Hance (Dupont-NEN, Boston, MA).

[79] U. K. Laemmli, *Nature (London)* **227**, 680 (1970).
[80] G. Dreyfus, S. A. Adam, and Y. D. Choi, *Mol. Cell. Biol.* **4**, 415 (1984).

Western Blot Analysis

The sensitivity of the enhanced chemiluminescence procedure enables the conditioned medium to be analyzed directly without purification. A time course of thrombin or factor Xa activation can be performed on total conditioned medium.

1. Conditioned medium is harvested and centrifuged at 500 rpm for 5 min to remove cellular debris.

2. The medium is transferred to a fresh tube.

3. To an aliquot or dilution, thrombin is added at 1 U/ml and the reaction is incubated at 25° or 37°.

4. At various time points an aliquot is removed for measurement of activity in a clotting assay. A second aliquot is diluted into 2.5% 2-mercaptoethanol and 0.5% SDS and the proteins are electrophoresed on a SDS-polyacrylamide gel.

5. The cleavage products are visualized after Western blotting and analyzed using a chemiluminescence detection system. The ECL system from Amersham (Arlington Heights, IL) gives high sensitivity.

Analysis of Tyrosine Sulfation

Tyrosine sulfation is a widespread posttranslational event that occurs on a number of proteins as they transit the trans-Golgi compartment. The addition of an O^4-sulfate ester to tyrosine may be the last event before the protein exits the cell.[81] Tyrosine sulfation is mediated by the enzyme protein-tyrosine sulfotransferase (EC 2.8.2.20), which utilizes the donor 5'-phosphoadenosine 3'-phosphosulfate (PAPS).[82]

Tyrosine is the only amino acid which is posttranslationally modified by the addition of sulfate.[81] Cells grown in the presence of $^{35}SO_4$ ion can incorporate the label into tyrosine or carbohydrate.[83] N-Glycanase (EC 3.5.1.52), O-glycanase, and inhibitors of glycosylation can be used to distinguish sulfate incorporated into carbohydrate versus tyrosine. In addition, SO_4 ion attached to tyrosine is sensitive to acid.[83] As a consequence, it is important to avoid low pH in lysis and gel buffers. The polyacrylamide gel fix should not contain trichloroacetic acid (TCA).[83]

1. Cells are grown to a density of $0.1–2 \times 10^7$ in a 100 mM plate, in complete medium supplemented with 10% dialyzed fetal bovine serum, 100 U/ml penicillin, 100 μg/ml streptomycin, 1 mM glutamine, and 0.1% aprotinin. [^{35}S]Sulfuric acid can be obtained from New England Nuclear

---

[81] W. B. Huttner and P. A. Baeuerle, *Mod. Cell Biol.* **6**, 97 (1988).

[82] C. Niehrs and W. B. Huttner, *EMBO J.* **9**, 35 (1990).

[83] W. B. Huttner, this series, Vol. 107, p. 200.

(NEX042, Boston, MA) or [^{35}S]sulfate can be obtained from Amersham (SJS.1A, Arlington Heights, IL). The medium used for labeling should be sulfate-free. Sulfate-free medium, alpha, or MEM can be special ordered from Hazelton Research Products (Lenexa, KS) or Specialty Media (Lavallette, NJ). For some cell types decreasing the methionine and cysteine to 1% of the concentration in the medium may increase the incorporation of ^{35}S. Medium which does not contain sulfate, methionine, and cysteine can be specially ordered from Specialty Media. For some proteins reducing the cysteine or methionine may affect the protein synthesis or secretion. This can be evaluated in a separate experiment, where the incorporation of [^{35}S]methionine into total protein can be monitored by precipitation with trichloroacetic acid.[77] Dialyzed fetal calf serum should be used to reduce the level of methionine, cysteine, and sulfate in the medium.

2. The medium is removed by aspiration. Cells are rinsed once with medium and incubated for 30 min at 37° in sulfate-free medium containing growth supplements and 10 mM HEPES (pH 7.3). The medium is removed and 5 ml of fresh sulfate-free medium containing 0.5 – 1 μCi ^{35}SO$_4$ ion and growth supplements is added. The cells are incubated overnight at 37°.

3. Conditioned medium is harvested as described previously (see *Radiolabeling Cells*). If the conditioned medium is to be assayed, no protease inhibitors are added. Incorporation of ^{35}SO$_4$ ion in total protein is monitored by precipitation of an aliquot of the conditioned medium with TCA. Prior to immunoprecipitation, protease inhibitors are added.

4. The conditioned medium is precipitated as described above. The immunoprecipitate may be subjected to thrombin digestion and treated with *N*-glycanase to remove N-linked carbohydrate or *O*-glycanase to remove O-linked carbohydrate before analysis by electrophoresis on a SDS-polyacrylamide gel.

Factor VIII and Factor V Assays

Several assays are available for both factor V and factor VIII. The amount secreted into the conditioned medium can be assayed as described in previous reviews.[5,84,85]

Method for Heteroduplex Plasmid Mutagenesis

The genetic manipulation of DNA is now a routine laboratory technique. Many excellent reviews cover numerous techniques of mutagenesis.

[84] M. J. Seghatchain, "Factor VIII-von Willebrand Factor," Vol. 1, p. 98. CRC Press, Boca Raton, FL, 1989.
[85] M. E. Nesheim, J. A. Katzman, P. B. Tracy, and K. G. Mann, this series, Vol. 80, p. 249.

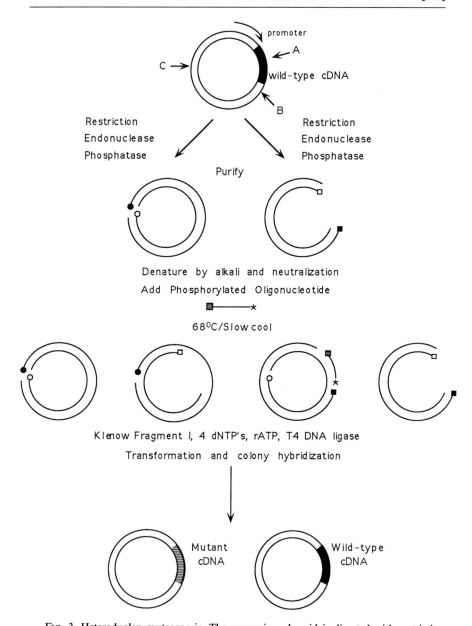

FIG. 3. Heteroduplex mutagenesis. The expression plasmid is digested with restriction enzyme C to linearize the plasmid. A separate aliquot is digested with restriction endonucleases A and B to form a gap around the site to be mutated. The vector is treated with phosphatase and purified. The DNA is denatured by alkali treatment and neutralization and

These techniques include the use of oligonucleotides to direct the introduction of the mutation,[86] enzymatic addition or removal of bases,[87] chemical modification,[88] the misincorporation of altered bases,[89-91] and the use of the polymerase chain reaction (PCR).[92] The most common method utilizes the single-stranded M13 vector as a template.[93] The major drawback of this method is that it is time consuming and involves shuttling the cDNA from the mutagenesis vector to the expression vector.

Some methods for mutagenesis can be performed in the expression vector containing the cDNA to be expressed. We have found this strategy to be most successful using a modification of the gapped heteroduplex procedure described by Morinaga (Fig. 3).[94] This method allows for direct mutagenesis of a sequence already inserted into an expression vector and is particularly advantageous for large cDNAs, such as factor VIII and factor V. This approach obviates the requirement for subcloning fragments between the vector used for mutagenesis and the expression vector.

Choice of Oligonucleotide. For heteroduplex mutagenesis two oligonucleotides are required. One is used to introduce the mutation into the DNA (mutagenic oligonucleotide) and the other is used to identify the clone harboring the correct mutation (screening oligonucleotide).

The design of the mutagenic oligonucleotide should allow it to produce a stable hybrid interaction around the site to be changed. Mutagenic oligonucleotides containing at least 12 bp of homology on each side of the mutation site are the most efficient. Less mispriming occurs with large oligonucleotides (40–60 bases), which are designed to form a G–C base pair at the 5' and 3' ends. The optimal size for a screening oligonucleotide is generally 14–17 bases, depending on the base composition. It is desirable to have a melting temperature (T_m) [in 5 × SSC (0.75 M NaCl, 0.075

[86] M. Smith, *Annu. Rev. Genet.* **19**, 423 (1985).
[87] J. Sambrook, E. F. Fritsch, and T. Maniatis, "Molecular Cloning: A Laboratory Manual." Cold Spring Harbor Lab., Cold Spring Harbor, NY, 1989.
[88] J. T. Kadonga and J. R. Knowles, *Nucleic Acids Res.* **13**, 1733 (1985).
[89] J. W. Taylor, J. Ott, and F. Eckstein, *Nucleic Acids Res.* **13**, 8765 (1985).
[90] T. A. Kunkel, J. D. Roberts, and R. A. Zakour, this series, Vol. 154, p. 367.
[91] M. Sugimoto, N. Esaki, H. Tanaka, and K. Soda, *Anal. Biochem.* **179**, 309 (1989).
[92] H. A. Erlich, "PCR Technology: Principles and Applications for DNA Amplification." Stocton Press, 1989.
[93] M. J. Zoller and M. Smith, this series, Vol. 100, p. 468.
[94] Y. T. Morinaga, T. Franceschini, S. Inouye, and M. Inouye, Biotechnology 2, 636 (1984).

the mutagenic oligonucleotide is added. The reaction is heated at 68° and is allowed to slowly cool. The mutagenesis reaction is performed and the resultant DNA is used to transform competent *E. coli.* Both wild-type and mutant DNAs are recovered from the reaction.

M sodium citrate, pH 7.0)] greater than 30°. When designing mutations it is important to avoid introducing DNA sequences for potential mRNA splice sites,[95,96] CpG dinucleotides,[97-99] and rare codons[100] which may interfere with expression.

Phosphorylation of Oligonucleotides

10× Kinase buffer
0.5 M Tris-HC1 (pH 7.6)
0.1 M MgCl$_2$
1 mM spermidine-HC1
1 mM EDTA (pH 8.0)

Phosphorylation of Mutagenic Oligonucleotide

1. To a 1.5-ml centrifuge tube, add 400 pmol of the mutagenic oligonucleotide to 2 μl 10× kinase buffer, 2 μl 50 mM dithiothreitol (DTT), 1 μl [γ-^{32}P]ATP (5000 Ci/mmol), 1 μl (10–14 U) T4 polynucleotide kinase (New England Nuclear, Boston, MA or BioLabs, Beverly, MA), and water to 20 μl.

2. Incubate the reaction at 37° for 10–20 min.

3. Add 2 μl of 20 mM adenosine phosphate (rATP), 1 μl T4 polynucleotide kinase, and incubate at 37° for 45 min.

4. To stop the reaction add 1 μl of 500 mM EDTA (pH 8.0), and heat at 65° for 5 min. Add 50 μl of 10 mM Tris-HC1 (pH 7.6) and 1mM EDTA (pH 8.0).

5. Extract once with an equal volume of phenol–chloroform, once with chloroform. Add 7.4 μl of sodium acetate (pH 5.2) to the aqueous phase and precipitate with 2 volumes of ethanol (v/v) at −20° for 20 min.

6. Centrifuge in a microcentrifuge, remove the supernatant fluid, and rinse the pellet once with 70% ethanol. Evaporate the ethanol.

7. Resuspend the precipitate in 80 μl of sterile 10 mM Tris-HC1 (pH 7.6) and 1mM EDTA (pH 8.0).

Phosphorylation of Screening Oligonucleotide

1. Mix 20 pmol of oligonucleotide with 2 μl 10 × kinase buffer, 2 μl 50 mM DTT, 1 μl T4 polynucleotide kinase (10–14 U/ml), 13.5 μl of [γ-^{32}P]ATP (5000 Ci/mmol), and add water to 20 μl.

[95] S. M. Mount, *Nucleic Acids Res.* **10**, 10 (1982).
[96] M. R. Green, *Annu. Rev. Genet.* **20**, 671 (1986).
[97] A. P. Bird, *Nucleic Acids Res.* **8**, 1499 (1980).
[98] D. Tonilo, G. Martin, B. R. Migeow, and R. Dono, *EMBO J.* **7**, 401 (1988).
[99] M. Alcalay and D. Tonilo, *Nucleic Acids Res.* **16**, 9527 (1988).
[100] R. Grantham, C. Gautier, M. Gouy, M. Jacobzone, and R. Mercier, *Nucleic Acids Res.* **9**, 43 (1981).

2. Incubate at 37° for 1 hr.

3. Add 1 μl 500 mM EDTA (pH 8.0), and heat at 65° for 5 min to inactivate the T4 polynucleotide kinase.

4. Dilute to 1 ml with 10 mM Tris-HCl (pH 7.6) and 1 mM EDTA (pH 8.0).

5. The efficiency of oligonucleotide phosphorylation is determined by chromatography on cellulose impregnated with polyethyleneimine (Polygram CEL 300 PEI, Brinkmann Instruments, Westbury, NY). Spot the radiolabeled oligonucleotide to the cellulose and allow the spot to dry. Develop 6 to 10 cm with a 0.75 M potassium phosphate buffer (pH 3.5). Dry the cellulose strip and expose to film at room temperature for several minutes. The free ATP migrates behind the aqueous front, and the oligonucleotide remains at the origin.

6. Generally, greater than 95% of the radioactivity is incorporated into the oligonucleotide. For most cases there is no need to separate the unincorporated label. If desired, unincorporated label may be separated by one of several methods.[87]

Mutagenesis

1. Linearize 10 μg of the plasmid at a unique restriction site distant from the mutagenesis target site. After the plasmid is digested, treat with 1–5 U of calf intestine phosphatase (Promega, Madison, WI) for 5 min at 37° to remove 5'-phosphate.

2. In a different tube digest 10 μg of the plasmid with one or more restriction enzymes which digest at two unique sites flanking the mutagenic target (Fig. 3).

3. Extract the DNA once with an equal volume of phenol–chloroform, once with equal volume of chloroform, and separate the linear DNA on an 0.8–1% agarose gel. Extract the DNA from the agarose with glass or other suitable agents.[87]

4. Mix 1 μg of each plasmid together, adjust volume to 18 μl with water, and add 2 μl of 2 N NaOH. Denature at room temperature for 10 min.

5. Neutralize with 180 μl of a solution of 0.1 M Tris-HCl (pH 8.0).

6. Remove 40 μl of the reaction mixture and add 5 μl of the phosphorylated mutagenic oligonucleotide.

7. Incubate at 68° for 90 min. Remove and allow the reaction to renature by slow cooling at room temperature. The remaining 160 μl of heteroduplex should be treated in the same manner. Analyze 10 μl of the 160-μl reaction on an 0.8% agarose gel and store the remaining mixture at −20° for future use. The presence of a new slower migrating band in addition to the starting plasmids indicates the successful formation of a heteroduplex.

8. To the 40 μl of heteroduplex/5 μl mutagenic oligonuclotide add 1 μl of a solution containing all four deoxyribonucleotide triphosphates at 5 mM each. Add 1 μl each of 20 mM rATP, 100 mM MgCl$_2$, 50 mM 2-mercaptoethanol, Klenow fragment of DNA polymerase I (3–4 U), and T4 DNA ligase (400 U).

9. Incubate at room temperature for 10 min, transfer to 16°, and incubate 8 hr to overnight.

10. Terminate the reaction by extracting sequentially with equal volumes of phenol–chloroform, and then chloroform. Ethanol precipitate the aqueous phase and resuspend the dried pellet in 10 μl of sterile Tris-HCl (pH 8.0) and 1 mM EDTA (pH 8.0).

11. Transform competent *Escherichia coli* DH-5 with 1–3 μl of the mutagenic reaction.[101]

12. Replica plate the colonies onto nitrocellulose using standard procedures.

13. Screen colonies with 1 × 10^6 cpm/ml of ^{32}P-labeled screening oligonucleotide in 5 × SSC (0.75 M NaCl, 0.075 M sodium citrate, pH 7.0), 0.1% sodium dodecyl sulfate (SDS), 5× Denhardt's reagent [50 × : 10 g Ficoll (Type 400, Pharmacia), 10 g polyvinylpyrrolidone, and 10 g bovine serum albumin (Fraction V; Sigma) in 1 liter of water],[102] 100 μg/ml denatured salmon sperm DNA.[87] Hybridize filters overnight 5° below the calculated T_m of the screening oligonucleotide estimated from the following formula: $T_m = 2°$ (A + T) + 4° (G + C).

14. Wash filters for 1–1.5 hr at 5° below the estimated T_m in 5 × SSC/ 0.1% SDS on a rotating shaker. Replace the wash every 10–15 min. Monitor filters and wash buffer with a handheld Geiger counter until little radioactivity remains in the wash. Transfer the filters to Whatman (Clifton, NJ) 3MM to remove excess moisture. Do not allow the filters to dry. Sandwich the filters between Saran wrap. Expose filters to film at −80°. Under some conditions it may be necessary to increase the temperature of the wash to distinguish the mutation from the wild-type sequence.

15. Isolate the DNA from positively hybridizing colonies using standard methods.[103,104] Use various restriction enzymes to digest the DNA and separate fragments on 0.8% agarose gel.

16. Transfer the DNA to a solid support such as nitrocellulose or nylon filter and screen by hybridization. Alternatively, a more efficient method is to dry the gel onto Whatman 541 or 3MM under vacuum and screen by direct hybridization (next section).

[101] D. Hanahan, this series, Vol. 100, p. 333.
[102] D. T. Denhardt, *Biochem. Biophys. Res. Commun.* **23,** 1645 (1966).
[103] D. Ish-Horowicz and J. F. Burke, *Nucleic Acids Res.* **9,** 2989 (1981).
[104] D. S. Holmes and M. Quigley, *Anal. Biochem.* **114,** 193 (1981).

17. Hybridize the filters as described in step 13. The appropriate restriction fragments containing the mutation should hybridize to the ^{32}P-labeled oligonucleotide probe.

18. Positively hybridizing colonies may contain a mixture of mutant and wild-type sequences. It is essential to retransform the positive hybridizing DNA and rescreen the colonies.

19. Repeat steps 15–17.

20. Prepare the mutant plasmid DNA by centrifugation to equilibrium in cesium chloride using standard procedures. It is best to work with at least two independent isolates through the preliminary analysis.

21. Sequence the mutant target region by the dideoxy-mediated chain-terminated method.[105] The DNA can be directly sequenced in the plasmid utilizing a unique oligonucleotide primer with the collapsed coil method of sequencing.[87]

22. Characterize the DNA by restriction endonuclease analysis using a number of different restriction enzymes, and Southern blot hybridization[106] using the radiolabeled oligonucleotide as a probe. The DNA should also be digested with several frequent cutting enzymes and analyzed on a polyacrylamide gel to ensure other rearrangements did not occur.

Direct Hybridization

This procedure is an effective method for screening mutations utilizing oligonucleotide probes of 14–18 bases. It is very rapid and less time-consuming than transferring the DNA from a gel to a solid support. DNA fragments from 1 to 50 kb are retained in the gel.[107] There is some loss of fragments smaller than 500 bp during hybridization. The gels are durable when dried down and can be rehybridized.

1. Resolve the DNA digest on a 0.8–1% 0.04 M Tris–acetate (pH 8.0) and 1 mM EDTA (pH 8.0) agarose gel. Stain the gel with 1 μg/ml ethidium bromide to visualize DNA fragments.

2. Place the gel on two sheets of Whatman 541 or 3MM, and cover the gel with Saran wrap. Dry the gel for 1–1.5 hr (or until gel is dry) on a protein gel dryer (Bio-Rad, Richmond, CA) with heat.

3. Wet the gel in water to remove the paper.

4. Denature the gel in 0.5 N NaOH/1.5 M NaCl for 15 min at room temperature.

5. Rinse once with 1 M Tris-HC1 (pH 8.0)/1.5 M NaCl and treat for 15 min at room temperature in the same buffer.

---

[105] F. Sanger, S. Nicklen, and A. R. Coulson, *Proc. Natl. Acad. Sci. U.S.A.* **74**, 5463 (1977).

[106] E. M. Southern, *J. Mol. Biol.* **98**, 503 (1975).

[107] T. M. Shinnick, E. Lund, O. Smithies, and F. R. Blattner, *Nucleic Acids Res.* **2**, 9527 (1975).

6. Hybridize the gel as described above for 4 hr overnight.

7. Wash as described above.

9. Stain the gel in water containing 1 μg/ml ethidium bromide for no more than 30 sec.

10. Remove excess liquid, place the gel on Saran wrap, and cover with another piece of Saran wrap. Expose the gel to film with an intensifying screen at $-80°$.

11. The autoradiograph can be overlaid on the gel placed on an ultraviolet light box.

Acknowledgments

We thank our colleagues K. Marquette, D. Michnick, J. Nove, and K. Tomkinson for protocols and critically reviewing the manuscript.

[15] Assembly of Prothrombinase Complex

By SRIRAM KRISHNASWAMY, MICHAEL E. NESHEIM, EDWARD L. G. PRYZDIAL, and KENNETH G. MANN

Introduction

The interaction of factor Xa with factor Va on membranes containing acidic phospholipids in the presence of calcium ions, results in the assembly of the macromolecular prothrombinase complex that efficiently converts prothrombin to α-thrombin by limited proteolysis.[1,2] Although the serine protease, factor Xa, is capable of catalyzing this reaction in the absence of the other constituents, the assembly of the enzyme complex dramatically improves the catalytic efficiency for this reaction by approximately 10^5-fold.[2] Thus, the assembly of prothrombinase is a key event in the rapid formation of thrombin following initiation of the coagulation cascade.

Factor Xa interacts reversibly with acidic phospholipid-containing membranes through interactions that are mediated via Ca^{2+} ions and γ-carboxyglutamic acid residues present at its amino terminus.[1,3] In contrast, the interaction of the heterodimeric cofactor, factor Va, with the same membranes does not require added Ca^{2+} ions.[4] Divalent metal ions

[1] C. M. Jackson and Y. Nemerson, *Annu. Rev. Biochem.* **49,** 765 (1980).

[2] K. G. Mann, R. J. Jenny, and S. Krishnaswamy, *Annu. Rev. Biochem.* **57,** 915 (1988).

[3] E. W. Davie and K. Fujikawa, *Annu. Rev. Biochem.* **44,** 799 (1975).

[4] J. W. Bloom, M. E. Nesheim, and K. G. Mann, *Biochemistry* **18,** 4419 (1979).

are, however, required for the intersubunit interactions within the cofactor and the maintenance of functional integrity.[5,6] The reversible and tight interaction between membrane-bound factors Xa and Va leads to prothrombinase assembly.[7] Thus, protein–metal ion, protein–membrane, and protein–protein interactions are involved in the assembly of this enzyme complex and the expression of its activity.

Although some quantitative information relating to the binary and ternary interactions within prothrombinase can be inferred from kinetic measurements of prothrombin activation,[8] a detailed understanding of the physical biochemistry of the assembly process has been aided by studies of complex assembly on synthetic phospholipid vesicles in the absence of prothrombin using fluorescent reporter groups that are specifically incorporated into one or more of the interacting macromolecules. This approach has permitted studies of the discrete binary and ternary interactions that lead to prothrombinase assembly and the expression of its activity by equilibrium and kinetic approaches.

In this article, we describe the methodology that has been developed for the studies of the ternary and binary interactions within the prothrombinase complex, ranging from the techniques for the preparation of fluorescent adducts of factors Xa and Va to their application in the dissection of the physical biochemistry of the various equilibria involved in complex assembly.

Proteins and Fluorescent Derivatives

Bovine factor Xa is prepared by the activation of factor X using the purified activator from Russell's viper venom,[9] followed by purification using benzamidine-Sepharose as previously described.[10] Procedures for the isolation of bovine factor V, factor Va, and the isolated chains of the cofactor have been described.[11,12] The concentrations of the proteins are determined using the following molecular weights and extinction coefficients: ($E_{280}^{0.1\%}$): Xa, 45,300 (1.24),[13,14] Va, 168,000 (1.74),[15,16] Va$_{HC}$, 92,300

[5] C. T. Esmon, *J. Biol. Chem.* **254**, 964 (1979).

[6] M. E. Nesheim and K. G. Mann, *J. Biol. Chem.* **254**, 1326 (1979).

[7] S. Krishnaswamy, K. C. Jones, and K. G. Mann, *J. Biol. Chem.* **263**, 3823 (1988).

[8] M. E. Nesheim, J. B. Taswell, and K. G. Mann, *J. Biol. Chem.* **254**, 10952 (1979).

[9] J. Jesty and Y. Nemerson, this series, Vol. 45, p. 95.

[10] S. Krishnaswamy, W. R. Church, M. E. Nesheim, and K. G. Mann, *J. Biol. Chem.* **262**, 3291 (1987).

[11] M. E. Nesheim, J. A. Katzmann, P. B. Tracy, and K. G. Mann, this series, Vol. 80, p. 249.

[12] M. Kalafatis, S. Krishnaswamy, M. D. Rand, and K. G. Mann, this volume [13].

[13] C. M. Jackson, T. F. Johnson, and D. J. Hanahan, *Biochemistry* **7**, 4492 (1968).

[14] K. Fujikawa, M. H. Coan, M. E. Legaz, and E. W. Davie, *Biochemistry* **13**, 5290 (1974).

$(1.24),^{15,16}$ and Va_{LC}, 82,500 $(2.23).^{15,16}$ Determinations of the concentration of fluorescent derivatives of factors Xa, Va, or the subunits are frequently complicated by spectral interference from the modifier to the UV spectrum of the protein. This problem is circumvented by determining protein concentrations colorimetrically using bicinchoninic acid (BCA assay reagent, Pierce Chemical Co., Rockford, IL). Protein concentrations are determined by interpolation from standard curves prepared using known concentrations of unmodified Xa, Va, Va_{HC}, or Va_{LC}. Alternatively, concentrations of fluorescent protein derivatives can be determined by quantitative amino acid analysis.[17] Small unilamellar vesicles composed of 75% (w/w) L-α-phosphatidylcholine and 25% (w/w) L-α-phosphatidylserine (PCPS) are prepared by sonication and differential centrifugation as previously described.[18] The concentration of phospholipid is determined following oxidation and colorimetric analysis of inorganic phosphate.[19] Phospholipid concentrations are expressed as the concentration of monomeric phospholipids.

Preparation of Fluorescent Derivatives of Factor Xa

The preparation of defined fluorescent adducts of factor Xa has been facilitated by the synthesis of fluorescent tripeptidyl chloromethyl ketones that react rapidly with the protease.[20,21] Treatment of factor Xa with such reagents results in its inactivation through the alkylation of the active site histidine,[22] with the concomitant incorporation of 1 mole fluorophore per mole of factor Xa. The resulting derivative is catalytically inactive but retains the ability to participate in the macromolecular interactions within prothrombinase in a manner that is indistinguishable from the native enzyme.[23] Thus, fluorescent derivatives of factor Xa prepared by this strategy serve as useful reporter groups for macromolecular interactions within the enzyme complex without the complicating effects of side reactions catalyzed by active factor Xa.

Because dansylglutamylglycinylarginyl chloromethyl ketone (DEGR-ck) has proved the most useful reagent to date for physical studies of

[15] T. M. Laue, A. E. Johnson, C. T. Esmon, and D. A. Yphantis, *Biochemistry* **23**, 1339 (1984).
[16] S. Krishnaswamy and K. G. Mann, *J. Biol. Chem.* **263**, 5714 (1988).
[17] P. Bohlen, this series, Vol. 91, p. 17.
[18] D. L. Higgins and K. G. Mann, *J. Biol. Chem.* **258**, 6503 (1983).
[19] G. Gomori, *J. Lab. Clin. Med.* **27**, 955 (1942).
[20] C. Kettner and E. Shaw, *Thromb. Res.* **22**, 645 (1981).
[21] C. Kettner, and E. Shaw, this series, Vol. 80, p. 826.
[22] G. Schoellmann and E. Shaw, *Biochemistry* **2**, 252 (1963).
[23] M. E. Nesheim, C. Kettner, E. Shaw, and K. G. Mann, *J. Biol. Chem.* **256**, 6537 (1981).

prothrombinase, a detailed description of the modification of factor Xa by this reagent is provided. The same procedure can also be applied to the inactivation of factor Xa by other fluorescent (and nonfluorescent) glutamylglycinylarginyl chloromethyl ketone derivatives that have more recently been described.[24-27]

DEGR-ck is prepared by the methods described[20] or is obtained commercially (Calbiochem, La Jolla, CA) and stored dry at −20°. The solid is hydrated in 10 mM HCl to achieve an approximate concentration of 2–10 mM, which is verified by acid hydrolysis of an aliquot containing norleucine as a recovery standard followed by quantitation of Glu and Gly by amino acid analysis. The hydrated inhibitor is stable for approximately 1 year when stored at −20° in the dark. Inactivation of factor Xa is achieved by the sequential addition of 0.1 M equivalents of DEGR-ck (~1 μM, 1–10 μl) to a solution of factor Xa (0.5 mg/ml, 11 μM, 1–5 mg) in 0.1 M Tris, 0.1 M HEPES, 0.15 M NaCl, pH 7.4, stirred at room temperature and protected from ambient light. Tris is included in the buffer to scavenge traces of contaminating dansyl chloride that may be present in the DEGR-ck preparations and thus preclude the nonspecific incorporation of fluorophore into factor Xa. Following each addition of inhibitor, the mixture is incubated for approximately 5 min and residual activity is assessed by initial velocity measurements of Spectrozyme Xa (methoxycarbonylcyclohexylglycylglycylarginine p-nitroanilide) hydrolysis [200 μM Spectrozyme in 20 mM HEPES, 0.15 M NaCl, 0.1% (w/w) polyethylene glycol (PEG) 8000, 1–5 nM Xa]. The procedure is repeated until less than 0.01% of the initial activity remains, typically resulting from the addition of 2–3 equivalents of DEGR-ck. Excess inhibitor is separated from modified protein by chromatography using Sephadex G-25-150 (1.5 × 60 cm) developed with 20 mM HEPES, 0.15 M NaCl, pH 7.4, at room temperature in the dark. Protein eluting in the void volume (~35 ml) is identified by absorbance and fluorescence measurements, pooled, and precipitated by the addition of solid ammonium sulfate to 80% saturation. Precipitated protein is collected by centrifugation (53,000 g, 20 min, 4°), dissolved in 50% (v/v) glycerol and stored at −20°.

Analysis of the resulting fluorescent adduct (DEGR–Xa) by SDS-PAGE followed by illumination with UV light indicates that fluorescence is exclusively localized in the heavy chain (protease domain) of factor Xa. The specificity of the labeling reaction is demonstrated by the single fluo-

[24] E. B. Williams, S. Krishnaswamy, and K. G. Mann, *J. Biol. Chem.* **264**, 7536 (1989).
[25] K. G. Mann, E. B. Williams, S. Krishnaswamy, W. R. Church, A. Giles, and R.P. Tracy, *Blood* **76**, 755 (1990).
[26] S. Krishnaswamy, E. B. Williams, and K. G. Mann, *J. Biol. Chem.* **261**, 9684 (1986).
[27] E. L. G. Pryzdial and K. G. Mann, *J. Biol. Chem.* **266**, 8969 (1991).

rescent peptide observed by reversed-phase HPLC following digestion of DEGR–Xa by lysyl endopeptidase and staphylococcal V8 protease. In addition, incorporation of fluorescence into factor Xa is completely obviated by its prior reaction with the nonfluorescent reagent, glutamylglycinylarginyl chloromethyl ketone (EGR-ck).

Modification of Factor Va with Sulfhydryl-Directed Fluorescent Probes

Factor Va or either of the isolated chains of the cofactor contain free thiols that can be quantitatively titrated using dithiobis(nitrobenzoic acid) (DTNB). Titration of factor Va under native conditions with DTNB indicates a single highly reactive thiol present in the heavy chain (Va_{HC}) that can be modified without any loss in the functional properties of the cofactor.[28] Separation of the subunits following treatment with EDTA by described methods[12] results in the exposure of a second reactive thiol in the light chain (Va_{LC}) that can also be readily modified by a variety of sulfhydryl-reactive compounds. However, unfolding of the polypeptide chain by SDS reveals two titratable thiols in each of the subunits of the cofactor.[28] Thus, factor Va can be modified by sulfhydryl-directed fluorescent reporter groups to yield a number of defined combinations of fluorescently labeled species that are useful in studies of structure–function relationships in the cofactor.

In our laboratories, we have prepared a variety of derivatives of factor Va by reacting fluorescent maleimides and iodoacetamides with the intact cofactor. Under nondenaturing conditions, these reactions generally result in the incorporation of the modifier specifically into Va_{HC} without any detectable modification of Va_{LC} or its derived peptides and without any loss in functional activity.

Factor Va (3–20 mg) is dialyzed into 20 mM HEPES, 0.15 M NaCl, 5 mM CaCl$_2$, pH 7.0, at 4°. If necessary, the dialyzed sample is further concentrated by centrifugal ultrafiltration (Centricon-30, Amicon, Danvers, MA) to achieve a concentration of between 30 and 100 μM. The protein solution is centrifuged (53,000 g, 10 min, 4°) to remove insolubles and is treated with a 10-fold molar excess of fluorescent maleimide prepared as a 10 mM stock solution in the same buffer. In cases in which the maleimide is sparingly soluble in aqueous buffer, stock solutions (20 mM) are prepared either in CH$_3$CN or DMSO and are added to the factor Va solution with vigorous vortexing. Insoluble maleimides such as N-(1-pyrene)maleimide can be incorporated into factor Va by the addition of the

[28] S. Krishnaswamy and K. G. Mann, J. Biol. Chem. **263**, 5714 (1988).

appropriate amount of reagent preadsorbed to Celite. The resulting mixture is incubated for 14 hr at 4° in the dark with periodic vortexing to suspend insoluble fluorophores uniformly. Insoluble material is removed by centrifugation (10,000 g, 10 min, 4°) and the clear supernatant is chromatographed using Sephadex G-25-150 in the same buffer to separate factor Va from excess, unreacted dye. Fractions containing protein are pooled and precipitated with $(NH_4)_2SO_4$ (80% saturation). Precipitated protein is collected by centrifugation (53,000 g, 30 min, 4°), dissolved in 20 mM HEPES, 5 mM $CaCl_2$, 50% (v/v) glycerol, pH 7.4, and stored in the dark at $-20°$.

The specificity and extent of modification of factor Va is evaluated by a variety of criteria: (1) absorbance measurements to determine protein concentration and the concentration of incorporated dye using published molar extinction coefficients; (2) the reactivity of the modified protein with DTNB in comparison to unmodified factor Va; (3) visualization of the modified species following separation of the peptides by SDS-PAGE and illumination of the gel with UV light; and (4) digestion of modified factor Va by lysyl endopeptidase (EC 3.4.21.50) followed by peptide separation by reversed-phase HPLC (RP-300, Brownlee Labs, Santa Clara, CA) and detection of fluorescence. In general, we observe the extent of modification to range between 0.85 and 1.1 mole dye incorporated per mole of Va, with incorporation exclusively confined to Va_{HC}. Further, essentially all the fluorescence (>85%) is confined to a single peptide separated by HPLC following proteolytic digestion.

We have used the same procedures to modify the isolated subunits of factor Va (Va_{HC} and Va_{LC}) with a variety of fluorescent maleimides. Modification of the separated subunits under the conditions described above also results in the incorporation of approximately 1 mole of modifier incorporated per mole of protein. Va_{LC}, however, appears to be extremely sensitive to NH_2-reactive impurities present in some commercially available maleimides. Side reactions by impurities or the reaction of maleimides with amino groups are minimized by reducing the pH of the buffer to 6.8 and/or including 0.1 M Tris during the modification reaction.

The consequences of covalent modification on protein function are evaluated by determining the specific activity of modified factor Va in either coagulation assays or in direct measurements of prothrombin activation.[11] Because either isolated subunit possesses minimal cofactor activity, the functional properties of modified Va_{HC} or Va_{LC} are determined by reassociating limiting concentrations of the modified subunit with a vast excess of unlabeled second subunit (below), followed by the characterization of the cofactor function of the product.

Ternary Interactions within Prothrombinase

Complex Assembly by Measurements of Fluorescence Anisotropy

Steady-state measurements of both fluorescence intensity and anisotropy using DEGR–Xa provide useful information regarding the interactions of the reporter group and factor Va on the membrane surface. Steady-state anisotropy measurements are performed in an SLM 8000 fluorescence spectrophotometer (SLM Instruments, Urbana, IL), equipped with Glan-Thompson polarizers and operated in T-format,[29] using $\lambda_{ex} =$ 330 nm (band pass 8 nm) and $\lambda_{em} = 545$ nm (band pass 16 nm). Long-pass filters (KV-500 Schott Glass, Duryea, PA) are placed in the emission paths to minimize scattered light artifacts. Reaction mixtures (2.0 ml) in 1×1-cm stirred quartz cuvettes typically contain 200 nM DEGR–Xa in 20 mM HEPES, 0.15 M NaCl, 2 mM CaCl$_2$, pH 7.4. Polarization is measured by integrating the signal at the two positions of the excitation polarizer for a period of 10 sec each and the final value is calculated by averaging 8–10 repeated measurements to provide a precision of 0.002 in the measurements. A systematic titration of this reaction mixture with PCPS results in a modest increase in the polarization from a value of 0.24 ($r = 0.174$) to 0.26 ($r = 0.19$) at saturating concentrations of PCPS, describing the reversible interaction of DEGR–Xa with PCPS (Fig. 1). A similar titration in the presence of saturating concentrations of factor Va results in a larger increase in polarization to a limiting value of 0.33 ($r = 0.245$) (Fig. 1). Thus, measurements of fluorescence polarization permit the differentiation of the interaction of DEGR–Xa with the membrane surface from its interaction with factor Va on the membrane surface. These approaches have been used to dissect successfully and to obtain quantitative information for both types of interactions in the formation of the ternary prothrombinase complex.[23,30,31]

Complex Assembly by Steady-State Fluorescence Intensity Measurements

An increase in steady-state fluorescence intensity also accompanies the interaction between DEGR–Xa and factor Va on the membrane surface (Fig. 2). This interaction results in an 18% enhancement of the fluorescence of the dansyl moiety ($\lambda_{ex} = 330$ nm) and a larger enhancement (51%) in the signal ($\lambda_{ex} = 280$ nm) arising due to energy transfer from aromatic residues to dansyl. The increase in fluorescence intensity of the

[29] G. Weber and B. Bablouzian, *J. Biol. Chem.* **241**, 2558 (1966).
[30] M. M. Tucker, M. E. Nesheim, and K. G. Mann, *Biochemistry* **22**, 4540 (1983).
[31] S. Krishnaswamy, *J. Biol. Chem.* **265**, 3708 (1990).

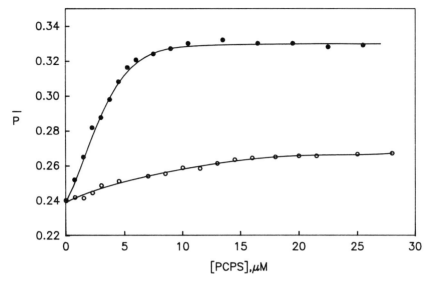

FIG. 1. Measurements of prothrombinase assembly by fluorescence polarization using DEGR–Xa. The interaction of DEGR–Xa (70 nM) with membranes was measured at the indicated concentrations of PCPS either in the absence (○) or presence (●) of 103 nM factor Va. Fluorescence polarization was measured at 22° in 20 mM Tris, 0.15 M NaCl, 2 mM CaCl$_2$, pH 7.4, using $\lambda_{ex} = 335$ nm and $\lambda_{em} = 565$ nm. Polarization was measured by integrating the signal for 30 sec and the data represent means of six repetitive measurements at each point. Adapted with permission from Nesheim et al.[23]

reporter group provides a more sensitive signal of complex assembly and differs from changes in anisotropy, in that the signal arises purely from the formation of the ternary complex and not from the possible binary interactions.[7,31] Thus, intensity measurements using DEGR–Xa provide a sensitive and direct tool for studies of prothrombinase assembly.

Equilibrium binding measurements of ternary complex formation (interaction between factor Va and DEGR–Xa on the membrane surface) are conducted using a fixed concentration of DEGR–Xa, a fixed and saturating concentration of PCPS, and increasing concentrations of factor Va with appropriate corrections for scattering and baseline fluorescence. Typical concentrations or reactants correspond to 15 nM DEGR–Xa, 100 μM PCPS in 20 mM HEPES, 0.15 M NaCl, 2 mM CaCl$_2$, pH 7.4, and factor Va varied from 0 to 50 nM. Stock solutions of the proteins (1–5 μM) are prepared in the same buffer by dialysis followed by centrifugation to remove particulates, and the buffer is filtered (0.45 μm) prior to use. Fluorescence intensity is measured using $\lambda_{ex} = 280$ nm (band pass 2 nm) and monitoring broadband fluorescence at $\lambda_{em} > 500$ nm with a long-pass

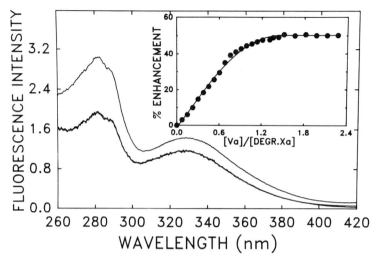

FIG. 2. Excitation spectra for DEGR–Xa in the presence and absence of PCPS and factor Va. Excitation spectra were collected over the indicated wavelengths (band pass 2 nm) using λ_{em} = 545 nm (band pass 4 nm). Reaction mixtures in 20 mM HEPES, 0.15 M NaCl, 2 mM CaCl$_2$, pH 7.4, contained 150 nM DEGR–Xa in the presence or absence of 90 μM PCPS (lower spectrum). The upper spectrum was collected using 150 nM DEGR–Xa, 90 μM PCPS, and 430 nM Va. The inset illustrates that fluorescence intensity monitored using λ_{ex} = 280 nm and λ_{em} = 545 nm increases saturably with increasing concentrations of factor Va to a limiting enhancement of 52% at a fixed concentrations of DEGR–Xa and a saturating concentration of PCPS. Reprinted with permission from Krishnaswamy *et al.*[7]

filter (Schott KV-500) in the emission beam. Three reaction mixtures (2.0 ml) in 1 × 1-cm stirred quartz cuvettes are used for these titrations: (A) DEGR–Xa, PCPS, and variable concentrations of Va; (B) DEGR–Xa, PCPS, and additions of buffer instead of factor Va; and (C) EGR–Xa (factor Xa inactivated with nonfluorescent peptidyl chloromethyl ketone), PCPS, and variable concentrations of factor Va. Fluorescence intensity is recorded for each sample following the addition of factor Va by averaging 32 readings over 10 sec. The experimental signal is provided by sample A, fluorescence baseline is provided by B, and the scattering contribution is provided by reaction C. The fractional fluorescence enhancement is calculated according to the following expression:

$$\frac{F}{F_0} = \frac{(F_A - F_C)}{(F_B - F_{C'})} \tag{1}$$

where F_A, F_B, and F_C are the fluorescence intensities recorded for reactions A–C and $F_{C'}$ represents the intensity recorded for reaction C in the absence of added factor Va. F/F_0 then yields the extent of fluorescence

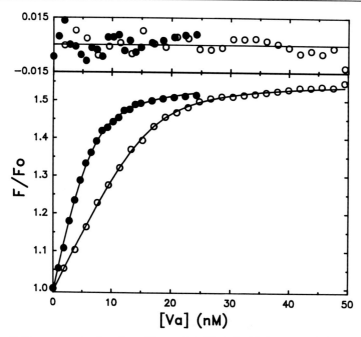

FIG. 3. Measurements of prothrombinase assembly at equilibrium by fluorescence intensity. Fluorescence intensity titrations were conducted using DEGR–Xa fixed at 6 nM (●) or 15 nM (○), 50 μM PCPS, and the indicated concentrations of Va in 20 mM HEPES, 0.15 M NaCl, 2 mM CaCl$_2$, pH 7.4. F/F_0 was measured using $\lambda_{ex} = 280$ nm and monitoring broadband fluorescence ($\lambda_{em} > 500$ nm). The data are analyzed to yield equilibrium parameters for the interaction between DEGR–Xa and factor Va on the membrane surface: $K_d = 1.0 \pm 0.2$ nM, $n = 1.13 \pm 0.02$ mol Va/mol Xa at saturation, and $F_{max}/F_0 = 1.56 \pm 0.02$. Adapted with permission from Krishnaswamy.[31]

enhancement relative to the signal observed with DEGR–Xa in solution or the DEGR–Xa–PCPS binary complex. Titration curves obtained by this procedure (Fig. 3) permit the calculation of the dissociation constant and stoichiometry for the interaction between DEGR–Xa and factor Va on the membrane surface, provided the concentration of PCPS is saturating.[31]

Rapid Kinetic Measurements of Prothrombinase Assembly

Stopped-flow measurements of prothrombinase assembly are conducted using an SLM-8000 fluorescence spectrophotometer equipped with a fluorescence stopped-flow box (Kinetic Instruments, Ann Arbor, MI) and controlled by data acquisition hardware and software from On-Line

Instrument Systems (Bogart, GA).[7] The temperature of the driving syringes and observation cell is regulated at 25° by a circulating water bath.

Protein solutions (DEGR–Xa and factor Va) are dialyzed into 20 mM HEPES, 0.15 M NaCl, 2 mM CaCl$_2$, pH 7.4, and further concentrated by centrifugal ultrafiltration if necessary prior to use. Because of the limitations of a two-syringe stopped flow, complex assembly is initiated in one of three configurations: (1) syringe A: DEGR–Xa, PCPS; syringe B: Va; (2) syringe A: DEGR–Xa, syringe B: PCPS, Va; and (3) syringe A: DEGR–Xa, Va, syringe B: PCPS. Typical final concentrations for these experiments range from 150 to 400 nM DEGR–Xa, 0–300 μM PCPS, and 0–2 μM factor Va. Complex assembly is measured (λ_{ex} = 280 nm, λ_{em} > 500 nm) by monitoring the fate of the reporter group (DEGR–Xa) following initiation by rapid mixing.

For each set of experimental conditions, the excitation slits are adjusted to maximize the sensitivity of the signal yet ensure that photobleaching over the course of the experiment does not complicate interpretations. The latter is verified by replicating traces at lower slit settings and comparing the amplitude of the observed fluorescence change to the expected values calculated from equilibrium measurements (above).

A typical stopped-flow experiment of prothrombinase assembly initiated by reacting the DEGR–Xa–PCPS binary complex (syringe A) with factor Va (syringe B) is illustrated in Fig. 4. The fluorescence change is reasonably described by a single exponential ($t_{1/2}$ ~ 11 msec) with an amplitude (F/F_0 = 1.51) that is consistent with "open-cuvette" experiments. Systematic studies of the dependence of the observed rate constant on reactant concentration along with stopped-flow light-scattering measurements have established that the rate-limiting step under these conditions corresponds to the initial interaction of factor Va with free sites on the vesicle surface prior to the assembly of the prothrombinase complex.

Binary Macromolecular Interactions within Prothrombinase

Fluorescence Anisotropy Measurements of Interaction of Factor Va with Phospholipid Vesicles

The reversible and tight interaction of factor Va with PCPS vesicles can be characterized using factor Va modified with N-(1-pyrene)maleimide on the heavy chain of the cofactor. These measurements rely on a change in fluorescence anisotropy of the reporter group when modified factor Va (Pyr-Va) binds to vesicles containing negatively charged phospholipid such as phosphatidylserine.[28]

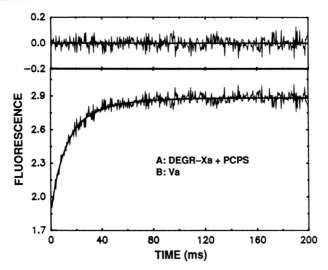

FIG. 4. Stopped-flow fluorescence measurements of prothrombinase assembly. Prothrombinase assembly was initiated by reacting equal volumes of syringe A (350 nM DEGR–Xa, 100 μM PCPS) and syringe B (760 nM factor Va in 20 mM HEPES, 0.15 M NaCl, 2 mM CaCl$_2$, pH 7.4). Prothrombinase assembly was monitored at 25° using λ_{ex} = 280 nm, λ_{em} > 500 nm. The data are fitted to a single exponential with the constants k_{obs} = 63.8 ± 1.7 sec^{-1}, F/F_0 (amplitude) = 1.51 ± 0.01, offset = 1.9. Adapted with permission from Krishnaswamy et al.[7]

Measurements are performed using an excitation wavelength of 330 nm (band pass 8 nm) and an emission wavelength of 396 nm (bandpass 16 nm) with long-pass filters (Schott KV-389) in the emission beams to minimize scattered-light artifacts. All buffers are filtered (0.45 μm), protein solutions are dialyzed into 20 mM HEPES, 0.15 M NaCl, 2 mM CaCl$_2$, pH 7.5, and centrifuged (10,000 g, 5 min, 4°) prior to use in the experiments. Reaction mixtures (2.0 ml) in 1 × 1-cm continuously stirred cuvettes contain a fixed concentration of Pyr-Va (10^{-7}–10^{-6} M) thermostatted at 25°. Titration curves are constructed by measuring polarization following incremental additions of PCPS vesicles to the reaction mixture. Polarization is measured approximately 20 sec following each addition by averaging 8–10 separate measurements obtained by integrating the signal at each position of the excitation polarizer for 10 sec. During the course of the titration, the anisotropy increases from a value of 0.155 in the absence of PCPS to a limiting value of 0.2 at saturating concentrations of PCPS. Systematic titritions obtained at two or more fixed concentrations of Pyr-Va are analyzed by the equations previously described,[28] to obtain equilibrium constants for the interaction of Va with phospholipid (Fig. 5).

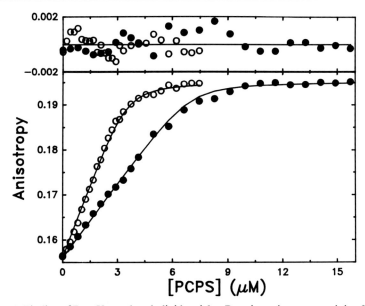

FIG. 5. Binding of Pyr–Va to phospholipid vesicles. Reaction mixtures containing 85 nM (O) or 170 nM (●) Pyr–Va in 20 mM HEPES, 0.15 M NaCl, 2 mM CaCl$_2$, pH 7.4, were titrated with increasing concentrations of PCPS. Fluorescence anisotropy was measured using λ_{ex} = 330 nm, λ_{em} = 396 nm. The lines are drawn according to the constants K_d = 2.7 ± 0.5 nM and stoichiometry = 42 ± 0.5 mol PCPS/mol Va at saturation. The residuals to the fitted line are shown in the upper panel. Adapted with permission from Krishnaswamy and Mann.[16]

The tight and well-defined interaction of Pyr-Va with PCPS membranes provides an ideal system for the evaluation of contributions of Va-derived peptides to the protein–membrane interaction or to infer the ability of other proteins to interact with PCPS vesicles through competition measurements. Typically, competition measurements are accomplished using low concentrations of Pyr-Va and subsaturating concentrations of PCPS (150 nM Pyr-Va, 4 μM PCPS) followed by anisotropy measurements (above) after incremental additions of the protein or peptide to be tested. Membrane binding of the competitor is inferred from its ability to reduce systematically the measured anisotropy to a limiting value corresponding to that observed with Pyr-Va in solution. Initial evidence of the displacement of Pyr-Va is further analyzed by repeating these experiments at two or more different fixed concentrations of PCPS, followed by analysis using equations previously described to extract the dissociation constant for the competitor–membrane interaction.[28] This approach is based on the assumption that the competitor and Pyr-Va compete for identical sites on the membrane surface and yields reasonable values when the equilibrium parameters for the competitor–PCPS interaction are comparable to those for factor Va. Qualitative evidence for the competitor–PCPS interaction can be obtained from such displacement curves when these conditions are not satisfied. Examples of both types of applications for these competition measurements are available in the literature.[28,32]

Rapid Kinetic Measurements of Interaction of Factor Va with Membranes

The binding of factor Va to small unilamellar PCPS vesicles results in a significant increase in vesicle radius with a concomitant increase in the intensity of scattered light.[4,33,34] Thus, the kinetics of the Va–membrane interaction can be continuously and conveniently monitored by changes in light-scattering intensity using fluorescence stopped-flow instrumentation.[7]

Prior to use, factor Va is dialyzed into 20 mM HEPES, 0.15 M NaCl, 2 mM CaCl$_2$, concentrated to values of 50 μM or greater by centrifugal ultrafiltration (Centricon 30, Amicon, Danvers, MA) and centrifuged (10,000 g, 10 min, 4°) to remove insolubles. Stopped-flow experiments are conducted in the same buffer, which is filtered (0.45 μm), degassed, and equilibrated at 25° prior to use. Appropriately diluted samples (2.0 ml, twice the final desired concentration) of factor Va and PCPS are prepared in Minisorp tubes (Nunc, Roskilde, Denmark) immediately before use and loaded into the driving syringes with care to prevent the introduction of air bubbles. The association of factor Va with PCPS membranes is then measured following initiation using $\lambda_{ex} = 320$ nm and monitoring broadband scattered light without an intervening emission monochromator or optical filters. Following preliminary shots, the excitation slit, channel gain, response time, and total collection time are adjusted to provide a ratiometric signal of reasonable amplitude and permit the collection of 200–400 data points over a time period corresponding to 10–20 half-lives of the observed change. Between six and eight replicate traces are collected for every set of experimental conditions.

Because of the rather large amplitude of the scattering change with factor Va, relatively noise-free data can be collected using final concentrations in the range of 0.1–0.4 μM factor Va and 10–250 μM PCPS with pseudo-first-order rate constants as high as 250 sec^{-1}. The results of a typical experiment illustrated in Fig. 6 demonstrate that when the concentration of PCPS is in excess, the association reaction can adequately be described by equations for a first-order process. Systematic studies of the

[32] M. A. Gendreau, S. Krishnaswamy, and K. G. Mann, *J. Biol. Chem.* **264,** 6972 (1989).
[33] M. L. Pusey and G. L. Nelsestuen, *Biochemistry* **23,** 6202 (1984).
[34] M. L. Pusey, L. D. Mayer, G. J. Wei, V. A. Bloomfield, and G. L. Nelsestuen, *Biochemistry* **21,** 5262 (1982).

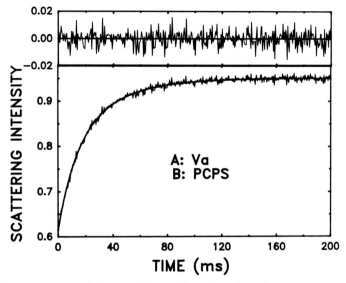

FIG. 6. Measurement of the association of factor Va with PCPS by stopped-flow light scattering. The binding of factor Va to PCPS was initiated by mixing equal volumes of 0.34 μM Va (syringe A) and 100 μM PCPS (syringe B) in 20 mM HEPES, 0.15 M NaCl, 2 mM CaCl$_2$, pH 7.4. The reaction was followed at 25° by measuring broadband scattered light using λ_{ex} = 320 nm. The data are analyzed according to a single exponential with the constants k_{obs} = 45.05 ± 0.45 sec^{-1}, amplitude = 0.319 ± 0.002, and offset = 0.629. Adapted with permission from Krishnaswamy et al.[7]

dependence of k_{obs} on PCPS concentrations permit the extraction of the site parameter (moles monomeric PCPS bound/moles Va) and the intrinsic second-order association rate constant expressed on a molar site basis.[7]

The dissociation of factor Va from PCPS membranes is also examined by stopped-flow light-scattering measurements by monitoring the decrease in light scattering following the mixing of the preformed Va-PCPS binary complex with an excess of the membrane-binding protein prothrombin fragment 1.[7] The basis for this approach is that prothrombin fragment 1 can displace factor Va from the membrane surface and the scattering intensity of the fragment 1–PCPS binary complex is substantially lower than that observed with factor Va bound to membranes.[7] For these experiments, reactions mixtures containing 0.2 μM Va and 20 μM PCPS are reacted with increasing concentrations of prothrombin fragment 1 (5–40 μM). The first-order decay in light-scattering intensity is demonstrated to be independent of the fragment 1 concentration to ensure that a limiting rate constant corresponding to the dissociation of factor Va from the membrane surface is achieved.[7]

While approaches for the study of the kinetics of the factor Va–membrane interaction are described, we have also used the same approach to examine the binding of vitamin K-dependent proteins to PCPS, including factor Xa, prothrombin, and factor X.[7]

Reassociation of Factor Va from Isolated Subunits

Fully functional cofactor activity can be regenerated from the isolated subunits by mixing Va_{HC} and Va_{LC} in the presence of calcium ions. Typical reassociation reaction mixtures contain 10^{-7}–10^{-6} M Va_{HC} and Va_{LC} mixed together in 20 mM HEPES, 0.15 M NaCl, 2 mM $CaCl_2$ and incubated at 37° in low-protein binding tubes (Minisorp tubes). The extent of reassociation is evaluated by withdrawing samples at various times and measuring cofactor activity by direct measurements of prothrombin activation. Typically, initial rates of prothrombin activation are measured by initiating reaction mixtures (2.0 ml) containing 1.4 μM prothrombin, 20 μM PCPS, 3 μM dansylarginine-N-(3-ethyl-1,5-pentanediyl)amide (DAPA), and 10 nM factor Xa in 20 mM HEPES, 0.15 M NaCl, 2 mM $CaCl_2$ with 1 nM reassociation mixture (based on the concentration of the limiting subunit). The initial velocity of prothrombin activation, determined by fluorescence measurements as previously described,[11] is linearly dependent on the concentration of the functional cofactor formed. In typical reaction mixtures, we observe that the maximum cofactor activity regained (based on the concentration of the limiting subunit) generally ranges between 95 and 105% of the activity of intact factor Va. The time course for reassociation can be accurately predicted using the second-order rate expression, the concentrations of Va_{HC}, Va_{LC}, and the second-order rate constant of 1.6×10^5 M^{-1} min^{-1} determined for these experimental conditions.[35]

In addition to indirect measurements of reassociation of factor Va based on cofactor activity, the physical interaction between Va_{HC} and Va_{LC} is also measured by fluorescence energy transfer using Va_{HC} modified with fluorescein-5-maleimide and Va_{LC} modified with 6-acryloyl-2-dimethylaminonaphthalene (acrylodan).[35] The energy transfer signal in this case arises purely from the physical interaction between the two subunits of the cofactor.

The interaction between the two subunits of factor Va is evaluated by measurements of quenching of donor fluorescence using 120 nM acrylodan–Va_{LC} (donor) and increasing concentrations of fluorescein–Va_{HC} (acceptor) in 20 mM HEPES, 0.15 M NaCl, 2 mM $CaCl_2$, pH 7.4. Energy transfer measurements are conducted using three different reaction

[35] S. Krishnaswamy, G. D. Russell, and K. G. Mann, *J. Biol. Chem.* **264**, 3160 (1989).

mixtures to correct for the fluorescence contributions of the individual species: (A) acceptor control, 120 nM Va$_{LC}$ with increasing concentrations of fluorescein–Va$_{HC}$; (B) donor control, 120 nM acrylodan–Va$_{LC}$ with increasing concentrations of Va$_{HC}$; and (C) experimental reaction, 120 nM acrylodan–Va$_{LC}$ with increasing concentrations of fluorescein–Va$_{HC}$. The reaction mixtures (2.2 ml each) are allowed to incubate for 15 hr at 37° in the dark to permit the achievement of equilibrium. Spectra are collected on the three reactions using identical instrument settings by excitation at 395 nm (band pass 2 nm) and scanning the emission monochromator between 420 and 620 nm (band pass 16 nm) with a KV-408 long-pass filter in the emission beam. The extent of donor quenching is calculated by integrating the experimental and control spectra between 420 and 485 nm. Representative spectra at saturating concentrations of fluorescein–Va$_{HC}$ (Fig. 7a) illustrate that the reassociation of the subunits to yield factor Va results in substantial donor quenching ($\sim 70\%$), which can be completely prevented by the addition of an excess of unmodified Va$_{HC}$ (Fig. 7b).

The reassociation of factor Va on a preparative scale is useful for the preparation of cofactor derivatives containing fluorescent reporter groups in Va$_{LC}$ or derivatives with different reporter groups on the two subunits. Reassociation is achieved as described above using 54 μM Va$_{HC}$ (10 mg, 5 mg/ml) and 61 μM Va$_{LC}$ (10 mg, 5 mg/ml) followed by incubation for 1 hr at 37°. The reaction mixture is then dialyzed for 4 hr versus 20 mM HEPES 5 mM CaCl$_2$, pH 7.4, at 4° and applied to a 2.5 × 20-cm column of Fast-Flow Q-Sepharose (Pharmacia, Piscataway, NJ) equilibrated in the same buffer. Elution of bound proteins with a linear gradient of increasing NaCl (0–0.6 M, 700 ml, 2 ml/min) in the same buffer results in the quantitative separation of free Va$_{LC}$ (elution at 0.05 M NaCl) from reconstituted Va (elution at 0.35 M NaCl). Fractions containing reconstituted cofactor are identified by coagulation assays[11] and are pooled and precipitated by the addition of solid ammonium sulfate (80% saturation). Precipitated protein is collected by centrifugation (53,000 g, 20 min, 4°), dissolved in 20 mM HEPES, 5 mM CaCl$_2$, 50% (v/v) glycerol, and stored at $-20°$.

Measurement of Solution-Phase Interaction between Factors Xa and Va by Analytical Ultracentrifugation

Measurements of the interaction between factors Va and Xa in the absence of membranes are complicated by the relatively low affinity for this interaction ($K_d \sim 10^{-6}\ M$), in comparison to their tight interaction ($K_d \sim 10^{-9}\ M$) on the membrane surface.[27,36] Thus, fluorescence approaches developed for physical measurements of prothrombinase assembly at low

[36] D. S. Boskovič, A. R. Giles, and M. E. Nesheim, *J. Biol. Chem.* **265**, 10497 (1990).

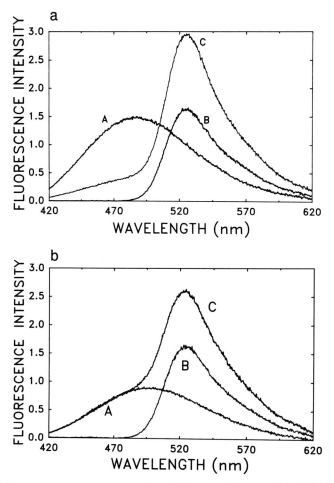

FIG. 7. Fluorescence resonance energy transfer between Va_{HC} and Va_{LC} following reassociation. (a) Corrected emission spectra were obtained using fluorescein–Va_{HC} (300 nM) and acrylodan–Va_{LC} (120 nM) reassociated to completion in 20 mM HEPES, 0.15 M NaCl, 2 mM CaCl$_2$, pH 7.4, at 37°. The excitation wavelength was 395 nm and spectra were obtained using identical settings: (A) acrylodan–Va_{LC} and unlabeled Va_{HC} (donor control); (B) fluorescein–Va_{HC} and unlabeled Va_{LC} (acceptor control); and (C) acrylodan–Va_{LC} and fluorescein–Va_{HC} (experimental spectrum). (b) The specificity of the observed signal was assessed by examining the ability of unlabeled Va_{HC} to reduce the extent of energy transfer. (A) 80 nM acrylodan–Va_{LC} plus 1.8 μM Va_{LC}; (B) 80 nM unlabeled Va_{LC}, 300 nM fluorescein–Va_{HC} plus 1.5 μM unlabeled Va_{HC}; and (C) 80 nM acrylodan–Va_{LC}, 300 nM fluorescein–Va_{HC}, and 1.5 μM unlabeled Va_{HC}. Reprinted with permission from Krishnaswamy et al.[35]

concentrations of reactants are not well suited for studies of the phospholipid-independent interaction between factors Xa and Va. However, the ability to incorporate chromophores selectively and differentially into factors Xa and Va permits direct measurements of the protein–protein interaction through sedimentation velocity measurements by taking advantage of the optical selectivity of the photoelectric scanning analytical ultracentrifuge. Using one labeled reactant and the appropriate detection wavelength, the interacting system is evaluated by monitoring the sedimentation velocity of a single chromophore-modified species. In this manner, the interpretation of multicomponent data is simplified by eliminating one of three superimposed sedimentation boundaries and the complication of several extinction coefficients. This approach has been useful for measurements of the interaction between Va_{HC} and Va_{LC} and for measurements of the interaction between factors Xa and Va in the absence of membranes.[27,35] The methodology for the latter measurements is provided below.

The solution-phase interaction between factors Xa and Va is evaluated by sedimentation velocity measurements using reaction mixtures containing (1) factor Xa and fluorescein-5-maleimide-modified factor Va (above) or (2) native factor Va and factor Xa modified with rhodamine X–glutamylglycinylarginyl chloromethyl ketone to yield rx–EGR–Xa. Because of the small change in sedimentation behavior of factor Va on interacting with factor Xa, the former approach yields qualitative information regarding this interaction. Quantitative information regarding the enzyme–cofactor interaction is obtained from the second approach, by taking advantage of the large change in the sedimentation velocity of rx–EGR–Xa following its interaction with unlabeled factor Va.

Centrifugation is performed with a Beckman Model E equipped with mirrored optics and a cylindrical lens to capture and focus stray incident monochromatic light. The improved sensitivity of the altered centrifuge optics (0.05 OD) and the high extinction coefficients of the chosen chromophores (70,000 M^{-1} cm^{-1}) permit measurements at 10^{-6} M concentrations of protein. Protein samples are dialyzed extensively into 20 mM HEPES, 0.15 M NaCl, 2 mM CaCl$_2$, pH 7.4, prior to use. Reaction mixtures (380 μl) containing the desired concentrations of reactants are loaded into double-sector cells housed with sapphire windows and centrifuged at 60,000 rpm (260,000 g) at 22°. The sedimentation of the chromophore-labeled protein is monitored by absorbance scans performed

[37] C. R. Cantor and P. R. Schimmel, "Biophysical Chemistry: Part II. Techniques for the Study of Biological Macromolecules." Freeman, San Francisco, 1980.
[38] R. J. Goldberg, J. Phys. Chem. 57, 194 (1953).

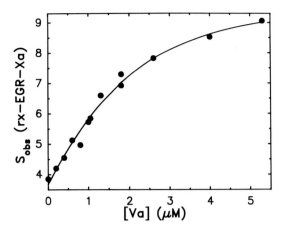

FIG. 8. Measurement of the phospholipid-independent interaction between rx–EGR–factor Xa and factor Va. The weight average sedimentation of 1.8 μM rx–EGR–factor Xa was monitored at 582 nm in the presence of various concentrations of factor Va. All samples were in 20 mM HEPES, 0.15 M NaCl, pH 7.4, 2 mM CaCl$_2$ at 20°. The line is drawn assuming a stoichiometry of 1 mol factor Va/mol rx–EGR–Xa at saturation, with the fitted parameters K_d = 0.8 ± 0.28 μM, S_{OBS} in the absence of added Va = 3.67 ± 0.13, and S_{OBS} at saturation = 10.14 ± 0.5. Adapted with permission from Pryzdial and Mann.[27]

at the absorption maxima of the chromophores ($\lambda = 495$ nm for fluorescein–Va or $\lambda = 572$ nm for rx–EGR–Xa). Scans are initiated immediately following the achievement of the desired rotor speed and repeated at 4-min intervals. The photomultiplier carriage position and absorbance output signals are digitized through an ISAAC data acquisition system (Cyborg). Weight average sedimentation coefficients of the labeled species are extracted from the digitized data by a linear regression analysis of a plot of the natural logarithm of the equivalent boundary position versus time.[37] The equivalent boundary position is calculated from the square root of the second moment of the individual scans.[38]

Sedimentation measurements as described above are repeated using a fixed concentration of rx–EGR–Xa and increasing concentrations of unlabeled factor Va, to yield the weight average sedimentation coefficient as a function of factor Va concentrations. Because this interaction is not significantly perturbed during differential mass transport in the cell, changes in the weight average sedimentation coefficient of rx–EGR–factor Xa are directly proportional to the fractional saturation of the rx–EGR–factor Xa at equilibrium.[27] Plots of the observed weight average sedimentation coefficient as a function of factor Va (Fig. 8) are analyzed to extract a dissociation constant for this interaction.[27]

TABLE I
CONSTANTS FOR TERNARY AND BINARY INTERACTIONS WITHIN PROTHROMBINASE

Interaction[a]	K_d (M)	n^b (mol/mol)	$k_{association}^c$ ($M^{-1} sec^{-1}$)	$k_{dissociation}$ (sec^{-1})	Ref.
Xa + nPCPS $\rightleftharpoons$ Xa–PCPS$_n$	1.1×10^{-7}	46	2.9×10^7	3.3	e
Va + nPCPS $\rightleftharpoons$ Va–PCPS$_n$	2.7×10^{-9}	42	5.7×10^7	0.17	e,f
Xa–L + nVa–L $\rightleftharpoons$ L–Xa–Va$_n$-L	1.0×10^{-9}	1.13	$\geq 1 \times 10^9$	ND^d	g,h
nVa$_{HC}$ + Va$_{LC}$ $\rightleftharpoons$ Va	5.9×10^{-9}	1	2.6×10^3	1.7×10^{-5}	i
Xa + nVa $\rightleftharpoons$ Xa–Va$_n$	8×10^{-7}	1	ND	ND	j,k

[a] The indicated reactions were measured in 20 mM HEPES, 0.15 M NaCl, 2 mM CaCl$_2$, pH 7.4, at 25°. The constants for the interaction between Va$_{HC}$ and Va$_{LC}$ were obtained at 37°.

[b] Stoichiometry for the observed interaction. For protein–membrane interactions, n indicates moles of monomeric phospholipid bound per mole of protein at saturation and L indicates protein-binding sites on the vesicle surface. For protein–protein interactions, n indicates moles of factor Va bound per mole of Xa at saturation.

[c] Association rate constant expressed as M sites^{-1} sec^{-1} for protein–PCPS interactions.

[d] ND, Not determined.

[e] S. Krishnaswamy, K. C. Jones, and K. G. Mann, J. Biol. Chem. **263**, 3823 (1988).

[f] S. Krishnaswamy and K. G. Mann, J. Biol. Chem. **263**, 5714 (1988).

[g] M. E. Nesheim, J. B. Taswell, and K. G. Mann, J. Biol. Chem. **254**, 10952 (1979).

[h] S. Krishnaswamy, J. Biol. Chem. **265**, 3708 (1990).

[i] S. Krishnaswamy, G. D. Russell, and K. G. Mann, J. Biol. Chem. **264**, 3160 (1989).

[j] D. S. Boskovič, A. R. Giles, and M. E. Nesheim, J. Biol. Chem. **265**, 10497 (1990).

[k] E. L. G. Pryzdial and K. G. Mann, J. Biol. Chem. **266**, 8969 (1991).

Summary

The approaches described in this article have resulted in an increased understanding of the reaction steps involved in the stabilization and assembly of the prothrombinase complex. Because prothrombinase is considered an archetype for some of the other coagulation complexes, the quantitative information derived from these studies (Table I) provides the framework for future studies of prothrombinase and suggests experimental approaches for studies of the other analogous coagulation reactions.

Acknowledgment

Supported by NIH Grants HL-38337 (S.K.), HL-34575, and HL-35058 (K.G.M.), Grant MA-9781 from the Medical Research Council of Canada (M.E.N.), and a postdoctoral fellowship from the Medical Research Council of Canada (E. P.).

[16] Procoagulant Activities Expressed by Peripheral Blood Mononuclear Cells

By PAULA B. TRACY, ROYCE A. ROBINSON, LAURA A. WORFOLK, and DEBRA H. ALLEN

Introduction

The ability of mononuclear cells, particularly monocytes and macrophages, to participate in the molecular events leading to thrombin formation is an integral part of their physiological and pathophysiological roles in wound repair, chronic inflammation, and atherosclerosis. There are several lines of investigation which support this concept. First, there is extensive fibrin deposition and an active fibroproliferative response accompanying these events. Also, because monocytes interact with vascular endothelium, platelets, and mesenchymal cells, and because all of these cells respond in a variety of ways to thrombin, it seems likely that the production of thrombin at the monocyte/macrophage membrane surface provides an important bioregulatory effector molecule at precise locations.

We and others have begun to explore the concept that the generation of procoagulant activities are an important mechanism by which mononuclear cells function. For example, it is clear that monocytes can provide the appropriate membrane surface for the assembly and function of virtually all the coagulation complexes involved in thrombin production. Monocytes express tissue factor at the membrane surface which binds factor VIIa, allowing the proteolytic activation of factor X to factor Xa, thereby initiating the extrinsic pathway of coagulation.[1] Propagation of the coagulant response is accomplished by the assembly and function of the prothrombinase complex, a stoichiometric complex of the cofactor factor Va and the enzyme factor Xa that, bound to the monocyte surface in the presence of calcium ions, effects the proteolytic conversion of prothrombin to thrombin.[2,3] Factor Va may be provided through the activation of the plasma procofactor factor V by a protease associated with the monocyte membrane.[4] Assembly of the prothrombinase complex may be further facilitated by the provision of additional factor Xa through the functional interactions of factors VIIIa and IXa with the monocyte surface.[5] Once

[1] R. L. Edwards and F. R. Rickles, *Prog. Hemostasis Thromb.* **7**, 183, (1984).
[2] P. B. Tracy, M. S. Rohrbäch, and K. G. Mann, *J. Biol. Chem.* **258**, 7264 (1983).
[3] P. B. Tracy, L. L. Eide, and K. G. Mann, *J. Biol. Chem.* **260**, 2119 (1985).
[4] D. H. Allen and P. B. Tracy, *Blood* **78**, Suppl. 1, Abstr. No. 240 (1991).
[5] M. P. McGee and L. C. Li, *J. Biol. Chem.* **266**, 8079 (1991).

generated, factor Xa not only participates in prothrombin activation, but also, when bound to the monocyte membrane independently of factor Va, will cleave factor IX to the activation intermediate, factor IXα.[6,7] Factor IXα then serves as a substrate for the assembled tissue factor/factor VIIa complex, leading to the generation of the active enzyme factor IX$\alpha\beta$.[7] All these reactions are required for the effective formation of thrombin. Thrombin, once generated, can bind to the monocyte membrane[8] and cleave bound fibrinogen to fibrin monomers,[9] which are subsequently cross-linked by a factor XIIIa-like molecule released from the cells.[10]

The ability of the monocyte to regulate the procoagulant response is supported by several observations. Initiation of the coagulant response is not achieved until the monocyte is stimulated to express tissue factor by an exogenous agonist.[1] The tissue factor expressed at the membrane surface is not fully functional, however, and requires additional cell stimulation to achieve maximal expression of cofactor activity.[11] Even though monocyte stimulation is not required for the functional interactions of factors Va and Xa[2,3] or factors VIIIa and IX$\alpha\beta$[5] with the monocyte surface, endotoxin stimulation significantly enhances prothrombin activation.[12] Yet, it is apparently without effect on the factor IXa-catalyzed activation of factor X.[5] This chapter details methods to assess several of these various procoagulant activities expressed by purified subpopulations of peripheral blood mononuclear cells, particularly monocytes and T lymphocytes.

Cell Isolation

The following protocols outline procedures for the purification of peripheral blood mononuclear cells and isolated subpopulations of monocytes and lymphocytes. New protocols have been developed, or published methods have been modified to prepare cell suspensions which have not been unintentionally stimulated with agonists (such as endotoxin) or damaged by the use of adherence techniques followed by various lifting protocols. Consequently, techniques are outlined which rely solely on density gradient centrifugation methods. All cell culture media and buffers are prepared with endotoxin-free water and stored in high-temperature (180°) baked glassware. All reagents, sterile plasticware, and glassware are

[6] L. Worfolk and P. B. Tracy, *Blood* **78,** Suppl. 1, Abstr. No. 221 (1991).
[7] L. A. Worfolk, R. A. Robinson, and P. B. Tracy, *Blood* **80,** 1989 (1992).
[8] L. T. Goodnough and H. Saito, *J. Lab. Clin. Med.* **99,** 873 (1982).
[9] S. R. Gonda and J. R. Shainoff, *Proc. Natl. Acad. Sci. USA* **79,** 4565 (1982).
[10] L. J. Weisberg, D. T. Shiu, P. R. Conkling, and M. A. Shuman, *Blood* **70,** 579 (1987).
[11] T. A. Drake, W. Ruf, J. H. Morrissey, and T. S. Edgington, *J. Cell Biol.* **109,** 389 (1989).
[12] R. A. Robinson, L. A. Worfolk, and P. B. Tracy, *Blood* **79,** 406 (1992).

screened for endotoxin contamination using the *Limulus* amebocyte lysate assay (Associates of Cape Cod, Woods Hole, MA), with a detection minimum of 0.06 ng/ml. Using methods detailed below as well, the cells are assayed for their ability to provide the appropriate membrane surface or the cofactors required for several described procoagulant reactions.

Mononuclear Cell Isolation

Mononuclear cells (MNCs), free of platelet contamination, are isolated from the buffy coat of one unit of blood using citrate–phosphate–dextrose–adenine as anticoagulant.[2,12] The buffy coat is obtained by centrifugation of the unit of blood at 1600 g for 4 min in a swinging-bucket rotor. Following expression of the platelet-rich plasma the buffy coat is subsequently expressed (35–50 ml) and diluted 1:4 with cold (4°) Ca^{2+}- and Mg^{2+}-free Hanks' balanced salt solution, pH 7.4, containing 20 mM EDTA (HBSS–EDTA). Aliquots of the diluted buffy coat (35 ml) are layered over 15 ml of cold Ficoll/Hypaque (1.076 g/cm³, Sigma Chemical Co., St. Louis, MO) in 50-ml polycarbonate centrifuge tubes and centrifuged at 400 g for 20 min at 4°. The MNCs are harvested from each interface, washed initially by a 5- to 10-fold dilution in cold HBSS–EDTA, and centrifuged at 200 g for 10 min. Following two additional washes with HBSS–EDTA the MNCs are resuspended in 2 ml of HBSS containing 1% endotoxin-free human serum albumin (American Red Cross, Burlington, VT) or low-endotoxin bovine serum albumin [Irvine Scientific, Santa Ana, CA (HBSS–albumin)] and counted using a hemocytometer to allow for the further purification of T lymphocytes and monocytes. If the MNCs are used in the assays described below they are resuspended in the appropriate buffer.

Mononuclear cell composition is distinguished using modified Wright's and myeloperoxidase stains. Typically, the final MNC preparations contain 25–35% monocytes, 65–75% lymphocytes, <1% neutrophils, and are devoid of platelets. Cell viability is typically >98% as determined by nigrosin or trypan blue dye exclusion.

T Lymphocyte Isolation

Purified suspensions of T lymphocytes are isolated from MNCs by rosette formation with sheep red blood cells (SRBCs) treated with 2-aminoethylisothiouronium bromide hydrobromide[12,13] (AET, Sigma Chemical Co., St. Louis, MO; see below). MNCs suspended in HBSS–albumin are

---

[13] M. A. Pellegrino, S. Ferrone, and A. N. Theofilopoulos, *J. Immunol. Methods* **11**, 273 (1976).

diluted to a cell count of 2×10^7 cells/ml and mixed with an equal volume of AET-treated SRBCs suspended in HBSS–albumin. After centrifugation at 200 g for 8 min at 4° to pellet the cells, they are then maintained at 4° for 30–60 min. Following resuspension, aliquots of the cells (30 ml) are layered over 15 ml of cold Ficoll/Hypaque and centrifuged at 400 g for 30 min at 4°. The interface consisting of monocytes and B lymphocytes is collected and used to purify monocytes (see below). The SRBC/T lymphocyte pellets are resuspended and washed once with HBSS (50 ml). The T lymphocytes are isolated by resuspension of the rosettes and lysis of the SRBCs at 4° by addition of 61 mM NH$_4$Cl, 11 mM KHCO$_3$ (50 ml), followed by centrifugation at 200 g for 10 min at ambient temperature, followed by two additional washes in HBSS. Based on modified Wright's and myeloperoxidase stains the T lymphocytes are 95–98% pure with >98% viability as determined by trypan blue dye exclusion.

AET Treatment of SRBCs

In this procedure[14] SRBCs are obtained commercially (Crane Laboratories, Syracuse, New York) suspended in Alsever's solution. The SRBCs (30 ml) are centrifuged at 200 g for 10 min and washed three times using 50 ml of HBSS. Following the final wash, 20 ml of 0.28 M AET, pH 8.5, is added to 5 ml of packed cells. A 1 : 4 ratio of packed cells to AET should be maintained. The cell suspension is incubated at 37° for 20 min, inverting occasionally to mix. The treated SRBCs are centrifuged at 200 g for 10 min and washed 4–5 times using HBSS until the supernatant is clear. The AET-treated SRBCs are stored as a 1% solution (1 ml of packed cells plus 99 ml of Alsever's solution) at 4° and used within 5 days. Prior to use in rosetting, the cells are washed once with HBSS and resuspended to their original volume in HBSS–albumin.

Monocyte Isolation

The monocyte and B lymphocyte interface obtained following SRBC rosette formation is washed two times with HBSS by centrifugation at 200 g for 10 min.[2] Following the second wash, the cells are suspended in HBSS and diluted to a cell concentration of 1×10^8 cells/ml. Monocytes are then purified by centrifugation through a continuous Percoll gradient $(1.03–1.12 \text{ g/cm}^3)$[15] in phosphate-buffered saline (PBS). Gradients are prepared by mixing 7 ml of Percoll (Pharmacia, Piscataway, NJ) with 6 ml

[14] M. A. Pellegrino, S. Ferrone, M. P. Dierich, and R. A. Reisfeld, *Clin. Immunol. Immunopathol.* **3**, 324 (1975).
[15] F. Gmelig-Meyling and T. A. Waldmann, *J. Imunol. Methods* **33**, 1 (1980).

of 25 mM sodium phosphate, 300 mM NaCl, pH 7.4, followed by centrifugation at 20,000 g for 10 min at room temperature using a fixed-angle rotor. Cell separation is achieved by overlaying 1 ml of cells (1 × 10^8/ml) on a 13-ml gradient followed by centrifugation at 1000 g for 20 min in a swinging-bucket rotor. The monocyte bands (1.06–1.065 g/cm^3) are combined, washed two times with HBSS, and collected by centrifugation at 400 g. The final monocyte preparations contain >85% monocytes with the remaining contamination predominantly lymphocytic. These preparations are <1% contaminated with neutrophils and devoid of platelets. Cells are >98% viable as determined by nigrosin or trypan blue dye exclusion.

When necessary, MNCs or monocytes are labeled with ^{51}Cr (Amersham, Arlington Heights, IL). Cells (1 × 10^7/ml) suspended in HBSS–albumin are incubated with ^{51}Cr (1 μCi/10^6 cells) for 20 min at 37°, followed by three washes (200 g, 10 min) to remove free chromium. Cells are resuspended, counted on a hemocytometer, and the amount of ^{51}Cr associated per cell is determined.

Cell Culture and Stimulation Protocols

Cells are suspended in siliconized glass or Teflon tubes (1–10 × 10^6/ml) in culture medium consisting of a 1:1 mixture of low-endotoxin RPMI 1640: Earle's medium 199 (Irvine Scientific, Santa Ana, CA) supplemented with 10% heat-inactivated, low-endotoxin fetal bovine serum (FBS; Irvine Scientific), penicillin (100 U/ml), streptomycin (100 μg/ml), and L-glutamine (2 mM), pH 7.4, and maintained at 37° in 5% CO$_2$.[2,12] When necessary, cells are stimulated with various concentrations of endotoxin [0.01–1 μg/ml (*Escherichia coli* lipopolysaccharide O55:B5; Sigma Chemical Co., St. Louis, MO)] for 1–24 hr at 37° in 5% CO$_2$.

Assays for Assembly and Function of Macromolecular Coagulant Enzyme Complexes on Monocytes and Lymphocytes

An appropriate membrane surface is required for the functional assembly of the protein constituents of the known macromolecular coagulant enzyme complexes.[16] Assays are described to assess how monocytes and lymphocytes participate in and potentially regulate both prothrombinase complex assembly and function, as well as the tissue factor/factor VIIa-catalyzed activation of factor X through the expression of monocyte-associated tissue factor.

---

[16] K. G. Mann, R. J. Jenny, and S. Krishnaswamy, *Annu. Rev. Biochem.* **57**, 915 (1988).

Prothrombinase Complex Assembly and Function on Cells in Suspension

An assay previously described to quantitate factor Va cofactor activity[17] has been modified to assess kinetically prothrombinase complex assembly and function at the bovine[18,19] and human platelet surface[2,19] and to describe the functional binding interactions of factors Va and Xa with both unstimulated[3,12] and stimulated[12] human monocytes and lymphocytes. This assay can be used also to describe changes in the cell membrane which modulate prothrombinase complex assembly and function.[12] Thrombin formation is monitored continuously by the change in fluorescence intensity of dansylarginine *N*-(3-ethyl-1,5-pentanediyl)amide (DAPA) as it interacts with thrombin, as described in detail previously.[17,20]

Materials. DAPA is synthesized as described[20] or obtained commercially (Haematologic Technologies, Essex Junction, VT). Human factor X and prothrombin are isolated as described by Bajaj *et al.*[21] Factor X is applied to an α-human protein C immunoaffinity column to remove trace protein C contamination undetectable by gel electrophoresis. Factor Xa is prepared as described by Jesty and Nemerson[22] using the factor X activator isolated from Russell's viper venom.[23] Taipan snake venom (Sigma, St. Louis, MO) is used to activate prothrombin to α-thrombin as previously described.[24] Factor V is isolated using immunoaffinity chromatography as described,[17,25] and activated with 3 units/ml of thrombin for 10 min at 37°.[17] Following factor V activation, thrombin is inhibited with 5 μM DAPA. Factors Va and Xa are prepared as 1 μM stock solutions in either 5 mM HEPES-buffered Tyrode's solution, pH 7.4 (0.137 M NaCl, 2.7 mM KCl, 12 mM NaHCO$_3$, 0.36 mM NaH$_2$PO$_4$, 1 mM MgCl$_2$, 2 mM CaCl$_2$, 5 mM dextrose), containing 0.35% recrystallized bovine serum albumin (HTA) or 20 mM HEPES, 0.15 M NaCl, 5 mM CaCl$_2$, pH 7.4, supplemented with albumin, as well. The stock solutions are kept on ice and remain stable for as long as 12 hr. Vesicles of 25% phosphatidylserine and 75% phosphatidylcholine (PCPS) are prepared as described by Barenholz *et al.*[26]

[17] M. E. Nesheim, J. A. Katzmann, P. B. Tracy, and K. G. Mann, this series, vol. 80, p. 249.
[18] P. B. Tracy, M. E. Nesheim, and K. G. Mann, *J. Biol. Chem.* **256**, 743 (1981).
[19] P. B. Tracy, M. E. Nesheim, and K. G. Mann, this series Vol. 215, p. 329.
[20] M. E. Nesheim, F. G. Prendergast, and K. G. Mann, *Biochemistry* **18**, 996 (1979).
[21] S. P. Bajaj, S. I. Rapapport, and C. Prodanos, *Prep. Biochem.* **11**, 394 (1981).
[22] J. Jesty and Y. Nemerson, this series, Vol. 45, p. 95.
[23] W. Kissiel, M. A. Hermodson, and E. W. Davie, *Biochemistry* **15**, 4901 (1976).
[24] W. G. Owen and C. M. Jackson, *Thromb. Res.* **3**, 705 (1973).
[25] J. A. Katzmann, M. E. Nesheim, L. S. Hibbard, and K. G. Mann, *Proc. Natl. Acad. Sci. USA* **78**, 162 (1981).
[26] Y. Barenholz, D. Gibbs, B. J. Littmann, J. Goll, E. Thompson, and F. D. Carlson, *Biochemistry* **16**, 2806 (1977).

Assay Procedure. Before samples can be assayed the maximum fluorescence change equal to total substrate conversion and representing 80–90% of full recorder scale is established as follows. A substrate solution consisting of prothrombin (0.2 mg/ml, 2.78 μM) and DAPA (6 μM) in 5 mM HEPES, 0.15 M NaCl, 5 mM Ca^{2+}, pH 7.4, is prepared and kept in the dark. The substrate (0.75 ml) is added to a cuvette containing 0.75 ml of the cells (1 – 10 × 10^6/ml) or cell-derived material to be assayed. Phospholipid vesicles (5 μM final concentration in an aliquot typically less than 10 μl) and factor Va [5 nM final concentration (7.5 μl of a 1 μM stock)] are added. The signal generated due to the intrinsic fluorescence of DAPA (λ_{em} 335 nm, λ_{ex} 565 nm) and light scattering of the cells is offset to zero. Thrombin generation is initiated by the addition of factor Xa [5 nM final concentration (7.5 μl of a 1 μM stock)]. Following completion of the reaction, i.e., no further change in the fluorescence signal, the signal is amplified by sensitivity adjustments on the fluorimeter such that 80–90% of full scale represents total substrate consumption. Test reactions are then analyzed without any further sensitivity adjustments on the fluorimeter. Phospholipid vesicles are omitted from the test reactions to ensure that only the cells or cell-derived material are providing the required membrane surface.

Following the addition of the substrate solution, cells or cell-derived material, and factor Va, any baseline fluorescence signal detected again is offset to zero. The reaction is initiated by factor Xa and the rate of thrombin formation is continuously recorded, allowing for calculation of initial rate data. In assays to monitor changes in the cell membrane, the cell concentration is held constant and rate-saturating concentrations of factor Va and Xa are used, as previously determined, to ensure that the membrane surface is the only limiting component. As shown in Fig. 1, this assay can be used to monitor endotoxin-induced changes in both the monocyte cell membrane and monocyte-derived vesicles over time, whereas lymphocytes remain unaffected.[12] Alternatively, this assay can be used to deduce the apparent functional binding interactions of factor Va (or factor Xa) with the monocyte membrane by varying the concentration of either ligand while the other is held constant at a rate-saturating, nonlimiting concentration (Fig. 2).[3]

Prothrombinase Complex Assembly and Function of Adherent Monocytes

Monocytes labeled with ^{51}Cr as described earlier are diluted into media containing 10% low-endotoxin fetal bovine serum to a cell concentration of 5 – 10 × 10^6 ml. Aliquots of the monocyte suspension (2 ml) are added to 12-well tissue culture plates and allowed to adhere for 1 hr at 37° in 5%

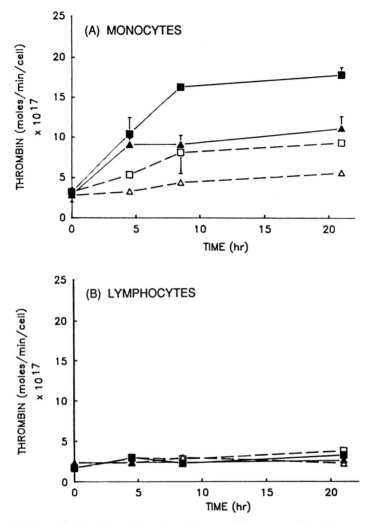

FIG. 1. Effect of endotoxin on the prothrombinase activity expressed by reisolated monocytes and T lymphocytes and their respective cell-free supernatants. Freshly isolated monocytes (A) and T lymphocytes (B) were diluted to 3.8×10^6 cells/ml and incubated over a 20-hr period at 37°, 5% CO_2, in the presence (■, ▲) and absence (□, △) of endotoxin (1 μg/ml). At the indicated time points, aliquots of the cell suspensions were removed and centrifuged at 550 g for 10 min. The cell-free supernatants (□, ■) were removed and assayed for prothrombinase activity. The cells (△, ▲) were resuspended to their original volume and assayed similarly. The data represent the mean and range of two experiments. In some instances, the error bars are within the symbol. Reproduced with permission from Robinson et al.[12]

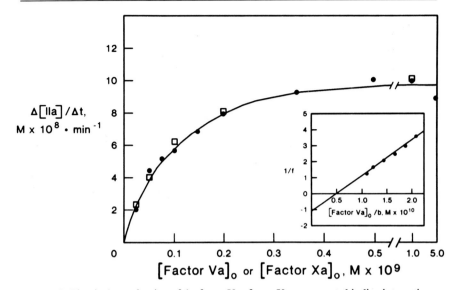

FIG. 2. Kinetic determination of the factor Va–factor Xa–monocyte binding interactions. In one set of assays, the factor Va concentration was varied as indicated and the reaction was initiated by addition of 5 nM factor Xa. Alternatively, factor Va (5 nM) was held constant and the reaction was initiated by varying the factor Xa concentration. The factor Va binding parameters were determined by double-reciprocal analysis of the saturation curve shown (●); apparent $K_d = 0.4 \times 10^{-10}$ M, $n = 12,000$, $k_{cat} = 34$ mol of thrombin/sec/mole of factor Va bound. The influence of factor Xa on thrombin generation in the presence of saturating amounts of factor Va (5 nM) is shown (□) and is identical to the curve obtained with varying amounts of factor Va, indicating that factor Va and factor Xa form a 1:1 stoichiometric complex on the monocyte surface. Reproduced with permission from Tracy et al.[3]

CO_2. Following incubation, the nonadherent cells are removed by gentle aspiration and the adherent cells are washed three times with 2 ml of 5 mM HEPES-buffered Tyrode's solution, pH 7.4, containing 0.35% bovine serum albumin (HTA) to remove loosely adherent cells. Following the third wash, the adherent cells are assayed for prothrombinase activity essentially as described previously for cultured endothelial cells.[27] The plated cells are overlayed with 2 ml of HTA containing 3 μM DAPA, to which is added equimolar concentrations of factors Va and Xa (5 nM final concentration, 10 μl of 1 μM stocks prepared in HTA); the cells are placed on an orbital shaker (1000 rpm). DAPA is included in the assay to prevent thrombin cleavage of the substrate to prethrombin-1, a poorer substrate for the prothrombinase complex. The reaction is initiated by the addition of

[27] M. R. Visser, P. B. Tracy, G. M. Vercellotti, J. L. Goodman, J. G. White, and H. S. Jacob, Proc. Natl. Acad. Sci. USA 85, 8227 (1988).

prothrombin (1.4 μM final concentration, typically 40 μl of a 70 μM stock). Multiple assays can be run simultaneously by adding substrate at 10-sec intervals. Aliquots (100 μl) of the activation mixtures are removed at 2, 4, 6, 8, and 10 min and added to 200 μl of cold (4°) 20 mM HEPES, 0.15 M NaCl, 0.05 M EDTA, 0.1% polyethylene glycol (PEG) 8000, pH 7.4 (quench buffer). The remaining reaction mixture in each well is removed by aspiration, followed by the addition of 1 N NaOH (1 ml) and vigorous trituration to lyse the cells. Following a 30-min incubation at ambient temperature, the lysate is removed. Each well is subsequently washed with an additional 1 ml of 1 N NaOH, which is combined with the original lysate and analyzed for ^{51}Cr activity by gamma counting. The number of cells present per well is determined based on the ^{51}Cr activity/cell calculated previously.

The rate of thrombin generation/well is calculated by determination of the thrombin concentration in the reaction subsamples removed from each well over time. Thrombin concentration is determined using the chromogenic substrate S2238 or Spectrozyme TH. Assays are done in a 96-well microtiter plate for analysis on a Molecular Devices (Menlo Park, CA) kinetic microplate reader. Assay mixtures contain 80 μl of substrate (0.4 mM) in quench buffer and 20 μl of the thrombin source. The rate of chromogenic substrate hydrolysis is determined and captured using software which calculates the thrombin concentration based on a standard curve (0–200 nM) prepared initially using purified enzyme. The rate of thrombin generated per cell can then be calculated.

Table I lists data comparing the thrombin generated, via the prothrombinase complex, by monocytes adhered to plastic or maintained in suspension prior to and following a 15-hr incubation with endotoxin. The adherent cells expressed considerably less prothrombinase activity then those maintained in suspension. These observations may reflect inaccessible membrane surface due to adherence. However, the adherent cells expressed a 3.6-fold increase in prothrombinase activity in response to endotoxin, whereas the cells in suspension expressed a 1.5-fold activity enhancement.

Assessment of Monocyte-Associated Tissue Factor Activity

At various times (0–24 hr) following endotoxin stimulation (detailed previously), a 1.2-ml aliquot of the cell suspension (0.5–1 × 10^7 cells/ml) previously labeled with ^{51}Cr is removed, subjected to centrifugation (550 g, 10 min), and resuspended in 1.2 ml of 20 mM HEPES, 0.15 M NaCl, 5 mM CaCl$_2$, pH 7.4 (assay buffer). Two 50-μl aliquots are removed and the ^{51}Cr activity is measured for accurate determination of the resuspended cell

TABLE I
PROTHROMBINASE ACTIVITY EXPRESSED BY ENDOTOXIN-STIMULATED MONOCYTES IN
SUSPENSION OR ADHERED TO PLASTIC

	Prothrombinase activity[a] [moles thrombin generated/min/cell ($\times 10^{17}$)]; cell-associated activity	
Cells	0 hr	15 hr
Suspension		
Control	12.2 ± 0.06	16.7 ± 2.56
Endotoxin	12.2 ± 0.06	25.2 ± 0.28
Adherent		
Control	1.8 ± 0.21	1.9 ± 0.13
Endotoxin	1.8 ± 0.21	6.9 ± 0.67

[a] Mean $\pm$ SD ($n = 3$).

concentration. Half of the remaining sample is subjected to three cycles of sonication on ice using a Heat Systems ultrasonic (W375) sonicator (Farmingdale, NY) equipped with a microtip. Each cycle consists of 15 sec of sonication followed by a 30-sec rest period. The tissue factor activity associated with the samples, prior to and following sonication, is determined by its ability to serve as the required cofactor for the factor VIIa-catalyzed activation of factor X as follows. Two separate reaction mixtures are prepared in assay buffer and allowed to incubate for 10 min at 37°. Tube A (0.5 ml) contains 20 nM recombinant factor VII/VIIa (NOVO Biolabs, Danbury, CT) and aliquots of the tissue factor source (≤ 400 μl intact or sonicated cells). Tube B (0.4 ml) contains 1 μM human factor X. The 10-min incubation allows for the tissue factor/factor VIIa-catalyzed activation of factor VII prior to substrate addition. After both tubes are incubated for 10 min and before factor X activation is initiated, a time 0 sample is prepared by adding 50 μl from tube A to 300 μl of assay buffer containing 25 mM EDTA, followed by 50 μl from tube B. Factor X activation is then initiated by adding 350 μl from tube A to the remaining 350 μl in tube B. At 1, 2, 4, 10, and 15 min, 100-μl aliquots are removed from the activation mixture and added to 300 μl of assay buffer containing 25 mM EDTA to quench the reaction.

Factor Xa concentration in the reaction subsamples is determined using the chromogenic substrate S2222 or Spectrozyme fXa. Assay mixtures contain 150 μl of the factor Xa source and 200 μl of substrate (0.2 μM) in 20 mM HEPES, 0.15 M NaCl, 25 mM EDTA, 0.1% polyethylene

TABLE II
TISSUE FACTOR EXPRESSED BY INTACT AND SONICATED MONONUCLEAR CELLS
STIMULATED WITH ENDOTOXIN[a]

Endotoxin concentration (μg/ml)	Tissue factor activity[c] (nM factor Xa/min/5 × 10^6 cells)	
	Intact cells	Sonicated cells
0	—[b]	0.06
0.1	0.7 ± 0.3	1.9 ± 0.5
1.0	3.1 ± 1.3	17.6 ± 5.7

[a] Mononuclear cells (5 × 10^6/ml) were stimulated with varying concentrations of endotoxin for 7 hr.
[b] Undetectable activity.
[c] Mean ± SEM.

glycol 8000, pH 7.4. The rate of chromogenic substrate hydrolysis is determined with a kinetic microplate reader equipped with the necessary software to calculate factor Xa concentrations based on a standard curve (0–200 nM) prepared initially using purified enzyme.

Table II lists data comparing the tissue factor activity associated with mononuclear cells subsequent to endotoxin stimulation, prior to and following cell sonication. As has been demonstrated by several investigators,[11,28] cell sonication substantially increases the amount of tissue factor activity expressed. Prior to sonication, and dependent on the endotoxin concentration used, intact cells express only 18–33% of the activity of the sonicated cells.

Assessment and Visualization of Coagulation Factor Cleavage at Monocyte Surface

The assays described in the previous section indicate that monocytes can participate in two reasonably well-characterized procoagulant reactions. Monitoring those reactions required only detection of enzymatic activity. In several instances it may be necessary to monitor a proteolytic event by visualizing the cleavage products directly, because no enzymatic activity may be apparent.[7] Consequently, in this section, methods are described for the autoradiographic visualization of radioiodinated coagulation proteins.

[28] R. Bach and D. B. Rifkin, *Proc. Natl. Acad. Sci. USA* **87,** 6995 (1990).

Assessment of Cleavage of ^{125}I-Labeled Factor IX by Factor Xa Bound to Monocytes

Materials. Human factor IX can be purified as described by Bajaj *et al.* and radioiodinated using the Iodogen transfer technique.[18,19] Na[125]I (~ 1 mCi/0.2 mg protein) is added to an Iodogen-coated tube (1 μg Iodogen/10 μg protein) in 100 μl of 0.02 M Tris, 0.15 M NaCl, pH 7.4. Following 5 min of gentle vortexing, the oxidized iodonium ion is transferred to a separate tube containing $\leq$0.5 ml factor IX (0.4 mg/ml) in 0.02 M Tris, 0.15 M NaCl containing 5 mM benzamidine, pH 7.4, and incubated on ice for 5 min. Labeled protein is separated from free isotope by gel filtration with a 10-ml Sephadex G-25-150 column (Sigma, St. Louis, MO) using 0.02 M Tris, 0.15 M NaCl buffer, pH 7.4, to develop the column. Labeled protein fractions (>95% precipitable with 10% trichloroacetic acid) are pooled, and dialyzed first against 0.02 M HEPES, 0.15 M NaCl, pH 7.4, to remove the benzamidine, and finally into 50% glycerol prior to storage at $-20°$. Specific radioactivities range from 4460 to 9677 cpm/ng of protein. To confirm that the labeled and unlabeled proteins are identical substrates, initial experiments are performed comparing the rate of cleavage of factor IX and ^{125}I-labeled factor IX by factor Xa as described below. No significant difference in cleavage should be observed between the labeled and unlabeled proteins.

Assay. All assays are performed in 0.02 M HEPES, 0.15 M NaCl, 5 mM CaCl$_2$, pH 7.4. Typically, 0.5–1.0 ml of freshly isolated peripheral blood monocytes ($0.01-2.0 \times 10^7$/ml) are incubated with factor Xa (10–200 nM) for 10 min at 37°, followed by the addition of ^{125}I-labeled factor IX (100 nM) to initiate the reaction.[7] At timed intervals, 50- to 100-μl aliquots are removed and added to 0.25 M EDTA, pH 7.4 (25 mM final concentration), to quench the reaction. The subsamples are microcentrifuged at 13,000 g for 30 sec. The supernatant is removed, frozen, and lyophilized. Control reaction mixtures include ^{125}I-labeled factor IX alone, monocytes plus ^{125}I-labeled factor IX (no factor Xa added), and ^{125}I-labeled factor IX plus factor Xa (at the indicated concentrations, without monocytes added).

The cleavage of factor IX is monitored by SDS-PAGE using 10–20% polyacrylamide gels under reducing conditions as described by Laemmli.[29] Following electrophoresis, dried gels are subjected to autoradiography at $-70°$ using Kodak (Rochester, NY) XR-1 film and Dupont "Lighting Plus" intensifying screens. Progress curves of the cleavage of factor IX by factor Xa bound to monocytes can be generated by densitometric analyses of the autoradiographs using a Microscan 1000 scanning densitometer (TRI, Inc., Nashville, TN) equipped with a solid-state linear diode array

[29] U. K. Laemmli, *Nature (London)* **227**, 680 (1970).

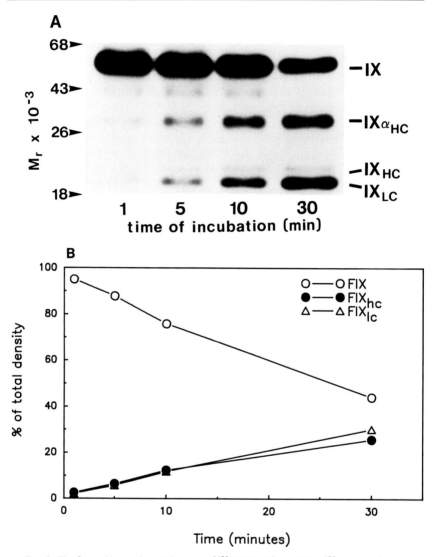

FIG. 3. The factor Xa-catalyzed cleavage of ^{125}I-labeled factor IX to ^{125}I-labeled factor IXα in the presence of monocytes. Monocytes (1×10^7/ml) suspended in 20 mM HEPES, 0.15 M NaCl, 5 mM CaCl$_2$, pH 7.4, were incubated with factor Xa (30 nM) for 10 min at 37° followed by the the addition of a plasma concentration of factor IX (100 nM). (A) At the time points indicated aliquots were removed and processed for autoradiography (see text for details). (B) The relative concentrations of factor IXα light and heavy chains formed and the factor IX consumed over time were determined by densitometric analysis of the autoradiograph, as detailed in the text. Each band was plotted as a percentage of total density per lane.

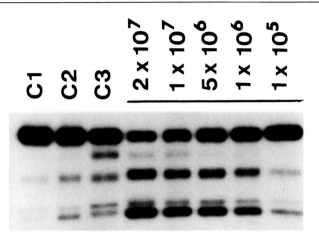

FIG. 4. The factor Xa-catalyzed cleavage of ^{125}I-labeled factor IX is dependent on the monocyte concentration. Varying concentrations of monocytes (as indicated) were incubated with factor Xa (100 nM) for 10 min at 37°, as described in Fig. 5, followed by the addition of a plasma concentration of ^{125}I-labeled factor IX (100 nM). At 5, 10, and 30 min, aliquots were removed and processed for autoradiography, as described in the text. Cleavage observed after a 30-min incubation is shown. Three control reactions were performed: C1, ^{125}I-labeled factor IX alone; C2, ^{125}I-labeled factor IX + factor Xa; and C3, ^{125}I-labeled factor IX + cells (2×10^7/ml).

camera to digitize images through a photographic lens. Resulting digitized images contain approximately 0.25×10^6 pixels, with each pixel measuring 0.16×0.16 mm. Linearity of this system is 0–2.5 absorbance units. Data are analyzed with a 80826-based computer equipped with a math coprocessor and software which allows for either automatic or manual background subtraction and full editing capability. Data are expressed as integrated volumes for each protein band using the arbitrary density units of the scanning system.

Representative data are shown in Fig. 3, where the incubation of factor Xa with monocytes and a plasma concentration of factor IX resulted in the generation of factor IXα. The time-dependent loss of factor IX and formation of the factor IXα heavy and light chains are visualized by autoradiography (Fig. 3A) and quantitated by scanning densitometry (Fig. 3B). Incubation of a fixed, high concentration of factor Xa with increasing concentrations of cells (Fig. 4) indicated that the extent of factor IX cleavage was dependent on the cell concentration. Furthermore, as indicated by the various control reactions, factor Xa in solution was an inef-

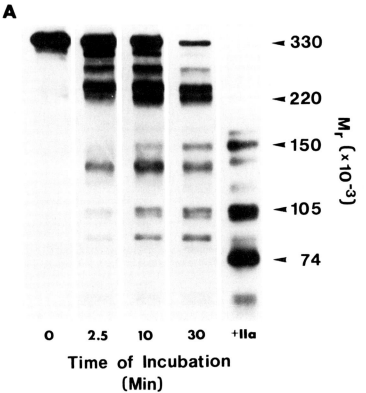

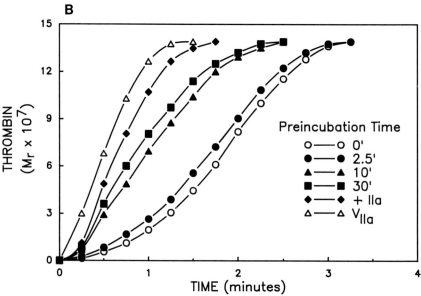

fective catalyst (C2) whereas the cells alone expressed some proteolytic activity toward factor IX in the absence of added factor Xa (C3).

Assay for Cleavage and Activation of Factor V by Monocyte-Associated Protease

Materials. Factor V is radioiodinated using the Iodogen transfer technique,[30] with a protocol similar to that described for factor IX. Na^{125}I (2 mCi/mg protein) is added to an Iodogen-coated plastic tube (1 μg Iodogen/10 μg proteins in 100 μl of 0.5 M phosphate buffer. Following 5 min of gentle vortexing, the oxidized iodonium ion is transferred to a separate vessel containing 0.5 ml of factor V (0.2–0.5 mg/ml) in 0.02 M Tris-HCl, 0.15 M NaCl, pH 7.4, which is then incubated on ice for 5 min with occasional mixing. The labeled protein is made 1.2 M in NaCl and applied to a 3-ml phenyl-Sepharose column equilibrated in 0.02 M imidazole hydrochloride, 1.2 M NaCl, pH 6.5. Free iodine is washed from the column with the same buffer. ^{125}I-Labeled factor V is eluted with the same buffer without NaCl. The protein fractions, which are typically >98% trichloroacetic acid (TCA)-precipitable, are pooled and dialyzed against 10 mM Tris/borate, 1 mM CaCl$_2$/50% glycerol, pH 6.5, and stored at $-20°$. Specific radioactivities typically range from 900 to 3100 cpm/ng protein (0.1–0.4 mol of ^{125}I/mol of protein). This labeling technique yields a radioiodinated molecule of relatively high specific radioactivity which retains 100% functional activity when cleaved to the cofactor, factor Va.

Assay. Reaction mixtures (0.5–1 ml), maintained at 37°, contain freshly isolated peripheral blood monocytes (1–10 × 10^6 cells/ml) suspended in 20 mM HEPES, 0.15 M NaCl, pH 7.4, and factor V (20–70 nM), containing trace ^{125}I-labeled factor V (500–1000 cpm/μl). At various

[30] D. D. Monkovič and P. B. Tracy, *Biochemistry* **29**, 1118 (1990).

Fig. 5. The time-dependent cleavage and activation of human factor V by a monocyte-associated protease. Monocytes (5 × 10^6/ml) were incubated with human factor Va (67 nM) containing trace ^{125}I-labeled factor V (1000 cpm/μl) in 20 mM HEPES, 0.15 M NaCl, 5 mM CaCl$_2$, pH 7.4, at 37°. At the time points indicated, aliquots were removed and centrifuged at 12,000 g for 10 sec. (A) The cell-free supernatants were processed for autoradiography as described in the text. (B) At select time points of incubation, total factor Va cofactor activity was assessed by measuring the ability of the factor V cleavage products to support the factor Xa-catalyzed activation of prothrombin, as detailed in the text. Over time of incubation, the decreased lag and increased initial rate of thrombin formation observed in B indicate that the cleavages observed in A result in the generation of factor Va cofactor activity.

times (30 sec–60 min) aliquots are removed and assayed for factor Va cofactor activity and/or prepared for gel electrophoresis. Factor Va cofactor activity is assessed using the previously detailed DAPA assay. Reaction subsamples containing ≤ 1 nM factor V/Va are used to initiate reactions containing DAPA (3 μM), prothrombin (1.4 μM), PCPS vesicles (20 μM), and factor Xa (5 nM) in 20 mM HEPES, 0.15 M NaCl, 5 mM CaCl$_2$, pH 7.4. Under these conditions the initial rate of thrombin formation is linear with respect to factor Va functional activity. Reaction subsamples to be used to visualize factor V cleavage are prepared for electrophoresis by one of several methods. Addition of glacial acetic acid (10% final concentration) followed by lyophilization and resuspension in Laemmli sample preparation buffer (SPB), or dilution with two volumes of SPB, allows visualization of the cleavage products in the total reaction mixtures. The cleavage products bound to the cell versus those remaining in solution are obtained by layering of an aliquot (100 μl) over an oil mixture [0.5 ml of 1 part Apiezon A oil (James G. Biddle, Plymouth Meeting, PA): 9 parts n-butyl phthalate] in a microcentrifuge tube followed by centrifugation at 10,000 g for 30 sec. A portion (50 μl) of the aqueous layer (containing the unbound cleavage products) is added to 5.5 μl of glacial acetic acid. The remaining aqueous layer and the oil layer are rapidly removed by aspiration and the cell pellet is solubilized in 10% acetic acid. The samples are lyophilized and resuspended in SPB. Factor V cleavage is assessed following slab gel electrophoresis and autoradiography of the dried gels.

Representative data are shown in Fig. 5, where the incubation of ^{125}I-labeled factor V with monocytes resulted in its time-dependent cleavage (Fig. 5A) that paralleled a concomitant increase in the expression of cofactor activity (Fig. 5B). The cofactor activity ultimately generated was nearly identical to that seen with factor Va produced by intentional activation with thrombin.

Summary

These combined data support the concept that the procoagulant response elicited by mononuclear cells, particularly monocytes, is accomplished through regulated binding site-mediated (or perhaps "receptor"-mediated) assembly of proteolytic activities at their membrane surface. Because the work of several laboratories indicate that the monocytes provide the appropriate membrane surface for the assembly and function of all the coagulation complexes required for thrombin production in $vivo$, monocytes may provide a unique opportunity to investigate how coagulant reactions are regulated on cell surfaces through both receptor-mediated events as well as by channeling a product of one reaction to serve as a mediator of a second reaction.

Acknowledgments

This work was supported by Grant HL34863 from the National Institutes of Health, an American Heart Established Investigator Award (P.B.T.), and the University of Vermont College of Medicine.

[17] Meizothrombin: Active Intermediate Formed during Prothrombinase-Catalyzed Activation of Prothrombin

By MARGARET F. DOYLE and PAUL E. HALEY

Introduction

The activation of prothrombin to thrombin proceeds via the cleavage of two peptide bonds in the prothrombin molecule. These bonds are cleaved by the prothrombinase enzyme complex, which is composed of the enzyme, factor Xa, and a cofactor, factor Va, assembled on a phospholipid surface in the presence of calcium ions. The activation of prothrombin (Fig. 1) can occur by two possible mechanisms: initial cleavage at the Arg^{273}-Thr^{274} bond,[1] giving rise to the intermediates fragment 1·2 and prethrombin-2, or initial cleavage at the Arg^{322}-Ile^{323} bond, giving rise to the intermediate meizothrombin. Further cleavage of either intermediate by factor Xa gives rise to the enzyme α-thrombin. The intermediate meizothrombin, which contains an active site similar to thrombin, can autoproteolyse to meizothrombin(desF1). Meizothrombin (desF1) is similar to meizothrombin but with the fragment 1 domain, which is responsible for phospholipid binding, removed. Human meizothrombin and meizothrombin (desF1) are substrates for further autoproteolysis at Arg^{286} to form the enzyme thrombin (des1–13). This species is the form of α-thrombin generated on activation of human prothrombin.

During the activation of prothrombin by factor Xa, multiple intermediate species which contain an active site may be formed.[2] These potential

---

[1] Amino acid numbering is based on human prothrombin. Human prothrombin consists of 581 residues, whereas bovine prothrombin contains 582 residues, due to the deletion of the fourth residue in the human molecule.

[2] M. F. Doyle and K. G. Mann, *J. Biol. Chem.* **265**, 10693 (1990).

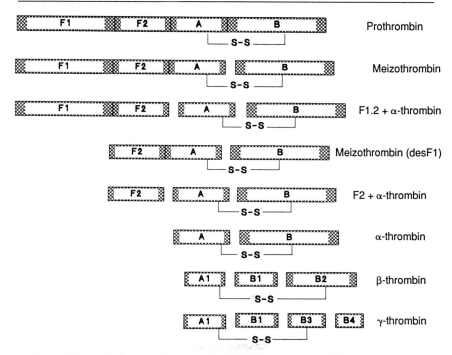

Fig. 1. Schematic diagram of prothrombin activation products which contain active sites. Prothrombin and its prothrombinase activation products are illustrated. The factor Xa cleavage sites in bovine prothrombin are at Arg-323 (between the A and B chains) and at Arg-274 (between fragment 2 and the A chain). The thrombin cleavage site on bovine prothrombin is at Arg-156 (between fragment 1 and fragment 2). Human prothrombin consists of one less amino acid due to a deletion of the fourth residue in the human molecule. Human α-thrombin is 13 amino acids smaller than bovine α-thrombin due to an additional thrombin cleavage site at Arg-286 in the human protein. Bovine β-thrombin has been cleaved at Lys-287 and Lys-305 in the A chain and at Lys-389 and Arg-396 in the B chain. Human β-thrombin is cleaved at Arg-384 and at Arg-395. Human γ-thrombin has an additional cleavage of the β-thrombin at Arg-445 and Lys-476. Reproduced from Doyle and Mann,[2] with permission.

species are depicted in Fig. 2. It is important to note that there are several significant differences between the bovine and human enzymes. First, the amino acid numbering used throughout this article is for human prothrombin, unless otherwise noted. Bovine prothrombin contains one more amino acid than human due to a deletion of the fourth residue in the human molecule. Second, there exists a thrombin cleavage site on human prothrombin at Arg[286] that does not exist in the bovine protein. Third, the autoproteolysis of α-thrombin to β- or γ-thrombin proceeds via different cleavages, as explained in the legend to Fig. 1.

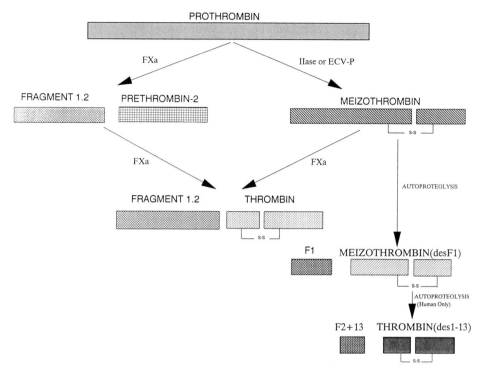

FIG. 2. Possible pathways of prothrombin activation. The possible pathways for prothrombin activation are illustrated and are described in the text. Abbreviations: FXa, factor Xa; IIase, prothrombinase complex; F1, fragment 1; F2 + 13, fragment 2 plus the first 13 amino acids from the thrombin A chain; ECV-P, prothrombin activator from *E. carinatus*.

Preliminary data on the activation of prothrombin indicated that the activation occurs solely via the prethrombin-2 intermediate.[3-6] These early experiments were performed with factor Xa alone, and not the entire prothrombinase complex. In 1986, Rosing *et al.*[7] utilized a fluorescent active-site-directed inhibitor, I-2581 [*N*-dansyl-(*p*-guanidino) phenylalanine-piperidine hydrochloride], to demonstrate that meizothrombin was

[3] C. M. Heldebrandt and K. G. Mann, *J. Biol. Chem.* **248**, 3642 (1973).
[4] C. M. Heldebrant, R. J. Butkowski, S. P. Bajaj, and K. G. Mann, *J. Biol. Chem.* **248**, 7149 (1973).
[5] C. T. Esmon, W. G. Owen, and C. M. Jackson, *J. Biol. Chem.* **249**, 8045 (1974).
[6] G. Tans, G. Rosing, G. van Diegen, and H. C. Hemker, *in* "The Regulation of Coagulation" (K. G. Mann and F. B. Taylor, eds.), p. 173. Elsevier/North-Holland, New York, 1980.
[7] J. Rosing, R. F. A. Zwaal, and G. Tans, *J. Biol. Chem.* **261**, 4224 (1986).

indeed an intermediate in the activation of prothrombin to thrombin.[8] At the same time, Krishnaswamy et al.[9,10] utilized a similar fluorescent thrombin inhibitor, dansylarginine-N-(3-ethyl-1,5-pentanediyl)amide (DAPA), to demonstrate that activation of prothrombin by the fully assembled prothrombinase complex proceeds in an ordered, sequential reaction via the intermediate meizothrombin. When DAPA binds the active site of thrombin, there is a concomitant increase in fluorescence intensity (excitation wavelength, 280 nm; emission wavelength, 540 nm), which can be used to monitor prothrombin activation.[11] In monitoring the activation of prothrombin, a transient increase in the fluorescent intensity can be seen. By removing aliquots from the reaction and running SDS-PAGE, it was clear that this increased fluorescence was due to the presence of the intermediate. Studies on the activation of prothrombin using the prothrombin activator purified from *Echis carinatus*,[12] which cleaves primarily the Arg^{322}-Ile^{323} bond, proved that meizothrombin displays a higher fluorescent intensity than thrombin at similar concentrations. The inclusion of DAPA also prevents the feedback cleavage of prothrombin by meizothrombin or thrombin, which can make interpretation of the data significantly more difficult. Further studies on the activation of human prothrombin indicate that the formation of meizothrombin as an intermediate is a consequence of the association of the cofactor, human factor Va, with the enzyme, human factor Xa, on the phospholipid surface. The absence of factor Va causes the activation to proceed via the prethrombin-2 intermediate.

The observation that meizothrombin binds DAPA indicates that the intermediate has expressed an active site, and therefore could possibly have activity similar to thrombin. In order to determine the activity, if any, of meizothrombin, it first becomes necessary to obtain the intermediate in a stable form suitable for study.

Preparation of Meizothrombin

Because meizothrombin is a transient intermediate in the activation of prothrombin, it is not yet possible to prepare the enzyme in an active form that is stable for long-term storage. For this reason, meizothrombin is

---

[8] Thrombin(des1–13) is thrombin lacking the first 13 amino acids due to a thrombin cleavage site in the human protein; prethrombin(des1–13) is prethrombin-2 lacking the first 13 amino acids due to a thrombin cleavage site in the human protein.

[9] S. Krishnaswamy, K. G. Mann, and M. E. Nesheim, *J. Biol. Chem.* **261**, 8977 (1986).

[10] S. Krishnaswamy, W. R. Church, M. E. Nesheim, and K. G. Mann, *J. Biol. Chem.* **262**, 3291 (1987).

[11] L. S. Hibbard, M. E. Nesheim, and K. G. Mann, *Biochemistry* **21**, 2285 (1982).

[12] T. Morita, S. Iwanaga, and T. Suzuki, *J. Biochem. (Tokyo)* **79**, 1089 (1976).

prepared immediately prior to use. Highly purified prothrombin is diluted to $1.4 \mu M$ in 20 mM Tris, pH 7.4, 0.15 M NaCl, 2 mM CaCl$_2$. DAPA is added to a final concentration of 3.0 μM. The reaction is started by the addition of *E. carinatus* prothrombin activator to a final concentration of 10 μg/ml. The progress of the reaction is monitored by following the increase in fluorescence at 545 nm (excitation wavelength, 280 nm) caused by the binding of DAPA to the active site of meizothrombin or thrombin. When the activation rate nears completion, the sample is put immediately on ice. Under these conditions, meizothrombin is stable [less than 10% conversion to meizothrombin (desF1)] for 15 min at 37°, 2 hr at 4°, and 2–3 days at −20°, as judged by SDS-polyacrylamide gel electrophoresis. At this concentration, however, the DAPA completely inhibits any activity meizothrombin might have. The advantage to using DAPA as the inhibitor is that it is reversible, so that diluting the meizothrombin to near physiologic concentrations (< 15 nM) dilutes the DAPA below its K_d of 3×10^{-8} M, thus removing the inhibitory effects of the DAPA, and creating a means for assessing meizothrombin activity toward physiologic substrates. Meizothrombin can also be prepared in a similar manner using the reversible thrombin inhibitor I-2581.[13]

To verify that no further cleavage of meizothrombin occurs during the course of study, prothrombin is radiolabeled[2] with ^{125}I prior to activation with the *E. carinatus* prothrombin activator. The radiolabeling has no effect on the activation of the prothrombin to meizothrombin. Because the meizothrombin contains both DAPA and *E. carinatus* prothrombin activator, controls are required to ensure that these compounds in no way affect the assays performed subsequent to the prothrombin activation.

To perform binding studies on meizothrombin, a stable intermediate can be obtained by reaction with a covalent active site inhibitor. Such a study has been reported utilizing the inhibitor dansylglutamylglycylarginyl chloromethyl ketone (DEGR-ck).[14] Because the rate of incorporation of this inhibitor into the active site is slow, further purification of the meizothrombin is necessary to remove autolytic breakdown products that are also obtained.

Assays of Meizothrombin Activity

Because meizothrombin has an active site similar to thrombin, the relative activity of meizothrombin toward various thrombin substrates can

[13] G. Tans, T. Janssen-Claessen, H. C. Hemker, R. F. A. Zwaal, and J. Rosing, *J. Biol. Chem.* **266,** 21864 (1991).
[14] S. A. Armstrong, E. J. Husten, C. T. Esmon, and A. E. Johnson, *J. Biol. Chem.* **265,** 6210 (1990).

be assessed. These results are of interest because meizothrombin has both the active site of thrombin as well as the lipid-binding domain of prothrombin. Assay procedures for the different substrates are as follows:

Active Site Titration

To determine the exact concentration of active sites in meizothrombin and thrombin, active site titrations are performed at three enzyme concentrations, using p-nitrophenyl-p'-guanidinobenzoate (NPGB), by the method of Chase and Shaw.[15]

Buffer. 0.1 M sodium veronal buffer, pH 8.3, 20 mM $CaCl_2$.

NPGB. 16.8 mg (0.05 nmol) is dissolved in 1 ml dimethylformamide, then diluted with 4 ml acetonitrile.

A reference cuvette is prepared which contains 2 ml of buffer and 5 μl of NPGB. A sample cuvette is prepared which contains 1.9–2.0 ml (depending on the volume of sample to be added) and 5 μl of NPGB. A baseline is recorded at 410 nm for 1 min. An appropriate volume of meizothrombin is added to the sample cuvette and the "burst" (sudden increase in absorbance at 410 nm) is recorded. The enzyme concentration is determined using an extinction coefficient at 410 nm for the p-nitrophenol of 16,595 M^{-1} cm^{-1}, using the equation $[E] = A_{410\,nm}/16,595$. This yields the concentration of active sites, which, when divided by the protein concentration, gives the percent titratable active sites.

Chromogenic Substrates

Chromogenic thrombin substrates, such as S2238 (phenylalanylpipecoylarginyl-p-nitroaniline) (Helena, Beaumont, TX) and Spectrozyme TH (H-D-HHT-alanylarginyl-p-nitroaniline) (American Diagnostica, Greenwich, CT), can be utilized to assess the integrity of the active site of meizothrombin, relative to thrombin. The assays are set up in a microtiter format as follows:

Meizothrombin or thrombin is prepared in 20 mM Tris, pH 7.4, 0.15 M NaCl, 2 mM $CaCl_2$, 0.1% polyethylene glycol (PEG) 8000. Aliquots of the enzyme (0.14 pmol) are placed in triplicate into 96-well assay plates (Corning, Corning, NY). Various concentrations (5–100 μM) of S2238 or Spectrozyme TH are prepared and the concentration is verified by determining the absorbance at 316 nm, using the extinction coefficient provided with each substrate. The reaction is started by the addition of chromogenic substrate (200 μl), and the plate is read immediately using a Molecular Devices (Menlo Park, CA) V_{max} kinetic plate reader. A sample of the hydrolyzed substrate is prepared for each concentration of substrate

---

[15] T. Chase, Jr. and E. Shaw, this series, Vol. 19, p. 20.

so that an extinction coefficient (4.3×10^3 M^{-1} at 405 nm for a 205-μl volume) can be determined. It is important to verify the extinction coefficient, as plates from different manufacturers may yield different results. Initial velocities are determined from the linear portion of the graphs (less than 15% substrate hydrolysis).

Fibrinogen Clotting Activity

The ability of meizothrombin to clot fibrinogen is assessed in a purified system.[16]

Fibrinogen. Bovine fibrinogen (3 g) (Sigma, St. Louis, MO) is dissolved in 100 ml of H_2O and the pH is adjusted to 7.2 with 0.5 M dibasic sodium phosphate. The volume is brought to 150 ml and stored in plastic vials at $-20°$. The fibrinogen is thawed at $37°$ immediately prior to use.

Imidazole Buffer. Imidazole (17.2 g) is dissolved in 900 ml of 0.1 N HCl and diluted to 1 liter with H_2O. The pH is then adjusted with HCl to 7.25.

Assay Mix 1.6 ml Imidazole buffer
14.4 ml 0.9% (w/v) NaCl
3.2 ml 4.95% (w/v) PEG 8000
4.8 ml Fibrinogen
Mix well and centrifuge to remove insoluble material

Assay. To 0.3 ml of assay mix add 1–10 μl thrombin or meizothrombin, and start a timer. The tube is gently rocked until a visible clot is formed, and the time is recorded. A standard curve, ranging from 0.2 to 2.0 NIH units, is prepared using thrombin of known specific activity. The relative activity of meizothrombin is determined from the standard curve.

Platelet Aggregation and Activation

The ability of meizothrombin to cause platelet aggregation and activation is determined using washed platelets and a lumiaggregometer (Chronolog, Havertown, PA). This machine allows for the simultaneous measurement of platelet aggregation, by monitoring the change in light scattering at 320 nm, and platelet activation, by monitoring ATP release as a function of photoluminescence using the luciferin-luciferase system. The luciferin-luciferase reaction solution is obtained directly from the manufacturers (Chronolog) and is used according to the directions supplied. Platelets are prepared from freshly drawn blood and washed by the method of Mustard.[17] The washed platelets are counted in a Coulter counter and the

[16] R. L. Lundblad, H. S. Kingdon, and K. G. Mann, this series, Vol. 45, p. 156.
[17] J. F. Mustard, D. W. Perry, N. G. Ardle, and M. A. Packham, *Br. J. Haematol.* **22,** 193 (1972).

concentration adjusted to 2×10^8 platelets/ml. The reactions are carried out at 37° using varying concentrations of enzyme.

Factor V Activation

The activation of purified factor V is monitored by both the increase in clotting activity using factor V-deficient plasma,[18,19] and the change in the sodium dodecyl sulfate-polyacrylamide gel electrophoresis pattern. The exact reaction conditions[2] are as follows:

A standard curve is prepared using factor V-deficient plasma, with normal pooled citrated plasma (diluted from 1/10 to 1/200) as the standard. For each assay, 50 μL of factor V-deficient plasma, 50 μl of thromboplastin, and 50 μl of normal citrated plasma or factor V sample are incubated for 1 min at 37°. Then 50 μl of 25 mM CaCl$_2$ is added and the time required for clot formation is recorded. To monitor factor V activation, purified factor V is diluted to 0.15 mg/ml (0.45 μM) in 20 mM Tris, pH 7.4, 0.1 M NaCl, and the initial activity of the factor V is determined. Meizothrombin (molar ratio of factor V to enzyme, 1000 : 1) is added and temporal aliquots are removed for assay and electrophoresis. The samples for electrophoresis are made (10 μM FPR-ck and 2% SDS) and heated at 90° for 5 min. Duplicate samples are prepared and brought to 2% 2-mercaptoethanol.

Protein C Activation

To assess the potential role of meizothrombin in the anticoagulant pathway, protein C activation by enzyme alone (meizothrombin), enzyme – cofactor (meizothrombin – thrombomodulin), or the enzyme – cofactor complex (meizothrombin – thrombomodulin – Ca^{2+} – phospholipid) can be determined. The general assay procedure[2] is as follows:

The buffer used for all assays is 20 mM Tris, pH 7.4, 0.15 M NaCl, 0.01% Lubrol-PX. Meizothrombin, either alone or in the presence of thrombomodulin or thrombomodulin – phospholipid, is preincubated for 10 min at 25°. Protein C is added and temporal aliquots (10 μl) are withdrawn and placed in a 96-well microtiter assay plate containing 90 μl of 3 μM antithrombin III. Controls are performed to ensure that this concentration is sufficient to ensure complete inhibition of the meizothrombin. Heparin is not used, as it has been reported to be an inefficient inhibitor of meizothrombin.[20] Chromogenic substrate is added (100 μl of 0.4 mM

[18] M. E. Nesheim, K. H. Myrmel, L. Hibbard, and K. G. Mann, *J. Biol. Chem.* **254**, 508 (1978).
[19] M. L. Lewis and A. G. Ware, *Proc. Soc. Exp. Biol. Med.* **84**, 640 (1953).
[20] P. Schoen and T. Lindhout, *J. Biol. Chem.* **262**, 11268 (1987).

S2238) and the absorbance change at 405 nm is monitored with time. The amount of APC generated is determined from a standard curve that compares the amount of chromogenic substrate hydrolysis versus the concentration of purified APC.

To determine an apparent dissociation constant and a stoichiometry of binding, titration experiments are performed by fixing the meizothrombin and varying the amount of thrombomodulin. Assays are performed as described above. Thrombomodulin shows a saturable increase in the activation of protein C by meizothrombin. The relative increase in rate can be used as an indication of the bound fraction, from which an apparent dissociation constant (K_d) can be determined.[21]

In order to obtain kinetic parameters for the activation of protein C at high substrate concentrations ($> 5 \mu M$), it is necessary to remove any traces of activated protein C activity from the substrate. This is accomplished by extensive dialysis of the protein C against 20 mM Tris, pH 7.4, 0.15 M NaCl, followed by chromatography on a benzamidine-Sepharose (1 × 5 cm) column. The flow through, which contains the protein C, is concentrated on a Centricon-10 microconcentrator (Amicon, Danvers, MA). The protein C prepared in this way contains <0.01% APC activity. A series of protein C samples (0.5–40 μM) are prepared and added to previously incubated enzyme–cofactor mixtures containing 20 mM thrombomodulin and 10 mM meizothrombin, and assayed as described above.

Contractile Activity in Rabbit Femoral Arteries

Femoral and brachial arteries are removed from white male rabbits and cleaned in cold buffer (see below) prior to slicing into 2-mm ring segments.[22] These segments are then mounted on two wires and placed in tissue chambers (3 ml) containing the following buffer mixture: 130 mM NaCl, 4.7 mM KCl, 14.9 mM NaHCO$_3$, 1.2 mM KH$_2$PO$_4$, 1.2 mM MgSO$_4$, 0.03 mM EDTA, 11 mM dextrose, and 1.6 mM CaCl$_2$. One wire is attached to a stationary base, while the other rests on a holder attached to a force transducer. Arteries are equilibrated for 1 to 2 hr in 37° buffer and aerated with 95% O$_2$ and 5% CO$_2$. After equilibration, segments are contracted with norepinephrine (10^{-5} M) to determine their maximal capacity to develop active force. After norepinephrine is washed out, a single dose of either thrombin or meizothrombin is added to each segment and the

[21] H. Gutfreund, "Enzymes: Physical Principals," p. 69. Wiley, New York, 1972.
[22] L. P. Thompson, M. F. Doyle, K. G. Mann, and J. A. Bevan, (1989) *J. Vasc. Med. Biol.* **1**, 347 (1989).

resultant responses are monitored with respect to the tissue maximum. In addition, the effect of cumulative dosing of thrombin or meizothrombin can be done on a single artery. The ability of thrombin or meizothrombin to cause artery relaxation can be tested by precontracting the segments with norepinephrine and then adding thrombin or meizothrombin. The relaxation ability is compared to that obtained by the addition of acetyl-choline, which is completely endothelium dependent and a good indicator of functional endothelium. The role of the endothelium in this contractile activity can be determined by mechanically removing the endothelium.

Polyacrylamide Gel Electrophoresis of Radiolabeled Meizothrombin

Because meizothrombin is an intermediate, the possibility that it has converted to meizothrombin(desF1) or thrombin during the course of the experiment must be ascertained. To this end, the experiments outlined above are also performed with radiolabeled meizothrombin. Experiments are done with both the labeled and unlabeled proteins to ensure that the process of radiolabeling has in no way altered the activity of the enzyme. At the conclusion of each experiment, an aliquot of the reaction mixture containing 6×10^5 cpm is removed and brought to 1 μM phenylalanylargi-nyl chloromethyl ketone (PPA-ck) (Calbiochem, La Jolla, CA) to prevent further autolysis. The sample is brought to 1% SDS. A duplicate sample is prepared in the presence of 1% 2-mercaptoethanol. The samples are heated at 90° for 5 min and then applied to 5–15% polyacrylamide gradient gels in the presence of SDS and subjected to electrophoresis. The gels are then dried and placed on Kodak (Rochester, NY) XAR-5 film for 14 hr at −70° prior to developing. Under all conditions tested, no more than 5–10% autolysis should occur, with some meizothrombin(desF1) visible, primarily in the procoagulant experiments (i.e., factor V activation, plate-let aggregation).

Amidolytic Assay for Meizothrombin Generation

In order to examine meizothrombin generation from prothrombin, Rosing et al.[7] developed an assay for meizothrombin activity utilizing the differences in heparin-mediated antithrombin III inhibition. The assay is performed by first determining the total amidolytic activity of the sample toward S2238, which measures both thrombin and meizothrombin activi-ties.[13] The sample is then incubated with 5 nM AT III in the presence of 20 μg/ml heparin, and the amidolytic activity is again determined. Under these conditions, the thrombin activity is theoretically inhibited so that the only activity measured is from meizothrombin or meizothrombin(desF1).

Characteristics of Meizothrombin

Meizothrombin and its inactive precursor, prothrombin, contain identical amino acid sequences, with the exception of a single bond cleavage at Arg-323. This single bond cleavage creates an active site similar to that found in thrombin. The role of this intermediate, which contains the active site of thrombin and the lipid-binding properties of prothrombin, had remained a mystery until the use of DAPA allowed a window for viewing the possible significance for this intermediate. Active site titrations of meizothrombin yielded an average of 93.6% titratable active sites. These results do not necessarily indicate the quantity of meizothrombin, because all the proteolysed forms of meizothrombin [i.e., meizothrombin(desF1) and thrombin] would have titratable active sites. This does confirm that, in general, 6–7% of the prothrombin was not activated, allowing for the calculation of a more correct value for k_{cat}. Analysis of the S2238 activity shows that meizothrombin ($K_m = 10.8 \pm 1.5 \ \mu M$, $k_{cat} = 74.9 \pm 4.7 \ \text{sec}^{-1}$) has activity that is very similar to that obtained with human thrombin ($K_m = 10.2 \pm 1.2 \ \mu M$, $k_{cat} = 59.8 \pm 3.2 \ \text{sec}^{-1}$). The slightly higher k_{cat} for meizothrombin may be significant in that a similar phenomenon was observed when fragment 2 associates with thrombin ($K_d = 7.7 \times 10^{-10}$ M).[23] This may indicate minor differences in the active site conformation between meizothrombin and thrombin.

Structural information on meizothrombin has been obtained using the fluorescent active-site-blocked DEGR–meizothrombin. The fluorescent dansyl dye, which is covalently attached to the active site of meizothrombin, is particularly sensitive to calcium ions. The addition of calcium causes a 24% decrease in the dansyl emission and a similar decrease in the fluorescence lifetime. The intensity change is half-maximal at 0.2 mM Ca^{2+}, which correlates well with the calcium-dependent conformational change observed in fragment 1 of prothrombin. This indicates that Ca^{2+} binding to meizothrombin via the membrane-binding domain elicits a conformational change in the active site. Fluorescence energy transfer studies indicate that the active site of meizothrombin is located 71 Å from the phospholipid bilayer.[14] The addition of factor Va to the membrane-bound meizothrombin causes an increase in the dansyl emission and a decrease in the distance of the active site of meizothrombin from the phospholipid bilayer to 67 Å. These data indicate that factor Va binds directly to meizothrombin, causing an alteration in the active site.

The most obvious activity of meizothrombin toward macromolecular substrates would have to be its ability to cleave prothrombin and mei-zothrombin.[7] In the absence of inhibitors, meizothrombin generation is

[23] K. H. Myrmel, R. L. Lundblad, and K. G. Mann, *Biochemistry* **15**, 1767 (1976).

followed almost instantaneously by its conversion to meizothrombin(desF1), and in the case of human meizothrombin, to thrombin(des1–13), due to the thrombin cleavage site 13 amino acids from the carboxyl terminus of the factor Xa cleavage site. The addition of catalytic amounts of meizothrombin to prothrombin results in the rapid conversion of prothrombin to prethrombin-1, and in the case of human prothrombin, to prethrombin-2(des1–13).[7] This would indicate that meizothrombin is capable of interacting with large macromolecular substrates.

The ability of meizothrombin to act on other physiologic substrates is somewhat limited.[2] Procoagulant substrates, such as fibrinogen, factor V, and platelets, proved to be' poor substrates for meizothrombin, showing approximately 2% activity relative to thrombin. In fact, this 2% activity may well represent trace amounts of meizothrombin(desF1) and/or thrombin present in the meizothrombin. This indicates that although the amino acids that comprise the active site of meizothrombin are identical to those in thrombin, the presence of the fragment 1 · 2 peptide greatly alters the enzyme specificity.

The contribution of meizothrombin to the anticoagulant pathway is far more relevant.[2] Meizothrombin activation of protein C proceeds at rates very similar to those observed with thrombin. In fact, kinetic data indicate that thrombomodulin binds meizothrombin ($K_d = 0.89$ nM) equally well as thrombin ($K_d = 0.97$ nM) with a 1 : 1 stoichiometry, and activates protein C with similar catalytic efficiency ($k_{cat} = 3.5$ μM^{-1} min^{-1} for thrombin/thrombomodulin; $k_{cat} = 2.7$ μM^{-1} min^{-1} for meizothrombin/thrombomodulin). The experiments were repeated with iodinated meizothrombin to verify that there was no significant proteolysis of the meizothrombin that might account for this activity.

Meizothrombin has been shown to be a potent agonist for vascular contraction in both rabbit femoral and brachial arteries.[22] Fivefold higher concentrations of thrombin are required to give similar reactivity to meizothrombin. Neither thrombin nor meizothrombin exhibits any relaxation properties, so the difference in reactivity is due to contractile ability. The interaction appears to be smooth muscle mediated, because removal of the endothelium had no effect on the thrombin or meizothrombin contractile effect.

Utilizing the amidolytic assay,[13] meizothrombin generation has been demonstrated under a wide variety of conditions. The results in this study were presented as the percentage of meizothrombin activity relative to the total activity. For this reason, no quantitation of prethrombin-1, and therefore of the alternate pathway of prothrombin activation, is assessed. The study did find that meizothrombin formation during initial rate experiments was not affected significantly, relative to thrombin formation, by pH, temperature, or NaCl concentration. They did find decreased

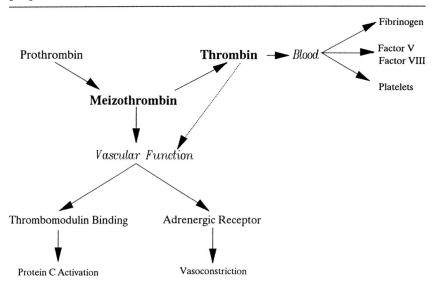

FIG. 3. Diagrammatic representation of the proposed trafficking of meizothrombin and thrombin.

percentages of meizothrombin formation when naturally occurring phospholipids (i.e., platelets, thromboplastin, or cephalin) are used instead of synthetic phospholipids (i.e., phosphatidylserine : phosphatidylcholine vesicles). One possible explanation for this may be that naturally occurring phospholipids contain binding sites for meizothrombin that could sequester the meizothrombin away from other substrates, alter the inhibition by antithrombin III and heparin, or inhibit the amidolytic activity of meizothrombin. Attempts to demonstrate meizothrombin generation in plasma have, to date, not been successful. It is important to note that no prethrombin-2 has been demonstrated either. Although the initial conclusion from these data would be that meizothrombin does not play an important physiologic role in blood, this may not be the case. Blood or plasma that is allowed to activate in a test tube does not mimic the type of reaction that is occurring in the blood vessel. The role of the endothelium and vascular smooth muscle cells, which are key elements in hemostasis, has been eliminated. This point is further illustrated when one examines the amount of thrombin generated in defibrinated human plasma *in vitro* (650 nM) compared to the amount estimated *in vivo* ($<$ 14 nM). The most likely reason for this is the lack of endothelial thrombomodulin, which serves as a cofactor for the activation of protein C, a key anticoagulant enzyme.

Thus, we envision the role of meizothrombin to be directed toward vascular function (see Fig. 3). In the event of vascular injury, where smooth muscle is exposed, meizothrombin, which is generated prior to the formation of thrombin, can react swiftly with vascular smooth muscle to cause vessel contraction and thus minimize blood loss. The fact that meizothrombin has no significant activity toward procoagulant substrates restricts it from substrate competition and targets it directly toward vascular functions. Any thrombin produced subsequent to this step would then be available to act on procoagulant substrates for clot formation. Meizothrombin also plays an important role in the event of small amounts of extraneous activation of the coagulation cascade. On initial activation of prothrombin to meizothrombin, the meizothrombin can bind to thrombomodulin and activate protein C. The activated protein C then inactivates factors Va and VIIIa, thus shutting down the coagulation cascade. The binding of meizothrombin to thrombomodulin is most important in the microvasculature, where thrombomodulin concentrations (500 nM) have been estimated to be well above the dissociation constant for thrombin (1.7 nM). Small amounts of thrombin formation would cause a fibrinogen clot that could restrict blood flow rather quickly. For this reason it is essential that procoagulant activity is minimized except in the case of severe injury to the vessels. Thus, the initial generation of meizothrombin, prior to the generation of thrombin, can immediately bind thrombomodulin and activate protein C. In the event of massive injury, prothrombin activation would go to completion, generating sufficient thrombin to promote clot formation.

[18] Identification and Characterization of Mutant Thrombins

By RUTH ANN HENRIKSEN

Introduction

Determination of the crystal structure for thrombin[1] has provided an increased opportunity for understanding structure–function relationships for both congenital mutants of thrombin and those prepared by site-specific mutagenesis. Congenitally dysfunctional prothrombins have been identified in fewer than 20 pedigrees.[2] Many of these defects appear to

[1] W. Bode, I. Mayr, U. Baumann, R. Huber, S.R. Stone, and J. Hofsteenge, *EMBO J.* **8**, 3467 (1989).
[2] H. R. Roberts and P. A. Foster, *In* (R. W. Colman, J. Hirsh, V. J. Marder, and E. W. Salzman, eds.), "Hemostasis and Thrombosis" 2nd ed., p. 162. Lippincott, Philadelphia, 1987.

result in altered activation of prothrombin without affecting the thrombin portion of the molecule. However, in six pedigrees, the defect results in production of a thrombinlike molecule that lacks normal fibrinogen clotting activity. These are summarized in Table I.[3-12] Although such congenitally dysfunctional forms of thrombin are a rare occurrence, their identification and characterization can provide valuable information on structure–function relationships in this molecule.

The gene for prothrombin is located on an autosomal chromosome, 11p11-q12, and individuals who are heterozygous for a normal and abnormal gene are not easily identified. There are two primary reasons for this. First, in the normal/abnormal heterozygote, where the functional prothrombin level is about 50% of normal, there is no identifiable clinical defect (phenotype) associated with expression of the abnormal gene. Second, the prothrombin time and partial thromboplastin time tests, which are used in the clinical laboratory to screen for blood coagulation abnormalities, are sensitive to factor deficiencies only below about 40% of the normal level.

Thus, phenotypic expression of a prothrombin abnormality, which may be detected by routine screening tests as a deficiency in fibrinogen clotting activity, occurs only when both genes are altered. Because these mutants are rare and distinct abnormalities, phenotypically affected individuals will frequently be heterozygous for two abnormal prothrombin genes. Reported cases of dysprothrombinemia include both individuals with two abnormal proteins and those with only one identifiable dysprothrombin, where a second gene product is either not detectable or is present at a markedly reduced concentration (hypoprothrombinemia). An approximate incidence of individuals heterozygous for dysprothrombinemia may be obtained from an estimate of the occurrence of individuals heterozygous for two prothrombin abnormalities. If the latter, a recessive trait, is roughly estimated at 1 in 100×10^6, then the incidence of hetero-

[3] S. J. F. Degen, R. T. A. MacGillivray, and E. W. Davie, *Biochemistry* **22**, 2087 (1983).
[4] T. Miyata, T. Morita, T. Inomoto, S. Kawauchi, A. Shirrakami, and S. Iwanaga, *Biochemistry* **26**, 1117 (1987).
[5] T. Inomoto, A. Shirakami, S. Kawauchi, T. Shigekiyo, S. Saito, K. Miyoshi, T. Morita, and S. Iwanaga, *Blood* **69**, 565 (1991).
[6] R. A. Henriksen and W. G. Owen, *J. Biol. Chem.* **262**, 4664 (1987).
[7] R. A. Henriksen and K. G. Mann, *Biochemistry* **27**, 9160 (1988).
[8] R. A. Henriksen and K. G. Mann, *Biochemistry* **28**, 2078 (1989).
[9] M.-J. Rabiet, M. Jandrot-Perrus, J. P. Boissel, and F. Josso, *Blood* **63**, 927 (1984).
[10] A. Bezeaud, J. Elion, and M.-C. Guillin, *Blood* **71**, 556 (1988).
[11] E. Morishita, M. Saito, I. Kumabashiri, H. Asakura, T. Matsuda, and K. Yamaguchi, *Blood* **80**, 2275 (1992).
[12] M.-J. Rabiet, *Thromb. Haemost.* **54**, 273 (abstr.) (1985).

TABLE I
CONGENITALLY MUTANT THROMBINS

Thrombin	Amino acid substitution[a]	Functional defect	Ref.
Tokushima	Arg-418(101) → Trp	Substituted residue is adjacent to Asp of catalytic triad residues. Mutant has reduced activity toward fibrinogen and a low-molecular-weight substrate for which there are both k_{cat} and K_m defects so that k_{cat}/K_m is 7% of that for thrombin	4,5
Quick I	Arg-382(67) → Cys	Substitution is within proposed fibrinogen binding groove, distal to catalytic site. Nearly normal activity toward low-molecular-weight substrates and antithrombin III. There is 1–2% of thrombin activity in release of fibrinopeptide A (k_{cat}/K_m) and stimulation of platelet aggregation	6,7
Quick II	Gly 558(226) → Val	Substitution is within primary substrate binding pocket. Less than 0.1% normal activity toward Arg substrates of thrombin, but binds DFP stoichiometrically and hydrolyzes a leucyl-p-nitroanilide with k_{cat}/K_m 15% of that for thrombin. Does not form complex with antithrombin III	8
Metz	n.d.[b]	Activity toward both biologic and low-molecular-weight substrates of thrombin decreased to less than 4% of normal. Increased K_m and decreased k_{cat} for hydrolysis of a chromogenic substrate. Also reduced heparin binding	9
Salakta	n.d.	Activity toward both biologic and low-molecular-weight substrates reduced. For the latter, defect is in K_m. Rate of reaction with antithrombin III is decreased to about 5% of normal	10
Himi I Himi II	Met 337(32) → Thr Arg 388(73) → His	Activation of dysprothrombin isolated from compound heterozygote yields dysthrombin with decreased fibrinogen clotting activity and nearly normal activity toward tripeptide-p-nitroanilide	11
Molise	n.d.	Dysthrombin has 10% of both amidolytic activity and fibrinogen clotting activity compared to normal thrombin	12

[a] Numbering for thrombin residues is based on prothrombin sequence[3] with corresponding chymotrypsinogen numbering[1] in parentheses.
[b] n.d., Not determined.

zygosity for dysprothrombinemia in the population would be 1 in 5000, which exceeds the incidence of the sex-linked coagulation disorder, classical hemophilia.

Reagents and Methods

Diagnosis of Congenital Dysprothrombinemia

The fundamental criterion for the diagnosis of congenital dysprothrombinemia is decreased functional prothrombin activity relative to the prothrombin concentration determined immunochemically. Because this finding can also occur in vitamin K deficiency or liver disease, it is important that these conditions be excluded as possible causes of dysprothrombinemia.

The concentration of prothrombin antigen in plasma, which is normally about 100 mg/liter, may be determined by immunoprecipitation methods such as the quantitative immunoelectrophoretic technique described by Laurell.[13] Polyclonal antisera to human prothrombin are available from Behring Diagnostica (Marburg, Germany) and from Dako (Carpinteria, California, or Glostrup, Denmark).

The functional prothrombin assay reported by Denson et al.[14] and described here uses taipan (Oxyuranus scutellatus scutellatus) venom (Sigma Chemical Company, St. Louis, MO) as a prothrombin activator. This venom activator, which is dependent on both phospholipid and calcium ions, has been used in identifying reported defects in prothrombin. The end point for this assay is fibrinogen clot formation, which may be detected by visual inspection and timed with a stopwatch or with instrumentation such as the Fibrometer (Becton, Dickinson and Company, Cockeysville, MD). A standard plasma pool is needed to calibrate the clotting-time dependence on prothrombin concentration. This may be prepared by obtaining blood samples, anticoagulated with 1 part 3.8% trisodium citrate to 9 parts whole blood, from 20 normal donors. The red cells are removed by centrifugation before pooling the plasma, which is then recentrifuged at 16,000 g for 30 min at 4°, aliquoted into plastic tubes, and frozen at −70°. Prompt processing of this plasma pool is necessary to preserve coagulation factor activity. Alternatively, standardized plasma pools may be obtained from suppliers of reagents for clinical coagulation laboratories. Taipan venom reagent is prepared by adding 0.1 mg/ml venom to 0.1 M NaCl, 0.04 M CaCl$_2$, 0.05 M Tris-HCl, pH 7.4,

[13] C.-B. Laurell and E. J. McKay, this series, Vol. 73, p. 339.
[14] K. W. E. Denson, R. Borrett, and R. Biggs, Br. J. Haematol. 21, 219 (1971).

0.1% in bovine serum albumin. This reagent may be stored below $-20°$. Rabbit brain cephalin (Sigma Chemical Company, St. Louis, MO) diluted as described by the manufacturer may be used as a source of phospholipid. Other cephalin preparations resuspended by sonication at about 0.2 mg/ml are also satisfactory. Fibrinogen may be isolated from plasma[15] or obtained from Kabi Diagnostica (Uppsala, Sweden or Helena Laboratories, Beaumont, TX) and is prepared as a stock solution at 3–4 mg/ml in 0.1 M NaCl, 0.05 M Tris-HCl, pH 7.4. The assay is performed by incubating 0.1 ml fibrinogen solution, 0.1 ml cephalin, and 0.1 ml plasma dilution together for 3 min at 37° before adding 0.1 ml of the taipan venom reagent to initiate clotting. The time from addition of the venom reagent to initial clot formation is recorded as the clotting time. To establish a standard curve, 1/10, 1/20, 1/50, 1/100, and 1/200 dilutions of the standard plasma pool are prepared in 0.1 M NaCl, 0.05 M Tris-HCl, pH 7.4, immediately before performing the assay. Clotting times are determined in duplicate for each of the standard dilutions and at dilutions of 1/10, 1/20, and 1/50 for the plasma being assayed. The standard curve is prepared as a linear plot of the clotting time in seconds versus the reciprocal of the standard plasma pool dilution. The concentration of prothrombin in the test plasma is determined from the standard curve as the standard plasma concentration corresponding to the test plasma clotting time. This concentration is multiplied by the plasma dilution × 100 to obtain percent of normal prothrombin activity in the test plasma. In the absence of an inhibitor, the results obtained on the three test plasma dilutions should agree within 10%.

Additional activators of prothrombin and alternate substrates have been used in prothrombin assays. As activators, staphylocoagulase and *E. carinatus* venom may not be sensitive to prothrombin abnormalities observed with other activators. Alternatively, assays that depend on factors Xa and Va avoid the problem of using a nonphysiologic activator that might not detect some prothrombin abnormalities. Because the specificity determinants for hydrolysis of low-molecular-weight synthetic substrates of thrombin generally differ from those of biologic substrates that interact with extended regions of the enzyme surface, mutant thrombins may have normal activity toward chromogenic substrates with significantly decreased activity toward physiologic substrates such as fibrinogen. For this reason chromogenic or fluorogenic substrate assays may not detect abnormalities that affect the biologic functions of thrombin. Further, some of the biologic activities may not have the same specificity determinants as for fibrinogn clotting. Therefore, the possibility of identifying congenital abnormalities that either do not affect or only minimally affect fibrinogen

[15] W. Straughn, III and R. H. Wagner, *Thromb. Diath. Haemorrh.* **16**, 198 (1966).

clotting activity should not be overlooked. Thus, one might assay for dysfunctional molecules that do not bind thrombomodulin or activate protein C normally. Factor XIII activation, heparin binding, or the stimulation of thromboxane production might be affected to a greater or lesser extent than fibrinogen clotting activity. Other abnormalities might relate to the chemotactic or mitogenic activities of thrombin.

An alternate approach to identifying mutant thrombins is by direct examination of the gene structure with methods such as restriction fragment length polymorphism (RFLP) analysis, as has been described for a polymorphism occurring in fragment 1 of prothrombin.[16] Combining this technique with analysis of single-strand conformation polymorphism may also be useful in characterizing mutant prothrombin genes.[11,17]

Isolation of Thrombin

Apheresis is probably the most satisfactory method for obtaining sufficient plasma to isolate the quantity of prothrombin needed for both structural and functional characterization of a mutant protein. For dysprothrombinemic individuals receiving regular replacement therapy, plasma should be obtained immediately prior to prothrombin infusion to minimize contamination with normal prothrombin. Plasma may also be obtained from individuals heterozygous for normal and abnormal prothrombin, provided a method can be devised to separate the normal and mutant proteins.

Small amounts of prothrombin may be isolated from plasma by adsorption to and elution from immobilized monoclonal antibodies.[18,19] This method should yield the best recovery of prothrombin from plasma if the epitope recognized by the antibody is not altered. However, monoclonal antibodies may not be readily available, and a more classical procedure for prothrombin isolation may be the method of choice. Experience in the isolation of prothrombin Quick[6] has shown that a procedure modified from that described by Comp *et al.*[20] for isolation of human vitamin K-dependent proteins consistently yields a good recovery of prothrombin, at approximately 50% of that present in the starting plasma.

In this procedure anionic proteins are first removed from diluted citrated plasma by adsorption to an anion-exchange resin. This appears to

[16] H. Iwahana, K. Yoshimoto, and M. Itakura, *Nucleic Acids Res.* **19**, 4309 (1991).
[17] H. Iwahana, K. Yoshimoto, T. Shigekiyo, A. Shirakami, S. Saito, and M. Itakura, *Int. J. Hematol.* **55**, 93 (1992).
[18] M.-J. Rabiet, B. C. Furie, and B. Furie, *J. Biol. Chem.* **261**, 15045 (1986).
[19] B. F. Le Bonniec and C. T. Esmon, *Proc. Natl. Acad. Sci. U.S.A.* **88**, 7371 (1991).
[20] P. C. Comp, R. R. Nixon, M. R. Cooper, and C. T. Esmon, *J. Clin. Invest.* **74**, 2082 (1984).

improve the yield of prothrombin, particularly from lipemic plasmas, when compared to adsorption of the vitamin K-dependent proteins directly from plasma by barium salts. Because the binding of prothrombin to anion-exchange resins is largely dependent on the fragment 1 domain containing γ-carboxyglutamic acid residues, separation of the two forms of the zymogen present in the plasma of heterozygous individuals may not be possible. If alterations occur in the fragment 1 region, the method for isolation of prothrombin may need to be modified.[21] Once isolated, the prothrombin is activated with taipan venom and the resultant thrombin is purified by cation-exchange chromatography with modification of the method described by Lundblad et al.[22]

The procedure that follows is appropriate for 250–500 ml (1–2 units) of plasma and may be further modified for larger or smaller plasma volumes. Plasma, collected by apheresis or low-speed centrifugation, is heparinized at 4 U/ml, recentrifuged at 12,000 g to remove residual platelets, and stored at −80° prior to isolation of prothrombin. A good recovery of prothrombin has been obtained from plasma stored in this manner for as long as 14 years. On the morning when prothrombin isolation is to begin, the plasma is thawed and diluted with an equal volume of 0.02 M Tris-HCl, 0.01 M benzamidine hydrochloride, pH 7.5. To the diluted plasma, 0.5–1.0 g QAE-Sephadex in 0.1 M NaCl, 0.02 M Tris-HCl, pH 7.5, is added with stirring, which is continued for 1 hr at 22°. The suspension is then allowed to settle for 30 min, after which the supernatant plasma is removed by siphon. The residual QAE-Sephadex, which is now blue-green in color, is transferred to a 1.5- to 2.5-cm column. A small flexible plastic spatula (the type ordinarily used in the kitchen) is helpful for ensuring maximum transfer of the QAE-Sephadex. After allowing the residual supernatant to drain, the QAE-Sephadex, contained in the column, is washed with 50 ml of 0.15 M NaCl, 0.02 M Tris-HCl, pH 7.5, and eluted with 50 ml of 0.4 M NaCl, 0.02 M Tris-HCl, pH 7.5. The eluate is brought to 10^{-4} M phenylmethylsulfonyl fluoride and diluted with 1.7 volumes 0.032 M sodium citrate, 5 mM benzamidine, 0.02 M Tris-HCl, pH 8.0. The citrate solution is chilled in an ice bath and 0.1 volume 1 M BaCl$_2$ is added with stirring. Stirring is continued for a minimum of 15 min. The suspension of barium citrate should be well chilled to minimize solubility of the barium salt. The vitamin K-dependent proteins adsorbed to barium citrate are removed by centrifugation at 12,000 g for 15 min at 4°. The pale blue supernatant is discarded. The precipitate is washed out of the centrifuge bottle with 125 ml chilled distilled water and transferred to a

[21] C. T. Esmon, G. A. Grant, and J. W. Suttie, Biochemistry 14, 1595 (1975).
[22] R. L. Lundblad, H. S. Kingdon, and K. G. Mann, this series, Vol. 45, p. 156.

Waring blender where it is homogenized for 2 min. The speed of the blender is reduced with a voltage regulator to avoid foaming. The homogenized suspension is transferred back to a centrifuge bottle, the blender is rinsed with a small amount of cold distilled water, and the suspension is again chilled in an ice bath for a minimum of 15 min prior to centrifugation at 12,000 g for 15 min at $0-4°$. The supernant is discarded and the precipitate is dissolved in 60 ml 0.2 M EDTA, pH 5.6, with stirring continued for 30 min, after which the dissolved barium salts and protein are dialyzed overnight against 0.25 M NaCl, 0.02 M Tris-HCl, 0.02 M EDTA, pH 7.2, at 4°. The dialyzed sample is applied to a 0.9 × 15-cm column of QAE-Sephadex equilibrated in the same buffer and the column is washed with the equilibration buffer to remove nonadhering proteins. Prothrombin is then eluted with a linear gradient to 0.5 M NaCl, 0.02 M Tris-HCl, 0.02 M EDTA, pH 7.2, with a total gradient volume of 60 ml (6 column volumes). The eluate is monitored for the presence of protein by ultraviolet absorbance at 280 nm. The human vitamin K-dependent proteins, including prothrombin, are not well resolved on this column and elute together as the only major protein peak. Prothrombin, a single-chain glycoprotein with a molecular weight of 72,000, migrates with a somewhat higher molecular weight on denaturing gel electrophoresis.

For activation, the prothrombin-containing fractions are pooled and dialyzed against 0.07 M NaCl, 0.05 M Tris-HCl, pH 7.5, at 4°. The dialyzed prothrombin solution (~ 1 mg/ml) is brought to 8 mM CaCl$_2$, and 20 μg/ml taipan venom is added. Rabbit brain cephalin (Sigma Chemical Co., St. Louis, MO) is resuspended according to the manufacturer's directions, and 1/40 volume of the concentrated stock suspension is added to the prothrombin solution. Other sources of phospholipid resuspended by sonication may be used at a final concentration of 20–30 μg/ml. Under these conditions, activation is complete in 20 to 40 min. When catalytically active forms of thrombin are formed, the extent of activation may be followed by chromogenic substrate assay of small samples taken from the activation mixture. When activation is complete, the reaction mixture is applied to a 0.6 × 15-cm column of sulfopropyl-Sephadex equilibrated in 0.07 M NaCl, 0.05 M Tris-HCl, pH 7.5. Following sample application and washing of the column with the same column buffer to remove the activation reagents and nonadhering proteins, a linear gradient to 0.45 M NaCl, 0.05 M Tris-HCl, pH 7.5, is started with 30 ml of buffer in each gradient vessel. Thrombin is eluted as the single major peak at about 0.3 M NaCl. If degradation of thrombin has occurred, the degraded forms, β- and/or γ-thrombin, will be eluted from the sulfopropyl-Sephadex column prior to the intact molecule, α-thrombin. The enzyme may be stored in small aliquots at $-80°$ for several years. Thrombin has a molecular weight of

36,000, and is composed of two disulfide-linked polypeptide chains, the larger of which (B chain) is glycosylated.

When plasma has been obtained from an individual heterozygous for two prothrombin genes, separation of two thrombin forms may be achieved by the gradient elution from sulfopropyl-Sephadex. This will be facilitated if the two forms differ in charge. Alternatively, affinity methods based on a functional defect identified in the mutant thrombin(s) might be used.

Functional Characterization of Mutant Thrombins

Interactions at Catalytic Site. Active site titration with the synthetic thrombin inhibitor, N^2-dansylarginine N-(3-ethyl-1,5-pentanediyl)amide (DAPA) (Haematologic Technologies Inc., Essex Junction, VT) can provide considerable information regarding the structure of the catalytic site of thrombin. The enhancement of fluorescence obtained following binding of this high-affinity inhibitor permits determination of K_d, and when compared to the enhancement of fluorescence for the native molecule may suggest changes in the molecular environment of the dansyl moiety. Further, when DAPA is bound to normal thrombin, enhancement of fluorescence can also occur as a result of energy transfer to the dansyl moiety following excitation in the absorption band for aromatic residues in the protein, 280–290 nm. If the fluorescence enhancement observed with energy transfer is similar for mutant and normal thrombins, similar conformational relationships for nearby aromatic residues are indicated.

For these titrations, 0.5 μM thrombin in 0.1 M NaCl, 0.05 M Tris-HCl is placed in a fluorescence cuvette and an aqueous solution of DAPA is added by using a syringe with repeating dispenser. The total volume of DAPA added should not exceed 2% of the total volume of protein solution to minimize concentration changes. The DAPA concentration may be determined from the absorbance at 330 nm, where the molar extinction coefficient is 4.01×10^3.[23] The fluorescence emission at 550 nm is determined after each addition of DAPA, with excitation at 340 and 290 nm. Fluorescence is also determined for DAPA added to the buffer only. This value is subtracted from the total protein fluorescence at each DAPA concentration, and the corrected titration data representing fluorescence enhancement resulting from fluorophore binding to the protein are fitted to the equation for single-site equilibrium binding by nonlinear least-squares analysis to determine maximum fluorescence, dissociation con-

[23] K. G. Mann, J. Elion, R. J. Butkowski, M. Downing, and M. E. Nesheim, this series, Vol. 80, p. 286.

stant, and concentration of ligand-binding sites. Results may be determined for data obtained from excitation at both 340 and 290 nm. Because of the high affinity of DAPA binding to thrombin, the precision of the K_d determination will be improved at thrombin concentrations of 0.5 μM or less.

Antithrombin III inhibition of thrombin also appears to result from interaction with residues located within close proximity to the catalytic site of thrombin. Thus, similar rates of inhibition by antithrombin III for mutant and normal molecules also indicates similarity in the catalytic site structure and a lack of functional alteration with respect to these residues, but only to the extent required for enzyme–inhibitor complex formation. To determine the rate of inhibition of thrombin by antithrombin III,[6] thrombin (to give a final concentration of 3–5 μg/ml) is added to excess inhibitor (final concentration, 75 μg/ml) in 0.1 M NaCl, 0.02 M Tris-HCl, pH 7.5, 0.1% albumin. Human antithrombin III may be prepared by affinity chromatography on heparin-Sepharose (Pharmacia LKB Uppsala, Sweden or Piscataway, NJ) as described previously.[24,25] At timed intervals, samples are removed from the reaction mixture and diluted into 0.1 M NaCl, 0.05 M Tris-HCl, pH 8.3, for determination of residual activity with a p-nitroanilide substrate such as tosyl-Gly-Pro-Arg-p-nitroanilide (Boehringer Mannheim, Mannheim, Germany or Indianapolis, IN). From these results second-order rate constants for inhibition can be calculated and compared.

Because of their small size and limited interaction with the enzyme surface, steady-state kinetic studies with low-molecular-weight substrates will be sensitive to structural and functional alterations only near the catalytic site. Full determination of steady-state kinetic parameters should permit detection of minor alterations in this region of the molecule. However, because of differences in rate-limiting steps in the catalysis of hydrolysis of different types of substrates, results for esters or anilides may not reflect the same changes in peptide bond hydrolysis.

Interaction at Secondary Specificity Sites. The normal release of fibrinopeptide A from fibrinogen involves an interaction with thrombin at both the catalytic site and a distal anion-binding or specificity site on the thrombin surface. Considerable information about the functional consequences of structural alterations may be obtained by determination of the steady-state kinetic parameters for fibrinopeptide A release. Because the mechanism for fibrinopeptide B release appears to differ from that for fibrinopeptide A release, determination of steady-state kinetic parameters

[24] W. G. Owen, *Biochim. Biophys. Acta* **405**, 380 (1975).
[25] P. S. Damus and R. D. Rosenberg, this series, Vol. 45, p. 653.

for release of fibrinopeptide B from fibrinogen or fibrin I may also be useful in characterizing the defect in a mutant thrombin. The use of analytical HPLC for quantitation of fibrinopeptide B has been described.[26]

Fibrinopeptide A may be quantitated by either a commercially available radioimmunoassay (Mallinckrodt, Inc., Radioisotopes Division, St. Louis, MO), described below,[6] or by analytical HPLC.[26] Human fibrinogen may be isolated from plasma[15] or obtained from Kabi Diagnostica (Uppsala, Sweden or Helena Laboratories, Beaumont, TX). To minimize the background level of fibrinopeptide A, the preparation is treated with diisopropyl fluorophosphate (DFP) and/or p-amidinophenylmethylsulfonyl fluoride and contaminating plasminogen is removed by affinity chromatography on lysine-Sepharose (Pharmacia LKB, Uppsala, Sweden or Piscataway, NJ). Residual inhibitors are removed and the fibrinogen is concentrated by precipitation with 25% saturated ammonium sulfate. The fibrinogen is then gel-filtered on Sephadex G-25 in 0.1 M NaCl, 0.05 M Tris-HCl, pH 7.5. For determination of rates of hydrolysis, 250 μl of a fibrinogen dilution is transferred to a polypropylene tube maintained at constant temperature in a water bath. To this is added 10 μl of thrombin. Aliquots of this reaction mixture are transferred to plastic tubes equilibrated at the reaction temperature and the reaction is stopped at timed intervals by the addition of p-amidinophenylmethylsulfonyl fluoride (4 mg/ml in dimethyl sulfoxide) to a final concentration of 2.6 mM. A final thrombin concentration of 30 pM with fibrinogen Aα chain concentrations ranging from 0.5 to 17 μM permits determination of steady-state kinetic parameters by this method. For the fibrinopeptide A radioimmunoassay, all volumes may be decreased to one-half of those specified by the manufacturer. Each sample is brought to an appropriate dilution for assay with 0.1 M NaCl, 0.05 M Tris-HCl, pH 7.5, 10 mg/ml bovine serum albumin before bentonite adsorption of fibrinogen. Substrate hydrolysis should not exceed 10% of the total fibrinogen present at each fibrinogen concentration. The rates of hydrolysis determined at various fibrinogen concentrations are fitted to the Michaelis–Menten equation by nonlinear regression to obtain the steady-state kinetic parameters K_m and k_{cat}.

Other Functional Alterations. Because thrombin has several biologic activities not involving the same structural domains of the protein, a particular mutant would be expected to display varying patterns of alteration. However, as indicated above, diagnosis of dysprothrombinemia has always been based on deficient fibrinogen clotting activity. For mutant thrombins, other functional activities of interest include the activation of factors V, VIII, and protein C as well as thrombomodulin binding, heparin binding, and the enhancement of inhibition by antithrombin III in the

---

[26] J. Hofsteenge, H. Taguchi, and S. R. Stone, *Biochem. J.* **237,** 243 (1986).

presence of heparin. Heparin cofactor II inhibition of thrombin in the presence of heparin involves different structural features of thrombin than for antithrombin III inhibition, making study of the interaction with heparin cofactor II also of interest with respect to understanding structure–function relationships. Platelet aggregation and thromboxane production appear to result from different mechanisms of platelet stimulation by thrombin, a discordance that might be further explored with mutant thrombins. The effects of alteration of thrombin structure on chemotaxis and mitogenesis, two additional cellular responses, have not been thoroughly explored.

Structural Characterization

Identification of Primary Structural Alterations. Because prothrombin is a relatively abundant plasma protein present in normal plasma at about 100 mg/liter, a few hundred milliliters of plasma should provide sufficient protein for initial functional characterization and identification of an altered amino acid. Less than 1 mg of each protein was needed for the complete primary structural characterizations of thrombin Quick I and thrombin Quick II (see Table I).

The basis for the structural characterization described here is essentially that of Miyata *et al.*[4] In the three studies reporting identification of the amino acid substitution in congenitally mutant thrombins, lysyl endopeptidase from *Achromobacter lyticus* has been used to prepare peptides for peptide mapping and sequencing studies. Of the 308 amino acids residues in thrombin, 22 are lysines distributed throughout the sequence. Therefore, lysyl endopeptidase hydrolysis yields a series of peptides, the longest of which, in normal thrombin, is 31 residues. For peptides of this size it should be possible to obtain complete sequence information by automated peptide sequencing.

Reduction and Alkylation. Thrombin contains four intramolecular disulfide bonds that need to be reduced and alkylated before proceeding with the enzymatic digestion. The choice of alkylating agent will depend on the particular capabilities of the system being used to identify the phenylthiohydantoin (PTH) amino acids obtained by Edman degradation. Generally, the use of vinylpyridine yields the most easily identifiable derivative of cysteine. The method for reduction and alkylation described below is modified[27] from that used in the characterization of thrombin Quick I[7] and thrombin Quick II.[8]

Prior to reduction, 1 mg of thrombin is dialyzed against 0.01 M acetic acid and lyophilized. The protein is then dissolved in 0.5 ml of 6 M

---

[27] A. S. Inglis, this series, Vol. 91, p. 26.

guanidine hydrochloride, 0.2 M Tris-HCl, 1 mM disodium ethylenedia-minetetraacetic acid, pH 8.6. Mercaptoethanol is added to 1% final concentration, and the mixture is incubated at 37° for 2 hr prior to the addition of a threefold excess of 4-vinylpyridine over total sulfhydryl groups and further incubation at room temperature for 2–3 hr. The solution is acidified to pH 3 with glacial acetic acid. The reduced and alkylated protein is transferred to Spectra/Por 6 dialysis tubing, molecular weight cutoff 2000 (Spectrum Medical Industries, Inc., Los Angeles, CA) to prevent loss of the thrombin A chain, dialyzed against 0.01 M acetic acid, and lyophilized.

Enzymatic Digestion. For enzymatic digestion, 1 mg of lyophilized, reduced, and alkylated protein is dissolved in 0.25 ml of 8 M urea, 0.05 M Tris-HCl, pH 9.0, then diluted with 0.25 ml of 0.05 M Tris-HCl, pH 9.0. To this solution is added 5 μl of a stock solution of lysyl endopeptidase (Wako Chemicals, Inc., Dallas, TX) at 10 activity U/ml in 0.15 M NaCl, 0.02 M Tris, pH 7.4, followed by incubation at 37° for 4 hr, after which an additional 5 μl of lysyl endopeptidase is added and incubation is continued at 37° for an additional 4 hr. The peptide digest may be stored at −20° prior to chromatography.

Peptide Maps. The use of reversed-phase columns with gradients of acetonitrile in dilute trifluoroacetic acid has yielded good separation of the peptides derived from thrombin by lysyl endopeptidase digestion. An ultraviolet absorbance detector is used to monitor the column at 214 nm. Simultaneous monitoring at 280 nm may be helpful in identifying peptides and determining peptide purity. To optimize peptide separation and identify a peptide with altered elution position, analytical-scale peptide maps may be prepared with 5–10 μg of the digested protein. The peptide maps obtained for thrombin and thrombin Quick I by this procedure are shown in Fig. 1. Varying elution parameters, including gradient patterns, eluants, and column supports, may assist in obtaining a satisfactory separation and in identifying an altered peptide. A change in pH may be particularly helpful in shifting the elution positions of peptides. The use of postcolumn derivatization with o-phthalaldehyde can assist in discriminating peptide from nonpeptide material eluting from the column. In the case of lysyl endopeptidase digestion each peptide contains only one lysine residue, and has roughly equivalent fluorescence so that examination of peak areas may assist in identifying regions of the chromatogram where two (or more) peptides are coeluting, as is suggested by the results shown in Fig. 1 (seen particularly in the A region). Peptides treated with o-phthalaldehyde cannot be used subsequently for sequencing. Therefore, once conditions are optimized for peptide separation, larger amounts of peptide material should be applied to the column and fractions collected while the effluent is monitored by UV absorbance or other nondestructive method. The

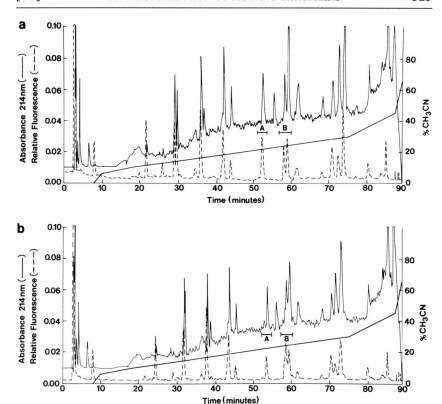

FIG. 1. Peptide maps for thrombin and thrombin Quick I. The chromatograms shown are for reversed-phase chromatography of 10 μg of peptides derived from a lysyl endopeptidase digest of S-carboxymethylated thrombin (a) or thrombin Quick I (b), applied to a Vydac 218TP54 C$_{18}$ column (The Sep/a/ra/tions Group, Hesparia, CA) and eluted with a gradient of acetonitrile in 0.1% trifluoroacetic acid at a flow rate of 1.0 ml/min. Fluorescence was monitored following postcolumn derivatization with *o*-phthalaldehyde. Peptide material is missing from thrombin Quick I in the region labeled A, and an additional shoulder appears in the region labeled B. Reprinted with permission from Henriksen and Mann,[7] copyright [1988] American Chemical Society.

fractions containing the peptide of interest are pooled and subjected to further chromatography if additional purification is needed. Finally peak fractions are concentrated by lyophilization prior to automated sequencing.

DNA Sequencing to Identify Defect. The development of the polymerase chain reaction provides an alternative approach for determination of the primary structural alteration in mutant thrombins.[11] This method may be used to amplify the prothrombin gene in DNA isolated from circulating white blood cells. Because the gene for prothrombin contains 14 exons, 7

TABLE II
SITE-SPECIFIC MUTANT THROMBINS

Mutation[a]	Functional defect	Ref.
Glu-345(39) → Lys	Activity on chromogenic substrates for thrombin is 50–200% of normal. Increased rates for inhibition by antithrombin III and for activation of protein C, primarily in the absence of thrombomodulin. The k_{cat}/K_m for fibrinopeptide A release is 50% of normal with increases in both k_{cat} and K_m	28
Lys-372(60F) → Glu	Normal chromogenic substrate activity; fibrinogen clotting activity 17% of normal. Increased activity in activation of protein C in the presence of thrombomodulin	29
Arg-388(73) → Glu	Normal chromogenic substrate activity; fibrinogen clotting activity and activation of protein C ± thrombomodulin are markedly reduced	29
Arg-390(75) → Glu	Normal chromogenic substrate and fibrinogen clotting activities. Activation of protein C in presence of thrombomodulin and Ca^{2+} is < 10% of normal	29
Glu-522(192) → Gln	Does not affect fibrinopeptide A release, but enhances rate of activation of protein C primarily in the absence of thrombomodulin by increasing k_{cat}	19
Ser-525(195) → Ala	Modification of Ser of catalytic triad. Lacks hydrolytic activity, but binds thrombomodulin normally	29,30

[a] Numbering for thrombin residues is based on prothrombin sequence[3] with corresponding chymotrypsinogen numbering[1] in parentheses.

of which code for thrombin, a maximum of 14 primers corresponding to the 7 thrombin exons are needed to determine the structural alteration in a mutant thrombin by this method. Further, the expected heterozygosity at the prothrombin locus in cases of congenitally mutant thrombins would require sequencing of multiple clones to identify both defects. Because of the relatively high levels of prothrombin in plasma, it may be as efficient to determine a primary structural defect by peptide sequencing as by DNA sequencing. In situations in which a cross-reacting protein is absent or present at very low concentration, however, identification of the defect would require DNA sequencing. The choice of technique to use will probably depend largely on the preference of the particular laboratory undertaking the project.

Tertiary Structural Consequences. An indication of the effects of an altered residue on the tertiary structure may be gained by examination of the crystal structure of normal thrombin,[1] the coordinates for which are available in the Brookhaven Protein Data Bank (Brookhaven National Laboratory, Upton, NY). The value of examining the known three-dimensional structure in relation to the substantial amount of literature describing structure–function relationships for thrombin cannot be overemphasized. With computerized molecular modeling systems, qualitative assessment of the effects of a mutation may be further enhanced by local energy minimization of the altered structure in the region of interest. Understanding of larger or more global structural effects will most probably require determination of the crystal structure.

Mutant Thrombins Obtained by Site-Specific Mutagenesis

In addition to the congenitally mutant thrombins, several other mutant thrombins have been prepared by site-specific mutagenesis and have been used to investigate specific questions regarding structure–function relationships in this enzyme. Mutations have been introduced into the cDNA for prothrombin by using the polymerase chain reaction to create single nucleotide substitutions or by oligonucleotide-directed mutagenesis. Various vectors have been used to express the mutant prothrombins in CV-1 cells, CHO cells, or baby hamster kidney cells. Prothrombin has been isolated from cell culture media by either adsorption to barium salts or affinity chromatography on a human prothrombin-specific monoclonal antibody. To obtain the mutant thrombin, recombinant prothrombins have been activated by prothrombinase (factor Xa, factor Va, phospholipid, and Ca^{2+}) and *E. carinatus* or taipan snake venoms, and the product isolated by cation-exchange chromatography. The isolation and characterization of these mutant forms relies on methods similar to those described above for congenitally mutant thrombins. These mutants are summarized in Table II.[19,28-30]

Acknowledgment

Support for preparation of this article has been provided by NIH Grant HL-45194 and a Senior International Fogarty Fellowship.

[28] B. F. Le Bonniec, R. T. A. MacGillivray, and C. T. Esmon, *J. Biol. Chem.* **266**, 13796 (1991).
[29] Q. Wu, J. P. Sheehan, M. Tsiang, S. R. Lentz, J. J. Birktoft, and J. E. Sadler, *Proc. Natl. Acad. Sci. U.S.A.* **88**, 6775 (1991).
[30] G. Pei, K. Baker, S. M. Emfinger, D. M. Fowlkes, and B. R. Lentz, *J. Biol. Chem.* **266**, 9598 (1991).

[19] Synthetic Selective Inhibitors of Thrombin

By Shosuke Okamoto and Akiko Hijikata-Okunomiya

Introduction

Experimental and clinical needs have made the search for synthetic selective inhibitors of thrombin extremely important. Thrombin, a major physiological enzyme in coagulation, affects the conversion of fibrinogen to fibrin, the activation of factor XIII (FXIII) to FXIIIa, and the aggregation of platelets. Other plasma proteins have been reported to be susceptible to this enzyme.[1]

In circulatory blood, thrombin is found as the zymogen, prothrombin, which is activated by intrinsic and extrinsic pathways. Once activated, thrombin is neutralized by plasma antithrombins, e.g., antithrombin III (ATIII). The result is that active thrombin is rarely found in circulatory blood. However, the activation of thrombin occurs in pathological states and such activation is undoubtedly hazardous. The synthetic selective inhibitors discussed in this chapter were designed based on biochemical knowledge regarding the conversion of fibrinogen by thrombin to fibrin.

Fibrinogen – Fibrin Conversion

Role of Phe and Arg in Fibrinogen Molecules

Studies of fibrinogen–fibrin conversion that are concerned with the basic and clinical aspects of coagulation have attracted the attention of many researchers. Some unique molecular aspects of fibrinogen–fibrin conversion have been reported especially by Blombäck and colleagues.[2,3] Their studies were extended in close cooperation with Edman, who developed the automatic amino acid sequence analyzer, and the results have been valuable in casting a new light on this field.

When thrombin acts on fibrinogen, it cleaves Arg^{16}-Gly^{17} in the Aα chain of fibrinogen and Arg^{14}-Gly^{15} in the Bβ chain, liberating fibrinopeptide A and B, respectively.[2,3] This action of thrombin is strikingly different

[1] J. W. Fenton, II, *Ann. N. Y. Acad. Sci.* **370**, 468 (1981).
[2] R. F. Doolittle and B. Blombäck, *Nature (London)* **202**, 147 (1964).
[3] B. Blombäck, M. Blombäck, B. Hessel, and S. Iwanaga, *Nature (London)* **215**, 1445 (1967).

from that of the digestive enzyme trypsin, which hydrolyzes nearly all of the peptide bonds located next to arginine and lysine.[4,5]

Comparative studies of the action of thrombin on fibrinogens from various species indicate that Phe is found consistently in position 9 from Arg-1 in fibrinopeptide A, despite other evolutionary differences. Particular attention was paid to the geometrical relationship between arginine and phenylalanine, and the following peptides were synthesized[6]: Phe-$(Val)_n$-Arg-COOCH$_3$. The number (n) of Val residues (or spacer amino acids) is significant in manifesting the thrombin inhibitory activity, i.e., the inhibitory activity becomes greater when n is 1 or 7 in this structure. Although they are hydrolyzed quickly, these kinds of ester compounds form the theoretical basis for designing synthetic substrates and inhibitors of thrombin.

Synthetic Selective Substrates for Thrombin

Although synthetic thrombin inhibitors prepared by B. Blombäck et al. were weak *in vitro* and *in vivo*[6] they opened a new field of study of synthetic selective chromogenic substrates for different kinds of proteases. Some of these substrates for thrombin include[7] H-D-Phe-pip-Arg-p-nitroanilide, boc-Val-Pro-Arg-4-methylcoumaryl-7-amide, and tosyl-Gly-Pro-Arg-p-nitroanilide.

Theoretical Basis for Selectivity of Thrombin

It is presumed that the mechanisms by which protease reads or "decodes" the information in the substrate are concentrated in the active center of the protease molecule.[8] The active center is somewhat lengthy (18 and 25 Å in carboxypeptidase A and papain, respectively), and has a stereogeometrical surface with elevated or lowered parts.[9,10] The charged side chain, hydrophobic binding site, and other hyperreactive sites may participate in decoding the information in the substrate molecule. The

[4] E. Mihalyi and J. E. Godfrey, *Biochim. Biophys. Acta* **67**, 73 (1963).
[5] H. A. Scheraga, in "Chemistry and Biology of Thrombin" (R. L. Lundblad, J. W. Fenton, II, and K. G. Mann, eds.), p. 145. An Arbor Sci. Publ., Ann Arbor, MI, 1977.
[6] B. Blombäck, M. Blombäck, P. Olsson, L. Svendsen, and G. Åberg, *Scand. J. Clin. Lab. Invest.* **107**, Suppl., 59 (1969).
[7] R. Lottenberg, U. Christensen, C. M. Jackson, and P. L. Coleman, this series, Vol. 80, p. 341.
[8] S. Okamoto and A. Hijikata, *Drug Des.* **6**, 143 (1975).
[9] I. Schechter and A. Berger, *Biochem. Biophys. Res. Commun.* **27**, 157 (1967).
[10] N. Abramowitz, I. Schechter, and A. Berger, *Biochem. Biophys. Res. Commun.* **29**, 862 (1967).

specificity code is "encoded" spatially in three-dimensional arrangements of amino acids within the substrate molecule. The inhibitor, having to a certain extent the same code as the substrate but with some differences, behaves as a "noise" in the system. This disturbs the reaction.

In the search for effective inhibitors we have been greatly stimulated by basic studies that enable us to discuss inhibitor design in terms of coding and decoding. It is even possible to encode signals, i.e., certain chemical groups, within the molecules to be designed.

N^α-Dansyl-L-arginine 4-Ethylpiperidineamide

Modification of Arginine

Studies on the fibrinopeptides described above suggest that thrombin recognizes the L-arginine residue, in particular its biophysical and geometrical arrangements. Therefore, we selected L-arginine as the most promising skeleton for designing thrombin inhibitors and studied the structure–function relationship between a number of arginine derivatives and their thrombin inhibitory activity.

Because the conversion of the L-Arg residue to other homologous amino acids or guanidino compounds decreased the thrombin inhibitory activity, we concluded that the L-Arg skeleton is essential.[11,12] Chemical modifications of the α-amino and the α-carboxyl groups changed the thrombin inhibitory activity as shown in Table I.[11,13,14] The α-amino group should not be removed, but it should be modified, e.g., with the dansyl group as one representative substituent. The carbonyl compounds are far less inhibitory than the sulfonyl compounds. The α-carboxyl group should not be removed, but it should be modified; n-butyl ester and n-butylamide are effective. Ring fixation of aliphatic chains in cyclic forms, e.g., piperidine and morpholine, have been successful. Furthermore, the introduction of methyl, ethyl, propyl, and phenyl groups into the piperidine ring indicated that the 4-ethylpiperidine derivative is the most potent.

Stepwise chemical modifications of L-arginine affecting the thrombin

[11] S. Okamoto, A. Hijikata, K. Kinjo, R. Kikumoto, K. Ohkubo, S. Tonomura, and Y. Tamao, *Kobe J. Med. Sci.* **21**, 43 (1975).

[12] S. Tonomura, R. Kikumoto, Y. Tamao, K. Ohkubo, S. Okamoto, K. Kinjo, and A. Hijikata, *Kobe J. Med. Sci.* **26**, 1 (1980).

[13] S. Okamoto, K. Kinjo, A. Hijikata, R. Kikumoto, Y. Tamao, K. Ohkubo, and S. Tonomura, *J. Med. Chem.* **23**, 827 (1980).

[14] R. Kikumoto, Y. Tamao, K. Ohkubo, T. Tezuka, S. Tonomura, S. Okamoto, Y. Funahara, and A. Hijikata, *J. Med. Chem.* **23**, 830 (1980).

TABLE I
ARGININE DERIVATIVES AND THROMBIN INHIBITORY ACTIVITY

$$\begin{array}{c} NH \\ \diagup \\ C-NH-(CH_2)_3-CH-CO-R_1 \\ \diagdown \\ NH_2 \qquad\qquad\quad NH \\ \qquad\qquad\qquad | \\ \qquad\qquad\qquad R_2 \end{array}$$

R_1	R_2	I_{50} for thrombin (M)
$-OCH_3$	$-SO_2-$⬡$-CH_3$	1.0×10^{-3}
$-OCH_3$	$-SO_2-$(naphthyl)$-N(CH_3)_2$ (Dansyl)	2.0×10^{-5}
$-OCH_2CH_3$	Dansyl	8.0×10^{-6}
$-OCH_2CH_2CH_3$	Dansyl	2.0×10^{-6}
$-OCH_2CH_2CH_2CH_3$	Dansyl	2.0×10^{-6}
$-OCH_2CH_2CH_2CH_2CH_3$	Dansyl	4.9×10^{-5}
$-NHCH_2CH_3$	Dansyl	1.0×10^{-4}
$-NHCH_2CH_2CH_3$	Dansyl	5.0×10^{-6}
$-NHCH_2CH_2CH_2CH_3$	Dansyl	2.1×10^{-6}
$-NHCH_2CH_2CH_2CH_2CH_3$	Dansyl	1.3×10^{-4}
$-NCH_2CH_2CH_2CH_3$ with CH_3	Dansyl	2.0×10^{-6}
$-N$(piperidine)	Dansyl	1.0×10^{-6}
$-N$(piperidine)$-CH_3$	Dansyl	3.0×10^{-7}
$-N$(piperidine)$-CH_2CH_3$	Dansyl	1.0×10^{-7}
$-N$(piperidine)$-$⬡	Dansyl	1.0×10^{-4}

inhibitory activity yielded a series of very potent thrombin inhibitors having the following tripod structure: (1) L-arginine skeleton, (2) α-amino substituent with a bulky aromatic group, and (3) α-carboxyl substituent with a hydrophobic chain or ring having a chain length of between 4 and 5 carbons. Accordingly, to recognize the tripods precisely, three binding sites

TABLE II
NATURE OF NO. 205

Parameter	Value	$K_i(M)$
Absorption maximum	330 nm	—
Extinction coefficient[a]	4400	—
Inhibition		
Thrombin		3.7×10^{-8}
Trypsin		1.0×10^{-5}
Plasmin		5.0×10^{-4b}
Factor Xa		5.7×10^{-4b}
Urokinase		1.7×10^{-4b}

[a] At 330 nm.
[b] Calculated from I_{50} value.

need to be identified in the thrombin molecule, namely, the negatively charged binding site, the aromatic binding site, and the hydrophobic binding site.

N^α-Dansyl-L-arginine 4-ethylpiperidineamide (No. 205) is one of the derivatives that fulfills the basic requirements described above. It inhibits thrombin competitively and very potently with a K_i value of $3.7 \times 10^{-8}M$. It inhibits trypsin only weakly[15] and scarcely inhibits plasmin, factor Xa, and urokinase (Table II). Therefore No. 205 is regarded as the most potent and selective thrombin inhibitor.

[15] A. Hijikata, S. Okamoto, R. Kikumoto, and Y. Tamao, *Thromb. Haemostasis* **42**, 1039 (1979).
[16] A. Hijikata, S. Okamoto, E. Mori, K. Kinjo, R. Kikumoto, S. Tonomura, Y. Tamao, and H. Hara, *Thromb. Res.* **8**, Suppl. II, 83 (1976).

Some animal experiments using No. 205 and its homologous inhibitor, N^α-dansyl-L-arginine 4-methylpiperidineamide (No. 189), indicate that these inhibitors are effective in preventing the action of thrombin *in vivo,* when they are infused carefully.[16]

Fluorescence Studies Using No. 205

Compound No. 205 becomes highly fluorescent when bound to thrombin.[17] The fluorescence emission wavelength maximum shifts from 578 to 555 nm in the presence of thrombin. The fluorescence intensity increases by a factor of ~3 on binding with the enzyme when an excitation wavelength of 350 nm is used. Compound No. 205 does not cause fluorescence interaction with prothrombin, making it a useful tool for the analysis of the activation process of prothrombin.[18] Special attention should be paid on the fact that No. 205 inhibits plasma cholinesterase (BuCHE) very strongly, as described below, and that it also causes fluorescence changes on interaction with BuChE.[19,20]

Reducing Toxicity of No. 205

Although No. 205 is effective in experimental systems, it is not applicable clinically because of its high toxicity: the LD_{50} is 80 mg/kg in mice when administered intraperitoneally. The ability of No. 205 to bind strongly to BuChE and tissue constituents seems to be closely related to its toxicity. To reduce toxicity, stepwise chemical modifications can be made on the N terminus and C terminus of L-arginine, leaving the arginine skeleton unchanged.

Very promising results have been obtained by introducing a carboxyl group into the C terminus of L-arginine. As shown in Table III, the thrombin inhibitory activity found in the carboxyl-containing amides follows the same rule as in the case of amide and ester compounds: The hydrophobic chains and rings corresponding to carbon chain lengths between C-4 and C-5 are the most potent.[21] However, the introduction of a carboxyl group into the C terminus of L-arginine decreases the acute

[17] M. E. Nesheim, F. G. Prendergast, and K. G. Mann, *Biochemistry* **18,** 996 (1979).
[18] L. S. Hibbard, M. E. Nesheim, and K. G. Mann, *Biochemistry* **21,** 2285 (1982).
[19] S. Nagano, S. Okamoto, K. Ikezawa, K. Mimura, A. Matsuoka, A. Hijikata, and Y. Tamao, *Thromb. Haemostasis* **46,** 45 (1981).
[20] S. Brimijoin, K. P. Mintz, and F. G. Prendergast, *Biochem. Pharmacol.* **32,** 699 (1983).
[21] R. Kikumoto, Y. Tamao, K. Ohkubo, T. Tezuka, S. Tonomura, S. Okamoto, and A. Hijikata, *J. Med. Chem.* **23,** 1293 (1980).

TABLE III
ARGININE DERIVATIVES AND THROMBIN INHIBITORY ACTIVITY

R	I_{50} for thrombin (M)
(−N with CH$_3$ and CH$_2$COOH)	$> 1.0 \times 10^{-4}$
(−N with CH$_2$CH$_2$CH$_3$ and CH$_2$COOH)	8.0×10^{-6}
(−N with CH$_2$CH$_2$CH$_2$CH$_3$ and CH$_2$COOH)	3.0×10^{-7}
(−N with CH$_2$CH$_2$CH$_2$CH$_2$CH$_3$ and CH$_2$COOH)	1.0×10^{-6}
(piperidine −N−CH$_3$, COOH)	4.0×10^{-7}
(piperidine −N−C$_2$H$_5$, COOH)	3.0×10^{-7}

toxicity, BuChE inhibitory activity, and tissue-binding affinity (Table IV).[22-24] After a series of studies on pharmacological and adverse effects, we finally found 4-methyl-1-N^2-[(3-methyl-1,2,3,4-tetrahydro-8-quinolinyl)sulfonyl]-L-arginyl)-2-piperidinecarboxylic acid very promising. Of

[22] S. Okamoto, A. Hijikata, R. Kikumoto, and Y. Tamao, *Abstr. Congr. Int. Soc. Haematol. 17th*, p. 301 (1978).
[23] K. Oda, K. Ohtsu, Y. Tamao, R. Kikumoto, A. Hijikata, K. Kinjo, and S. Okamoto, *Kobe J. Med. Sci.* **26**, 11 (1980).

the four stereoisomers at the 2- and 4-positions of the piperidine ring, the most potent is the (2R,4R)-isomer, which is approximately 20,000 times stronger than the (2S,4S)-isomer. The (2R,4R)-isomers is also called No. 805, MD805, and MC19038, and is now available clinically as argatroban.[25,26]

Nature of Argatroban

The structure and some physicochemical and enzymatic properties of argatroban are shown in Table V. Argatroban inhibits thrombin competitively, with a K_i value of $1.9-3.8 \times 10^{-8} M$. No other enzyme has ever been reported to be inhibited so potently, and argatroban inhibits other related enzymes only weakly.[26]

Argatroban inhibits all the reactions involving the active site of thrombin directly and without any cofactor: fibrinogen–fibrin conversion, FXIII activation, platelet aggregation, protein C activation, thrombin–ATIII complex formation, endothelin release, vessel wall contraction, and plasminogen activator release from vessel walls.[25–33] Argatroban effectively inhibits platelet aggregation caused by thrombin; it scarecely inhibits platelet aggregation caused by ADP and collagen. At concentrations lower than 1 μM in plasma, argatroban prevents thrombus formation in various animal models.[34–37]

[24] A. Hijikata-Okunomiya, S. Okamoto, Y. Tamao, and R. Kikumoto, J. Biol. Chem. 263, 11269 (1988).

[25] S. Okamoto, A. Hijikata, R. Kikumoto, S. Tonomura, H. Hara, K. Ninomiya, A. Maruyama, M. Sugano, and Y. Tamao, Biochem. Biophys. Res. Commun. 101, 440 (1981).

[26] R. Kikumoto, Y. Tamao, T. Tezuka, S. Tonomura, H. Hara, K. Ninomiya, A. Hijikata, and S. Okamoto, Biochemistry 23, 85 (1984).

[27] Y. Tamao, T. Yamamoto, R. Kikumoto, H. Hara, J. Itoh, T. Hirata, K. Mineo, and S. Okamoto, Thromb. Haemostasis 56, 28 (1986).

[28] H. Hara, Y. Tamao, and R. Kikumoto, Jpn. Pharmacol. Ther. 14, Suppl. 5, 875 (1986).

[29] K. Suzuki, H. Kusumoto, and S. Hashimoto, Biochim. Biophys. Acta 882, 343 (1986).

[30] A. Hijikata-Okunomiya, S. Okamoto, and K. Wanaka, Thromb. Res. 59, 967 (1990).

[31] I. Maruyama, Jpn. J. Clin. Hematol. 31, 776 (1990).

[32] K. Nakamura, Y. Hatano, and K. Mori, Thromb. Res. 40, 715 (1985).

[33] A. Hijikata-Okunomiya, H. Kitaguchi, and M. Hirata, Thromb. Res. 45, 699 (1987).

[34] T. Yamamoto, T. Hirata, M. Inagaki, J. Itoh, R. Kikumoto, and Y. Tamao, Jpn. Pharmacol. Ther. 14, Suppl. 5, 887 (1986).

[35] M. Iwamoto, T. Hara, H. Ogawa, and M. Tomikawa, Jpn. Pharmacol. Ther. 14, Suppl. 5, 903 (1986).

[36] H. Ikoma, K. Ohtsu, Y. Tamao, R. Kikumoto, and S. Okamoto, Blood Vessel 13, 72 (1982).

[37] H. Ikoma, H. Hara, Y. Tamao, and R. Kikumoto, Jpn. Pharmacol. Ther. 14, Suppl. 5, 883 (1986).

TABLE IV

ENZYME SELECTIVITY AND TOXICITY OF ARGININE DERIVATIVES

$$\begin{array}{c} NH \\ \| \\ C-NH-(CH_2)_3-CH-CO-R_1 \\ \end{array} \quad \begin{array}{c} \\ \\ NH \\ | \\ R_2 \end{array}$$

NH₂ group: $\underset{NH_2}{\overset{NH}{C}}$

Compound	R₁	R₂	I₅₀ (M)		LD₅₀ (mg/kg in mice)
			Thrombin	BuChE	
No. 205	—N⟨piperidine⟩—CH₂CH₃	—SO₂⟨naphthyl⟩—N(CH₃)₂	1.0×10^{-7}	1.6×10^{-7}	80 i.p.
No. 407	—N(CH₂CH₂OCH₃)(CH₂COOH)	—SO₂⟨naphthyl⟩—OCH₃, OCH₃	3.0×10^{-7}	$>2.5 \times 10^{-3}$	890 i.p., 377 i.v.
Argatroban	—N⟨piperidine⟩—CH₃, COOH (2R,4R)	—SO₂⟨tetrahydroquinolinyl, HN⟩—CH₃	3.2×10^{-8}	$>1.5 \times 10^{-4}$	550 i.p., 81 i.v.

TABLE V
NATURE OF ARGATROBAN

Parameter	Value	K_i (M)
Absorption maximum	333 nm	—
Extinction coefficient[a]	5000	—
Inhibition		
Thrombin		
Bovine		1.9×10^{-8}
Human		3.8×10^{-8}
Trypsin		5.0×10^{-5}
Plasmin		8.0×10^{-4}
Factor Xa		2.1×10^{-4}
Plasma kallikrein		1.5×10^{-3}

[a] At 333 nm.

Argatroban is less toxic than No. 205: in mice the LD_{50} is 550 mg/kg intraperitoneally and 81 mg/kg intravenously. After venous infusion, the plasma concentration in humans decreases biphasically with half-lives of 7 and 40 min.[38] Although the solubility of argatroban in water is rather low, a 1 mg/ml solution can be obtained by dissolving argatroban in 0.1 N HCl, followed by neutralization.

Relation to Antithrombin III and Heparin

There are two naturally occurring thrombin inhibitors: antithrombin III and heparin. ATIII, the major thrombin inhibitor in plasma, binds to the active site of thrombin, but its inhibition is time dependent. Heparin

[38] O. Izawa, M. Katsuki, T. Komatsu, and S. Iida, *Jpn. Pharmacol. Ther.* **14**, Suppl. 5, 1113 (1986).

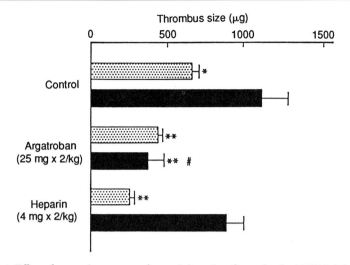

FIG. 1. Effect of argatroban on experimental thrombus formation in ATIII-deficient rats. *Stippled bars:* thrombus size (μg) in control and in argatroban-treated and heparin-treated rats with normal ATIII values. *Solid bars:* thrombus size (μg) in control and in argatroban-treated and heparin-treated rats with low ATIII values. *, $p < 0.05$; **, $p < 0.01$ vs. normal control; #, $p < 0.01$ vs. ATIII-deficient control. (Data used were cited from Ref. 39.)

accelerates the reaction. Thus, ATIII and heparin depend on each other in manifesting their inhibitory activity, indicating that the actions of these two inhibitors are rather complicated in controlling the anticoagulant activity. In contrast, argatroban inhibits thrombin instantaneously without any cofactor. Therefore, argatroban is expected to exhibit fewer individual differences in its efficacy. In fact, the antithrombotic effect of argatroban is independent of the amount of ATIII present, indicating a stable pharmacological effect (Fig. 1).[39]

ATIII interacts with thrombin at the active site, whereas argatroban competes with the same site, preventing the consumption of ATIII.[30] Based on the fact that argatroban inhibits the formation of the thrombin–ATIII complex, argatroban is useful for stable measurement of prothrombin or thrombin–ATIII complex in human plasma.[40,41]

Thrombi Are "Alive"

Tissue Plasminogen Activator and Argatroban

The thrombolytic action of tissue plasminogen activator (t-PA) is accelerated by the addition of argatroban when compared with the addition

[39] T. Kumada and Y. Abiko, *Thromb. Res.* **24**, 285 (1981).
[40] A. Hijikata-Okunomiya, *Thromb. Res.* **57**, 705 (1990).

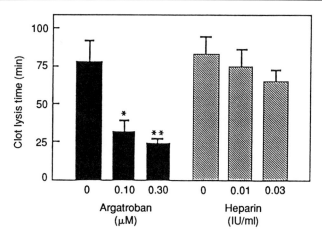

FIG. 2. Effect of argatroban on fibrinolysis by t-PA. *, $p < 0.05$; **, $p < 0.01$. (Data used were cited from Ref. 27.)

of heparin (Fig. 2).[27] We have concluded that argatroban has an adjuvant effect on t-PA. Such an effect was also reported *in vivo* in animal experiments. Reocclusion with reformed thrombi was prevented when argatroban was administered with t-PA in an animal model of arterial thrombosis.[42,43] These results suggest the necessity of studies to establish the clinical usage of t-PA with argatroban as a new thrombolytic therapy.

Fibrinopeptide A Released from Thrombi in Vivo

Extensive clinical studies on the application of argatroban have been made in Japan for the treatment of chronic arterial occlusions, including Buerger's disease; a number of patients were treated with a venous infusion of argatroban, and favorable results and no side effects have been reported.[44] Further, increased amounts of fibrinopeptide A in the circulating blood were reported on blood samples taken within 3–4 weeks after thrombi formation (Fig. 3).[45] This indicates that thrombin is still active possibly within the network of fibrin, and thrombin within the thrombi is effectively inhibited by the low-molecular-weight synthetic thrombin inhibitor.[46] Fibrin white thread obtained by extensive washing of blood clots

[41] T. Yamashita, K. Kouzai, N. Inoue, K. Mimura, and A. Matsuoka, *J. Jpn. Soc. Thromb. Hemostasis* **3**, 49 (1992).

[42] D. J. Fitzgerald and G. A. Fitzgerald, *Proc. Natl. Acad. Sci. U. S. A.* **86**, 7585 (1989).

[43] I. K. Jang, H. K. Gold, R. C. Leinbach, J. T. Fallon, and D. Collen, *Circ. Res.* **67**, 1552 (1990).

[44] T. Tanabe, Y. Mishima, K. Furukawa, S. Sakaguchi, K. Kamiya, S. Shionoya, T. Katsumura, and A. Kusaba, *J. Clin. Ther. Med.* **2**, 1645 (1986).

[45] W. M. Feinberg, D. C. Bruck, M. E. Ring, and J. J. Corrigan, Jr., *Stroke* **20**, 592 (1989).

[46] J. I. Weitz, M. Hudoba, D. Massel, J. Maraganore, and J. Hirsh, *J. Clin. Invest.* **86**, 385 (1990).

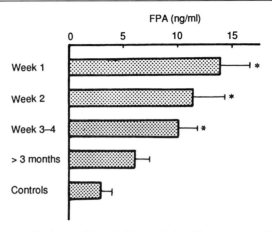

FIG. 3. Variation of fibrinopeptide A in blood obtained from stroke patients. *, $p < 0.05$ vs. controls. (Data used were cited from Ref. 45.)

with water contains a considerable amount of active thrombin. Such thrombin within the thrombi may induce further growth of the thrombi, which has been observed clinically.

Extensive clinical trials on the use of argatroban in cerebral thrombosis are now being carried out in Japan. The results obtained, e.g., through double-blind tests, indicate that argatroban is effective and useful in patients with cerebral thrombosis.[47,48] A problem confronting us may be to determine the kind of mechanism or the action of argatroban in ameliorating cerebral thrombosis. It is known that (1) argatroban accelerates fibrinolysis by t-PA, (2) argatroban inhibits the activation of FXIII, (3) argatroban may inhibit the activity of the remaining thrombin within the thrombi, and (4) argatroban does not require ATIII. Further studies are undoubtedly required.

Acknowledgment

The series of L-arginine derivatives mentioned in this article were synthesized as described in previous papers.[13,14,17,21,25] The synthesis of No. 205 was described by R. Kikumoto et al.[14] and M. E. Nesheim et al.,[17] and the synthesis of argatroban was described by S. Okamoto et al.[25]

[47] Y. Yonekawa, H. Handa, S. Okamoto, Y. Kamijo, Y. Oda, J. Ishikawa, H. Tsuda, Y. Shimizu, M. Satoh, T. Yamaguchi, I. Yano, Y. Horikawa, and E. Tsuda, *Arch. Jpn. Chir.* **55**, 711 (1986).
[48] S. Kobayaski, M. Kitani, S. Yamaguchi, T. Suzuki, K. Okada, and T. Tsunematsu, *Thromb. Res.* **53**, 305 (1989).

[20] Quantifying Thrombin-Catalyzed Release of Fibrinopeptides from Fibrinogen Using High-Performance Liquid Chromatography

By Assunta S. Ng, Sidney D. Lewis, and Jules A. Shafer

Introduction

Fibrinogen, the soluble precursor to the insoluble fibrin matrix of blood clots, is present in normal human blood plasma at approximately 2.9 mg/ml.[1,2] Figure 1 illustrates the structural features of this asymmetric glycoprotein, which is comprised of 2Aα, 2Bβ, and 2γ polypeptide chains linked together by disulfide bonds.[3,4] The horizontal arrows in Scheme I delimit the predominant pathway for the conversion of fibrinogen to the insoluble fibrin matrix of blood clots.[5] This process is initiated by α-thrombin-catalyzed hydrolysis at Arg-Aα16, to yield fibrin I and fibrinopeptide A (FPA), a 16-aminoacyl residue peptide from the amino terminus of each Aα chain. The newly exposed N termini at the end of the resulting α chains of fibrin I intermolecularly associate with preexisting polymerization sites in the D domains of fibrin to form fibrin I protofibrils (Fig. 2). Protofibril formation is followed by α-thrombin-catalyzed hydrolysis at Arg-Bβ14, releasing fibrinopeptide B (FPB), a 14-aminoacyl residue peptide from the amino terminus of each Bβ chain to form fibrin II protofibrils. Release of FPB increases the propensity of fibrin to undergo lateral interactions that result in formation of insoluble fibrin fibers.[6,7] The fibrin fibers are subsequently mechanically stabilized by cross-links that are formed via factor XIII-catalyzed reactions between certain lysyl and glutamyl residues in the α and γ chains of fibrin. (See Lorand *et al.*[8] for a review of this process.)

[1] T. W. Meade, M. Brozovič, R. R. Chakrabarti, A. P. Haines, J. D. Imeson, S. Mellows, G. J. Miller, W. R. S. North, Y. Stirling, and S. G. Thompson, *Lancet* 2, 533 (1986).

[2] N. S. Cook and D. Ubben, *Trends Pharmacol. Sci.* 11, 444 (1990).

[3] J. A. Shafer and D. L. Higgins, *CRC Crit. Rev. Clin. Lab. Sci.* 26, 1 (1988).

[4] R. R. Hantgan, C. W. Francis, H. A. Scheraga, and V. J. Marder, *in* "Hemostasis and Thrombosis: Basic Principles and Clinical Practice" (R. W. Colman, J. Hirsh, V. J. Marder, and E. W. Salzman, eds.), p. 269. Lippincott, Philadelphia, 1987.

[5] S. D. Lewis, P. P. Shields, and J. A. Shafer, *J. Biol. Chem.* 260, 10192 (1985).

[6] R. R. Hantgan and J. Hermans, *J. Biol. Chem.* 254, 11272 (1979).

[7] R. R. Hantgan, J. McDonagh, and J. Hermans, *Ann. N.Y. Acad. Sci.* 408, 344 (1983).

[8] L. Lorand, M. S. Losowsky, and K. J. M. Miloszewski, *Prog. Hemostasis Thromb.* 5, 245 (1980).

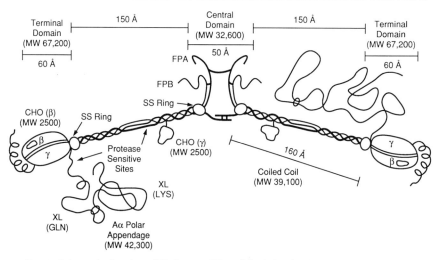

FIG. 1. Schematic drawing of fibrinogen. The original drawing has been modified slightly to illustrate the antiparallel alignment of the N-terminal disulfide junction of the γ chains. XL denotes cross-linking sites. Sites labeled in only one-half of the molecule are also present at corresponding positions in the other half of the molecule. Reproduced with permission from *Annu. Rev. Biochem,* Volume 53, © 1984 by Annual Reviews Inc.

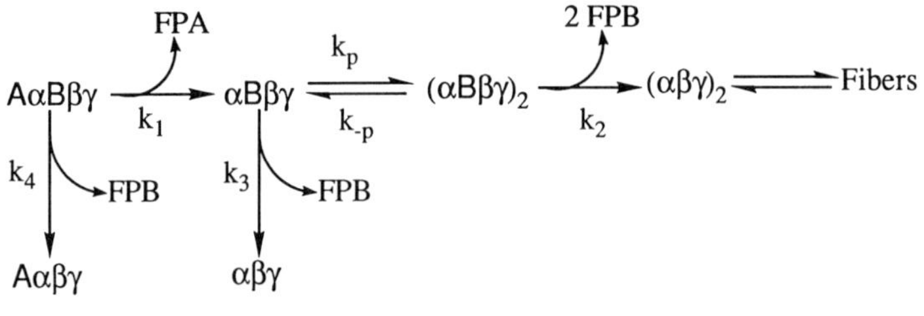

SCHEME I. Reaction pathway for release of fibrinopeptides during fibrin assembly. Values for the numbered rate constants are products of specificity constants (k_{cat}/K_m) and the α-thrombin concentration [e]. At 37°, pH 7.4, $\Gamma/2 = 0.15$, $k_1/[e] = 1.1 \times 10^7 \ M^{-1} \ sec^{-1}$; $k_2/[e] = 4.2 \times 10^6 \ M^{-1} \ sec^{-1}$; $k_3/[e] = 6.5 \times 10^5 \ M^{-1} \ sec^{-1}$; $k_4/[e] < 3.7 \times 10^5 \ M^{-1} \ sec^{-1}$; $k_p = 1 \times 10^6 \ M^{-1} \ sec^{-1}$; and $k_{-p} = 0.064 \ sec^{-1}$. The predominant pathway is that described in the horizontal direction. Reproduced with permission from Lewis *et al.*[5]

Theory

Kinetic Parameters for Release of Fibrinopeptides

Kinetic studies of thrombin-catalyzed release of fibrinopeptides from bovine[9] and human[5,10-13] fibrinogen yielded observations consistent with

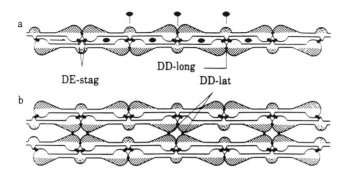

FIG. 2. (a) Model of the two-stranded protofibril. The two contacts involved in protofibril formation are identified as DD-long and DE-stag. (b) Lateral association of protofibrils to make the fibrin fibers. This association must involve a different type of lateral contact, probably between D nodules, identified as DD-lat. Reproduced with permission from W. E. Fowler, H. P. Hantgan, J. Hermans, and H. P. Erickson, *Proc. Natl. Acad. Sci. U.S.A.* **78,** 4872 (1981).

the contention that Scheme I accurately represents the reaction pathway for α-thrombin-catalyzed conversion of fibrinogen to fibrin in the presence of α-thrombin. The sequentiality of fibrinopeptide release is dependent on the value of k_1/k_4. At pH 7.4 $\Gamma/2 = 0.15$, and 37°, $k_1/k_4 \geq 30$ for α-thrombin-catalyzed release of fibrinopeptides from normal human fibrinogen.[14] Consistent with the pathway depicted in Scheme I: (1) Release of FPB from fibrinogen occurs with a time lag which is not observed for the release of FPB from fibrin I polymer.[10,15] (2) The value of the specificity constant for release of FPB subsequent to release of FPA and polymerization (i.e., $k_2/[e]$ of Scheme I, where [e] is enzyme concentration) determined in studies of the action of α-thrombin on fibrinogen is similar to the value obtained from independent studies of the release of FPB from purified fibrin I polymer.[10,15] (3) The rate constant determined for k_p from a kinetic analysis of α-thrombin-catalyzed release of fibrinopeptides from fibrinogen is in reasonable agreement with the directly determined value for the rate constant for bimolecular association of fibrin I monomers from light-scattering measurements.[7] (4) Inhibitors of fibrin polymerization in-

[9] E. Mihalyi, *Biochemistry* **27,** 976 (1988).
[10] M. C. Naski and J. A. Shafer, *J. Biol. Chem.* **265,** 1401 (1990).
[11] M. C. Naski and J. A. Shafer, *J. Biol. Chem.* **266,** 13003 (1991).
[12] R. DeCristofaro and M. Castagnola, *Haemostasis* **21,** 85 (1991).
[13] J. Hofsteenge, H. Taguchi, and S. R. Stone, *Biochem. J.* **237,** 243 (1986).
[14] D. L. Higgins, S. D. Lewis, and J. A. Shafer, *J. Biol. Chem.* **258,** 9276 (1983).
[15] S. D. Lewis, T. J. Janus, L. Lorand, and J. A. Shafer, *Biochemistry* **24,** 6772 (1985).

hibit release of FPB but not FPA.[5,14,16] (5) Fibrinogen-Detroit (R → SAα19), a fibrinogen variant which exhibits delayed polymerization and a decreased rate of FPB release, exhibits a near normal rate of FPA release.[17]

The relative values for the specificity constants for release of FPB from fibrinogen ($k_4/[e]$), fibrin I monomer (k_3/e), and fibrin II polymer ($k_2/[e]$) (see Scheme I) are consistent with the view that conformational rearrangements occur on release of FPA and polymerization of fibrin I that increase the accessibility of the scissile bond at R-Bβ14 to α-thrombin. The rate constants listed in Scheme I may be a function of the pH, temperature, and ionic strength. Thus, use of different reaction conditions (25°, pH 8, Γ/2 = 0.3) may in part account for the observation by Hanna *et al.*[18] that FPA is released only about 6-fold faster than FPB (rather than 30-fold as indicated in Scheme I). It is interesting to note that in fibrinogen-Petoskey (R → H-Aα16) release of FPA(R → H16) is 160-fold lower than release of normal FPA from fibrinogen[14] and that release of FPA(R → H16) from the variant fibrinogen is slower than the release of FPB.[19,20] Neglecting the effects of differences in experimental conditions on rate constant ratios, these observations are not inconsistent with the finding that $k_1/[e] \geqslant 30$ $k_4/[e]$ for normal human fibrinogen, because for the variant fibrinogen, where the value of $k_1/[e]$ is decreased 160-fold, $k_1/[e]$ may have been lowered to a value less than $k_4/[e]$.

HPLC Determination of Release of Fibrinopeptides

The initial report of Martinelli and Scheraga[21] documenting the use of HPLC to separate and assay bovine fibrinopeptides heralded studies of the use of HPLC for assaying fibrinopeptides and derivatives thereof from various species.[22-27] Kinetic studies of α-thrombin-catalyzed release of

[16] A. Hurlet-Jensen, H. Z. Cummins, H. L. Nossel, and C. Y. Liu, *Thromb. Res.* **27**, 419 (1982).

[17] B. Blombäck, B. Hessel, D. Hogg, and L. Therkildsen, *Nature (London)* **275**, 501 (1978).

[18] L. S. Hanna, H. A. Scheraga, C. W. Francis, and V. J. Marder, *Biochemistry* **23**, 4681 (1984).

[19] C. Southan, D. A. Lane, W. Bode, and A. Henschen, *Eur. J. Biochem.* **147**, 593 (1985).

[20] R. F. Ebert, W. E. Schreiber, and W. R. Bell, *Thromb. Res.* **43**, 7 (1986).

[21] R. A. Martinelli and H. A. Scheraga, *Anal. Biochem.* **96**, 246 (1979).

[22] J. A. Koehn and R. E. Canfield, *Anal. Biochem.* **116**, 349 (1981).

[23] J. P. Sellers and H. G. Clark, *Thromb. Res.* **23**, 91 (1981).

[24] D. L. Higgins and J. A. Shafer, *J. Biol. Chem.* **256**, 12013 (1981).

[25] M. Kehl, F. Lottspeich, and A. Henschen, *Hoppe-Seyler's Z. Physiol. Chem.* **362**, 1661 (1981).

[26] M. Kehl, F. Lottspeich, and A. Henschen, *in* "Fibrinogen—Recent Biochemical and Medical Aspects" (A. Henschen, H. Graeff, and F. Lottspeich, eds.), p. 217. de Gruyter, Berlin, 1982.

[27] R. F. Ebert and W. R. Bell, *Anal. Biochem.* **148**, 70 (1985).

fibrinopeptides from bovine[28] and human[14] fibrinogen were the first to demonstrate HPLC as a useful method for the determination of kinetic parameters for thrombin-catalyzed release of fibrinopeptides.

In this chapter we show several methods of how HPLC analysis of fibrinopeptides can yield (1) initial rates of thrombin-catalyzed release of fibrinopeptides from fibrinogen, (2) the Michaelis–Menten parameters k_{cat} and K_m for α-thrombin-catalyzed release of FPA and FPB from fibrinogen and fibrin I, (3) the specificity constant k_{cat}/K_m for both FPA and FPB at concentrations of fibrinogen below K_m, and (4) the inhibition constant for hirugen, an exosite inhibitor of α-thrombin. We also show the HPLC profile of coeluting fibrinopeptides and activation peptides from factor XIII. We also suggest a method for determining the concentration of α-thrombin that is catalytically competent to process fibrinogen.

Materials and Methods

Preparation of Fibrinogen

Laboratory-prepared fibrinogen is purified from outdated plasma as previously described.[29] Fibrinogen from Kabi and Sigma (St. Louis, MO) are also used. Solutions of fibrinogen are dialyzed against 0.3 M NaCl, pH 6–8, followed by centrifugation. The concentration of fibrinogen is determined using an $E^{1\%}_{280\,nm}$ of 15.1 and an M_r of 340,000.[30] All fibrinogen solutions are stored at $-70°$ prior to use.

Preparation of α-Thrombin

Human α-thrombin is prepared according to the method of Fenton *et al.*[31] Human α-thrombin from Sigma (3010 NIH units/mg of protein) was also used by DeCristofaro and Castagnola.[12] The concentration of α-thrombin is determined from its absorbance at 280 nm using an $E^{1\%}_{280\,nm}$ of 18.3 in 0.1 M NaOH and M_r of 36,500.[31] α-Thrombin is greater than 93% active by active site titration.[32] To prevent nonspecific adsorption of thrombin on surfaces, the microfuge tubes used in assays are soaked overnight with 1% polyethylene glycol (PEG) 20,000 (w/v), and air dried

[28] R. A. Martinelli and H. A. Scheraga, *Biochemistry* **19**, 2343 (1980).
[29] S. D. Lewis and J. A. Shafer, *Thromb. Res.* **35**, 111 (1984).
[30] E. Mihalyi, *Biochemistry* **7**, 208 (1968).
[31] J. W. Fenton, II, M. J. Fasco, A. B. Stackrow, D. L. Aronson, A. M. Young, and J. S. Finlayson, *J. Biol. Chem.* **252**, 3587 (1977).
[32] T. Chase, Jr. and E. Shaw, this series, Vol. 19, p. 20.

without rinsing. Additionally, thrombin stock solutions are diluted with buffer containing PEG 8000.[33]

Preparation of Fibrin I

Reptilase R (34 mg lyophilized powder) purchased from Abbott Laboratories (North Chicago, IL) is reconstituted by addition of 1 ml of water. Fibrinogen (5 ml, 0.6 mg/ml) in 9.5 mM sodium phosphate buffer, pH 7.4, 0.14 M NaCl is incubated at 37° for 4 hr with 0.01 ml of a solution of reptilase R. The mixture is then cooled in an ice bath. The resulting clot is first collected on a glass rod, then rinsed with water, and dissolved in 0.4 ml of 0.02 M acetic acid.[15] The dissolved fibrin is precipitated twice in 80 mM sodium phosphate, pH 6.3. It is then dissolved in a minimum volume of 0.02 M acetic acid. The concentration of fibrin I is calculated using an $E_{280\ nm}^{1\%}$ of 14.0 and an M_r of 340,000.[10]

Preparation of Factor XIII

Factor XIII is isolated from human plasma according to the procedure of Lorand et al.[34] Factor XIII is stored at 4° in 50 mM Tris-HCl, pH 7.5, 1 mM EDTA and dialyzed into reaction buffer just before use.[35] The concentration of ab protomeric units of factor XIII is determined from the absorbance at 280 nm using an $E_{280\ nm}^{1\%}$ value of 13.8 and an equivalent weight of 160,000 for the ab protomeric unit according to Schwartz et al.[36]

Sample Preparation for FPA, FPB, and Activation Peptide Release

Time-Dependent FPA and FPB Release from Fibrinogen at [Aα] ≤ 0.1K$_{mA}$. The time dependence of FPA release from fibrinogen is followed in 0.04 M sodium phosphate, pH 7.4, 0.084 M NaCl, 0.1% PEG 8000.[14] Aliquots of 0.9 ml of 0.39 μM fibrinogen are added to 1.8-ml microfuge tubes. After warming to 37° in a water bath, 0.1 ml of buffer (the zero-time control) or 0.1 ml of 2.0–5.0 nM α-thrombin is added with mixing. Reaction mixtures are quenched at the specified times by addition (with mixing) of 0.1 ml of 3 M perchloric acid and 0.3 ml of water. A high concentration of α-thrombin (0.1 ml of 20–50 nM) is added to one sample that is quenched at 30 min for determination of the maximum release of

[33] W. Wasiewski, M. J. Fasco, B. M. Martin, T. C. Detwiler, and J. W. Fenton, II, *Thromb. Res.* **8**, 881 (1976).
[34] L. Lorand, R. B. Credo, and T. J. Janus, this series, Vol. 55, p. 333.
[35] T. J. Janus, S. D. Lewis, L. Lorand, and J. A. Shafer, *Biochemistry* **22**, 6269 (1983).
[36] M. L. Schwartz, S. V. Pizzo, R. L. Hill, and P. A. McKee, *J. Biol. Chem.* **246**, 5851 (1971).

fibrinopeptides. The quenched samples are centrifuged at 10,000 g for 5 min at room temperature and 1.0 ml of supernatant solution is analyzed by HPLC (about 0.5 nmol of FPA and FPB if 100% released).

Initial rates of FPA Release from Fibrinogen. The time dependence of FPA release from fibrinogen is followed in 9.47 mM sodium phosphate, pH 7.4, 0.137 M NaCl, 2.5 mM KCl, 0.9 mM CaCl$_2$, 0.5 mM MgCl$_2$, 0.1% PEG 8000.[14] Aliquots of fibrinogen (2–40 μM based on AαB$\beta\gamma$ protomer units) are added to 1.8-ml microfuge tubes. After warming to 37° in a water bath, buffer (the zero-time control) or α-thrombin is added with mixing. Reaction mixtures are quenched at the specified times by addition (with mixing) of 3 M perchloric acid (final concentration of acid 0.3 M). A high concentration of α-thrombin is added for 30 min to determine the maximum release of fibrinopeptides. The volume of the quenched samples is brought to 1.4 ml with water and centrifuged at 10,000 g for 5 min at room temperature and 1.0 ml of supernatant is analyzed by HPLC. Initial rates are calculated from the slopes of linear plots of FPA concentration versus time under conditions where release of FPA is <10%.

Initial rates of FPB Release from Fibrin I. The time dependence of FPB release from fibrin I is followed in 30 mM sodium phosphate, pH 7.4, 0.10 M NaCl, 4.8 mM acetate, 0.1% PEG 8000.[10] A sample of α-thrombin is equilibrated at 37° with sufficient buffering so that the pH is 7.4 after addition of fibrin I dissolved in 0.02 M acetic acid. The kinetics are started with addition of enough fibrin I to give the desired concentration in 0.5 ml. The fibrin I concentration varied from 0 to 24 μM αB$\beta\gamma$ ensembles (i.e., fibrin I protomers). The reaction is quenched with 75 μl of 14% perchloric acid (v/v), precipitated protein removed by centrifugation, and fibrinopeptides in the supernatant solution quantified by HPLC. Initial rates are calculated from the slopes of linear plots of FPB concentration versus time under conditions where the release of FPB is <10%.

Initial rates of FPA Release from Fibrinogen in Presence of Hirugen. The time dependence of FPA release from fibrinogen is followed in 0.04 M sodium phosphate, pH 7.4, 0.084 M NaCl, 0.1% PEG 8000.[37] Aliquots of 0.5 ml of fibrinogen (2–80 μM based on AαB$\beta\gamma$ protomer unit) and hirugen (0–50 μM, prepared as described in Maraganore et al.[38]) are equilibrated at 37°, then α-thrombin is added with mixing. Reactions are quenched at specified times with 75 μl of 14% (v/v) perchloric acid. The samples are then centrifuged and the supernatant solution quantified by

[37] M. C. Naski, J. W. Fenton, II, J. M. Maraganore, S. T. Olson, and J. A. Shafer, *J. Biol. Chem.* **265,** 13484 (1990).
[38] J. M. Maraganore, B. Chao, M. L. Joseph, J. Jablonski, and K. L. Ramachandran, *J. Biol. Chem.* **265,** 8692 (1989).

HPLC. Initial rates of FPA release are calculated from the slopes of linear plots of FPA concentration versus time under condition where the release of FPA is <10%.

Activation Peptide Release from Factor XIII. The time dependence of FPA, FPB, AP (a 36-amino acid major activation peptide), and AP' (a minor activation peptide) is followed in 9.47 mM sodium phosphate, pH 7.4, 0.137 M NaCl, 2.5 mM KCl, 0.1% PEG.[35] Fibrinogen and factor XIII are stirred at 37° in a polypropylene tube. A sample is removed for a zero-time point. Then an aliquot of thrombin is added. Within 2 min, aliquots of 0.8 ml of the reaction mixture are added to 1.8-ml microfuge tubes, which are prewarmed to 37° in a water bath. Reaction mixtures are quenched at the specified times by addition (with mixing) of 0.1 ml of 3 M perchloric acid and 0.4 ml of water. A high concentration of α-thrombin (0.1 ml of 20–50 nM) is added to one sample that is quenched at 30 min for determination of the maximum release of fibrinopeptides. The quenched samples are centrifuged at 10,000 g for 5 min at room temperature and 1.0 ml of supernatant solution is analyzed by HPLC.

HPLC Conditions for FPA and FPB Detection

Fibrinopeptides are separated using a reversed-phase C$_{18}$ column (0.46 × 25 cm, Spherisorb S50 ODS2 from Phase Sep), eluted at room temperature with a gradient of 0.08 M sodium phosphate buffer (pH 3.1) and acetonitrile and detected at a wavelength of 205 nm. A two-solvent gradient system is used at a flow rate of 1 ml/min. Solvent A is 0.08 M sodium phosphate [HPLC grade H$_3$PO$_4$ (Fischer, Pittsburgh, PA) diluted into HPLC water (Burdick & Jackson, Muskegon, MI)] adjusted to pH 3.1 with NaOH pellets and filtered through 0.22-μm filters. Solvent B is 100% acetonitrile (Burdick & Jackson). Optimal separation of FPA-P, FPA, and FPB is obtained using the following program for elution: 0 to 15 min, 17% B; then a 10-min linear gradient to 40% B.

A typical HPLC profile is depicted in Fig. 3, where FPA-P (FPA phosphorylated at Ser-3) elutes first, followed by FPA and FPB. Recovery of FPA and FPB is >90% as determined by amino acid analysis. The method can detect a lower limit of FPA and FPB of approximately 25 pmol. Fibrinopeptide A and B have molar absorptivities (205 nm) of 4.4 × 10^4 and 5.12 × 10^4 cm^{-1} M^{-1}, respectively.[24]

HPLC Conditions for Detection of Activation Peptide Release from Factor XIII

Fibrinopeptides are separated using a reversed-phase C$_{18}$ column (0.45 × 25 cm, Partisil 10 ODS 3 from Whatman, Clifton, NJ), eluted at room

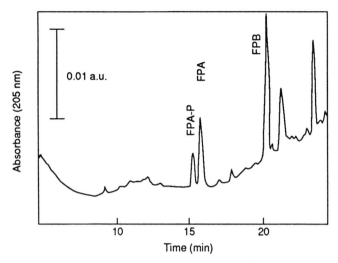

FIG. 3. HPLC elution profile of fibrinopeptides (100 pmol) from laboratory-prepared fibrinogen. HPLC conditions and peak identification are described in the text. A peak corresponding to the void volume (not shown) is present in the first 5 min of the profile. Unlabeled peaks are present in a buffer control.

temperature with a gradient of 0.08 M sodium phosphate buffer (pH 3.1) and acetonitrile and detected at a wavelength of 205 nm.[35] A two-solvent gradient system is used at a flow rate of 1 ml/min. Solvent A is 90% 0.08 M sodium phosphate, pH 3.1, and 10% (v/v) acetonitrile. Solvent B is 60% 0.08 M sodium phosphate, pH 3.1, and 40% (v/v) acetonitrile. The program for elution is 0 to 10 min, 15% B; then a 50-min linear gradient to 90% B.

A typical HPLC profile is depicted in Fig. 4, where FPA elutes first, followed by FPB, AP, and AP'. AP and AP' have molar absorptivities (205 nm) of 1.12×10^5 cm^{-1} M^{-1}.[35]

Determination of k_{catA} and K_{mA} for Thrombin-Catalyzed Release of FPA from Fibrinogen

The α-thrombin-catalyzed release of FPA from fibrinogen follows simple Michaelis–Menten kinetics [Eq. (1)],

$$d[\text{FPA}]/dt = -d[\text{A}\alpha]/dt = k_{\text{catA}}[\text{e}][\text{A}\alpha]/(K_{\text{mA}} + [\text{A}\alpha]) \qquad (1)$$

where [FPA], [Aα], and [e] represent the concentrations of FPA, fibrinogen Aα chains, and thrombin, respectively. Transformation of Eq. (1) yields Eq. (2). The steady-state kinetic parameters k_{catA} and K_{mA} are obtained by

$$[\text{A}\alpha]/V = K_{\text{mA}}/k_{\text{catA}}[\text{e}] + [\text{A}\alpha]/k_{\text{catA}}[\text{e}] \qquad (2)$$

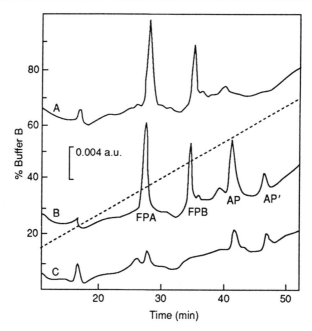

FIG. 4. HPLC elution profiles for fibrinopeptides and for factor XIII activation peptides. Aliquots (0.8 ml) of reaction mixtures containing 0.27 nM thrombin at 37°, pH 7.4, $\Gamma/2 =$ 0.17, 0.1% PEG were quenched at the times specified with 0.1 ml of 3 M HClO₄. Precipitated protein was removed by centrifugation. The supernatant solution was adjusted to 1.3 ml with water and 1 ml of the resulting solution was subjected to HPLC. (A) 0.2 μM fibrinogen, after 15 min of reaction; (B) 0.2 μM fibrinogen and 0.2 μM factor XIII protomer after 18 min of reaction; (C) 0.4 μM factor XIII protomer after 66 min of reaction. The initial part of the chromatograms (not shown) contained a peak at the void volume. The dashed line indicates the percentage of buffer B in the gradient. Reprinted with permission from Janus et al.,[35] copyright [1983] American Chemical Society.

fitting the dependence of the initial rate of hydrolysis (V) of Aα chains on their concentration to Eq. (2) as shown in Fig. 5. Values for the steady-state kinetic parameters k_{catA} and K_{mA} obtained were 84 sec⁻¹ and 7.2 μM, respectively.[14]

Determination of k_{catB} and K_{mB} for Thrombin-Catalyzed Release of FPB from Fibrin I

Fibrin I was evaluated as a substrate for α-thrombin. Under reaction conditions, polymerization of fibrin I was rapid and the fibrin I gel was uniformly distributed throughout the solution.[10] Assuming equal accessa-

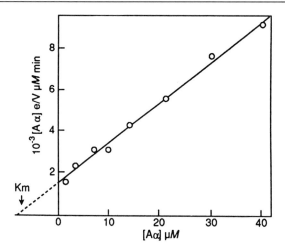

FIG. 5. Determination of Michaelis–Menten kinetic parameters for the release of FPA from fibrinogen by 0.041 nM thrombin (2.87 NIH units/μg) at pH 7.4, 37°, $\Gamma/2 = 0.15, 0.1\%$ PEG. The line, the best fit of the data to Eq. (2), yielded values of 84 ± 4 sec^{-1} and 7.2 ± 0.9 μM for k_{catA} and K_{mA}, respectively. Initial rates (V) of release of FPA were determined from slopes of plots of FPA versus time prior to release of 10% of the FPA. Reproduced with permission from Higgins et al.[14]

bility of Arg-Bβ14 in αB$\beta\gamma$ and $(\alpha$B$\beta\gamma)_2$ (see Scheme I) for α-thrombin, the α-thrombin-catalyzed release of FPB from fibrin I follows simple Michaelis–Menten kinetics. Substituting [αB$\beta\gamma$] for [Aα] in Eq. (1) and (2), the dependence of the initial velocity of FPB release on fibrin I concentration (see Fig. 6) yielded values of 49 sec^{-1} and 7.5 μM for k_{catB} and K_{mB}, respectively.[10]

Direct Evaluation of Specificity Constant, k_{catA}/K_{mA}, for Release of FPA under Conditions Where [Aα] $\leq 0.1K_{mA}$

The α-thrombin-catalyzed release of FPA from fibrinogen follows simple Michaelis–Menten kinetics [see Eq. (1) and (2)]. At pH 7.4, 37°, $\Gamma/2 = 0.15$, the Michaelis constant K_{mA} for release of FPA is 7.2 μM based on concentration of Aα chains.[14] Under reaction conditions, [Aα] $\leq 0.1K_{mA}$, Eq. (1) simplifies to

$$-d[A\alpha]/dt = k_{catA}[e][A\alpha]/K_{mA} \qquad (3)$$

Integration of Eq. (3), observing the identity

$$\{([FPA]_f - [FPA])/[FPA]_f\} = [A\alpha]/[A\alpha]_0 \qquad (4)$$

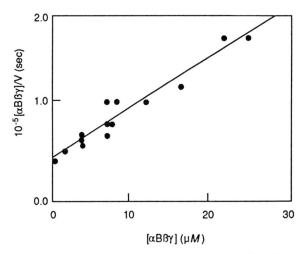

FIG. 6. Determination of Michaelis–Menten kinetic parameters for α-thrombin-catalyzed hydrolysis of fibrin I. Reaction conditions were 37°, pH 7.4, $\Gamma/2 = 0.17$, 0.1% PEG, 3.55 pM α-thrombin. The line, the best fit of the data to Eq. (2), yielded values of 48 ± 5 sec^{-1} and $7.2 \pm 1.5\ \mu M$ for k_{catB} and K_{mB}, respectively, where [$\alpha B\beta\gamma$], the concentration of fibrin I protomers, is substituted for [Aα] in Eq. (2). Reproduced with permission from Naski and Shafer.[10]

where the subscripts 0 and f denote initial and final values, yields Eq. (5)

$$\ln\{([FPA]_f - [FPA])/[FPA]_f\} = -k_1 t \tag{5}$$

where

$$k_1 = k_{catA}[e]/K_{mA} \tag{6}$$

A plot of $\ln\{([FPA]_f - [FPA])/[FPA]_f\}$ versus time is depicted in Fig. 7. The slope (k_1) of the plot in Fig. 7, together with the concentration of active α-thrombin, yielded a value of k_{catA}/K_{mA} of approximately 11.4 × $10^6\ M^{-1}$ sec^{-1}. The value of the specificity constant k_{catA}/K_{mA} is consistent with the individually reported values of k_{catA} and K_{mA} of 84 s^{-1} and 7.2 μM, respectively.[14] At pH 7.4, $\Gamma/2 = 0.15$, 37°, a similar value of k_{catA}/K_{mA} is obtained when the time-dependent appearance of FPA + FPA-P is analyzed using HPLC conditions, which do not separate FPA and FPA-P.

Direct Evaluation of Specificity Constant, k_{catB}/K_{mB}, for Release of FPB under Conditions Where [Aα] $\leq$ 0.1K_{mA}

At low concentrations of thrombin (<2 nM) and the fibrinogen concentration used in these studies, polymerization has been shown to be

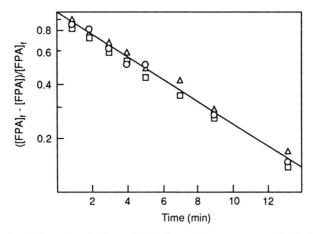

FIG. 7. Thrombin-mediated release of FPA from different sources of 0.35 μM fibrinogen by 0.213 nM α-thrombin: laboratory prepared (O); Kabi, grade L (□); Sigma #F-3879 (△). Reaction conditions are described in the text. The line was the best fit to Eq. (5) and yielded a value for k_1 of 2.43 × 10^{-3} sec^{-1}. Using this value of k_1 and the α-thrombin concentration of 0.213 nM in Eq. (6) yielded the specificity constant k_{catA}/K_{mA} of 11.4 × 10^6 M^{-1} sec^{-1}. Reprinted with permission from Lewis and Shafer,[29] copyright 1984, Pergamon Press plc.

rapid (relative to catalysis) and complete.[5,14] Thus, the time-dependent release of FPB as depicted in Scheme I can be represented by Eq. (7),[14]

$$[FPB]/[FPB]_f = 1 + [k_2/(k_1-k_2)] \, e^{-k_1 t} - [k_1/(k_1-k_2)] \, e^{-k_2 t} \qquad (7)$$

where the subscript f denotes final value. The rate constants k_1 and k_2 are equivalent to the product of the thrombin concentration and the specificity constant for FPA [see Eq. (6)] and FPB release, respectively. Equation (7) is the standard equation for time-dependent product release in a simple first-order reaction of the type A → B → C. Interestingly, more complicated equations that take into account slow FPB release from AαB$\beta\gamma$ and αB$\beta\gamma$ are not necessary under the above reaction conditions.[5,11,14] In fact, it has been shown that Eq. (7) can also be used to represent the time-dependent appearance of fibrinopeptides from bovine[9] and human fibrinogen,[11] even when the concentration of Aα and Bβ chains is not less than K_m. The ability of Eq. (7) to represent accurately the time course for the release of fibrinopeptides, regardless of the initial concentration of Aα and Bβ chains, is a fortuitous consequence of the fact that the K_m values for FPA and FPB release are similar and that fibrin II is a competitive inhibitor of FPA and FPB release with a K_i that is similar to the K_m values for release of FPA and FPB.[9,10]

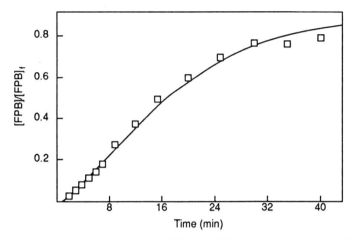

FIG. 8. Thrombin-mediated release of FPB from 0.35 μM fibrinogen by 0.266 nM α-thrombin. Reaction conditions are described in the text. The line was the best fit of the data to Eq. (7), fixing k_1 to the previously determined value of 3.7×10^{-3} sec^{-1}, and yielded a value for k_2 of 1.07×10^{-3} sec^{-1}. The specificity constant for FPB release was calculated to be 4.0×10^6 M^{-1} sec^{-1}. Reproduced with permission from Higgins et al.[14]

We routinely solve for k_1 by fitting the time dependency of FPA release to Eq. (5), and then solve for k_2 by a nonlinear least-squares fit of the time dependence of FPB to Eq. (7), using the value determined for k_1. The value of k_2 obtained is divided by the α-thrombin concentration to give the specificity constant k_{catB}/K_{mB} for FPB release. Plots of the type shown in Fig. 8 typically yield specificity constants of approximately 4×10^6 M^{-1} sec^{-1} for FPB release. This value is in reasonable agreement with the value of 6.5×10^6 M^{-1} sec^{-1} determined from the individually determined values of k_{catB} and K_{mB} from the initial release of FPB from fibrin I.[10] Negative deviation from the fit of Eq. (7) to the experimental data is usually observed at higher extents of FPB release.[14] This effect is probably due to inaccessibility of a fraction of Bβ chains to α-thrombin after clot formation. To minimize errors, data after the release of 70% FPB are not used to determine kinetic constants.

Determination of Inhibition Constant of Hirugen for Thrombin-Catalyzed Release of FPA from Fibrinogen

Hirugen, the synthetic N-acetylated COOH-terminal dodecapeptide of hirudin, was shown to behave as a pure competitive inhibitor of human α-thrombin-catalyzed release of FPA from fibrinogen.[37] In contrast to this inhibitor activity, hirugen had a modest effect on thrombin activity toward small substrates. These observations suggest that hirugen binds to α-

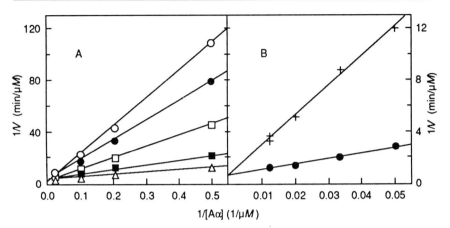

FIG. 9. Double-reciprocal plot for the dependence of the initial velocity of α-thrombin-catalyzed release of FPA from fibrinogen on the concentration of fibrinogen and hirugen. Reactions were performed under standard conditions (37°, pH 7.4, $\Gamma/2 = 0.17$, 0.1% PEG) with 55.5 pM α-thrombin, and 2–38.8 μM AαB$\beta\gamma$ fibrinogen promoter (A). Each line represents a linear least-squares fit of the data at 0.0 ($\triangle$), 0.5 ($\blacksquare$), 1.47 ($\square$), 2.94 ($\bullet$), and 5.0 ($\bigcirc$) μM hirugen to Eq. (8). [Aα] and V represent the fibrinogen protomer concentration and initial rate of FPA release, respectively. (B) Data at 50 (+) and 5 ($\bullet$) μM hirugen with 20–80 μM AαB$\beta\gamma$ using 0.33 nM α-thrombin. Reproduced with permission from Naski et al.[37]

thrombin at an exosite distinct from the active site. The dependence of initial velocity on the concentration of hirugen and substrate was fit to Eqs. (8) and (9).

$$1/V = (1/k_{catA} [e]) + [K_{mA(app)}/(k_{catA} [e] [A\alpha])] \qquad (8)$$

$$K_{m(app)} = K_{mA} + (K_{mA} [H]/K_i) \qquad (9)$$

where [H] and K_i are the hirugen concentration and the inhibition constant for hirugen and thrombin, respectively. Figure 9 shows a plot of $1/V$ versus $1/[A\alpha]$ at several hirugen concentrations. These lines intersect at a common point on the ordinate, showing that hirugen is a competitive inhibitor of α-thrombin. A nonlinear least-squares fit of Eq. (8) for each hirugen concentration in Fig. 9 yields a value for the apparent Michaelis constant, $K_{mA(app)}$. The dependence of $K_{mA(app)}$ on hirugen concentration yielded a value of 0.54 μM for K_i using an independently determined value of 9.1 μM for K_{mA} (Fig. 10).[37]

Activation Peptide Release from Factor XIII

HPLC-based studies of α-thrombin-catalyzed release of fibrinopeptides and activation peptide (a 37-amino acid peptide from the amino terminus

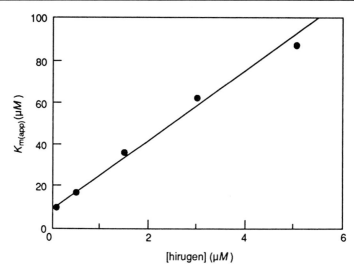

FIG. 10. Dependence of the $K_{m(app)}$ for the α-thrombin-catalyzed release of FPA from fibrinogen on the concentration of hirugen. The line represents a nonlinear least-squares fit of the dependence of $K_{m(app)}$ on the concentration of hirugen to Eq. (9) with the values of K_{mA} set to 9.1 μM. The best fit yielded a value for K_i of 0.54 ± 0.2 μM. Reproduced with permission from Naski et al.[37]

of the a chain of factor XIII) from factor XIII have provided evidence for a reaction pathway wherein α-thrombin proteolysis of factor XIII proceeds through the formation of a fibrin–factor XIII complex.[15,35,39] Figure 4 illustrates the conditions for HPLC separation of FPA, FPB, and activation peptide in a single chromatogram.[35]

Determination of Concentration of α-Thrombin That Is Competent to Process Fibrinogen

Thrombin is frequently assayed with a peptidyl chromogenic substrate.[40] This assay, however, yields little information regarding the putative fibrin(ogen)-binding exosite of thrombin. For example, γ-thrombin, an autolyzed derivative of α-thrombin containing proteolytic nicks that compromise interactions involving the fibrin(ogen)-binding exosite of thrombin, releases FPA 2400-fold slower than does α-thrombin, although the efficiency of γ-thrombin-catalyzed hydrolysis of small chromogenic sub-

[39] M. C. Naski, L. Lorand, and J. A. Shafer, *Biochemistry* **30**, 934 (1991).
[40] R. Lottenberg, V. Christensen, C. M. Jackson, and P. C. Coleman, this series, Vol. 80, p. 341.

strates is similar to that of α-thrombin.[41] In this regard an alternative method has been proposed for determining the concentration of α-thrombin based on the competency of α-thrombin to hydrolyze FPA from fibrinogen under standard conditions.

The slope of the plot in Fig. 7 for FPA release yields a value of k_1 which is related to the thrombin concentration by Eq. (6). Thus, the ratio of the measured value of k_1 (determined with ≤ 0.35 μM, $> 90\%$ clottable fibrinogen when the assay is carried out at $37°$, $\Gamma/2 = 0.15$, pH 7.4) and the specificity constant (11.4×10^6 M^{-1} sec^{-1}) yields the concentration of α-thrombin that is competent to hydrolyze fibrinogen. The data in Fig. 7 reveal that the time dependence for the release of FPA is essentially independent of the source of fibrinogen. Other methods used to assay fibrinogen that are based on the time to clot fibrinogen are sensitive to the quality of the fibrinogen substrate and the presence of inhibitors of fibrin polymerization.[14,29] The concentration of thrombin determined from measurements of k_1 can be related to the commonly used NIH unit, because 1 NIH unit, as determined in clotting-time assays of thrombin, is equivalent to $\sim 10^{-11}$ mol of thrombin.[29]

Immunochemical Detection of Fibrinopeptides

Although HPLC studies of fibrinopeptide release are usually performed in buffers with purified fibrinogen, HPLC can also be used to analyze fibrinopeptides directly from plasma.[42,43] Radioimmunoassay[44-47] and enzyme immunoassay[48] methods for detecting fibrinopeptides are convenient, because FPA antisera are commercially available,[48] and have better sensitivity ($\sim 0.35 - 1$ pmol)[44,47] than HPLC methods (~ 25 pmol). However, the HPLC method described here is more specific for identifying each type of fibrinopeptide, and does not suffer from interference resulting from cross-reactivity with other fibrinopeptides and contaminating fibrinogen inadvertently left in solution.[48,49] Antibody sensitivity toward FPA depends

[41] S. D. Lewis, L. Lorand, J. W. Fenton, II, and J. A. Shafer, *Biochemistry* **26,** 7597 (1987).
[42] H. H. Seydewitz and I. Witt, *Thromb. Res.* **40,** 29 (1985).
[43] C. Southan, E. Thompson, and D. A. Lane, *Thromb. Res.* **43,** 195 (1986).
[44] H. L. Nossel, L. R. Younger, G. D. Wilner, T. Procupez, R. E. Canfield, and V. P. Butler, Jr., *Proc. Natl. Acad. Sci. U.S.A.* **68,** 2350 (1971).
[45] H. L. Nossel, I. Yudelman, R. E. Canfield, V. P. Butler, Jr., K. Spanondis, G. D. Wilner, and G. D. Qureshi, *J. Clin. Invest.* **54,** 43 (1974).
[46] S. B. Bilezikan, H. L. Nossel, V. P. Butler, Jr., and R. E. Canfield, *J. Clin. Invest.* **56,** 438 (1975).
[47] V. Hoffman and P. W. Straub, *Thromb. Res.* **11,** 171 (1977).
[48] J. Soria, C. Soria, and J. J. Ryckewaert, *Thromb. Res.* **20,** 425 (1980).
[49] J. Owen, *Thromb. Haemostasis* **62,** 807 (1989).

on the degree of FPA phosphorylation (FPA-P),[49,50] which can increase sharply in patients undergoing acute therapy.[42,51] HPLC clearly separates FPA-P from FPA.

Other Applications of HPLC Analysis of Fibrinopeptides

HPLC analysis of fibrinopeptides has facilitated structural characterization of mutationally altered fibrinopeptides[24,52–55] as well as the posttranslationally modified forms of fibrinopeptides.[53,54,56] Normal human FPA has been reported to exist in three major forms, namely, the unmodified form (~70%), which has the sequence ADSGEGDFLAEGGGVR; a form (~20%) wherein the hydroxyl group of Ser-3 is phosphorylated (FPA-P); and an N-terminal truncated form (~10%) wherein the alanyl residue is absent (FPA-Y). Normal human FPB has the sequence ZGVNDNEEGFFSAR and exists mainly in the unmodified form (~95%). A variable amount of FPB which has lost its C-terminal arginyl residue (des-Arg FPB) has been reported.[22,26,27,53] The des-Arg FPB probably arises via the action of a carboxypeptidase contaminant in the assay sample.[22,27] HPLC has been reported to detect a form of FPB which has lost the first two N-terminal residues (i.e., a peptide with the $B\beta3-14$ sequence of fibrinogen)[27] as well as peptides corresponding to $A\alpha1-21$ of fibrinogen (arising from the action of elastase on fibrinogen),[57] $A\alpha1-19$ of fibrinogen (arising from the action of trypsin or elastase together with plasma peptidases on fibrinogen),[41,57] and $B\beta1-42$ of fibrinogen (arising from the action of plasmin on fibrinogen).[58]

[50] D. Prisc, *Res. Clin. Lab.* **20**, 217 (1990).
[51] O. C. Leeksma, F. Meijer-Huizinga, E. A. Stoepman-van Dalen, C. J. W. van Ginkel, W. G. van Aken, and J. A. van Mourik, *Blood* **67**, 1460 (1986).
[52] D. L. Higgins, S. D. Lewis, J. A. Penner, and J. A. Shafer, *Thromb. Haemostasis* **48**, 182 (1982).
[53] M. Kehl, F. Lottspeich, and A. Henschen, *in* "Fibrinogen: Structure, Functional Aspects and Metabolism" (F. Haverkate, A. Henschen, W. Nieuwenhuizen, and P. W. Staub, eds.), p. 183. de Gruyter, Berlin, 1983.
[54] A. Henschen, M. Kehl, and C. Southan, *in* "Current Problems in Clinical Biochemistry (E. A. Beck and M. Furlan, eds.), p. 273. Huber, Berne, 1984.
[55] D. A. Lane, E. Thompson, and C. Southan, *in* "Fibrinogen 2: Biochemistry, Physiology and Clinical Relevance" (G. D. O. Lowe, J. T. Douglas, C. D. Forbes, and A. Henschen, eds.), p. 71. Elsevier, Amsterdam, 1987.
[56] J. Lucas and A. Henschen, *J. Chromatogr.* **369**, 357 (1986).
[57] J. I. Weitz, *Ann. N.Y. Acad. Sci.* **624**, 154 (1991).
[58] J. A. Koehn, A. Hurlet-Jensen, H. L. Nossel, and R. E. Canfield, *Anal. Biochem.* **133**, 502 (1983).

[21] Protein C Activation

By Charles T. Esmon, Naomi L. Esmon, Bernard F. Le Bonniec,
and Arthur E. Johnson

Introduction

Physiologically relevant protein C activation occurs on the surface of the endothelium and is catalyzed by a complex of thrombin bound reversibly to a cell surface receptor, thrombomodulin (TM).[1] This complex exhibits specificities that differ from those of free thrombin. Complex formation accelerates the activation of protein C to form the anticoagulant, activated protein C, while inhibiting the ability of thrombin to promote clotting.[1-6] In this chapter, emphasis will be placed on approaches that have been taken to elucidate the mechanisms of protein C activation and to understand how interaction of thrombin with TM alters the specificity of thrombin. Other reviews on the structure and biology of the system are also available.[1,7-14]

Insights into the mechanisms that control thrombin specificity and regulate protein C activation have been facilitated by analysis of the structure–function relationships in the activation complex.

[1] C. T. Esmon, J. Biol. Chem. 264, 4743 (1989).
[2] C. T. Esmon, N. L. Esmon, and K. W. Harris, J. Biol. Chem. 257, 7944 (1982).
[3] J. Hofsteenge, H. Taguchi, and S. R. Stone, Biochem. J. 237, 243 (1986).
[4] H. V. Jakubowski, M. D. Kline, and W. G. Owen, J. Biol. Chem. 261, 3876 (1986).
[5] M.-C. Bourin, A.-K. Öhlin, D. A. Lane, J. Stenflo, and U. Lindahl, J. Biol. Chem. 263, 8044 (1988).
[6] J. F. Parkinson, B. W. Grinnell, R. E. Moore, J. Hoskins, C. J. Vlahos, and N. U. Bang, J. Biol. Chem. 265, 12602 (1990).
[7] J. Stenflo, A.-K. Öhlin, E. Persson, C. Valcarce, and J. Astermark, Ann. N.Y. Acad. Sci. 614, 11 (1991).
[8] N. L. Esmon, Prog. Hemostasis Thromb. 9, 29 (1988).
[9] J. Stenflo, Semin. Thromb. Hemostasis 10, 109 (1984).
[10] W. Kisiel and E. W. Davie, this series, Vol. 80, p. 320.
[11] B. Dahlbäck, P. Fernlund, and J. Stenflo, in "Blood Coagulation" (R. F. A. Zwaal and H. C. Hemker, eds.), p. 285. Elsevier, Amsterdam, 1986.
[12] W. A. Dittman and P. W. Majerus, Blood 75, 329 (1990).
[13] B. Dahlbäck, Thromb Haemostasis 66, 49 (1991).
[14] K. A. Bauer and R. D. Rosenberg, Blood 70, 343 (1987).

Structural Information

Protein C

Human protein C circulates as a single chain ($\approx$ 20%) and as a two-chain molecule. Linking the chains is a Lys-Arg dipeptide that appears to be cleaved during secretion.[15,16] In addition, the human protein C heavy chain undergoes variable glycosylation, giving rise to a doublet on SDS-gel electrophoresis.[15,16] All forms can be activated, although the glycosylation variants have been reported to have differences in activation rates and anticoagulant activity, but no differences in rates of reaction with inhibitors.[17]

Protein C is composed of several domains (Fig. 1). The amino-terminal domain contains the vitamin K-dependent γ-carboxyglutamic acid (Gla) residues involved in membrane binding.[18] This domain is required for both optimal activation and expression of anticoagulant activity.[19-21] The Gla domain is connected by a "hydrophobic stack" to two epidermal growth factor (EGF)-like domains and these domains are followed by the protease domain.

The protease domain has sequence similarity with chymotrypsin and contains the catalytic triad.[22] Chemically, human protein C activation requires only the release of a 12-residue peptide from the amino terminus of the heavy chain,[23] as depicted in Fig. 1.

Thrombomodulin

Thrombomodulin is an integral membrane protein[24-27] composed of an N-terminal region of no known function that has remote sequence

[15] J. P. Miletich and G. J. Broze, Jr., *J. Biol. Chem.* **265**, 11397 (1990).

[16] B. W. Grinnell, J. D. Walls, D. Gerlitz, D. T. Berg, D. B. McClure, H. Ehrlich, N. U. Bang, and S. B. Yan, *Adv. Appl. Biotechnol. Ser.* **11**, 29 (1991).

[17] B. W. Grinnell, J. D. Walls, and B. Gerlitz, *J. Biol. Chem.* **226**, 9778 (1991).

[18] G. L. Nelsestuen, W. Kisiel, and R. G. DiScipio, *Biochemistry* **17**, 2134 (1978).

[19] S. V. D'Angelo, P. C. Comp, C. T. Esmon, and A. D'Angelo, *J. Clin. Invest.* **77**, 416 (1986).

[20] J. B. Galvin, S. Kurosawa, K. Moore, C. T. Esmon, and N. L. Esmon, *J. Biol. Chem.* **262**, 2199 (1987).

[21] N. L. Esmon, L. E. DeBault, and C. T. Esmon, *J. Biol. Chem.* **258**, 5548 (1983).

[22] J. Stenflo and P. Fernlund, *J. Biol. Chem.* **257**, 12180 (1982).

[23] W. Kisiel, *J. Clin. Invest.* **64**, 761 (1979).

[24] T. E. Petersen, *FEBS Lett.* **231**, 51 (1988).

[25] R. W. Jackman, D. L. Beeler, L. Fritze, G. Soff, and R. D. Rosenberg, *Proc. Natl. Acad. Sci. U.S.A.* **84**, 6425 (1987).

[26] D. Wen, W. A. Dittman, R. D. Ye, L. L. Deaven, P. W. Majerus, and J. E. Sadler, *Biochemistry* **26**, 4350 (1987).

[27] K. Suzuki, H. Kusumoto, Y. Deyashiki, J. Nishioka, I. Maruyama, M. Zushi, S. Kawahara, G. Honda, S. Yamamoto, and S. Horiguchi, *EMBO J.* **6**, 1891 (1987).

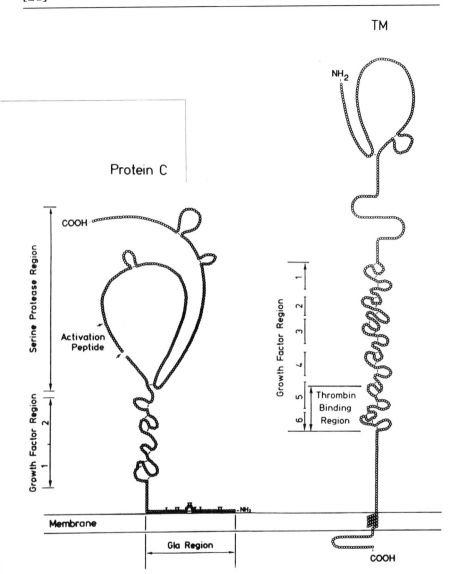

FIG. 1. Schematic representation of protein C and TM (see text for discussion) based on primary structure and homology with other proteins. The Gla residues (γ-carboxyglutamic acid) are indicated as small Y-shaped symbols on protein C. These residues are required for biological activity and require vitamin K for their biosynthesis. Adapted from Esmon.[1]

similarity to the asialoglycoprotein receptor.[24] This region is followed by six EGF-like domains, an O-linked sugar region that contains the attachment site for a chondroitin sulfate moiety,[5,6] a transmembrane region, and a short cytoplasmic tail that contains a potential phosphorylation site.[28] The major thrombin binding site is located in the EGF repeats,[29,30] specifically EGF domain 5 and domain 6,[31,32] but the ability to form near-optimal activation complexes requires at least EGF domains 4–6.[33,34] Trace activity is resident in EGF 4–5[32] and a peptide from EGF 5 has been shown to bind to thrombin and displace TM.[32,35] This peptide and EGF 5–6[36] appear to bind in the anion-binding exosite of thrombin.[35]

Thrombin

Thrombin is a serine protease with an overall structure similar to chymotrypsin. Descriptions of the three-dimensional structure of thrombin with the covalent inhibitor D-Phe-Pro-Arg-chloromethyl ketone (FPR-ck)[37] and the noncovalent leech inhibitor, hirudin,[38-40] have significantly advanced our understanding of the mechanisms involved in substrate recognition by thrombin. These insights provide useful models from which to formulate specific experiments related to how TM functions. The salient features of thrombin structure critical to protein C activation are illustrated in Fig. 2. The structure differs from that of the digestive serine proteases primarily in the size and extent of the substrate binding cleft and the

[28] W. A. Dittman, T. Kumada, J. E. Sadler, and P. W. Majerus, *J. Biol. Chem.* **263**, 15815 (1988).

[29] S. Kurosawa, J. B. Galvin, N. L. Esmon, and C. T. Esmon, *J. Biol. Chem.* **262**, 2206 (1987).

[30] K. Suzuki, T. Hayashi, J. Nishioka, Y. Kosaka, M. Zushi, G. Honda, and S. Yamamoto, *J. Biol. Chem.* **264**, 4872 (1989).

[31] S. Kurosawa, D. J. Stearns, K. W. Jackson, and C. T. Esmon, *J. Biol. Chem.* **263**, 5993 (1988).

[32] T. Hayashi, M. Zushi, S. Yamamoto, and K. Suzuki, *J. Biol. Chem.* **265**, 20156 (1990).

[33] D. J. Stearns, S. Kurosawa, and C. T. Esmon, *J. Biol. Chem.* **264**, 3352 (1988).

[34] M. Zushi, K. Gomi, S. Yamamoto, I. Maruyama, T. Hayashi, and K. Suzuki, *J. Biol. Chem.* **264**, 10351 (1989).

[35] M. Tsiang, S. R. Lentz, W. A. Dittman, D. Wen, E. M. Scarpati, and J. E. Sadler, *Biochemistry* **29**, 10602 (1990).

[36] J. Ye, L.-W. Liu, C. T. Esmon, and A. E. Johnson, *J. Biol. Chem.* **267**, 11023 (1992).

[37] W. Bode, I. Mayr, U. Baumann, R. Huber, S. R. Stone, and J. Hofsteenge, *EMBO J.* **8**, 3467 (1989).

[38] T. J. Rydel, K. G. Ravichandran, A. Tulinsky, W. Bode, R. Huber, C. Roitsch, and J. W. Fenton, II, *Science* **249**, 277 (1990).

[39] M. G. Grutter, J. P. Priestle, J. Rahuel, H. Grossenbacher, W. Bode, J. Hofsteenge, and S. R. Stone, *EMBO J.* **9**, 2361 (1990).

[40] T. J. Rydel, A. Tulinsky, W. Bode, and R. Huber, *J. Mol. Biol.* **221**, 583 (1991).

Thrombin

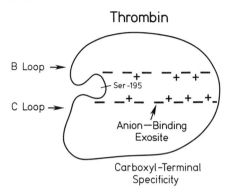

FIG. 2. Schematic representation of thrombin. The model reflects the restricted access to the active site due to the insertion loops in thrombin relative to trypsin, indicated as B and C. The other major feature is the extended substrate-binding site that is rich in basic residues, which has led to its description as the anion-binding exosite. In addition to several substrates, receptors and thrombomodulin appear to bind in this region. See text for details and references.

presence of several insertions that narrow access to the binding cleft. The active site cleft is primarily lined with hydrophobic residues and the base of the cleft contains many polar/charged side chains. Beyond the S2′ subsite, the active site cleft continues as a long groove with many basic residues. The basic nature of this region has led to its description as the anion-binding exosite.

Isolation of Protein C

Classical purification procedures for protein C have been summarized by Kisiel and Davie in a previous paper in this series.[10] During the ensuing years, the major advance with respect to isolation from plasma has involved the use of monoclonal antibodies.

Immunization and Screening for Hybridoma Antibodies to Protein C

Of particular utility are antibodies that recognize protein C only in the presence of Ca^{2+}. At least one such antibody can usually be obtained from three fusions. It is our impression that the chances of obtaining such Ca^{2+}-dependent antibodies are improved by incubating protein C with Ca^{2+} prior to injection. BALB/c mice (Jackson Labs, Bar Harbor, ME) are immunized intraperitoneally three times with 100 μg of protein C in 0.1 M NaCl, 0.02 M Tris-HCl, pH 7.5, and 5 mM $CaCl_2$. The first injection is in Freund's complete adjuvant and subsequent injections follow at 2-week intervals with Freund's incomplete adjuvant. The mice are injected 2 months later with 50 μg of protein C and the fusion is performed 3 days later. A myeloma cell line (P3-X-63-Ag8-653, 1.5×10^8 cells) is mixed with a fivefold excess of spleen cells from the immunized mouse. The cells are fused by standard techniques using polyethylene glycol (PEG) 1500 (Boehringer Mannheim). Metal-dependent antibodies can be detected simply by screening with standard ELISA assays in the presence of 5 mM Ca^{2+} and 5 mM EDTA. The EDTA is useful because most laboratory buffers contain sufficient Ca^{2+} to allow recognition of epitopes expressed due to occupancy of the high-affinity site. Alternatively, nearly one-third of the monoclonal antibodies we have characterized against protein C can be used for isolation and eluted rather gently with 50 or 80% ethylene glycol (v/v), 1 mM MES-HCl, pH 6.0.

Isolation of Protein C with Ca²⁺-Dependent Antibody

The advantage of the monoclonal antibody approach in general, and metal-dependent antibodies in particular, is especially great when working with human plasma and with proteins (such as protein C) that are present at low concentrations (3–5 μg/ml). The inherent problems of handling barium salts in the classic procedures are difficult to reconcile with safety issues related to viral pathogens carried in human plasma. Risks are minimized by concentrating the vitamin K-dependent factors on QAE-Sephadex in the first step.[19,41] The purification is performed at room temperature. Human plasma (30 liters) is diluted with an equal volume of 0.02 M Tris-HCl, pH 7.5, 10 mM benzamidine, 1 U/ml heparin. All buffers contain 0.02% sodium azide. QAE-Sephadex A 50 (30 g swollen in 0.1 M NaCl) is gently stirred into the diluted plasma and mixed with a broad, slow-moving mechanical paddle for 1 hr. The QAE with bound protein is extremely dense and settles 30 min after stirring is halted. The supernatant is removed, and the QAE is packed into a 10×60-cm column, washed with 4 liters of 0.15 M NaCl, 0.02 M Tris-HCl, pH 7.5, 10 mM benzamidine hydrochloride, and eluted with 0.5M NaCl in the above buffer. The eluate containing protein C is approximately 0.6–1 liter and can be followed conveniently by the bright blue-green color which coelutes. Heparin is added to 10 U/ml, Ca^{2+} to 10 mM, and the Ca^{2+}-dependent antibody ($\approx$ 100 ml at 5 mg/ml on Affi-Gel 10) is stirred into the concentrate for 1 hr, allowed to settle, packed into a 2.5×30-cm column, and washed with 200 ml of 0.5 M NaCl, 0.02 M Tris-HCl, pH 7.5, 2 mM Ca^{2+}, 10 mM benzamidine hydrochloride, followed by 100 ml of 0.1 M NaCl, 0.02 M Tris-HCl, 2 mM Ca^{2+} before eluting with 0.1 M NaCl, 0.02 M Tris-HCl, pH 7.5, 2 mM EDTA. Depending on the monoclonal antibody employed,

[41] C. T. Esmon, F. B. Taylor, Jr., L. B. Hinshaw, A. Chang, P. C. Comp, G. Ferrell, and N. L. Esmon, *Dev. Biol. Stand.* **67**, 51 (1987).

the protein C is usually eluted in 100 ml or less. As mentioned above, the real utility of this approach is safety and speed. Isolation of 30 mg of protein C is readily achieved in 2 days. The eluate usually needs to be purified further to remove contaminating serum amyloid P, a plasma protein that binds to Sephadex in a Ca^{2+}-dependent fashion independent of the presence of the antibody. For small amounts of material (5–10 mg), the protein C can be separated from the serum amyloid P on a Mono Q column (HR5/5, Pharmacia, Piscataway, NJ) developed with a 20-ml linear gradient from 0.1–0.6 M NaCl in either 0.02 M Tris-HCl, pH 7.5, or 0.05 M MES-HCl, pH 6.0. DEAE-Sepharose and QAE-Sephadex columns (0.9 × 60 cm) eluted under the same conditions with a 200-ml linear gradient are also effective for up to 40 mg of protein C.

Comments. QAE-Sephadex and possibly other resins often have some contaminating substance that appears to interact with protein C and inhibit activation by the thrombin–TM complex. It is usually not dialyzable, does not seem to block protein C activation by high levels of thrombin alone, and does not inhibit amidolytic activity. With QAE-Sephadex, this problem can largely be eliminated by extensive washing of the swollen resin with 4 M NaCl prior to equilibration. This is especially important in the final stages of isolation.

Isolation of Domains of Protein C for Structure–Function Studies

Proteolytic Formation and Isolation of Gla-Domainless Protein C. Ca^{2+} has major positive and negative influences on protein C activation. Ca^{2+} enhances activation by the thrombin–TM complex > 60-fold[42] and inhibits activation by thrombin alone comparably.[42,43] A useful reagent can be generated by the proteolytic removal of the Gla domain from either bovine or human protein C.[19,21] The resultant "Gla-domainless protein C" can still be converted to activated protein C by either thrombin or the thrombin–TM complex. Removal of the Gla domain reduces the number of Ca^{2+} binding sites from approximately 10 to 1,[42,43] greatly simplifying structure–function analysis.

Human or bovine protein C (2 mg/ml) is incubated with chymotrypsin (5 μg/ml) in 0.1 M NaCl, 0.02 M Tris-HCl, pH 7.5, for 10 min at 37°. The reaction is stopped by addition of 1 mM diisopropyl fluorophosphate (DFP). Gla-domainless protein C is separated from intact protein C, the Gla peptide and chymotrypsin by ion-exchange chromatography on either QAE-Sephadex Q 50 or a Mono Q (HR 5/5) column (Pharmacia). Typi-

[42] A. E. Johnson, N. L. Esmon, T. M. Laue, and C. T. Esmon, *J. Biol. Chem.* **258**, 5554 (1983).
[43] G. W. Amphlett, W. Kisiel, and F. J. Castellino, *Biochemistry* **20**, 2156 (1981).

cally, 25 mg of reaction mixture is chromatographed on a 0.9 × 30-cm column of QAE-Sephadex. The column is developed with a 50-ml linear gradient from 0.1–0.6 M NaCl in 0.02 M Tris-HCl, pH 7.5, at room temperature. Gla-domainless protein C elutes with approximately 0.2 M NaCl.

Proteolytic Formation and Isolation of the EGF-like Domains of Protein C. A Ca^{2+} binding site has been localized to the first EGF-like domain of protein C.[44-46] This domain contains a β-hydroxyaspartic acid residue with properties consistent with a role in Ca^{2+} ligation. No direct documentation of a functional role for the hydroxylation is known at this time, though mutation of the Asp to Glu did result in loss of anticoagulant activity.[45] In factor IX, a structurally similar protein, blocking the hydroxylation reaction did not alter the biological activity.[47]

Gla-domainless bovine protein C (10 mg/ml) in 0.1 M NaCl, 50 mM Tris-HCl, pH 7.5, 2 mM EDTA is incubated with 2% trypsin (w/w) for 30 min at 37°.[46] The reaction is terminated by addition of 10 mM DFP. The EGF-like fragment is separated from the other reaction products by chromatography on an AcA 44 Ultrogel (Pharmacia) column (1.6 × 94 cm) in 0.1 M NaHCO$_3$. In addition to the two EGF-like repeats, a 24-residue peptide from the heavy chain remains linked through a disulfide bond.

Modification of this method allows isolation of the EGF-like repeats with the Gla domain covalently attached.[48] Bovine protein C (20 mg/ 10 ml) in 0.1 M Na$_2$B$_4$O$_7$ is reacted with citraconic anhydride (three times with 50 μl per addition with stirring) at room temperature with the pH maintained at 8.7 by addition of NaOH throughout the reaction. The citraconylated protein C is then digested with trypsin (2%, w/w) for 5 min at 37° in 0.1 M NaCl, 5 mM EDTA, 0.05 M Tris-HCl, pH 8.0. The reaction is terminated by addition of 10 mM DFP. The fragment containing the covalently associated Gla domain and EGF-like domains is isolated by chromatography on a Q-Sepharose Fast Flow ion-exchange column (1.6 × 16 cm) by sequential washing with 50 mM Tris-HCl, 1 mM EDTA, pH 8.0, the same Tris buffer containing 0.1 M NaCl, then 0.2 M NaCl before elution with 0.2 M NaCl, 20 mM CaCl$_2$.

[44] A.-K. Öhlin and J. Stenflo, *J. Biol. Chem.* **262**, 13798 (1987).
[45] A.-K. Öhlin, G. Landes, P. Bourdon, C. Oppenheimer, R. Wydro, and J. Stenflo, *J. Biol. Chem.* **263**, 19240 (1988).
[46] A.-K. Öhlin, S. Linse, and J. Stenflo, *J. Biol. Chem.* **263**, 7411 (1988).
[47] C. K. Derian, W. J. VanDusen, C. J. Przsiecki, P. N. Walsh, K. L. Berker, and P. A. Friedman, *J. Biol. Chem.* **264**, 6615 (1989).
[48] A.-K. Öhlin, I. Björk, and J. Stenflo, *Biochemistry* **29**, 644 (1990).

Comments. The fragment containing the Gla domain covalently attached to the EGF-like repeats is a competitive inhibitor of protein C activation and hence appears to contain many of the sites responsible for recognition by the activation complex. Neither the isolated Gla domain nor the EGF-like repeats were effective inhibitors.[49] To date, no reports have appeared on the isolation of functional protein C protease domain by either proteolytic or recombinant approaches.

Recombinant Protein C

Recombinant protein C has been expressed in functional form in mammalian cells in culture. Several expression vectors have been employed[16,17,45] and nearly complete carboxylation has been achieved in human 293 cells.[16] The proteolytic processing to two-chain protein C is also more complete with the 293 cells than with BHK-AD cells.[16] Activated protein C has been expressed directly by inserting the insulin receptor cleavage sequence (PRPSRKRR) in place of the normal protein C activation peptide.[50] Secretion levels of 8–12 μg/ml have been obtained with this mutant. These recombinant systems are amenable to site-directed mutagenesis, which will allow considerable structure–function analysis in the future.

Separation of Partially Carboxylated from Functional Protein C. Expression of recombinant protein C presents special problems because some of the material is not totally carboxylated and hence has lower specific activities in protein C assays that are dependent on membrane binding of protein C or activated protein C. A useful approach to this dilemma was developed by Yan and co-workers,[16] who noted that completely or nearly completely carboxylated protein C and undercarboxylated forms exhibited different properties on anion exchange chromatography on Fast Flow Q. Specifically, it was noted that 85% of the protein C from "good" carboxylating cell lines could be eluted from the column when 10 mM Ca^{2+} was added to 0.15 M NaCl, 0.02 M Tris-HCl, pH 7.4, whereas in excess of 0.3 M NaCl is required in the absence of Ca^{2+}. Less effectively carboxylated protein C forms were not eluted from the column in a Ca^{2+}-dependent fashion, but eluted essentially normally in the salt gradient. Presumably this difference is due to the role of the Gla region in determining the binding properties of protein C to this column and to the fact that fully

[49] P. J. Hogg, A.-K. Öhlin, and J. Stenflo, *J. Biol. Chem.* **267,** 703 (1992).
[50] H. J. Ehrlich, S. R. Jaskunas, B. W. Grinnell, S. B. Yan, and N. U. Bang, *J. Biol. Chem.* **264,** 14298 (1989).

carboxylated protein C binds Ca^{2+} more effectively and the bound Ca^{2+} blocks the binding to the column.

Isolation of Thrombomodulin

Isolation of TM has been accomplished from human placenta[51] and rabbit,[20,52] rat,[53] and bovine[4] lung. The basic approach takes advantage of the integral membrane nature of TM and the fact that the TM–thrombin interaction is extremely ionic strength dependent, but not dependent on the membrane interaction. Lungs from small animals are an especially rich source. Either fresh perfused lungs or frozen lungs are acceptable starting materials, with less degradation and a significantly greater yield per lung being obtained from fresh material. Fresh lungs are processed immediately without freezing. With frozen lungs, 80 pairs of young rabbit lungs are homogenized in a commercial meat grinder while still frozen. Unless otherwise specified, all steps are performed at 4°. Tissue is passed through the grinder three times, then suspended in 8 liters of 0.25 M sucrose, 0.02 M Tris-HCl, pH 7.5, and 5 mM benzamidine hydrochloride. The tissue is collected by centrifugation at 4800 g for 30 min. The pellet is again passed through the tissue grinder three times and washed a total of three times by repeated centrifugation and resuspension. The tissue is homogenized in a large Dounce homogenizer driven by a commercial drill press or in a Waring blender in 2.4 liters of sucrose buffer containing 2% Triton X-100. The cellular debris is separated from the solubilized TM by centrifugation at 4800 g for 40 min. Although not strictly necessary, filtering the eluate through a Sephadex G-50 column (5 × 5 cm) and adsorbing and eluting TM from QAE-Sephadex before proceeding with affinity chromatography confer the advantage of minimizing damage to the affinity columns from residual debris, adhesive molecules, and proteases in the crude extract. TM in the flow-through from the Sephadex G-50 column is diluted with an equal volume of 0.4 M NaCl, 0.02 M Tris-HCl, 5 mM benzamidine containing 20 g of swollen QAE-Sephadex and allowed to adsorb with gentle mixing for 40 min. The resin is then allowed to settle, packed into a column (5 × 30 cm) and washed with 1 liter of 0.2 M NaCl, 0.02 M Tris-HCl, pH 7.5, 5 mM benzamidine, 0.5% Lubrol PX (v/v). The resin is removed from the column, suspended vigorously with a mechanical stir-

[51] H. H. Salem, I. Maruyama, H. Ishii, and P. W. Majerus, *J. Biol. Chem.* **259**, 12246 (1984).
[52] N. L. Esmon, W. G. Owen, and C. T. Esmon, *J. Biol. Chem.* **257**, 859 (1982).
[53] C. T. Esmon, N. L. Esmon, J. Saugstad, and W. G. Owen, *in* Pathobiology of the Endothelial Cell (H. L. Nossel and H. J. Vogel, eds.), p. 121. Academic Press, New York, 1982.

ring bar in 300 ml of 1 M NaCl in the above buffer for 30 min, and then packed back into the column. The flow-through is collected and the column further eluted with 300 ml of the 1 M salt solution, allowing the resin to run dry. The reason for removing the ion-exchange material for the initial elution is that the extract often causes the resin to clump and the column can channel as a result. This leads to incomplete recovery that is avoided by the batch elution procedure.

Final purification involves affinity chromatography on either a thrombin affinity matrix or on a monoclonal antibody. For laboratories with a commitment to analysis of protein C activation, the monoclonal antibody approach is highly recommended, because it is more specific, and adsorption can be achieved directly from the QAE eluate. Because these procedures are antibody dependent, they are not presented here but have been described elsewhere.[20] Hybridoma antibodies to TM have been consistently obtained with the immunization and fusion schedules described for protein C above except that the antigen did not contain Ca^{2+}.

If a thrombin affinity column step is employed, the eluate is dialyzed or desalted into 0.1 M NaCl, 0.02 M Tris-HCl, pH 7.5, 0.5 mM $CaCl_2$, 10 mM benzamidine hydrochloride before application to the affinity resin. Thrombin (1 mg/ml of resin) linked to BioGel A-15m by the CNBr method[52] has provided the most reproducible affinity matrix. For unknown reasons, the adsorption and elution works much more efficiently if the enzyme is linked to the resin before inactivation with DFP rather than the reverse order. Inactivation of the thrombin is important because formation of thrombin complexes with inhibitors like antithrombin III results in loss of the capacity to bind to TM. The QAE eluate is applied to a 2.5 × 10-cm thrombin agarose column at room temperature, washed with 2 liters of 0.2 M NaCl, 0.02 M Tris-HCl, 0.5 mM $CaCl_2$, 1 mM benzamidine, 0.5% Lubrol PX, pH 7.5 and eluted with 1 M NaCl, 0.1 mM EDTA, 1 mM benzamidine, 0.02 M Tris-HCl, 0.05% Lubrol PX, pH 7.5. The eluate is then dialyzed into the 0.1 M NaCl buffer as above and reapplied to a 0.9 × 30-cm column of thrombin–agarose. The column is washed with 10 ml of 0.4 M NaCl, 0.02 M Tris-HCl, 0.5 mM $CaCl_2$, 1 mM benzamidine hydrochloride, 0.05% Lubrol PX, pH 7.5, and subsequently eluted with a linear gradient from 0.4 to 1 M NaCl in the above buffer with 0.1 mM EDTA replacing the Ca^{2+}.

Some species exhibit much more proteolytic degradation during the isolation procedure. A nice method to limit this degradation was developed by Jakubowski et al.[4] The lungs are suspended in the sucrose buffer made with 20 mM CAPS, pH 10.5. The procedure is carried out essentially as described for the rabbit lungs except that after Triton extraction, the pH is decreased to 7.5 and the extract made 25% in methanol. The methanol precipitates some proteins, and most proteases are not active at pH 10.5. The ability to utilize these methods to isolate TM from multiple species is of utility in the preparation of immunological reagents for studies of TM modulation on endothelial cells.

Preparation of Thrombomodulin Fragments

TM is stable from pH 2–10, in 1% SDS and in 6M guanidine.[52] The activity is rapidly lost in the presence of reducing agents. These remarkable stability characteristics allow the generation of functional proteolytic and CNBr fragments of TM. A useful soluble derivative can be formed by incubation of TM (1.16 mg/ml) for 10 min at 37° in 0.1 M NaCl, 0.02 M Tris-HCl, pH 7.5, 0.02% Lubrol PX with porcine pancreatic elastase (0.116 mg/ml).[29] The reaction is terminated with 10 mM DFP. The resultant active TM fragment is composed essentially of the six EGF-like repeats (Fig. 1). It can be separated from other reaction products and uncleaved TM by chromatography on a Mono Q HR 5/5 column. The column is developed with a 20-ml linear gradient from 0.1 to 2 M NaCl in 0.02 M Tris-HCl, pH 7.5, 0.02% Lubrol PX at room temperature. The unreacted TM is eluted from the column with NaCl concentrations greater than 1 M, apparently due to the presence of the chondroitin sulfate moiety. The elastase-modified TM fragment lacks the transmembrane region and the amino terminal domain and is thus soluble in the absence of detergents. For biophysical studies, this proteolytic derivative constitutes a major advantage compared to detergent-solubilized intact TM.

The EGF domain can be divided further by CNBr cleavage. Two methionines are strategically located in the elastase fragment: one midway through the second EGF domain, the other between EGF 4 and 5 (Fig. 1). CNBr cleavage[31] involves reacting TM (1 mg/ml) in 70% formic acid with 50 mg/ml CNBr for 24 hr at room temperature. Cleavage is incomplete and yields all possible combinations of the EGF domains. To obtain a larger percentage of the bigger fragments, reduce the CNBr to 10 mg. Fragments containing EGF 5–6 bind to immobilized thrombin and can be affinity purified as described above for intact TM. These fragments can then be separated by gel filtration on a TSK-125 column (Biorad, Richmond, CA) in 3 M guanidine hydrochloride, 0.02 M Tris-HCl, pH 7.5.[31] It is of particular interest that the EGF 5–6 fragment of TM binds to thrombin and blocks fibrinogen clotting activity, but does not accelerate protein C activation in the presence of Ca^{2+}. The latter property can be exploited in studies of the molecular mechanisms of TM function (see later, discussion of fluorescence and spectroscopic characterization of interactions).

Recombinant Thrombomodulin

Intact TM has been expressed in CV-1 cells,[35] which do not normally express TM. Several soluble forms of TM have been expressed by recombinant approaches in human 293, hamster AV12-644,[6] and COS-1 cells.[34] A form truncated at the membrane-spanning domain is synthesized either with or without chondroitin sulfate attached.[6] The variants can be separated on anion-exchange chromatography essentially as described above for proteolytic fragments of TM.

EGF domains 1–6, 4–6, 4–5, and 5–6 have been expressed and isolated.[34] In general, the properties of the recombinant proteins have agreed well with those observed with natural TM. The soluble TM derivatives have been expressed at relatively high levels (0.2–2 μg/ml).

Role of Chondroitin Sulfate in TM Function

Chondroitin sulfate can be removed from TM with chondroitinase ABC. Intact TM accelerates the inactivation of thrombin by antithrombin III.[5] Removal of the chondroitin sulfate eliminates this activity. With chondroitin sulfate present, TM accelerates the activation of factor XI by thrombin, an activity that is also lost on removal of the chondroitin.[54] Removal of the chondroitin decreases the apparent affinity of TM for thrombin about 5- to 10-fold based on kinetic analysis.[6,29] Chondroitin-free forms are less effective at blocking thrombin's clotting activity.[6,29] Whether this is due primarily to decreased thrombin–TM affinity, an additional steric hindrance contribution by the chondroitin, and/or a unique thrombin conformation induced by chondroitin remains unknown. Forms lacking the chondroitin sulfate exhibit an extremely unusual Ca^{2+} dependence with respect to protein C activation. There is a sharp optimum at approximately 300 μM Ca^{2+} and a rapid decline in activation rate at higher Ca^{2+} concentrations.[6,29] Sedimentation analysis reveals that high Ca^{2+} concentrations disrupt ternary complex formation by displacing the substrate.[55] Forms containing the chondroitin sulfate exhibit a simple saturable Ca^{2+} dependence,[6,29] presumably because the chondroitin prevents dissociation of the substrate within the ternary complex, though this has not been demonstrated directly. The affinity of the chondroitin-free TM derivatives for protein C is considerably greater at the optimal Ca^{2+} concentration than that observed with intact TM, but similar to the values obtained with membrane-reconstituted TM (Table I).

---

[54] D. Gailani and G. J. Broze, Jr., *Science* **253,** 909 (1991).
[55] P. H. Olsen, N. L. Esmon, C. T. Esmon, and T. M. Laue, *Biochemistry* **31,** 746 (1992).

TABLE I
PROPERTIES OF PROTEIN C ACTIVATORS

Activator	K_d (nM)	K_{cat} (M/min)	K_m (μM) $-Ca^{2+}$	K_m (μM) $+Ca^{2+}$	Ca^{2+}	Ref.
Thrombin alone	—	≈2	1–2	≈60	Hyperbolic inhibition, $\frac{1}{2}$ max, 250 μM	21
With TM	≈1	≈250	≈60	≈8	Hyperbolic acceleration, $\frac{1}{2}$ max, 250 μM	21
With Chondroitin-free TM	≈10	250	—	≈0.7	Enhances to ≈300 μM, inhibition above 300 μM	6,29
TM in neutral membranes	≈1	200	—	0.7	Positive cooperativity	20
TM in negatively charged membranes	≈1	200	—	0.1	—	20
On endothelium	0.5	—	—	0.7	Required	57

Endothelial cells grown under different conditions appear to add chondroitin to a variable extent,[56] which likely accounts for the range of K_d apparent observed over endothelial cells from 0.7 to 5–8 nM.[35,56,57] Growth of cells in β-D-xyloside appears to influence the extent of chondroitin attachment.[56]

Removal of Chondroitin Sulfate from TM. Rabbit TM (75 μg/ml) is incubated with 0.05 U/ml chondroitin ABC lyase from *Proteus vulgaris* (Seikaguku, Tokyo, Japan) in 0.1 M NaCl, 0.05 M Tris-HCl, 30 μM sodium acetate, pH 7.5 containing 2.5 mM o-phenanthroline, 10 μg/ml pepstatin, 20 μg/ml leupeptin, and 0.2% Nonidet P-40. Removal of the chondroitin renders the TM much less acidic and therefore the chondroitin-free TM can be separated from the unreacted material by ion exchange as described above for elastase-digested TM.

[56] J. F. Parkinson, J. G. N. Garcia, and N. U. Bang, *Biochem. Biophys. Res. Commun.* **169**, 177 (1990).
[57] W. G. Owen and C. T. Esmon, *J. Biol. Chem.* **256**, 5532 (1981).

Membrane Involvement in Protein C Activation

The activation of protein C by the thrombin–TM complex is greatly affected by the presence of phospholipids. Simply adding phospholipid[52,58] or preformed liposomes to TM solutions[59] can affect the reaction. However, incorporation into liposomes rather than adsorption would most likely mimic the physiologically relevant interactions on the endothelial cell surface.[20] The location of TM on the luminal surface of the endothelium makes it unlikely that the TM–thrombin complex activates protein C in the presence of a large pool of negatively charged phospholipids, because these are known to promote clot formation.

Method of TM Incorporation into Phospholipids

We have used the 1-palmitoyl-2-oleoyl forms of phosphatidylcholine and phosphatidylserine. These are stored dried in glass tubes under nitrogen. The lipid is dissolved to 20 mg/ml in 0.1 M NaCl, 0.02 M Tris-HCl, pH 7.5, 405 mM (119 mg/ml) octylglucoside, yielding a phospholipid:detergent molar ratio of 1:16. Once the lipids are fully dissolved with the aid of occasional vortexing, combinations at the preferred ratios can be made by simple mixing. A small aliquot of TM containing 5–50 μg protein is added to 0.2 ml of the phospholipid–octylglucoside solution. (It is important that the Lubrol PX concentration in the phospholipid mixture is <0.001% for vesicle formation to occur. If necessary, the detergent can be exchanged by binding the TM to a Mono Q column, washing slowly with ≈ 100 ml 15 mM octylglucoside in 0.1 M NaCl, 0.02 M Tris-HCl, pH 7.5, and elution in 2 M NaCl, 0.02 M Tris-HCl, pH 7.5, 15 mM octylglucoside.) A trace level of [^{14}C]phosphatidylcholine can be incorporated at this step to facilitate monitoring of the phospholipid distribution, if desired. The mixture is dialyzed against 3 × 1 liter 0.1 M NaCl, 0.02 M Tris-HCl, pH 7.5, at room temperature over 72 hr. The samples are then applied to a sucrose density gradient formed as follows for a 13-ml gradient in a SW41Ti rotor (Beckman, Palo Alto, CA): (a) 2.5 ml 30% sucrose, (b) liposome sample adjusted to 0.5 ml with buffer + 0.5 ml 50% sucrose, (c) 8 ml 20% sucrose, (d) 0.5–2 ml buffer to fill tube. Samples are spun at 26,000 rpm at 22° for 18–24 hr. Depending on the composition of the liposomes, they will be present at the 20% sucrose/buffer interface, in the buffer, or at the air/buffer interface. The distribution of protein and lipid after centrifugation can be determined by assay and by liquid scintillation

[58] J.-M. Freyssinet, B. Brami, J. Gauchy, and J.-P. Cazenave, *Thromb Haemostasis* **55,** 112 (1986).
[59] J.-M. Freyssinet, J. Gauchy, and J.-P. Cazenave, *Biochem. J.* **238,** 151 (1986).

counting the ^{14}C label, respectively. When comparing preparations of TM in liposomes of different phospholipid compositions or with unincorporated TM, it is useful to use Gla-domainless protein C as substrate rather than the native protein. Gla-domainless protein C activation is unaffected by lipid, whereas native protein C activation is affected differently by different lipid environments.[20] When activation of the native protein is being addressed, an additional safeguard is to neutralize any calcium present in the activation mixes before assaying for activated protein C in the chromogenic assays described elsewhere. Calcium-dependent binding of the activated protein C to the various phospholipid vesicles can affect the chromogenic activity of the enzyme.[59]

Comments. TM incorporates into membranes simply by addition to preformed liposomes.[59] It is unclear whether this method is equivalent to the method described above. With rabbit TM reconstituted into phosphatidylcholine vesicles by the dialysis method, the TM orientation is random.[20] Incorporation results in a decrease in the K_m for protein C from 8 to approximately 0.7 μM, a value virtually identical to that observed on the endothelial cell surface. Inclusion of 20% phosphatidylserine further decreases the K_m to 0.1 μM, presumably reflecting further protein C binding. With human TM, simple addition to preformed liposomes is essentially without influence on activation kinetics unless negatively charged lipids are present.[59] Whether this difference is related to species differences or to reconstitution methodologies remains to be determined.

Assays of Protein C Activation

General Considerations

The conditions for activation of protein C are extremely dependent on the activator used, be it thrombin alone, thrombin plus full-length TM, a fragment of TM, or liposome-incorporated TM. Therefore, the choice of activation conditions must be adjusted accordingly. The kinetic constants provided in Table I are useful for this purpose. Of major importance are the differences in K_d apparent observed with different TM forms, and the different Ca^{2+} dependencies observed with different activators. At saturating thrombin concentrations, the initial rate of protein C activation is linearly related to TM concentration because the reaction rate by free thrombin is less than 1% that of the complex. As a precaution, the value obtained with thrombin in the absence of added TM should be measured routinely and should be negligible under most conditions. As mentioned previously, the thrombin–TM interaction is very ionic strength dependent, and changes from 0.1 to 0.2 M NaCl can result in a twofold or more change in activation rate under conditions in which the thrombin concentration is near the K_d.[8]

Activation of Protein C

Protein C is incubated with thrombin in the presence or absence of TM and other modulators. The reaction is allowed to proceed for a set time interval and stopped by addition of antithrombin III (0.1 mg/ml) or hirudin (in fivefold excess over thrombin), neither of which inhibits activated protein C.

Determination of Activated Protein C Concentration

With purified protein C as the substrate, activated protein C formation can be measured with chromogenic substrates. Because activated protein C chromogenic activity is influenced by Ca^{2+} and other variables, it is advisable to prepare a standard curve each day to be certain that reaction components do not influence activity. Quantitative protein C activation can be obtained by incubating protein C (0.1 mg/ml) with 5 μg/ml thrombin for 3 hr at 37° at physiologic ionic strength and pH in 2 mM EDTA or by incubating the same concentration of protein C with 20 nM thrombin, 30 nM TM, 3 mM Ca^{2+} for 60 min at 37°.

Satisfactory substrates in microtiter plate assays of activated protein C are 0.4 mM Spectrozyme PCa (American Diagnostica, Greenwich, CT), 0.4 mM S2238, or 0.5 mM S2366 (Chromogenix, Franklin, OH). Total sample volumes of 100 μl 0.1 M NaCl, 0.02 M Tris-HCl, pH 7.5, containing 10 mg/ml gelatin or bovine serum albumin (BSA) are appropriate buffers and the assay sensitivity is approximately 0.5–6 nM activated protein C in the well when measured kinetically at 405 nm.

Thrombin Specificity

The sequence of protein C near the scissile bond is EDQVDP**R**LIDG, with cleavage occurring at the Arg residue typeset in boldface. Protein C activation by thrombin is very slow and TM functions to increase that rate to near the rate of cleavage of fibrinogen by free thrombin. This implies that residues in protein C inhibit thrombin–protein C interaction. Analysis of synthetic substrates and peptidyl chloromethyl ketone inhibitors reveal that the protein C sequence is compatible at the P1 site (Arg) and the P2 site (Pro) with the best substrates and inhibitors.[60,61] Studies with pep-

[60] R. Lottenberg, U. Christensen, C. M. Jackson, and P. L. Coleman, this series, Vol. 80, p. 341.
[61] C. Kettner and E. Shaw, this series, Vol. 80, p. 826.

tides comprising residues on both sides of the scissile bond demonstrate that cleavage is inhibited by the acidic residues in protein C at P3 and P3'.[62-64]

Influence of Anion-Binding Exosite Occupancy on Thrombin Activity toward Synthetic Substrates

The anion-binding exosite is relatively nonspecific with respect to sequence requirements for occupancy.[65,66] TM as well as peptides derived from hirudin (hirugen), heparin cofactor II, and the thrombin receptor all interact with thrombin's anion-binding exosite and alter peptidyl chromogenic and fluorogenic substrate hydrolysis, and/or the inhibition constant of peptidyl chloromethyl ketones.[3,36,66-68] In addition, heparin and other polyanions, such as dextran sulfate, which are believed to interact with thrombin outside of the anion-binding exosite, also influence thrombin catalysis toward p-nitroanilide substrates.[69] Although the changes in amidolytic activity are relatively small, they can be used to monitor complex formation simply by plotting change in rate versus TM concentration.

Proteolytic Derivatives of Thrombin

The proteolytic derivative, γ-thrombin, binds TM weakly.[70] Antibodies directed to the proteolytic fragment released from thrombin during autolysis to γ-thrombin also inhibit the ability of TM to accelerate protein C activation.[71] Elastase-digested thrombin, cleaved at residue 147 in the human thrombin sequence, has decreased affinity for TM.[72] A peptide corresponding to Thr-147 to Asp-175 also inhibits TM binding, and a monoclonal antibody with its epitope in this region inhibits TM binding.[72]

[62] J.-Y. Chang, Eur. J. Biochem. 151, 217 (1985).
[63] B. F. Le Bonniec, R. T. A. MacGillivray, and C. T. Esmon, J. Biol. Chem. 266, 13796 (1991).
[64] B. F. Le Bonniec and C. T. Esmon, Proc. Natl. Acad. Sci. U.S.A. 88, 7371 (1991).
[65] T.-K. H. Vu, V. I. Wheaton, D. T. Hung, I. Charo, and S. R. Coughlin, Nature (London) 353, 674 (1991).
[66] L.-W. Liu, T.-K. H. Vu, C. T. Esmon, and S. R. Coughlin, J. Biol. Chem. 266, 16977 (1991).
[67] G. L. Hortin and B. L. Trimpe, J. Biol. Chem. 266, 6866 (1991).
[68] G. L. Hortin, D. M. Tollefsen, and B. M. Benutto, J. Biol. Chem. 264, 13979 (1989).
[69] L. C. Petersen, Eur. J. Biochem. 137, 531 (1983).
[70] A. Bezeaud, M.-H. Denninger, and M.-C. Guillin, Eur. J. Biochem. 153, 491 (1985).
[71] G. Noé, J. Hofsteenge, G. Rovelli, and S. R. Stone, J. Biol. Chem. 263, 11729 (1988).
[72] K. Suzuki, J. Nishioka, and T. Hayashi, J. Biol. Chem. 265, 13263 (1990).

Thrombin Mutants

The approaches that have been taken involve molecular modeling of thrombin, prediction of interaction sites with TM, fibrinogen and protein C, and site mutagenesis to test the predictions. The details of this approach are beyond the scope of the present chapter, but examples of this work can be found in Refs. 16, 64, and 73–75. In essence, this work has demonstrated that the P3 and P3′ residues of protein C inhibit activation by thrombin, most probably by interacting at or near Glu residues 192 and 39 of thrombin, and that TM interaction appears to involve the anion-binding exosite of thrombin.

Spectroscopic Characterization of Interactions

Spectroscopy can be used to monitor the interactions involved in protein C activation. This approach provides information about conformational changes that result from complex formation, the affinities that govern the association of various proteins and ligands, and the topography of the complexes (i.e., the distances between different sites in the complexes). Furthermore, spectral data complement the data obtained from activity assays because one can spectroscopically detect, directly monitor, and characterize intermediate states that do not lead to product formation, such as thrombin–TM complex formation in the absence of substrate or calcium, or interactions between species that are not active in stimulating protein C activation.

Fluorescence spectroscopy requires only small concentrations (as little as 1 nM, depending on the dye) of fluorescent-labeled protein to obtain interpretable data. A wide range of information is provided by a complete analysis of the fluorescence signal. The emission properties of a fluorescent moiety (intensity, lifetime, energy/emission wavelength, and polarization/anisotropy) are sensitive to its environment, so a fluorescent molecule in a particular conformation and environment will emit a characteristic fluorescence signal when excited. In principle, a given signal can be correlated with both a specific conformation and a particular functional state (determined by a bioassay), thereby directly relating structure and function. The distribution of the molecules in different states (e.g., free or bound) can be determined if the characteristic signal for each state is known, and from

[73] Q. Wu, J. P. Sheehan, M. Tsiang, S. R. Lentz, J. J. Birktoft, and J. E. Sadler, *Proc. Natl. Acad. Sci. U.S.A.* **88,** 6775 (1991).

[74] R. A. Henriksen and K. G. Mann, *Biochemistry* **27,** 9160 (1988).

[75] H. J. Ehrlich, B. W. Grinnell, S. R. Jaskunas, C. T. Esmon, S. B. Yan, and N. U. Bang, *EMBO J.* **9,** 2367 (1990).

such data, equilibrium constants can be determined. Furthermore, once a binding constant has been measured using a fluorescent-labeled derivative, the binding constant of an unmodified, nonfluorescent species can be determined by competition experiments.[76]

Dansyl–Thrombin

A fluorescent probe can be covalently attached to the active site serine (Ser-195) of human thrombin using dansyl fluoride to form dansyl–thrombin. This probe is useful for detecting the association of thrombin with TM or one of its fragments because the binding of TM to dansyl–thrombin reduces its emission intensity by nearly 40%.[77]

Preparation of Dansyl–Thrombin. Dansyl–thrombin is synthesized essentially as described.[77,78] Human (but not bovine) thrombin (3 mg/ml in 0.07 M sodium phosphate, pH 7.2, 0.2 M NaCl) is mixed with 0.05 volumes of dioxane containing dansyl fluoride at 2.5 mg/ml, and is incubated for 7–10 hr at 25° and then for 3 days at 4°. The labeling efficiency is improved if the dioxane is passed through alumina immediately prior to solubilizing the dansyl fluoride. Free dansyl is removed at the end of the reaction by dialysis against 0.02 M Tris-HCl, pH 7.5, 0.1 M NaCl overnight at 4°. In our experience, the reaction does not go to completion. Active thrombin is removed from the sample by adding an excess of antithrombin III to the sample (2.5 mg antithrombin III/mg active thrombin), and then applying it slowly to an immobilized TM column (0.6 × 12 cm; 2 mg TM/ml of Affi-Gel 15).[77] Because dansyl–thrombin binds to TM and does not react with antithrombin III, while neither antithrombin III nor the complex between thrombin and antithrombin III will bind to TM, only dansyl–thrombin binds to the affinity column. After a 10-ml wash, dansyl–thrombin is eluted at 3 ml/hr with 20 mM Tris-HCl, pH 7.5, 2.0 M NaCl, dialyzed, and stored in small aliquots at −75°. The residual activity of the dansyl–thrombin prepared in this way is less than 1% of unmodified thrombin. A sham reaction with dansyl fluoride using DFP-treated thrombin shows that essentially all of the dansyl dyes are covalently attached to thrombin via Ser-195.[77]

Fluorescence Measurements. Depending on the instrument and light source, a dansyl–thrombin concentration of 50–100 nM is required in experiments. The excitation maximum of the dansyl dye is near 340 nm, but the emission intensity of dansyl–thrombin is greater when excited at

[76] J. K. Abrahamson, T. M. Laue, D. L. Miller, and A. E. Johnson, *Biochemistry* **24,** 692 (1985).
[77] J. Ye, N. L. Esmon, C. T. Esmon, and A. E. Johnson, *J. Biol. Chem.* **266,** 23016 (1991).
[78] L. J. Berliner and Y. Y. L. Shen, *Thromb. Res.* **12,** 15 (1977).

280 nm because of fluorescence energy transfer from thrombin trypto-
phan(s) to dansyl. Because the dansyl–thrombin emission maximum is
near 550 nm, a filter that adsorbs light below 300–400 nm must be placed
in the emission light path if one excites with 280-nm light; this is required
to eliminate any second-order excitation light from contributing to the
observed emission.

Emission intensity is monitored near 550 nm as a function of the
concentration of thrombin-binding species (TM, TM fragment, hirudin
fragment, etc.). The signal of a parallel dye-free blank or titrant-free blank
must be subtracted from the sample signal after each addition of titrant to
control for protein adsorption and contaminants in the titrant solution.
The signal of a parallel sample titrated with buffer instead of titrant will
control for photobleaching. The resulting net emission intensity must also
be corrected for dilution due to the addition of titrant. The fraction of
dansyl–thrombin bound to, e.g., TM in a sample is then given by
$(F_0 - F)/(F_0 - F_{min})$, where F_0, F, and F_{min} are the net dilution-corrected
dansyl emission intensities in the absence of TM, in the presence of a
particular amount of TM, and in the presence of saturating TM, respectively. The dissociation constant for the complex is then calculated using a
nonlinear regression analysis program (e.g., ENZFITTER, Imperial College of Science and Technology, London) to fit the data to the Scatchard
equation.

To reduce the amount of sample needed per experiment, 4 × 4-mm
quartz microcells (Starna Cells, Atascadero, CA) can be used with microcell holders (Perkin-Elmer Corp., Norwalk, CT) that insert into the cell
holder of the instrument. This permits the use of volumes as small as
200 μl in an SLM 8000C instrument (SLM Instruments, Inc., Urbana, IL).
To avoid light-scattering artifacts and obtain reproducible measurements,
extreme care must be taken in positioning these cells in the light path.
Mixing can be accomplished with a 2 × 2-mm Teflon-coated stirring bar in
the microcell and vertical stroking of the outside of the cell with a large
magnetic bar. Cleaning is facilitated by filling and aspirating the wash
solution at least 50 times through a 5-cm length of Teflon tubing that fits
snugly on the end of a Pasteur pipette.

Cautions. Dansyl–thrombin is somewhat unstable both chemically and
photochemically. Dansyl is hydrolyzed almost completely from the enzyme after 48 hr at pH 7.5 and room temperature; the hydrolysis rate is
much slower at 4°. In addition, the emission intensity of a sample decreases within minutes when excited continuously with a strong light
source, apparently because of photodegradation of the sample. Thus, it is
desirable to perform experiments as quickly as possible and to keep the
instrument excitation shutters closed except during measurements.

Thrombin adsorbs readily to the walls of cuvettes, thereby causing a time-dependent decrease in observed fluorescence intensity. This problem can be largely eliminated by coating the cells with 420 μl of 450–500 μM dioleoylphosphatidylcholine in vesicles, prepared by sonication as described previously,[77] for 4–20 hr at room temperature, and removing the phospholipid solution immediately before use.

Fluorescein–FPR–Thrombin and Similar Derivatives

Fluorescent probes can also be positioned at other locations in the active site by reacting various chloromethyl ketone affinity-labeling reagents with the active site histidine of thrombin. For example, dyes that are covalently linked to the active site histidine via tripeptide tethers, as in fluorescein–FPR–thrombin, are located in the active site groove at least 15 Å from the active site serine.[77] The sensitivity of these probes to the association of thrombin with TM and other ligands is, not surprisingly, significantly different from that of dansyl–thrombin. Thus, the data obtained with the different fluorescent derivatives are complementary, and have in fact provided a means to distinguish between TM-dependent conformational changes in the thrombin active site that do or do not correlate with cofactor function, specifically the TM stimulation of protein C activation.[77]

Preparation of Fluorescent Thrombin Derivatives with Dye Covalently Attached to Active Site Histidine via Tripeptide Tether. Fluorescent tripeptidyl chloromethyl ketone derivatives can either be purchased [dansyl-Glu-Gly-Arg-(DEGR)chloromethyl ketone; Calbiochem, La Jolla, CA], synthesized *de novo*,[61,79] or prepared from commercially available tripeptidyl chloromethyl ketones [(FPR) or D-Phe-L-Phe-Arg-(FFR)chloromethyl ketones; Calbiochem].[80] In the last approach, 10 mg of the FPR- or FFR-chloromethyl ketone in 2 ml of 50 mM sodium phosphate, pH 7.0, is mixed with 2 ml of methanol containing 25 mg of succinimidyl (acetylthio)acetate (Molecular Probes, Eugene, OR). The resulting N^{α}-[(acetylthio)acetyl]-FPR-(or FFR-)chloromethyl ketone is purified by chromatography over sulfopropyl-Sephadex C-25 and then Sephadex G-10 as described in detail by Bock.[80] After lyophilization and resuspension in 1 mM HCl, these reagents are stored in small aliquots at $-75°$ until use.

Modification of thrombin with chloromethyl ketone reagents can be done efficiently in various solvents. For example, thrombin (15–30 μM) in 20 mM Tris-HCl, pH 7.5, 0.1 M NaCl is incubated with 70 μM DEGR-chloromethyl ketone at room temperature until the amidolytic activity of

[79] E. B. Williams, S. Krishnaswamy, and K. G. Mann, *J. Biol. Chem.* **264,** 7536 (1989).
[80] P. E. Bock, *Biochemistry* **27,** 6633 (1988).

the enzyme is reduced to $<1\%$. Excess reagent is removed by dialysis at 4° to yield DEGR-thrombin.[81] Alternatively, thrombin (30 μM) is incubated with 70 μM N^α-[(acetylthio)acetyl]-FPR-chloromethyl ketone in 0.1 M HEPES, pH 7.0, 0.3 M NaCl, 1 mM EDTA until no activity is detected (<1 hr), and the excess reagent is removed by dialysis.[77] After hydrolysis of the terminal acetyl group to yield a free sulfhydryl group, any sulfhydryl-specific fluorescent reagent can be used to attach a dye covalently to the tripeptide tether in the active site of thrombin. For example, N^α-[(acetylthio)acetyl]-FPR–thrombin (13–16 μM) is incubated at room temperature with 75 μM 5-(iodoacetamido)fluorescein in 0.1 M HEPES, 0.3 M NaCl, 1 mM EDTA, and 0.1 M hydroxylamine, adjusted to pH 7.0. After 1 hr, the reaction mixture is passed through a Sephadex G-25 column (1.5 × 20 cm) in the same buffer lacking hydroxylamine to terminate the reaction and remove unreacted reagent, and then dialyzed to yield fluorescein–FPR–thrombin. Using this approach, we have also prepared ANS–FPR–thrombin.[77] Each of these fluorescent thrombins can be purified further by chromatography over a column containing immobilized TM as described above. The extent of labeling is very close to one dye per thrombin. Labeling at sites other than at the active site is very low ($<5\%$).

Fluorescent Measurements. Spectral analyses are carried out as described above for dansyl–thrombin, except that the excitation and emission wavelengths are 490 and 520 nm for fluorescein, and 328 and 454 nm for ANS.

The binding of ligands to thrombin can also be detected using anisotropy. This is necessary, for example, in the case of DEGR-thrombin because the binding of TM alters the anisotropy, but not the emission intensity, of the dansyl probe.[81] The emission intensity measured when a sample is excited by vertical plane-polarized light and the emission detected through a horizontal polarizer, designated as I_{VH}, I_{HH}, I_{HV}, and I_{VV}, are defined analogously. The component intensities of a dye-free blank sample containing the same concentrations of unmodified thrombin and ligand are subtracted from the corresponding component intensities of the sample to obtain the net emission intensities of the fluorescent thrombin species in a sample. The steady-state fluorescence anisotropy, r, is then calculated:

$$r = (I_{VV} - GI_{VH})/(I_{VV} + 2GI_{VH})$$

where the grating factor G is I_{HV}/I_{HH}. K_d values can be calculated from the ligand concentration dependence of the fluorescence-detected binding of ligand to the fluorescent thrombin species because the fraction of, e.g.,

[81] R. Lu, N. L. Esmon, C. T. Esmon, and A. E. Johnson, *J. Biol. Chem.* **264,** 12956 (1989).

DEGR-thrombin bound to a ligand at any point in a titration is given by $(r - r_0)/(r_{max}-r_0)$ when there is no change in emission intensity on complex formation. The maximum value r_{max} of anisotropy is observed in the presence of an excess of ligand, while r_0 is the anisotropy in the absence of ligand.

TM-Dependent Spectral Changes. The binding of EGF 1–6 to these fluorescent derivatives of thrombin, each with a dye located more than 15 Å from the active site serine, elicits very different spectral responses when samples are analyzed as described above for dansyl–thrombin. The EGF 1–6-dependent emission intensity change is +18% for ANS–FPR-thrombin, −13% for fluorescein–FPR-thrombin, and close to 0% for DEGR-thrombin.[77,81] Thus, it is clearly worthwhile to examine different dyes and tethers when searching for the optimal probe to detect and monitor a given interaction. Bock's method makes this relatively easy to do. It should also be noted that EGF 5–6[77] and hirugen,[36] the C-terminal fragment of hirudin, each elicit fluorescence changes in dansyl–thrombin, but not in the fluorescein or ANS derivatives of thrombin with dyes positioned 15 Å from Ser-195. Therefore, dyes positioned at different locations in the thrombin active site detect different cofactor-dependent structural changes. Only one of the two distinguishable TM-dependent conformational changes identified by fluorescence in the thrombin active site correlates with the TM-dependent stimulation of protein C activation.[77]

Distance Measurements by Fluorescence Energy Transfer

Fluorescent dyes can transfer their excited-state energy to appropriate nearby chromophores. This phenomenon, termed singlet–singlet energy transfer, can be utilized to measure distances on a molecular scale. After excitation, a fluorescent dye (the donor, D) nonradiatively transfers energy to a second chromophore or fluorophore (the acceptor, A). The efficiency of transfer depends on, among other things, the extent of overlap of D emission and A absorption spectra, the relative orientation of the transition dipoles of D and A, and the distance between D and A. By the proper choice of D and A, distances of 20–100 Å can be measured.

Fluorescence energy transfer was originally used to make point-to-point measurements between a single D location and a single A location in a molecule or complex. Energy transfer between randomly and uniformly distributed donors in one plane and randomly and uniformly distributed acceptors in a parallel plane has been employed experimentally.[81–83] This variation of the energy transfer technique provides a measurement of the

[82] E. J. Husten, C. T. Esmon, and A. E. Johnson, *J. Biol. Chem.* **262**, 12953 (1987).
[83] B. Snyder and G. G. Hammes, *Biochemistry* **23**, 5787 (1984).

distance of closest approach between donor dyes localized in one plane and acceptor dyes localized in the other plane, and hence is particularly useful for investigating the interactions of proteins with membranes.

Energy Transfer Experiments. Fluorescence energy transfer from donor dyes to acceptor dyes results in a decrease in donor fluorescence. The efficiency of energy transfer is usually determined by comparing the emission intensity of the donor dyes in the presence and absence of the acceptor dyes. Energy transfer experiments therefore require four parallel samples: D (donor only); DA (donor + acceptor); A (acceptor only) to correct for any emission caused by direct excitation of A; and B (blank) to correct for background fluorescence and light scattering. The background fluorescence can be a problem when using dansyl dyes because of their relatively weak fluorescence and the frequent occurrence of contaminants with spectral characteristics similar to those of dansyl. For example, "ultrapure" HEPES buffer samples obtained from various sources differ by a factor of five in their content of fluorescent impurities. It is generally best to sample different lots and different suppliers to identify the batch with the lowest background, and then to buy large quantities of that lot. Fluorescent contaminants in octylglucoside can also be a problem in vesicle reconstitution experiments.

The initial net emission intensity in each sample is obtained by subtraction of the signal of the B sample from those of the D, DA, and A samples to yield $(F_D)_0$, $(F_{DA})_0$, and $(F_A)_0$. After each addition of vesicles, the signal of B is subtracted from that of the other three, and the resulting net intensities are then multiplied by the dilution factor to yield the net dilution-corrected fluorescence intensity values F_D, F_{DA}, and F_A. The donor emission in the absence of acceptor is given by F_D, while donor emission intensity in the presence of acceptor is given by $(F_{DA} - F_A)$. Before the addition of vesicles, the initial intensities of the D and DA samples should, in principle, be equal, but they generally differ by a small amount. To take account of this difference, each subsequent intensity measurement of a sample is normalized by dividing it by the initial intensity value of that sample. When the samples are completely homogeneous, the ratio of the quantum yields in the D and DA samples is then given by

$$Q_D/Q_{DA} = [F_D/(F_D)_0]/\{(F_{DA} - F_A)/[(F_{DA})_0 - (F_A)_0]\}$$

For energy transfer between dyes localized in two planes (e.g., donor dyes at the active site of the thrombin–TM complex and acceptor dyes at the membrane surface), the distance of closest approach between the acceptor and donor dyes can be determined using the above expression and the first term of an approximate series solution for this situation[84] if the efficiency

[84] B. S. Isaacs, E. J. Husten, C. T. Esmon, and A. E. Johnson, *Biochemistry* **25**, 4958 (1986).

of energy transfer is low (i.e., the separation between the acceptor and donor dyes is large).[83] If the measured distance is less than about 1.4 R_0, more complicated analysis routines must be employed to determine the distance between the planes.[84,85]

Comments. An important point, too often overlooked, is that the interpretation of an energy transfer measurement relies entirely on the quality of the biochemical and chemical characterization of the sample. The spectral measurement is most accurate when the donor and acceptor populations are homogeneous. If heterogeneous, the fraction of donor and acceptor dyes in different states must be quantified both biochemically and spectroscopically. In the absence of a biochemical determination of the fraction of dyes in a particular functional and spectral state, distance calculations are meaningless.

Topology of Membrane-Bound Thrombin–TM Complex. In addition to altering the conformation of the active site of thrombin, TM also localizes the enzyme at the membrane surface, presumably positioning the thrombin at the optimal height above the surface to activate the membrane-bound substrate, protein C. The molecular architecture of the enzyme–cofactor complex can be discerned by using singlet–singlet energy transfer to provide information on the arrangement of the macromolecules and their domains in the complex. To determine the location of the thrombin active site in the thrombin–TM membrane complex, TM is reconstituted into large unilamellar phosphatidylcholine vesicles by dialysis, as described above, in two parallel samples, one of which contains the acceptor (e.g., octadecylrhodamine, or OR). The concentration of TM on the outer surface of these vesicles is determined enzymatically using Gla-domainless protein C as a substrate. The surface density of the OR is determined by measuring the absorbance at 564 nm and estimating the recovery of phospholipid in the purified vesicles by the recovery of radioactive phospholipid, which is added to the sample before dialysis, and making the assumptions detailed elsewhere.[81] DEGR-thrombin is added to the D and DA microcells, while an equivalent amount of DFP-treated thrombin is added to the A and B cuvettes. TM in vesicles lacking OR is then titrated into samples D and B, while TM in vesicles containing OR is titrated into samples DA and A. After an excess of TM is added to the samples, the net dilution-corrected donor intensities will be constant because all of the thrombin is bound to TM. Following this procedure and assuming a random orientation of the dyes ($\kappa^2 = 2/3$), the distance of closest approach between a dansyl in the active site of thrombin bound to TM and an OR at the membrane surface was calculated to be 66 Å.[81]

[85] T. G. Dewey and G. G. Hammes, *Biophys. J.* **32,** 1023 (1980).

Because thrombin is a globular protein with a diameter near 45 Å, it does not contact the membrane, but is located far above the membrane surface when bound to TM. TM therefore constitutes a platform to position the thrombin active site 66 Å above the membrane surface.

Conclusions

Current research into the nature of the protein C activation complex will require the integration of functional assays with physical measurements of the conformation and topography of the components of the activation complex. Crystallography and site-specific mutagenesis promise to contribute significantly to defining the structural basis for the acceleration of protein C activation elicited by thrombomodulin.

Acknowledgments

The research discussed herein was supported in part by grants awarded by the National Heart, Lung, and Blood Institute of the National Institutes of Health (Grant Nos. R37-HL30340 (to CTE) and R01-HL32934 (to AEJ). CTE is an investigator of the Howard Hughes Medical Institute.

[22] Protein C Inhibitor

By Koji Suzuki

Introduction

Protein C inhibitor (PCI) is a heparin-dependent inhibitor of activated protein C (APC), the most potent anticoagulant enzyme.[1,2] PCI was isolated first from human plasma[1] and then from bovine plasma.[3] In addition to APC, PCI inhibits some blood coagulation enzymes such as thrombin,[2,3] factor Xa,[2,3] plasma kallikrein,[3,4] and factor XIa.[3,4] Because PCI also in-

[1] K. Suzuki, J. Nishioka, and S. Hashimoto, J. Biol. Chem. 258, 163 (1983).
[2] K. Suzuki, J. Nishioka, H. Kusumoto, and S. Hashimoto, J. Biochem. (Tokyo) 95, 187 (1984).
[3] K. Suzuki, H. Kusumoto, J. Nishioka, and Y. Komiyama, J. Biochem. (Tokyo) 107, 381 (1990).
[4] J. C. M. Meijers, D. H. A. Kanters, R. A. A. Vlooswijk, H. E. van Erp, M. Hessing, and B. N. Bouma Biochemistry 27, 4231 (1988).

hibits tissue and urinary plasminogen activators (t-PA and u-PA), it is also known as plasminogen activator inhibitor-3 (PAI-3).[5,6]

In 1980, Marlar and Griffin found PCI in human plasma and speculated that the combined deficiency of factors V and VIII in some hemophiliacs was caused by a congenital deficiency of PCI.[7] However, plasma PCI antigen and PCI activity levels in such patients were similar to those of healthy controls, negating their hypothesis.[8,9] In repeatedly frozen and thawed plasma[10] or in outdated plasma,[11] PCI is proteolytically inactivated.

The normal concentrations of PCI in plasma and urine are 3.3–6.8 μg/ml (mean 5.3 μg/ml) and 0.5–1.0 μg/ml (mean 0.75 μg/ml), respectively.[12,13] The level of plasma PCI decreases in patients with liver dysfunction or disseminated intravascular coagulation (DIC).[12-16] In patients with DIC, or in those undergoing cardiopulmonary bypass with valve replacement or aortocoronary bypass, plasma PCI levels decrease nearly in parallel with that of protein C, and the APC–PCI complex is formed.[17] These findings suggest that PCI regulates the protein C anticoagulation system as a physiological inhibitor. α_1-Antitrypsin (α_1-AT) may also be a heparin-independent inhibitor of APC.[18-20] The plasma of patients with intravascular coagulation contains both APC–PCI and APC–α_1-AT complexes.[19] The half-life of human PCI administered to rabbits is 23.4 hr and that of the APC–PCI complex 19.6 min.[21] In urine, PCI occurs not only as free

[5] T. W. Stief, K.-P. Radtke, and N. Heimburger, *Biol. Chem. Hoppe-Seyler* **368**, 1427 (1987).
[6] M. J. Heeb, F. Espana, M. Geiger, D. Collen, D. C. Stump, and J. H. Griffin, *J. Biol. Chem.* **263**, 15813 (1987).
[7] R. A. Marlar and J. H. Griffin, *J. Clin. Invest.* **66**, 1186 (1980).
[8] W. M. Canfield and W. Kisiel, *J. Clin. Invest.* **70**, 1260 (1982).
[9] K. Suzuki, J. Nishioka, S. Hashimoto, T. Kamiya, and H. Saito, *Blood* **62**, 1266 (1983).
[10] J. E. Gardiner and J. H. Griffin, *Thromb. Res.* **36**, 197 (1984).
[11] M. Laurell and J. Stenflo, *Thromb. Haemostasis* **62**, 885 (1989).
[12] K. Suzuki, *Semin. Thromb. Hemostasis* **10**, 154 (1984).
[13] F. Espana and J. H. Griffin, *Thromb. Res.* **55**, 671 (1989).
[14] R. B. Francis and W. Thomas, *Thromb. Haemostasis* **52**, 71 (1984).
[15] R. A. Marlar, J. Endres-Brooks, and C. Miller, *Blood* **66**, 59 (1985).
[16] F. Espana, V. Vicente, I. Scharrer, D. Tabrnero, and J. H. Griffin, *Thromb. Res.* **59**, 593 (1990).
[17] K. Suzuki, Y. Deyashiki, J. Nishioka, and K. Toma, *Thromb. Haemostasis* **61**, 337 (1989).
[18] F. J. M. Van der Meer, N. H. van Tilburg, I. K. van der Linden, E. Briet, and R. M. Bertina, *Thromb. Haemostasis* **58**, 277 (1987).
[19] M. J. Heeb and J. H. Griffin, *J. Biol. Chem.* **263**, 11613 (1988).
[20] J. C. M. Meijers, R. A. A. Vlooswijk, D. H. A. J. Kanters, M. Hessing, and B. N. Bouma, *Blood* **72**, 1401 (1988).
[21] M. Laurell, J. Stenflo, and T. H. Carlson, *Blood* **76**, 2290 (1990).

form but also as a complex with u-PA.[22] PCI has been found in seminal plasma (150–220 μg/ml) at levels more than 40 times higher than that in blood plasma, and some of it is complexed with prostate-specific antigen (PSA).[23] PCI is also present in the testis, epididymis glands, prostate, Graafian follicle, and synovial fluids, and is interestingly absent from seminal plasma of patients with dysfunctional seminal vesicles.[24] These findings suggest that PCI plays a role in the regulation of the reproductive system. No congenital deficiencies of PCI have been reported.

Assay Methods

The assay is based on the inhibition of APC activity resulting from complex formation during preincubation with the inhibitor. Although plasma contains other inhibitors, that can inactivate APC, such as α_1-AT and perhaps α_2-macroglobulin, it can be assayed because the inhibitory rate of PCI for APC is specifically accelerated by heparin or negatively charged dextran sulfate.[2,3,25]

Activated Protein C Inhibitory Assay

Reagents

Buffer for substrate. 0.05 M Tris-HCl, pH 8.0, 0.10 M CsCl, 2 mM CaCl$_2$, and 5 μM of thrombin-specific inhibitor, MD-805 (argatroban)[26] [(2R,4R)-4-methyl-1-(N^2-(3-methyl-1,2,3,4-tetrahydro-8-quinolinyl) sul-fonyl)-L-arginyl)-2-piperidine carboxylic acid] or DAPA[27] [dansylarginine-N-(3-ethyl-1,5-pentanediyl)amide].

Substrate. Fluorogenic substrate; Boc-Leu-Ser-Thr-Arg-MCA (Protein Research Foundation, Osaka, Japan) or chromogenic substrate; L-pyroGlu-Pro-Arg-pNA (S-2366; Kabi, Stockholm, Sweden).

Enzyme. Human or bovine APC, from which protein C zymogen is purified from human[28] or bovine[29] plasma, then activated by thrombin

[22] D. C. Stump, M. Thienpont, and D. Collen, *J. Biol. Chem.* **261**, 1267 (1986).

[23] F. Espana, J. Gilabert, A. Estelles, A. Romeu, J. Aznar, and A. Cabo, *Thromb. Res.* **64**, 309 (1991).

[24] M. Laurell, A. Christensson, P.-A. Abrahamsson, J. Stenflo, and H. Lilja, *J. Clin. Invest.* **89**, 1094 (1992).

[25] F. Espana, M. Berrettini, and J. H. Griffin, *Thromb. Res.* **55**, 369 (1989).

[26] R. Kikumoto, Y. Tamao, T. Tezuka, S. Tonomura, H. Hara, K. Ninomiya, A. Hijikata, and S. Okamoto, *Biochemistry* **23**, 85 (1984).

[27] K. G. Mann, J. Elion, R. J. Butkowski, M. Downing, and M. E. Nesheim, this series, Vol. 80, p. 286.

[28] K. Suzuki, J. Stenflo, B. Dahlback, and B. Teodorsson, *J. Biol. Chem.* **258**, 1914 (1983).

[29] J. Stenflo, *J. Biol. Chem.* **251**, 355 (1976).

with thrombomodulin[30] in the presence of Ca^{2+}, by thrombin alone in the presence of EDTA,[28] or by snake venom protein C activator, Protac.[31] A stock solution of APC (5 μg/ml) is prepared in 0.05 M Tris-HCl, pH 8.0, 0.15 M NaCl, 0.1% BSA, and stored at $-80°$ until use.

Procedure. To 80 μl of buffer consisting of 0.05 M Tris-HCl, pH 8.0, 0.15 M NaCl, 2 mM EDTA, and 0.1% bovine serum albumin (BSA) containing heparin (5 U/ml), add 10 μl of either serially diluted plasma or samples containing PCI, and 10 μl of APC. After the mixture is incubated at 37° for 30 min, 50 μl of the reaction mixture is withdrawn and added to 450 μl of 0.2 mM fluorogenic substrate in substrate buffer in a cuvette. Changes in the intensity of fluorescence of 7-amino-4-methylcoumarin (AMC) liberated from the substrate are measured by fluorescence spectrophotometry with excitation at 380 nm and emission at 440 nm. When S-2366 is used as a substrate, 50 μl of the reaction mixture is added to 450 μl of 0.2 mM S-2366 in the substrate buffer in a cuvette, and changes in the absorbance of p-nitroaniline (pNA) liberated are measured at 405 nm. The specific activity of APC ranges from 2.2 to 2.4 μmol AMC or 1.7 to 1.8 μmol pNA liberated min^{-1} mg^{-1} of enzyme under these conditions. The PCI activity is expressed in terms of units of activity found in 1 ml of pooled normal plasma.

Purification Procedure

PCI is isolated from human and bovine plasma basically by the same procedure. The isolation of human PCI described here is a modification of the original procedure described by Suzuki *et al.*[1] All steps in the purification procedure are performed at 4°.

Starting Material. To freshly frozen human plasma (4 liters), add benzamidine chloride (10 mM final), diisopropyl fluorophosphate (DFP) (1 mM), phenylmethylsulfonyl fluoride (PMSF) (1 mM), soybean trypsin inhibitor (SBTI) (50 μg/ml), and aprotinin (100 U/ml). These abundant protease inhibitors are required in order to protect PCI from degradation mainly by plasma kallikrein.[11,20,32,33]

Barium Citrate Adsorption. Add 80 ml of 1 M $BaCl_2$ dropwise per liter of plasma with stirring for 30 min, then stir the mixture for 1 hr. The barium citrate precipitate is removed by centrifugation at 5000 rpm for 15 min. The barium citrate pellet is stored at $-80°$ as a source of vitamin K-dependent proteins involving protein C.

[30] N. L. Esmon, W. G. Owen, and C. T. Esmon, *J. Biol. Chem.* **257,** 859 (1982).
[31] C. L. Orthner, P. Bhattacharya, and D. K. Strickland, *Biochemistry* **27,** 2558 (1988).
[32] F. Espana, A. Estelles, J. H. Griffin, and J. Aznar, *Thromb. Haemostasis* **65,** 46 (1991).
[33] K.-P. Padtke, T. W. Stief, and N. Heimburger, *Biol. Chem. Hoppe-Seyler* **369,** 965 (1988).

Polyethyleneglycol Fractionation. Solid polyethylene glycol (PEG) 6000 is added (60 g/liter) to the supernatant resulting from the previous step, and the mixture is stirred for 1 hr. The precipitate is removed by centrifugation at 5000 rpm for 15 min. Solid PEG 6000 (60 g) is further added to each liter of supernatant. After the mixture is stirred for 1 hr, the 6–12% PEG precipitate is collected by centrifugation at 5000 rpm for 30 min.

Ammonium Sulfate Fractionation. The 6–12% PEG precipitate is dissolved in 500 ml of chilled 0.05 M Tris-HCl, 0.1 M NH$_4$Cl, pH 7.5, containing 10 mM benzamidine chloride, 1 mM DFP, 1 mM PMSF, 50 μg/ml SBTI, and 100 U/ml aprotinin. Solid ammonium sulfate is added to the solution to 50% saturation. After stirring for 1 hr, the mixture is centrifuged at 8000 rpm for 15 min. Solid ammonium sulfate is then added to the supernatant to make 70% saturation. The solution is then stirred for 1 hr and the precipitate is collected by centrifugation at 8000 rpm for 30 min.

Dextran Sulfate-Sepharose Chromatography. The precipitate from the previous step is dissolved in 0.05 M Tris-HCl, 0.1 M NaCl, pH 7.0, containing 1 mM benzamidine chloride, 1 mM DFP, 1 mM PMSF, 50 μg/ml SBTI, and 100 U/ml aprotinin and dialyzed overnight against the same buffer. The dialyzate is applied to a column of dextran sulfate-Sepharose (2.6 × 60 cm) equilibrated with the same buffer used for the dialysis. After washing the column with the same buffer, PCI is eluted with a linear gradient of NaCl (0.1–0.8 M in 1 liter) in the equilibration buffer at a flow rate of 50 ml/hr (Fig. 1). Activity of PCI or immunoreactive material against anti-PCI–IgG is determined and the proteins in the pooled fraction containing PCI activity or antigen are concentrated by solid ammonium sulfate (70% saturation).

Gel Chromatography on Ultrogel AcA 44. The precipitate from the previous step is collected by centrifugation at 10,000 rpm for 15 min and dissolved in a minimum volume of 0.05 M Tris-HCl, 0.15 M NaCl, pH 7.5, containing 1 mM benzamidine chloride, 1 mM DFP, 1 mM PMSF, 50 μg/ml SBTI, and 100 U/ml aprotinin. The sample is applied to a column of Ultrogel AcA 44 (1.5 × 80 cm) equilibrated with the same buffer used for dissolution, and eluted with the same buffer (Fig. 2). The flow rate is maintained at 4 ml/hr, and 1-ml fractions are collected. The fractions containing PCI activity or antigen are pooled and dialyzed overnight against 0.05 M Tris-HCl, 0.05 M NaCl, pH 7.5, containing 1 mM benzamidine chloride, 1 mM DFP, and 1 mM PMSF.

Heparin-Sepharose Chromatography. The dialyzate from the previous step is applied to a column of heparin-Sepharose CL-6B (0.9 × 10 cm) equilibrated with the same buffer for the dialysis. The column is washed with 10 ml of 0.05 M Tris-HCl, 0.2 M NaCl, pH 7.5, containing 1 mM

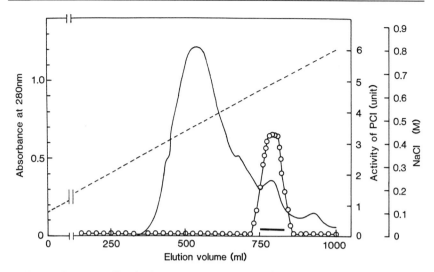

FIG. 1. Dextran sulfate-Sepharose chromatography of the 50–70% ammonium sulfate precipitate. Absorbance at 280 nm (—), activity of PCI (—○—), and NaCl concentration (---) were monitored. Fractions indicated by horizontal bar are pooled.

benzamidine chloride, 1 mM DFP, and 1 mM PMSF, and then with 10 ml of 0.05 M Tris-HCl, 0.2 M NaCl, pH 7.5, without protease inhibitors. PCI is then eluted with 0.05 M Tris-HCl, 0.5 M NaCl, pH 7.5 at a flow rate of 10 ml/hr, and 1 ml fractions are collected (Fig. 3). The fractions containing PCI activity or antigen are pooled and stored at −80° until use.

The purification of human PCI is summarized in Table I.

Similar purification procedures are presented by Meijers et al.,[20] Espana et al.,[25] and Padtke et al.[33] It is also possible to isolate PCI directly from plasma by affinity chromatography using monoclonal anti-PCI–IgG–Sepharose followed by gel filtration[34] or heparin-Sepharose chromatography.[11]

Properties

Physical Properties

Human PCI is a single-chain glycoprotein with a M_r of 57,000.[1] The absorption coefficient $E^{1\%}$ 280 nm is 14.1, and the isoelectric point (pI) is 4.5–6.0, suggesting microheterogeneity. Electrophoretically, human PCI is

[34] M. Laurrell, T. Carlson, and J. Stenflo, Thromb. Haemostasis 60, 334 (1988).

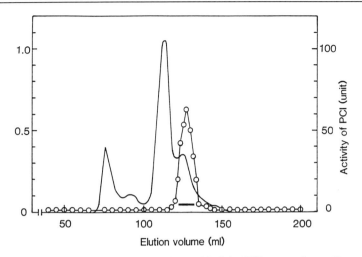

Fig. 2. Gel chromatography on Ultrogel AcA 44 of the 70% ammonium sulfate precipitate of fractions containing PCI obtained by dextran sulfate-Sepharose chromatography. Absorbance at 280 nm (—) and activity of PCI (—○—) are monitored. Fractions indicated by horizontal bar are pooled.

an α_1-α_2-globulin and differs in antigenicity, M_r, and enzyme inhibitory characteristics from other plasma serine protease inhibitors (serpin) such as antithrombin III (ATIII), heparin cofactor II (HCII), α_1-AT, α_1-antichymotrypsin (α_1-ACT), C1 esterase inhibitor, α_2-plasmin inhibitor, and plasminogen activator inhibitor-1 (PAI-1). PCI has affinity for heparin and dissociates from it in the presence of > 0.4 M NaCl.[17]

Kinetic Properties

PCI inhibits APC by forming a 1 : 1 complex. This complex formation is promoted by heparin (1–5 U/ml). It is not affected by other mucopolysaccharides such as chondroitin sulfate A and C, dermatan sulfate, heparan sulfate, and hyaluronic acid, but is markedly promoted by synthetic dextran sulfate (containing 18% sulfur, M_r 7500–200,000).[17] In the urinary tract, dermatan sulfate-containing glycosaminoglycans is suggested to stimulate uPA inhibitory activity of PCI.[35] PCI inhibits neither factor XIIa nor plasmin. Kinetic studies[2,20,25] show that the second-order rate constants (M^{-1} sec^{-1}) of human PCI for inhibition of (1) APC, (2) plasma kallikrein, (3) factor XIa, (4) factor Xa, (5) thrombin, (6) u-PA, and (7) t-PA are (1) 0.65×10^4, (2) 6.5×10^4, (3) 9.03×10^4, (4) 2.01×10^4, (5) 0.61×10^4, (6)

[35] M. Geiger, U. Prigliner, J. H. Griffin, and B. R. Binder, *J. Biol. Chem.* **266,** 11851 (1991).

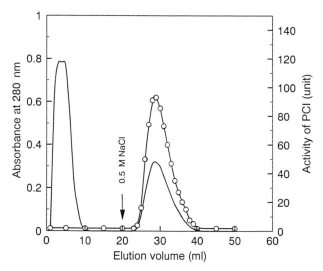

FIG. 3. Heparin-Sepharose chromatography of the dialyzate of pooled fractions containing PCI obtained by gel chromatography. Absorbance at 280 nm (—) and activity of PCI (—○—) are monitored. PCI is eluted with 0.5 M NaCl in 0.05 M Tris-HCl, pH 7.5.

0.22×10^4, and (7) 0.08×10^4 M^{-1} sec^{-1}, respectively, in the absence of heparin, and (1) 1.58×10^6, (2) 0.18×10^6, (3) 0.74×10^6, (4) 0.09×10^6, (5) 0.52×10^6, (6) 0.43×10^6, and (7) 0.03×10^6, respectively, in the presence of optimal concentration of heparin. Bovine PCI has similar second-order rate constants for the enzymes described above in the absence or presence of heparin.[3]

The mechanism of inhibition of APC by PCI is evaluated by SDS-polyacrylamide gel electrophoresis, followed by immunoblotting or autoradiography. Reacting PCI with APC results in the formation of a complex with a M_r of 102,000 and a proteolytic product of PCI with a M_r of 54,000. As is observed in other serpins, the reactive site of PCI is cleaved by the enzyme, and the peptide on the COOH-terminal side is released. Following its reaction with thrombin, factor Xa, u-PA, t-PA, and trypsin, PCI forms complexes with each factor and a proteolytic product of PCI with a M_r of 54,000. Plasma kallikrein and factor XIa form the respective complex and two modified PCIs with M_r of 54,000 and 52,000.[25] Two additional cleavage sites, besides a reactive site (Arg^{354}-Ser^{355}) have been determined in modified PCI, i.e., Arg^{357}-Leu^{358} and Arg^{362}-Leu^{363}.[11]

The enzyme–PCI complex is dissociated by nucleophilic reagents such as ammonia and hydroxylamine into the active enzyme and a proteolytic

TABLE I
PURIFICATION OF PROTEIN C INHIBITOR

Step	Volume (ml)	Total protein[a] (mg)	Total activity (units)	Specific activity[b] (units/mg)	Yield[b] (%)	Purification[b] (-fold)
Starting plasma	4000	240,000	4000	0.017	100	1
Barium citrate supernatant	4220	225,000	4000	0.018	100	1
6–12% PEG 6000 precipitate	910	60,000	2800	0.046	70	2.7
50–70% ammonium sulfate precipitate	1200	45,000	2600	0.058	65	3.4
Dextran sulfate–Sepharose	88	32.0	1020	31.9	26	1876
Ultrogel AcA 44	22	6.2	880	142.0	22	8353
Heparin-Sepharose	6	3.0	620	206.7	15.5	12,159

[a] Estimated from the absorbance at 280 nm, assuming $E_{1\,cm}^{1\%} = 10$.
[b] Calculated from the value of total protein and total activity.

product of PCI.[2] The bond between these factors may be an acyl bond such as that observed between ATIII and factor Xa or thrombin.[36]

Although equimolar concentrations of PCI and APC result in complete inhibition of APC activity, only 40% of the total PCI is found in the form of the APC–PCI complex and the remaining is in the form of a proteolytic product with a M_r of 54,000 when detected by immunoblotting after SDS-polyacrylamide gel electrophoresis and also by sandwich enzyme immunoassay specific for the complex.[17]

Structure of PCI

Amino Acid Sequence of PCI

The amino acid sequence of human PCI is deduced from the nucleotide sequence of PCI cDNA.[37] The cDNA structure and genomic DNA structure[38] suggest that the PCI mRNA is about 2.2 kilobase pairs and that the mature protein consists of 387 amino acid residues with a signal peptide of 19 amino acid residues (Fig. 4). As with α_1-AT and α_1-ACT, no disulfide

[36] I. Bork, C. M. Jackson, H. Jornvall, K. K. Lavine, K. Nordling, and W. J. Salsgiver, J. Biol. Chem. 257, 2408 (1982).
[37] K. Suzuki, Y. Deyashiki, J. Nishioka, K. Kurachi, M. Akira, S. Yamamoto, and S. Hashimoto, J. Biol. Chem. 262, 611 (1987).
[38] J. C. Meijers and D. C. Chung, J. Biol. Chem. 266, 15028 (1991).

```
                                                                   -19
CGAGCTCTGTGACCTTATGCTCCACACTAACTCTGGCAGAGCCTCCGTTTCCTCATAGAACAAAGAACAGCCACC  Met Gln Leu Phe Leu Leu
                                                                              ATG CAG CTC TTC CTC CTC    93

     -10                                          -1  +1                          10
Leu Cys Leu Val Leu Leu Ser Pro Gln Gly Ala Ser Leu  Ser Ser Arg His His Pro Arg Glu Met Lys Arg Val
TTG TGC CTG GTG CTT CTC AGC CCT CAG GGG GCC TCC CTT  AGC AGC AGG CAC CAC CCC CGG GAG ATG AAG AGA GTC   168

                                20                          30
Glu Asp Leu His Val Gly Ala Thr Val Ala Pro Ser Arg Asp Ser Arg Asp Thr Phe Asp Leu Tyr Arg Ala Leu
GAG GAC CTC CAT GTA GGT GCC ACG GTG GCC CCC AGC AGG GAC AGC AGA GAC ACC TTT GAC CTC TAC AGG GCC TTG   243

        40                              50                      60
Ala Ser Ala Ala Pro Ser Gln Asn Ile Phe Phe Ser Pro Val Ser Pro Val Ser Ile Met Ser Met Leu Ser Leu
GCT TCC GCT GCC CCC AGC CAG AAC ATC TTC TTC TCC CCT GTG TCC CCT GTG AGC ATC ATG TCC ATG CTC TCC CTG   318

                                70                      80
Gly Ala Gly Ser Ser Thr Lys Lys Gln Ile Leu Met Gly Phe Gln Leu Glu Gly Leu Leu Gln Gly Ser Lys Glu
GGG GCT GGG TCC AGC ACA AAG AAA CAG ATC CTC ATG GGC TTT CAG CTG GAG GGC CTC CTG CAG GGC TCA AAG GAG   393

            90                              100                     110
Leu His Arg Gly Phe Gln Leu Asn Gln Glu Leu Gln Asn Gln Pro Arg Asp Gly Phe Gln Leu Ser Leu Gly Asn
CTG CAC AGA GGC TTT CAG CTC AAC CAG GAA CTC CAG AAC CAG CCC AGA GAT GGC TTC CAG CTG AGC CTC GGC AAT   468

                                120                     130
Ala Leu Phe Thr Asp Val Leu Val Asp Gln Thr Phe Val Ser Ala Met Lys Thr Leu Tyr Leu Ala Asp
GCC CTT TTC ACC GAC GTA CTG GTA GAC CAG ACC TTC GTA AGT GCC ATG AAG ACG CTG TAC CTG GCA GAC          543

                                140                     150                     160
Thr Phe Pro Thr Asn Phe Arg Asp Ser Ala Gly Ala Met Lys Gln Ile Asn Asp Tyr Val Ala Lys Gln Thr Lys
ACT TTC CCC ACC AAC TTT AGG GAC TCT GCA GGG GCC ATG AAG CAG ATC AAT GAT TAT GTG GCA AAG CAA ACG AAG   618

                                170                     180
Gly Lys Ile Val Asp Leu Leu Asp Lys Leu Asn Lys Leu Asp Ser Asn Ala Val Val Ile Met Asn Tyr Ile Phe Lys
GGC AAG ATT GTG GAT CTC TTG GAC AAG CTT AAC AAG CTC GAT AGC AAT GCG GTG GTC ATC ATG AAT TAC ATC TTT AAA  693
```

394

```
                 190                          200                          210
Ala Lys Glu Thr Ser Phe Asn His Lys Gly Thr Gln Glu Gln Asp Phe Tyr Val Thr Ser Glu Thr Val Val
GCT AAG GAG ACA AGC TTC AAC CAC AAA GGC ACC CAA GAG CAA GAC TTC TAC GTG ACC TCG GAG ACT GTG GTG   768

                              220              230
Arg Val Pro Met Met Ser Arg Glu Asp Gln Tyr His Tyr Leu Leu Asp Arg Asn Leu Ser Cys Arg Val Val Gly
CGG GTA CCC ATG ATG AGC CGC GAG GAT CAG TAT CAC TAC CTC CTG GAC     AAC CTC TCC TGC AGG GTG GTG GGG   843

             240                     250                     260
Val Pro Tyr Gln Gly Asn Ala Thr Ala Leu Phe Ile Leu Pro Ser Glu Gly Lys Met Gln Gln Val Glu Asn Gly
GTC CCC TAC CAA GGC AAT GCC ACG GCT TTG ATT CTC CCC AGT GAG GGA AAG ATG CAG CAG GTG GAG AAT GGA   918

                 270                     280
Leu Ser Glu Lys Thr Leu Arg Lys Trp Leu Lys Met Phe Lys Arg Gln Leu Glu Leu Tyr Leu Pro Lys Phe
CTG AGT GAG AAA ACG CTG AGG AAG TGG CTT AAG ATG TTC AAG AGG CAG CTC GAG CTC TAC CTT CCC AAA TTC   993

             290                     300                     310
Ser Ile Glu Gly Ser Tyr Gln Leu Glu Lys Val Leu Pro Ser Leu Gly Ile Ser Glu Asn Val Phe Thr Ser His Ala
TCC ATT GAG GGC TCC TAT CAG CTG GAG AAA GTC CTC CCC AGT CTG GGG ATC AGT AAC GTC TTC ACC TCC CAT GCT   1068

                 320              330
Asp Leu Ser Gly Ile Ser Asn His Ser Asn Ile Gln Val Ser Glu Met Val His Lys Ala Val Glu Val Asp
GAT CTG TCC GGC ATC AGC AAC CAC TCA AAT ATC CAG GTG TCT GAG ATG GTG CAC AAA GCT GTG GAG GTG GAC   1143·

             340                     350                 →   360
Glu Ser Gly Thr Arg Ala Ala Ala Thr Gly Thr Ile Phe Thr Phe Arg Ser Ala Arg Leu Asn Ser Gln Arg
GAG TCG GGA ACC AGA GCG GCA GCC ACG GGG ACA ATC TTC ACT TTC AGG TCG GCC CGC CTG AAC TCT CAG AGC   1218

             370                     380          387
Leu Val Phe Asn Arg Pro Phe Leu Met Phe Ile Leu Val Asp Asn Asn Ile Leu Gly Lys Val Asn Arg Pro
CTA GTG TTC AAC AGG CCC TTT CTG ATG TTC ATT CTC GTG GAT AAC AAC ATC CTC GGC AAA GTG AAC CGC CCC   1293
```

Fig. 4. *(Continues)*

395

```
***
TGA GGTGGGGCTTCTCCTGAAATCTACAGGCCTCAGGGTGGGAGATGAAGGGGGCTATGCTATGGCCCATCTGTATGCTGGTAGCTAGTGATTTACAC    1391

AGGTTTAGTTGACTAATGAGGCATTACAAATAATATTACTCTCTATGATGATTGCTTCCACCCACACGACTGCAACATACAGGTGCCTTGGGGAAATGTGG    1490

AGAACATTCAATCTTGCCGTCACTATTCATCAATGAAGATTAGCACTGAGATCCAGAGAGGCTGGATGACTTGCTCAAGTTCACCAGCATGGTAGTGGC    1589

AAAGAGAGGTCCAGAGTCCTGGCCCTTGATGCCCAGCTCAGTGCCACAAAGCTCAGTAGGAGGGATGTTCCAGTGATGAGGGCCACCAGGAAGCACAG    1688

GTCCAAGGCTGGTCCCACACTTATCAGCAGCAACAACTGTCATCCTGCAGTTCATCCTGCAGTTCTGCAGTTCATCCTGCATGGGAAAAATGTTGGAATGGGAGTCTGAAATGGGGCTACTGTTTC    1787

AGTCCTAACGTGCTGTGTGACATTGGGACAACACTTTCCCTCTCTGGACCTCAGTTTCCCTCTGTATACAAGGATCAGATTCTTGCTGTGTGÁCCCAAGAA    1886

CTCCTGAAATCATATAGAAAGGCTGGGGTGGGCCCTGTCATTCGTGGTTGATTTCAATACACTCAAGTGCCATTCATCCTTTAAGAAAAACATCTGGAT    1985

ATCAAGGTGGAAATGGCCCATTTAATGATTGATTATATCATTTGTGGATATAGTTATAATCTGATGGGCCTGGCTGGGAGTGGAAGAGGGGAAGCCTT    2084

TTGCAAATAGTAGAGTGTCAGTTGCCAGTGCCAATGACTAACTTTTTGAATTCTATGTTGGCATTAACAATAAAGCATTTTGCAAACACTG    2175
```

Fig. 4. The complete nucleotide sequence of the cDNA of human PCI and the amino acid sequence deduced from it[37] and the genomic DNA sequence.[38] The amino acid sequence residues in the mature protein are numbered 1 to 387, and those in the putative signal peptide are − 1 to − 18; the initiation codon for Met is number −19. The numbering of the nucleotide sequence starts with the initiation codon of the first exon of the human PCI gene (T. Hayashi and K. Suzuki, 1992), and the AATAAA polyadenylation signal is double-underlined. The positions of the five potential glycosylation sites starting at amino acids 20, 39, 230, 243, and 319 are underlined. The reactive site peptide bond is indicated by an arrow.

bond is detected and only one cysteine residue is present per molecule. Three Asn-X-Ser (or Thr) sequences and two Thr (or Ser)-X-X-Pro sequences, which are possible binding sites for carbohydrate chains, are present. Because the M_r of the entire polypeptide is 43759, carbohydrate chains with a total M_r of 13,200 may be bound to some of the above binding sites for carbohydrate chains. The signal peptide domain is highly hydrophobic, and mature PCI protein may be synthesized by cleavage of the Leu-His bond. The sites of cleavage of the signal peptide of other inhibitors are Cys-His in ATIII,[39] Ala-Glu in α_1-AT,[40] and Pro-Asn in α_1-ACT,[41] suggesting that the cleavage site peptide bonds of plasma serpins are not constant.

The reactive site of PCI is located at Arg^{354}-Ser^{355} near the COOH terminus[37]; those of other serpin proteins are in the COOH terminus. The amino acid sequences around the reactive site of various serpin superfamily proteins are shown in Table II. The phenomenon that PCI inhibits not only APC but also thrombin may be caused by the same reactive site structure (Arg-Ser) as that of ATIII or the abnormal α_1-AT detected in α_1-AT Pittsburgh patients who have a tendency to bleed.[42]

Heparin-binding site in PCI is identified in the region involving residues 264–283 relatively near the COOH-terminus of the molecule in comparison to the site in ATIII or HCII.[43]

The amino-terminal sequence of bovine PCI[3] has a high degree of homology with that of human PCI (Fig. 5).

Structural Homology of PCI with Serpin Superfamily Proteins

Plasma serpins, except for α_2-macroglobulin, are structurally highly homologous. Chicken ovalbumin (OVALB),[44] angiotensinogen (ANGIO),[45] and thyroxine-binding globulin (TBG)[46] are also structurally highly homologous with serpins (> 28%), and are thus called members of the serpin superfamily. Based on the alignment of the amino acid se-

[39] T. Chandra, R. Stackhouse, V. J. Kidd, and S. L. C. Woo, *Proc. Natl. Acad. Sci. U. S. A.* **80,** 1845 (1983).
[40] G. L. Long, T. Chandra, S. L. C. Woo, E. W. Davie, and K. Kurachi, *Biochemistry* **23,** 4828 (1984).
[41] T. Chandra, R. Stackhouse, V. J. Kidd, K. J. H. Robson, and S. L. C. Woo, *Biochemistry* **22,** 5055 (1983).
[42] M. C. Owen, S. O. Brennan, J. H. Lewis, and R. W. Carrell, *N. Engl. J. Med.* **309,** 694 (1983).
[43] C. W. Pratt and F. C. Church, *J. Biol. Chem.* **267,** 8789 (1992).
[44] L. McReynolds, B. W. O'Malley, A. D. Nisbett, J. E. Fothergill, D. Givol, S. Fields, M. Robertson, and G. G. Brownlee, *Nature (London)* **273,** 723 (1978).
[45] R. Kageyama, H. Ohkubo, and S. Nakanishi, *Biochemistry* **23,** 3603 (1984).

TABLE II

Amino Acid Sequence around the Reactive Site in Serpin Superfamily Proteins

Species	Superfamily proteins	Target enzymes	Residues around reactive site								Ref.[a]
			P4	P3	P2	P1	P1'	P2'	P3'	P4'	
Human	Protein C inhibitor	Activated protein C, thrombin, kallikrein	Phe	Thr	Phe	Arg	Ser	Ala	Arg	Leu	1
Human	Antithrombin III	Thrombin, factor Xa	Ile	Ala	Gly	Arg	Ser	Leu	Asn	Pro	2
Human	Heparin cofactor II	Thrombin	Phe	Met	Pro	Leu	Ser	Thr	Gln	Val	3
Human	α_1-Antitrypsin	Elastase factor XIa, activated protein C	Ala	Ile	Pro	Met	Ser	Ile	Pro	Pro	4
Human	α_1-antichymotrypsin	Elastase	Ile	Thr	Leu	Leu	Ser	Ala	Leu	Val	5
Human	α_1-plasmin inhibitor	Plasmin	Ala	Met	Ser	Arg	Met	Ser	Leu	Ser	6
Human	Tissue plasminogen activator inhibitor-1	Tissue plasminogen activator, urokinase	Val	Ser	Ala	Arg	Met	Ala	Phe	Glu	7
Human	C1 inhibitor	Factor XIIa, C1 esterase	Ser	Val	Ala	Arg	Thr	Leu	Leu	Val	8
Human	Thyroxine-binding globulin	—	Glu	Val	Glu	Leu	Ser	Asp	Gln	Glu	9
Human	Angiotensinogen	—	Gln	Gln	Leu	Asn	Lys	Pro	Glu	Val	10
Baboon	α_1-Antitrypsin	Elastase, factor XIa	Ala	Ile	Pro	Met	Ser	Ile	Pro	Pro	11
Mouse	Contrapsin	Trypsin	Gly	Gly	Ile	Arg	Lys	Ala	Ile	Leu	12
Rat	Angiotensinogen	—	Gln	Gln	Pro	Gly	Ser	Pro	Glu	Val	13
Chicken	Ovalbumin	—	Val	Asp	Ala	Ala	Ser	Val	Ser	Glu	14
Barley	Protein Z	—	Gly	Val	Ala	Met	Ser	Met	Pro	Leu	15

[a] Key to references: (1) K. Suzuki, Y. Deyashiki, J. Nishioka, K. Kurachi, M. Akira, S. Yamamoto, and S. Hashimoto, *J. Biol. Chem.* **262**, 611 (1987). (2) T. Chandra, R. Stackhouse, V. J. Kidd, S. L. C. Woo, E. W. Davie, and K. Kurachi, *Proc. Natl. Acad. Sci. U.S.A.* **80**, 1845 (1983). (3) H. Ragg, *Nucleic Acids Res.* **14**, 1073 (1986). (4) G. L. Long, T. Chandra, S. L. C. Woo, E. W. Davie, and K. Kurachi, *Biochemistry* **23**, 4828 (1984). (5) T. Chandra, R. Stackhouse, V. J. Kidd, K. J. H. Robson, and S. L. C. Woo, *Biochemistry* **22**, 5055 (1983). (6) W. E. Holmes, L. Nelles, H. R. Lijnen, and D. Collen, *J. Biol. Chem.* **262**, 1659 (1987). (7) T. Ny, M. Sawdey, D. Lawrence, J. L. Millan, and D. J. Loskutoff, *Proc. Natl. Acad. Sci. U.S.A.* **83**, 6776 (1986). (8) A. E. Davis, III, A. S. Whitehead, R. A. Harrison, A. Dauphinais, G. A. P. Bruns, M. Cicardi, and F. S. Rosen, *Proc. Natl. Acad. Sci. U.S.A.* **83**, 3161 (1986). (9) I. L. Flink, T. J. Bailey, T. A. Gustafson, B. E. Markham, and E. Morkin, *Proc. Natl. Acad. Sci. U.S.A.* **83**, 7708 (1986). (10) R. Kageyama, H. Ohkubo, and S. Nakanishi, *Biochemistry* **23**, 3603 (1984). (11) K. Kurachi, T. Chandra, S. J. Friezner-Degen, T. T. White, T. T. White, S. L. C. Woo, and E. W. Davie, *Proc. Natl. Acad. Sci. U.S.A.* **78**, 6826 (1981). (12) R. E. Hill, R. K. Shaw, P. A. Boyd, H. Baumann, and N. D. Hastie, *Nature (London)* **311**, 175 (1984). (13) H. Ohkubo, R. Kageyama, M. Ujihara, T. Hirose, S. Inayama, and S. Nakanishi, *Proc. Natl. Acad. Sci. U.S.A.* **80**, 2196 (1983). (14) L. McReynolds, B. W. O'Malley, A. D. Nisbett, J. E. Fothergill, D. Givol, S. Fields, M. Robertson, and G. G. Brownlee, *Nature (London)* ... (15) I. Heigaard, S. K. Rasmussen, A. Brandt, and J. Svendson, *FEBS Lett.* **180**, 89 (1985).

```
            1       5          10          15          20            25
Bovine      R-R-S-Q-K-K-K-I-Q-E-V-P-P-A-V-T-T-A-P-P-G-S-R-D-F-
            *       * *                        *   * *       * * *
Human     H-R-H-H-P-R-E-M-K-K-R-V-E-D-L-H-V-G-A-T-V-A-P-S-S-R-R-D-F-
          1       5          10          15          20         25

            26        30           35          40          45        49
Bovine      V-F-D-L-Y-R-A-L-A-A-A-A-P-A-Q-N-I-F-F-X-P-L-X-I
            * * * * * * *   * * *   * * * * *   *       *
Human       Y-F-D-L-Y-R-A-L-A-S-A-A-P-S-Q-N-I-F-F-S-P-V-S-I
            30          35          40          45          50      53
```

FIG. 5. The NH$_2$-terminal amino acid sequences of bovine PCI and that of human PCI, for comparison.[3] The X indicates an amino acid that was not determined. Asterisks indicate positions where amino acids are identical in bovine and human PCI.

quences of proteins in this superfamily, in which each protein is aligned so that the number of gaps in the amino acid sequence is minimal, the degree of homology between PCI and other proteins in the serpin super family evaluated as the number of homologous amino acid residues is 42.3% for α_1-ACT, 41.6% for α_1-AT, 27.9% for ATIII, 26.4% for HCII, 26.9% for OVALB, and 23.2% for ANGIO. PCI appears to be a protein closely associated with α_1-ACT and α_1-AT.

The PCI gene is 13.0 kilobase pairs long and consists of five exons and four introns.[38] The same gene organization is observed in the genes of α_1-AT,[47] α_1-ACT,[48] HCII,[49] and ANGIO.[50] This suggests that PCI branched from a common ancestral protein involving these serpin superfamily proteins.

Acknowledgments

We thank Mr. J. Nishioka and Dr. T. Hayashi for their efforts in preparing this manuscript.

[46] I. L. Flink, T. J. Bailey, T. A. Gustafson, B. E. Markham, and E. Morkin, *Proc. Natl. Acad. Sci. U. S. A.* **83,** 7708 (1986).
[47] G. L. Long, T. Chandra, S. L. C. Woo, E. W. Davie, and K. Kurachi, *Biochemistry* **23,** 4828 (1984).
[48] J. Bao, R. N. Sifers, V. J. Kidd, F. D. Ledley, and S. L. C. Woo, *Biochemistry* **26,** 7755 (1987).
[49] H. Ragg and G. Preibisch, *J. Biol. Chem.* **263,** 12129 (1988).
[50] T. Tanaka, H. Ohkubo, and S. Nakanishi, *J. Biol. Chem.* **259,** 8063 (1984).

[23] Immunochemical Techniques for Studying Coagulation Proteins

By Richard J. Jenny, Terri L. Messier, Laurie A. Ouellette, and William R. Church

Introduction

Twenty years after the work by Anfinsen[1] on the folding of ribonuclease, antibodies are now routinely used as probes of protein structure. Specific antibodies can be prepared to any protein domain or peptide sequence and new technologies, including monoclonal antibodies, antibodies to synthetic peptides, and the genetic engineering of antibody idiotypes, have expanded the use of antibodies to all aspects of protein research. In blood coagulation and fibrinolysis, antibodies are versatile and powerful agents for immunoaffinity purification of proteins, for protein quantitation by radioimmunoassays and enzyme-linked immunosorbent assays (ELISA), as site-specific probes of metal ion- and activation-dependent epitopes, and as reaction-specific regulators of the macromolecular catalytic complexes of blood. Technical details for the production of monoclonal antibodies have been described elsewhere in this series.[2]

The purpose of this chapter is to provide and discuss some basic immunochemistry techniques which are applicable to the isolation and characterization of plasma proteins, more specifically, the plasma coagulation proteins. The following discussion will focus on three major topics: (1) selection and purification of an appropriate antibody; (2) immunoaffinity purification of coagulation proteins; (3) several antibody–antigen binding assays that have proven useful in the study of blood plasma proteins.

Selecting and Purifying Appropriate Antibody

Monoclonal versus Polyclonal

Generally, the type and amount of antibody required depends on the experimental approach and factors such as antibody and antigen availability, design of the immunoassay, and the limitations of cost and technical expertise. Because of the unlimited availability of a single-idiotype antibody, monoclonal antibodies are preferred for use in antigen purification, in studies on the conformational properties of the antigen, and in studies

---

[1] C. B. Anfinsen, *Science* **181**, 223 (1973).
[2] P. Parham, this series, vol. 92, p. 110.

on the functional activity of an enzyme. Polyclonal antisera are useful in quantitative immunoassays and in techniques such as blotting, where the conformational properties of antigenic determinants recognized by monoclonal antibodies are variable. Murine monoclonal and rabbit polyclonal antibodies produced to synthetic peptides frequently fail to recognize native antigen and are limited to studies where the conformational properties of the epitope are exposed and not masked. Polyclonal antibodies are the least desirable for immunoaffinity techniques, primarily due to their relative lack of specificity for a single epitope. In all cases, the use of purified antibodies is highly recommended and they should be used whenever possible. Immunoglobulin purification is straightforward and the use of purified antibody simplifies both the experimental design and interpretation of experimental results. Unfractionated antiserum or ascites fluid should be used with caution because contaminating rabbit or murine serum components could alter the outcome of experiments. Antibodies from nonimmune animals or antibodies to irrelevant antigens should also be used routinely as controls in all immunoassays.

Several factors influence the performance of an antibody with respect to immunochemical applications: (1) the affinity of the antibody–antigen complex; (2) the ability of the antibody to bind the native antigen under conditions of the binding reaction during immunoaffinity purification; and (3) the ability to dissociate the antibody–antigen complex under a set of conditions which are nondestructive to both the antibody and the antigen.

Purification of Murine Immunoglobulin from Ascites Fluid

The procedure described by Parham[2] with the following modifications is used for purifying murine immunoglobulin. Ascites fluid is centrifuged at 1000 g for 15 min at room temperature after harvesting to remove the cells and stored frozen at $-20°$ until used. After thawing, the crude ascites is centrifuged at 20,000 g for 20 min at 4° and the supernatant adjusted to 50% ammonium sulfate by addition of solid ammonium sulfate (29.1 g/100 ml of ascites). The solution is stirred for 1 hr at 4° and then centrifuged at 20,000 g for 20 min at 4°. The pellet is dissolved in a minimal volume of 20 mM Tris, 0.15 M NaCl, pH 7.4 buffer (TBS), centrifuged if necessary to remove insoluble material, and applied to a gel-filtration column (Sephacryl S-200, Pharmacia Piscataway, NJ; or Spectra/Gel AcA 34, Spectrum, Los Angeles, CA) equilibrated with TBS containing 0.02% (w/v) sodium azide at room temperature. Up to 50 ml of ascites is routinely chromatographed in a single run on a 2.5 × 100-cm column. For less than 10 ml of ascites, the 50% ammonium sulfate step is eliminated and ascites fluid is applied directly to the column following centrifugation. Volumes of 50 to 100 ml of ascites can be chromatographed on a larger column (5 × 100 cm) following ammonium sulfate precipitation. A typical column profile of absorbance at 280 nm versus fraction number is shown in Fig. 1A. Immunoglobulin-containing fractions are identified by immunoassay and sodium dodecyl sulfate-polyacrylamide gel electrophoresis (SDS-PAGE). For the chromatogram in Fig. 1A, fractions 35–40 contain immunoglobulin. Often the purity at this step exceeds 90% and for some applications is adequate. The immunoglobulin-containing fractions are pooled and the protein precipitated by addition of solid ammonium sulfate to 70% saturation (43.6 g/100 ml solution), stirred for 1 hr at 4°, and centrifuged at 20,000 g for 20 min at 4°. The pellet is dissolved in 5 mM Tris (pH 7.4) buffer and dialyzed overnight at 4° against the same buffer. The solution is centrifuged to remove insoluble material and applied at room temperature to a column of DEAE-cellulose (approximately 10 ml packed resin/100 mg protein) equilibrated with 5 mM Tris buffer. The column is washed until the absorbance is less than 0.01 and then eluted by gradient elution with 0 to 0.1 M NaCl in 5 mM Tris (pH 7.4) buffer. For a 10-ml column volume, a typical gradient is 150 ml of 5 mM Tris buffer in reservoir A and 150 ml 0.1 M NaCl, 5 mM Tris in reservoir B with a fraction size of 4–5 ml. The immunoglobulin peak elutes first and the albumin is removed by batch elution using 0.3 M NaCl in 5 mM Tris buffer. A typical elution profile is shown in Fig. 1B. The DEAE-cellulose column is washed with 1 M NaCl in 10 mM Tris (pH 7.4) buffer and stored at 4° in water containing of 0.02% sodium azide. The antibody is concentrated by adding solid ammonium sulfate to 70% saturation, the precipitate is centrifuged, and the pellet is redissolved in 50% (v/v) glycerol/water and stored at −20°. An alternative method for storing antibodies is lyophilization following dialysis versus 10 mM sodium phosphate, 0.15 M NaCl (pH 7.0) buffer. Aliquots of lyophilized antibodies are dissolved in water or water containing 0.02% (w/v) sodium azide before use. It is recommended that the stability of each antibody to lyophilization be checked first. Analysis of the purified immunoglobulin from the hybridoma αHFVIIa-13 by SDS-PAGE is shown in Fig. 1C. Antibody αHVIIa-13 is a murine IgG and shows a typical electrophoretic pattern of 150 kDa nonreduced (lane 2, Fig. 1C), and under reducing conditions a heavy-chain polypeptide of 50 kDa and a light-chain polypeptide of 25 kDa (lane 3, Fig. 1C).

For the occasional monoclonal antibody that elutes from DEAE-cellulose coincident with serum albumin, chromatography on Cibacron Blue resin (Sigma, St. Louis, MO; R-2507) is a simple alternative to anion-exchange chromatography. Immunoglobulin-containing fractions from the gel-filtration column are pooled, diluted with an equal volume of water,

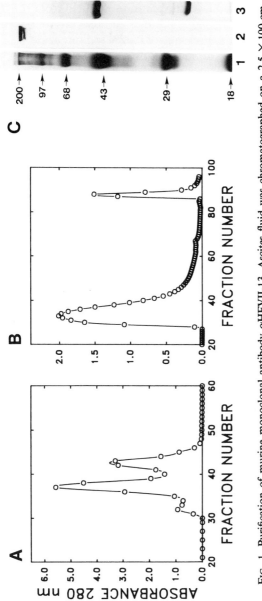

FIG. 1. Purification of murine monoclonal antibody αHFVII-13. Ascites fluid was chromatographed on a 2.5 × 100-cm Sephacryl S-200 column at room temperature (A). Fraction size was 6 ml. Immunoglobulin-containing fractions (fractions 35–40) were pooled, dialyzed versus 5 mM Tris (pH 7.4) buffer, and applied to a 2.5 × 5-cm DEAE-cellulose column equilibrated in 5 mM Tris at room temperature. (B) The column was eluted with a linear gradient of 0–0.1 M NaCl in 5 mM Tris (pH 7.4) buffer. Total gradient volume was 300 ml and 5-ml fractions were collected. The peak at fractions 25–45 is immunoglobulin. Murine albumin was eluted with 0.3 M NaCl in 5 mM Tris buffer (fractions 88–92). Purity of the monoclonal antibody was determined by sodium dodecyl sulfate-polyacrylamide gel electrophoresis and staining with Coomassie blue (C). Lane 1, molecular weight standards (GIBCO BRL, Gaithersburg, MD); lane 2, 10 μg purified immunoglobulin, nonreduced; lane 3, 10 μg purified immunoglobulin, reduced with 2% 2-mercaptoethanol.

and passed through a column of Cibacron Blue-Sephadex at room temperature. The flow-through contains immunoglobulin and is concentrated by ammonium sulfate precipitation. Murine albumin is eluted from the resin with 0.5 M NaCl in 10 mM Tris (pH 7.4) buffer and the column is washed with 1 M NaCl and stored in the presence of 0.02% sodium azide.

Purification of monoclonal antibody on protein A-Sepharose is sometimes a practical alternative, but use of protein A is generally limited to small amounts of antibody and appropriate immunoglobulin heavy-chain isotypes. Most monoclonal antibodies are IgG$_1$ isotype and do not bind well to protein A.[3]

Additional Points

The purification scheme described above is for total murine immunoglobulin. Depending on the relative amounts of hybridoma to normal mouse immunoglobulin present in the starting ascites fluid, the final purified immunoglobulin will contain a varying percentage of nonspecific antibody. This point may be important when quantitative measurements such as affinity constants are determined. If the specificity of the immunoglobulin is critical, antigen affinity columns are perhaps the best choice. For the case of the blood clotting proteins, prothrombin-Sepharose would be relatively easy to prepare, but factor VII-Sepharose would not be easy to prepare for most laboratories due to its limited availability. Unless there is a compelling reason, fractionation of murine immunoglobulin into specific and nonspecific antibody populations is of little practical value.

Polyclonal antibody preparations that have not been affinity purified are likely to contain subpopulations of antibodies that bind various immunoglobulins, or bind other antigen-related proteins. Furthermore, unless either affinity purified or raised against a small peptide antigen, polyclonal antibody preparations are likely to contain antibodies that react with a variety of epitopes within a single antigen. Reaction with multiple epitopes is unacceptable for immunoaffinity chromatography because it is nearly impossible to develop efficient elution techniques that release all bound antigen and preserve the integrity of both the antibody and the antigen. If a polyclonal antibody preparation is to be used in antigen purification, passing the sera over an antigen or peptide column and eluting the antibody increases the amount of specific antibody.

Antibody concentration is most conveniently determined by absorbance measurement at 280 nm and using an extinction coefficient ($E_{280 \ nm}^{1\%}$) of 14.0.

[3] P. C. Ey, S. J. Prowse, and C. R. Jenkin, *Immunochemistry* 15, 429 (1978).

Immunoaffinity Purification

For developing an antigen purification scheme, the most practical approach is to perform small-scale pilot studies. The information gained from a well-designed pilot study will not only determine if binding occurs, but can also provide information on the capacity of the resin, as well as efficient elution conditions. Scaled-down pilot studies can then be scaled up for large- or industrial-scale antigen production.

Immobilization of Antibody on Solid Support

A variety of techniques exist for immobilization of antibodies to a solid support such as agarose. There are many conflicting opinions as to which techniques yield the most stable and efficient product. By far, the most popular and most cited method of immobilization is the coupling of proteins to CNBr-activated matrices.[4] The popularity of this method is due to the fact that it is technically straightforward, inexpensive, and efficient. Critics of this method challenge the stability of the isourea bond which links the protein to the solid support.[5] Although this bond may be subject to nucleophilic attack in the presence of such agents as Tris or sodium azide, this only becomes a problem when there is prolonged exposure to strong alkaline conditions.

Most consistent success in antigen purification is achieved with antibody immobilized to CNBr-activated Sepharose CL-4B. Despite the reported instability of the isourea bond joining the protein to the solid support, we have observed minimal leeching of the antibody from the solid support even with extremely heavy use. Furthermore, we have noticed no degradation due to the activity of nucleophiles such as Tris and NaN_3, as long as neutral pH is maintained.

An alternative method of immobilization utilizes tosyl chloride-activated Sepharose.[6] Although less common than CNBr activation, this would be a method of choice if extremely harsh elution conditions, such as high pH (i.e., > 10), are employed.

Activation of Sepharose CL-4B with Cyanogen Bromide

The activation of agarose using cyanogen bromide is described in detail in the literature,[4] and will not be described here. Because CNBr is volatile at room temperature, all activation steps should be performed in a properly

[4] S. C. March, I. Parikh, and P. Cuatrecasas, *Anal. Biochem.* **60**, 149 (1974).
[5] J. Lasch and R. Koelsch, *Eur. J. Biochem.* **82**, 181 (1978).
[6] K. Nilsson and K. Mosbach, *Eur. J. Biochem.* **112**, 397 (1980).

ventilated fume hood. Preactivated Sepharose is also commercially available.

Coupling of Antibody to the Activated Support

The coupling of an antibody to the washed CNBr-activated matrix requires a free nucleophilic amino group and therefore the most extensive coupling occurs at a pH above the pK of the amino groups. However, to avoid losing antibody reactivity due to a loss of essential amino groups, overcoupling must be avoided. Overcoupling can be decreased by lowering the pH during the reaction, and thus minimizing the number of amino groups available for coupling. A suitable coupling buffer for most antibodies is 0.1 M sodium citrate, pH 6.5. Under these coupling conditions, a coupling efficiency of 80 to 90% is generally observed.

To initiate the coupling reaction, the antibody, which has been dialyzed into coupling buffer, is added to the activated resin, and the slurry is gently mixed at 4° for a minimum of 5 hr or overnight using a rotator or rocker platform. After mixing, the resin is allowed to settle, and an aliquot (approximately 1 ml) is removed and centrifuged for 1 min at 10,000 g at room temperature. The 280-nm absorbance of the supernatant is determined and the amount of antibody in the supernatant is calculated. The percent coupling efficiency can be determined. Productive coupling is usually represented by an 80 to 90% coupling efficiency. Reactive groups remaining on the resin are blocked by adding 0.1 volume of 1 M glycine, pH 8.5, or 0.1 volume of 1 M ethanolamine, pH 8.5, and allowing the slurry to mix for one additional hour at 4°. After blocking the remaining reactive groups, the resin is transferred to an appropriate chromatography column. Prior to using the immunoaffinity column for the first time, the column must be preeluted and reequilibrated. Because the appropriate elution conditions may not have yet been determined, and the stability of the affinity matrix to various elution conditions is unknown, the preelution is accomplished by applying several volumes of 0.02 M Tris, 1.5 M NaCl, pH 7.4. The column is then reequilibrated in the buffer of choice, which for most applications is TBS.

Preparation and Loading of Protein Sample

The conditions under which a sample is applied to an immunoaffinity column depend largely on the affinity of the antibody–antigen complex and hence the coupling density of the antibody as well as the concentration of the antigen in the sample medium. Furthermore, with diffusion-limited kinetics and an immobilized antibody, the interaction of an antigen with an immobilized antibody will be much slower than the corresponding solution-phase interaction. In theory, the sample concentration should be an order of magnitude higher than the dissociation constant of the antigen–antibody complex to bind effectively approximately 90% of the antigen. However, in practice, efficient binding to immobilized antibody often requires antigen concentrations that are 100 times that of the dissociation constant, mostly due to the kinetics of binding to the immobilized antibody.

The flow rate at which a sample is applied to an immunoaffinity column directly affects the efficiency of removing an antigen from the sample medium. Application rates of about 0.5 ml/min are generally a good starting point.

Protein samples for immunoaffinity chromatography should be contained in a buffer which mimics physiological pH and ionic strength, and should be free of insoluble materials or materials which could precipitate when applied to the affinity column. Buffers of choice include TBS, PBS (0.01 M Na_2HPO_4, 0.002 M KH_2PO_4, 0.14 M NaCl, 0.003 M KCl, pH 7.2) and HBS (0.02 M HEPES, 0.15 M NaCl, pH 7.4). If properties governing the antigen–antibody interaction are known, the composition of the loading buffer can be tailored to facilitate this interaction. Tailoring a buffer often requires changing the pH, changing the ionic strength, adding metal ion, or adding a chelating agent. These tailoring aspects are again best determined by small-scale empirical studies.

Adequate washing of the affinity column prior to elution is the single most important factor contributing to a homogeneous eluate. Too often this step is rushed or terminated too quickly, and nonspecifically adsorbed material is found in the sample eluate. Nonspecific binding can most often be eliminated by increasing the ionic strength of the wash buffer, by increasing the wash volume, and, if the sample loading is done at less than ambient temperature, by increasing the temperature of the wash step to near ambient. In general, most antigen–antibody complexes will not dissociate in the presence of 0.5 M NaCl, whereas most nonspecific ionic interactions will. Furthermore, relatively insoluble globular proteins that may be entrapped on the column are likely to be removed with elevated temperature and extensive washing.

Elution of Bound Antigen

Dissociation of the antigen–antibody complex, while maintaining the activity and/or integrity of both species, is the most challenging aspect of immunoaffinity isolation. Unfortunately, the development of an efficient elution technique is predominantly accomplished by trial and error. There are a variety of dissociation methods that one can experiment with, which range from very gentle to extremely harsh conditions with respect to their

TABLE I
ELUTION CONDITONS FOR IMMUNOAFFINITY CHROMATOGRAPHY

Type of reagent	Reagent composition[a]
High ionic strength	1.5 M Sodium chloride in 0.02 M Tris, pH 7.4
	3.0 M Sodium chloride in 0.02 M Tris, pH 7.4
	3.0 M Magnesium chloride in 0.02 M Tris, pH 7.4
Low ionic strength	Deionized water
Denaturant	3.0 M Sodium thiocyanate in 0.02 M Tris, pH 7.4
	2.0 M Guanidine hydrochloride in 0.02 M Tris, pH 7.4
	2.0 M Urea in 0.02 M Tris, pH 7.4
Organic solvent	50% Ethylene glycol
Low pH	0.1 M Glycine, pH 2.5
High pH	0.1 M Triethylamine, pH 11.5

[a] Indicates the highest concentration recommended, but does not ensure antigen or antibody stability.

effects on both the antigen and the affinity matrix. Techniques include increasing or decreasing the pH, increasing the ionic strength, adding a chelator or particular metal ion, adding a denaturant or chaotropic agent, or eluting with an organic solvent. Some of the most common elution techniques are listed in Table I.

The stability of both the antigen and the affinity matrix to a particular set of elution conditions may depend upon the duration of exposure to the elution buffer. To minimize prolonged exposure of the antigen several techniques can be employed. For example, when eluting under alkaline or acidic conditions, the effluent can be rapidly neutralized by collecting fractions into tubes containing a concentrated buffer such as 0.2 M Tris, pH 7.4. If a denaturant or chaotropic agent is used, immediate dialysis may allow the protein to regain a native conformation.

The vitamin K-dependent proteins provide an excellent example of proteins which are easily eluted and recovered from immunoaffinity chromatography.[7] These proteins may be eluted from affinity columns with 3 M NaSCN, and full activity is readily recovered following dialysis to remove the NaSCN. Figure 2 illustrates a Coomassie blue-stained gel of vitamin K-dependent proteins that were purified using immunoaffinity chromatography.

Immunoaffinity techniques have also proved useful for purification of recombinant coagulation proteins. Media from a transfected Chinese

[7] R. Jenny, W. Church, B. Odegaard, R. Litwiller, and K. Mann, *Prep. Biochem.* **16(3)**, 227 (1986).

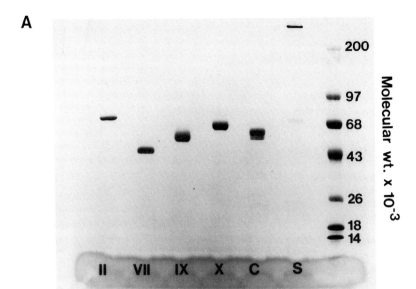

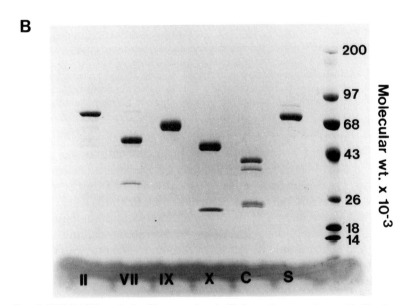

Fig. 2. SDS-PAGE analysis of human vitamin K-dependent proteins purified by immunoaffinity chromatography. Proteins eluted from immunoaffinity columns with 3 M NaSCN were dialyzed against TBS to remove the NaSCN, and samples (8 μg) were electrophoresed on SDS-PAGE with disulfide bonds intact (A), and following reduction with 2-mercaptoethanol (B). Proteins were then visualized by staining with Coomassie blue. The proteins include (left to right) prothrombin, factor VII, factor IX, factor X, protein C, and the complex of protein S with C4b binding protein. Reprinted from Jenny et al.[7] by courtesy of Marcel Dekker Inc.

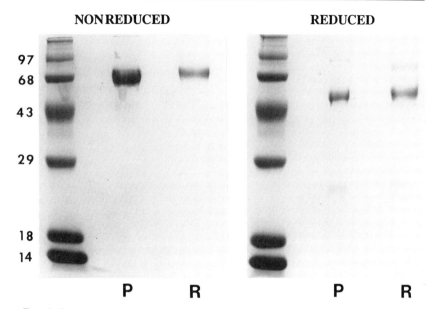

FIG. 3. Purified recombinant factor X analyzed by SDS-PAGE. Samples (2 μg) were electrophoresed on 10% polyacrylamide minigels and the gels were stained with Coomassie blue. The recombinant human factor X (lane R) purified from the CHO cell media appeared similar to immunoaffinity-purified plasma factor X (lane P) and consisted of a 45-kDa heavy-chain polypeptide disulfide bonded to a 17-kDa light-chain peptide. The recombinant factor X also contained a 74-kDa species previously shown to be single-chain factor X.

hamster ovary (CHO) cell line[8] was concentrated using a Millipore (Bedford, MA) Minitan concentration unit. The concentrate was chromatographed in 20 mM HEPES (pH 7.4) buffer on Pharmacia (Piscataway, NJ) Fast Flow QAE-Sepharose and eluted with HEPES buffer containing 0.5 M NaCl, 5 mM CaCl$_2$. Factor X-containing fractions were pooled and applied to an antifactor X monoclonal antibody immunoaffinity column. The factor X protein was eluted from the antibody column using 3 M NaSCN and concentrated by ammonium sulfate. The purity of the recombinant factor X was analyzed by sodium dodecyl sulfate-polyacrylamide gel electrophoresis (Fig. 3).

Additional points

Antibody columns are reusable many times, and some columns are several years old and have been used in excess of 50 times. Proper care and

---

[8] D. L. Wolf, U. Sinha, T. Hancock, P.-H. Lin, T. L. Messier, C. T. Esmon, and W. R. Church, *J. Biol. Chem.* **266**, 8384 (1991).

storage of an immunoaffinity column will greatly prolong its life span. By immediately reequilibrating an immunoaffinity column following the elution of bound antigen, prolonged exposure to harsh elution conditions can be avoided. Columns are stored in 0.02 M Tris, 0.15 M NaCl, pH 7.4, containing 0.02% sodium azide.

Immunoassays

Antibody–antigen interactions are dependent on the assay type and conditions used to measure the binding reaction. In the following section, three examples of immunoassays are described: (1) a solid-phase ELISA with antigen-coated microtiter plates for determining the cross-reactivity of a murine monoclonal antibody to factor VII with other vitamin K-dependent proteins; (2) a solid-phase competitive ELISA for quantitating human factor X; (3) and a solution-phase assay used with an antikringle monoclonal antibody produced to a synthetic peptide in prothrombin kringle 2.

Materials

Polystyrene 96-well microtiter plates (Corning 25801)
12-Channel pipettor, ELISA plate washer, and ELISA plate spectrophotometer
Murine or rabbit anticoagulation protein antibody (primary antibody); ascites fluid, serum, or purified antibody
Rabbit or goat antimurine or antirabbit immunoglobulin antibody (secondary antibody); divide into one-use aliquots, store frozen, thaw when needed
Sulfosuccinimidyl-6-(biotinamido)hexanoate (NHS-LC-biotin) (Pierce, Rockford, IL; 21335G)
Avidin peroxidase (Pierce 21123 G)

Solutions

50 mM Sodium carbonate buffer: 0.82 g Na_2CO_3, 1.43 g $NaHCO_3$ in 500 ml water; pH approximately 9.5
TBS: 10 mM Tris, 0.15 M NaCl (pH 7.4) buffer
1% BSA–TBS: 1 g bovine serum albumin in 100 ml TBS
0.1% BSA–TBS: 1/10 dilution 1% BSA–TBS in TBS
50 mM Citrate–phosphate buffer, pH 5.0: mix 0.1 M citric acid (24 ml) and 0.2 M sodium dibasic phosphate (add to pH 5.0, approximately 20–25 ml), add water to final volume of 100 ml
Chromogenic substrate: 4 mg o-phenylenediamine dihydrochloride (Sigma Chemical Co., P-1526) and 4 μl 30% hydrogen peroxide in 10 ml citrate–phosphate buffer

Cross-Reactivity of Monoclonal Antibody αHFVIIa-13 with Other
Vitamin K-Dependent Proteins

Due to the sequence homology of the vitamin K-dependent and plasma cofactor proteins of blood coagulation, the possibility exists that an antibody prepared against one purified blood clotting protein can bind a homologous epitope on other related proteins. Examples of this are described in the literature for both polyclonal and monoclonal antibodies.[9,10] Caution must be exercised when such cross-reactivity is observed to establish that other factors, such as contamination of antigen preparations with other proteins or the nonclonality of the hybridoma cell line, are not contributing to the observed antibody binding to antigen.

Aliquot 100 μl of protein antigen [10 μg/ml in carbonate (pH 9.6) buffer] into each well of 96-well microtiter plates. Let plates sit overnight at 4° or at room temperature for 2 hr. A concentration of 10 μg/ml is adequate for most proteins, but initial experiments using different concentrations of protein for coating should be performed for optimum results. Aspirate wells and wash once with TBS. Add 200 μl of 1% BSA–TBS to each well and let sit for 2 hr at room temperature to block nonspecific protein binding sites. If the antigen is stable, the plates can be frozen with 1% BSA–TBS for future use. Store at −20°. Serial dilute antibody into 0.1% BSA–TBS using a starting dilution for ascites fluid of 1/1000 (v/v) and for purified antibody of 20 μg/ml. Make 10–12 1 × 1 dilutions for each antibody. If solution conditions are to be varied, for example, by adding either 10 mM CaCl$_2$ or EDTA, dilute 1 M stock solutions into the 0.1% BSA–TBS solution. Remove the BSA blocking solution from the wells and wash once with TBS. Add 100 μl/well of the appropriate antibody dilution to duplicate or triplicate wells and let plate sit at room temperature for 1–4 hr or overnight at 4°. Optimum incubation time and temperature should be determined for each antibody–antigen pair. After incubation, aspirate wells and wash three times with TBS. Dilute second antibody into 0.1% BSA–TBS and add 100 μl of second antibody solution to each well. Optimum dilution of secondary antibody should be determined for each lot of reagent, but a typical working dilution is 1/2000–1/ 4000 (v/v). Let plate sit at room temperature for 1 hr, wash five times with TBS, and add the chromogenic substrate. Stop the reaction by adding 50 μl/well of 4 N H$_2$SO$_4$. Read the absorbance at 490 nm of each well using a microplate spectrophotometer and average the absorbances of identical wells. Plot A_{490} versus log of the dilution or the log of the antibody

[9] B. Furie and B. C. Furie, *J. Biol. Chem.* **254**, 9766 (1979).
[10] W. R. Church, T. Messier, P. R. Howard, J. Amiral, D. Meyer, and K. G. Mann, *J. Biol. Chem.* **263**, 6259 (1988).

concentration. Representative data for an antifactor VII monoclonal antibody are summarized in Fig. 4A. The antibody in this assay binds plasma factor VII and recombinant VIIa, but not prothrombin, factor X, or factor IX.

Controls should include wells with primary and secondary antibodies but no coated antigen; secondary antibody alone, no primary antibody or antigen; and antigen and secondary antibody but no primary antibody. Negative controls should include an ascites fluid or purified antibody to an irrelevant antigen. Do not use an antibody to another related plasma protein if possible. If rabbit polyclonal antisera are available for the antigen under consideration, serial dilutions (starting at 1/100 for antisera) can be used as a positive control. If no color develops in any of the wells, check the secondary antibody solution by mixing an aliquot directly with the chromogenic substrate. If color develops in all the wells or in the negative control wells, repeat the assay, changing one or more of the conditions, such as the adsorbed antigen concentration, the BSA concentration in the blocking buffer, or the primary or secondary antibody concentration.

This assay is convenient, reliable, and can readily accommodate a large number of samples. The assay is also useful for some antibody–synthetic peptide combinations using a coating concentration of 2 μM peptide in carbonate buffer. The assay is not appropriate for all peptide antigens and for those protein antigens that undergo plastic- or surface-dependent conformational changes. Adsorption of the protein antigen onto polystyrene can alter the expression of some epitopes.

Quantitative ELISA for Human Factor X

This assay uses biotin-labeled factor X and an immobilized monoclonal antibody to factor X. Biotin labeling of the antigen eliminates the use of radioactive material, but radioiodinated protein can be substituted if necessary. As with any chemical modification, the effect of labeling the protein must be a consideration and may influence the results.

Purified protein is labeled with NHS-LC-biotin. Dialyze 250 μg protein overnight at 4° versus 0.1 M sodium bicarbonate (pH 7.8) buffer. Remove from dialysis and add 0.1 M sodium bicarbonate buffer to a final volume of 1.0 ml. Dissolve the NHS-LC-biotin in water to a final concentration of 1 mg/ml and add 100 μl to the protein solution and let sit on ice for 2 hr. Dialyze overnight at 4° versus 0.1 M sodium phosphate (pH 7.0) buffer. Remove biotin-labeled protein from dialysis and dilute with 1% BSA–TBS to final BSA concentration of 0.1% BSA, or dilute 1:1 with glycerol and store at −20° in aliquots. In preliminary experiments, dilute the biotin-labeled protein to determine an optimum dilution to use in the immunoassay. Determination of the optimum antibody concentration to coat the

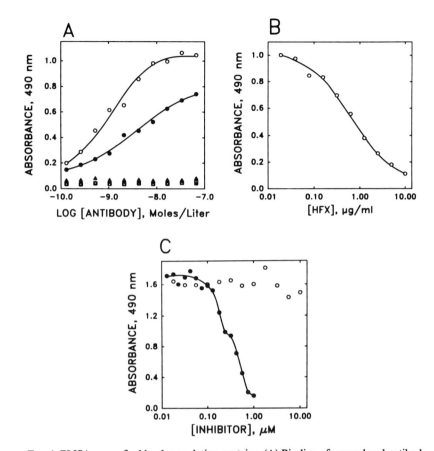

FIG. 4. ELISA assays for blood coagulation proteins. (A) Binding of monoclonal antibody αHFVIIa-13 to recombinant factor VIIa (O), plasma factor VII (●), human prothrombin (□), human factor IX (△), and human factor X (▲). Dilutions of purified monoclonal antibody were incubated in wells of 96-well microtiter plates coated with purified proteins. Following incubation and washing, antigen–antibody complexes were detected by a second incubation with goat antimurine immunoglobulin conjugated with horseradish peroxidase, followed by development with chromogenic substrate. The presence of bound antibody was detected by measuring the absorbance at 490 nm using a V_{max} spectrophotometer. (B) Binding of monoclonal antibody αHFX-27 to biotin-labeled human factor X in the presence of unlabeled factor X. Varying concentrations of unlabeled factor X (indicated on the abscissa) were mixed with a constant amount of biotin-labeled factor X. Aliquots of the mixture were incubated in wells of 96-wells tissue culture plates coated with antibody αHFX-27. The wells were washed and the presence of antigen–antibody complexes were determined by a second incubation with avidin–peroxidase followed by chromogenic substrate. (C) Binding of antibody αHII-5 to prothrombin and plasminogen in solution. Varying concentrations of prothrombin (●) and plasminogen (O) were incubated with antibody of αHII-5 at room temperature. Aliquots of the antigen–antibody mixture were transferred to wells of 96-well microtiter plates coated with prothrombin. After a 20-min incubation at room temperature the wells were washed and bound antibody was determined by a second incubation with goat antimurine imunoglobulin conjugated with horseradish peroxidase and chromogenic substrate. The data are plotted as absorbance due to chromogenic substrate hydrolysis *versus* the concentration of prothrombin or plasminogen in the first incubation.

96-well plates is recommended. Typical working dilutions are 10 μg/ml antibody concentration and 1/5000 dilution of the biotin-labeled protein. Dilute purified monoclonal antibody into carbonate buffer to a final concentration of 10 μg/ml and pipette 100 μl/well in 96-well microtiter plates. Let the plate sit for 2 hr at room temperature or overnight at 4°. Aspirate the antibody solution, rinse the wells once with TBS, and fill the well with 1% BSA–TBS for 2 hr at room temperature. Using Nunc Minisorp tubes (Thomas Scientific, Swedesboro, NJ, 6107-G15) dilute purified antigen, plasma, or other antigen-containing sample in 0.1% BSA–TBS and add an equal volume of biotin-labeled protein at twice the desired final concentration in 0.1% BSA–TBS. Aspirate albumin–blocking solution and add 0.1 ml of each antigen solution into duplicate antibody–coated wells. Incubate from 2 to 4 hr at room temperature or 1 hr at 37°. Wash wells 5 times with TBS and add avidin peroxidase (0.2 μg/ml in 0.1% BSA–TBS). Incubate at room temperature for 45 min, wash 3 times with TBS, and then add chromogenic substrate. After 10 min stop the reaction by adding 50 μl/well 4 N H$_2$SO$_4$. Average duplicate wells and plot absorbance at 490 nm versus log of the inhibitor concentration. The linear portion of the curve is used for extrapolation of unknown antigen-containing samples. Representative data for an antihuman factor X monoclonal antibody is shown in Fig. 4B.

Solution-Phase Immunoassay for Anti-kringle Monoclonal Antibody

Some of the limitations of the solid-phase ELISA are overcome using assay conditions wherein both antigen and antibody are soluble and binding occurs in solution.[11] This assay is useful when the competing antigen is a small protein fragment or synthetic peptide and for comparing the relative binding of homologous proteins. This type of assay can also be used for quantitative measurements of antigen concentration. The assay consists of two stages. In the first stage, a constant amount of antibody is mixed and incubated with varying concentration of protein antigen. Aliquots of the antigen–antibody mixture are then transferred to antigen-coated wells of a microtiter plate, incubated for a relatively short period of time, and then the presence of antibody–antigen complexes determined using secondary antibody and chromogenic substrate. The signal in this assay corresponds with free antibody from the solution antigen–antibody incubation. In the following example, a monoclonal antibody produced against a synthetic peptide representing prothrombin kringle 2 is shown to bind native prothrombin, but not plasminogen, in solution.[12]

[11] B. Friguet, A. F. Chaffotte, L. Djavadi-Ohaniance, and M. E. Goldberg, *J. Immunol. Methods* **77**, 305 (1985).
[12] W. R. Church, L. A. Ouellette, and T. L. Messier, *J. Biol. Chem.* **266**, 8384 (1991).

Coat wells of a 96-well microtiter plate with protein antigen (2 μg/ml in carbonate buffer, 130 μl/well). Let the plate sit at 4° overnight. Aspirate wells and block nonspecific protein binding sites with 2% BSA – TBS (250 μl/well) for 2 hr at room temperature. Using Nunc Minisorp tubes, dilute protein antigen into antibody solution. Optimum concentration of antibody must be determined; for the data summarized in Fig. 4C, the antibody concentration was 0.5 nM. Both antibody and antigen dilutions are conveniently made from stock solutions for which the protein concentrations are determined by absorbance measurements at 280 nm. The antigen concentration range must also be determined experimentally and a convenient starting concentration is 10 μM. Serial dilutions are made into antibody – containing 0.1% BSA – TBS. Following a 3-hr incubation at room temperature, 100 μl of antibody – antigen solution is transferred to duplicate or triplicate wells of antigen-coated microtiter plates. Aspirate the 2% BSA – TBS solution from the wells before using and pipette the samples rapidly. Let the plate sit for 30 min at room temperature; wash the wells 4 times with TBS. Add horseradish peroxidase-conjugated second antibody and incubate for 30 min at room temperature; wash 4 times with TBS. Add chromogenic substrate. Allow the reaction to develop for 5 – 10 min and stop the reaction by adding 50 μl 4 N H$_2$SO$_4$ to each well. Read the absorbance of each well at 490 nm, average the duplicate samples, and plot absorbance at 490 nm versus log of inhibitor concentration. Representative data are shown in Fig. 4C for the antiprothrombin monoclonal antibody αHII-5.[12] The antibody is to an epitope on prothrombin kringle 2 and binds prothrombin but does not bind the kringle-containing protein plasminogen in solution.

[24] Isolation of Intact Modules from Noncatalytic Parts of Vitamin K-Dependent Coagulation Factors IX and X and Protein C

By CARMEN VALCARCE, EGON PERSSON, JAN ASTERMARK, ANN-KRISTIN ÖHLIN, and JOHAN STENFLO

Introduction

Factors IX and X and protein C are zymogens of vitamin K-dependent serine proteases. The active form of factor IX, factor IXa, is part of the macromolecular complex (factor IXa, factor VIIIa, phospholipid, and calcium ions) that activates factor X, and factor Xa is part of the macromolecular complex (factor Xa, factor Va, phospholipid, and calcium ions) that activates prothrombin.[1-5] The active form of protein C functions as an anticoagulant enzyme by selectively inactivating the active forms of the cofactors, factors Va and VIIIa.[6] The three zymogens have an identical modular structure containing a carboxy-terminal serine protease that constitutes approximately two-thirds of the molecule. The noncatalytic third of each molecule consists of an amino-terminal, vitamin K-dependent γ-carboxyglutamic acid (Gla)-containing module, followed by two modules that are homologous to the epidermal growth factor (EGF). The amino-terminal EGF module of all three proteins contains one β-hydroxyaspartic acid (Hya) residue, and in factor IX this module also contains a unique O-linked di- or trisaccharide side chain containing xylose.[7]

The involvement of the Gla modules in calcium ion and phospholipid interactions is characteristic of vitamin K-dependent proteins.[8-10] In contrast, little is known about the function of the EGF modules.[7] Experiments to elucidate the structure and function of EGF modules are advancing along four lines of inquiry. First, studies of mutant factor IX molecules from patients with hemophilia B, with a biological activity from < 1 to 10% of normal, have led to the identification of several point mutations in the amino-terminal EGF module.[7,11] Some of these mutations seem to affect a Gla-independent calcium ion binding site. Second, intact recombinant coagulation factors, either with point mutations or with entire modules exchanged, have been expressed and their properties studied.[12-15] For instance, recombinant factor IX with the amino-terminal EGF module replaced by the corresponding module in factor X has been studied. Third, individual EGF modules have either been expressed in tissue culture or

[1] C. M. Jackson and Y. Nemerson, *Annu. Rev. Biochem.* **49**, 765 (1980).
[2] K. G. Mann, R. J. Jenny, and S. Krishnaswamy, *Annu. Rev. Biochem.* **57**, 915 (1988).
[3] K. G. Mann, M. E. Nesheim, W. R. Church, P. Haley, and S. Krishnaswamy, *Blood* **76**, 1 (1990).
[4] Y. Nemerson, *Blood* **71**, 1 (1988).
[5] E. W. Davie, K. Fujikawa, and W. Kisiel, *Biochemistry* **30**, 10363 (1991).
[6] C. T. Esmon, *J. Biol. Chem.* **264**, 4743 (1989).
[7] J. Stenflo, *Blood* **78**, 1637 (1991).
[8] J. Stenflo and J. W. Suttie, *Annu. Rev. Biochem.* **46**, 157 (1977).
[9] J. W. Suttie, *Annu. Rev. Biochem.* **54**, 459 (1985).
[10] B. Furie and B. C. Furie, *Blood* **75**, 1753 (1990).
[11] F. Giannelli, P. M. Green, K. A. High, S. Sommer, J. N. Lozier, D. P. Lillicrap, M. Ludwig, K. Olek, P. H. Reitsma, M. Goossens, A. Yoshioka, and G. G. Brownlee, *Nucleic Acids Res.* **19**, 2193 (1991).
[12] D. J. G. Rees, I. M. Jones, P. A. Handford, S. J. Walter, M. P. Esnouf, K. J. Smith, and G. G. Brownlee, *EMBO J.* **7**, 2053 (1988).
[13] S. W. Lin, K. J. Smith, D. Welsch, and D. W. Stafford, *J. Biol. Chem.* **265**, 144 (1990).
[14] W. F. Cheung, D. L. Straight, K. J. Smith, S. W. Lin, H. R. Roberts, and D. W. Stafford, *J. Biol. Chem.* **266**, 8797 (1991).
[15] J. R. Toomey, K. J. Smith, and D. W. Stafford, *J. Biol. Chem.* **266**, 19198 (1991).

chemically synthesized, and used in structural and metal ion binding studies.[16-18] Fourth, vitamin K-dependent proteins have been subjected to proteolytic cleavage and intact EGF modules have been isolated and studied. Initially, chymotrypsin was used to cleave between the Gla module and the amino-terminal EGF module in factor X[19] and later to remove the Gla module selectively from factors VII[20] and IX,[21] and proteins C[22] and Z.[23] The enzymatic approach has now been adopted to isolate individual modules from the noncatalytic parts of vitamin K-dependent coagulation factors containing EGF modules.[24-30]

Intact EGF modules can be isolated from controlled proteolytic digests of factors X (Fig. 1)[24,25] and IX (Fig. 2)[26,27] and protein C.[28-30] It is also possible to obtain the EGF modules linked to the Gla module. In some instances reversible chemical modification of Lys residues is required prior to digestion. With this approach, in contrast to methods that use expression of the modules in prokaryotes or chemical synthesis of individual modules, the postribosomal amino acid modifications do not pose a problem. The fragments that contain one or two EGF modules linked to the Gla module retain the calcium and phospholipid binding properties, which is useful in studies aimed at elucidating the function of the EGF modules.

Bovine factors IX[31] and X[32] and protein C[33,34] can be purified in large amounts by standard procedures. All operations described below are performed at room temperature unless otherwise stated.

[16] P. A. Handford, M. Baron, M. Mayhew, A. Willis, T. Beesley, G. G. Brownlee, and I. D. Campbell, *EMBO J.* **9,** 475 (1990).
[17] L. H. Huang, X. H. Ke, W. Sweeney, and J. P. Tam, *Biochem. Biophys. Res. Commun.* **160,** 133 (1989).
[18] P. A. Handford, M. Mayhew, M. Baron, P. R. Winship, I. D. Campbell, and G. G. Brownlee, *Nature (London)* **351,** 184 (1991).
[19] T. Morita and C. M. Jackson, *J. Biol. Chem.* **261,** 4015 (1986).
[20] T. Sakai, T. Lund-Hansen, L. Thim, and W. Kisiel, *J. Biol. Chem.* **265,** 1890 (1990).
[21] T. Morita, B. S. Isaacs, C. T. Esmon, and A. E. Johnson, *J. Biol. Chem.* **259,** 5698 (1984).
[22] N. L. Esmon, L. E. DeBault, and C. T. Esmon, *J. Biol. Chem.* **258,** 5548 (1983).
[23] T. Morita, H. Kaetsu, J. Mizuguchi, S. I. Kawabata, and S. Iwanaga, *J. Biochem. (Tokyo)* **104,** 368 (1988).
[24] E. Persson, M. Selander, S. Linse, T. Drakenberg, A.-K. Öhlin, and J. Stenflo, *J. Biol. Chem.* **264,** 16897 (1989).
[25] E. Persson, I. Björk, and J. Stenflo, *J. Biol. Chem.* **266,** 2444 (1991).
[26] J. Astermark, I. Björk, A.-K. Öhlin, and J. Stenflo, *J. Biol. Chem.* **266,** 2430 (1991).
[27] J. Astermark, P. J. Hogg, I. Björk, and J. Stenflo, *J. Biol. Chem.* **267,** 3249 (1992).
[28] A.-K. Öhlin and J. Stenflo, *J. Biol. Chem.* **262,** 13798 (1987).
[29] A.-K. Öhlin, S. Linse, and J. Stenflo, *J. Biol. Chem.* **263,** 7411 (1988).
[30] A.-K. Öhlin, I. Björk, and J. Stenflo, *Biochemistry* **29,** 644 (1990).
[31] K. Fujikawa and E. W. Davie, this series, Vol. 45, p. 74.
[32] K. Fujikawa and E. W. Davie, this series, Vol. 45, p. 89.
[33] J. Stenflo, *J. Biol. Chem.* **251,** 355 (1976).
[34] W. Kisiel and E. W. Davie, this series, Vol. 80, p. 320.

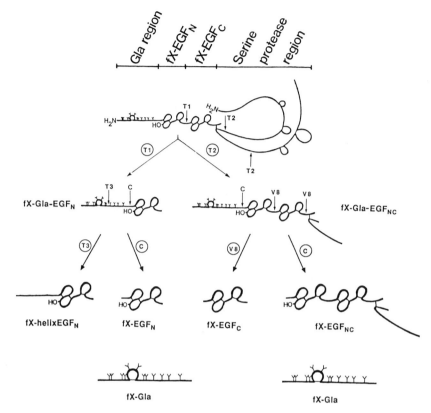

FIG. 1. Schematic illustration of the procedure for isolation of fragments from bovine factor X. The intact protein is citraconylated prior to tryptic digestion to avoid cleavage at Lys residues. High trypsin concentrations and long digestion times result in a cleavage between the two EGF modules with the production of fX-Gla-EGF$_N$ (T1). At lower trypsin concentrations and shorter digestion times, fX-Gla-ECF$_{NC}$ is obtained (T2). These fragments are isolated and the lysine blocking groups removed. Chymotryptic digestion (C) results in the removal of the Gla module from either fragment and the production of fX-EGF$_N$ and fX-EGF$_{NC}$, respectively. fX-EGF$_C$ is obtained by digestion of fX-Gla-EGF$_{NC}$ with V8 protease (V8). Decarboxylation and citraconylation of fX-Gla-EGF$_N$ followed by tryptic digestion (T3) result in the isolation of fX-helix-EGF$_N$.

Isolation and Characterization of EGF Module-Containing Fragments from Factor X

Isolation of Two EGF Modules Linked to Gla Module (fX-Gla-EGF$_{NC}$)

Careful tryptic cleavage of citraconylated factor X (fX) yields fX-Gla-EGF$_{NC}$. The Gla module remains intact, as trypsin does not cleave Arg-Gla peptide bonds. Prior to the tryptic cleavage, the lysine residues must be reversibly modified with citraconic anhydride[35] to avoid internal cleavages (see Fig. 1).

Citraconylation of Factor X. Bovine factor X (≈ 1 mg/ml) is dialyzed against 0.1 M borate buffer, pH 8.2, to remove Tris and any other compounds containing primary amino groups. After adjusting pH to 8.7 with 2 M NaOH, citraconic anhydride (250 μl/100 mg protein) is added in 3 aliquots over a period of 10 min. The pH of the stirred solution is kept at 8.7 by small additions of 2 M NaOH. After about 40 min, the sample is dialyzed against 50 mM Tris-HCl, pH 8.0, containing 0.1 M NaCl and 5 mM EDTA.

Tryptic Digestion. The citraconylated protein (≈ 1 mg/ml) is digested with 5 μg trypsin (EC 3.4.21.4, type XIII, Sigma, St. Louis, MO) per milligram of factor X for 10 min. The reaction is terminated by adding diisopropyl fluorophosphate (DFP) to a final concentration of 2 mM. Longer digestion time results in an internal cleavage in the second EGF module (between Arg-104 and Ser-105).[25] This cleavage does not result in an altered electrophoretic mobility of the fragment, as the carboxy-terminal part remains linked by a disulfide bridge, but a new amino terminus appears in the sequence analysis.

Ion-Exchange Chromatography. The tryptic digest is chromatographed on a Q-Sepharose Fast Flow column (1.6 $\times$ 13 cm) (Pharmacia, Uppsala, Sweden), on top of which there is a 1-cm layer of Sepharose 4B with immobilized soybean trypsin inhibitor (≈ 6 mg soybean trypsin inhibitor per milliliter of gel). The column is equilibrated with 50 mM Tris-HCl, pH 8.0, containing 0.1 M NaCl, and eluted with a linear gradient from 0.1 to 0.7 M NaCl (2 $\times$ 250 ml) in the same buffer. The flow rate is 35 ml/hr, and fractions of 7 ml are collected. fX-Gla-EGF$_{NC}$ is identified by SDS-PAGE and by Gla and Hya measurements after alkaline and acid hydrolysis, respectively.[36,37] The fragment elutes with about 0.45 M NaCl. Fractions containing fX-Gla-EGF$_{NC}$ are pooled, and the pH is lowered to 3.0 with glacial acetic acid to remove the lysine blocking groups. The sample is stirred slowly overnight. The protein precipitates at low pH, but dissolves readily when pH is adjusted to 8.0 before gel filtration.

Gel Filtration. The pool containing fX-Gla-EGF$_{NC}$ is concentrated to 2 ml using a Filtron Novacell (Filtron Technology Corp., Clinton, MA) (cutoff level 8 kDa), and loaded on a Sephadex G-75 Superfine column (1.5 $\times$ 97 cm) equilibrated either with 0.1 M NH$_4$HCO$_3$ or with 50 mM

[35] M. Z. Atassi and F. S. A. Habeeb, this series, Vol. 25, p. 546.
[36] P. Fernlund and J. Stenflo, *J. Biol. Chem.* **250,** 6125 (1975).
[37] P. Fernlund and J. Stenflo, *J. Biol. Chem.* **258,** 12509 (1983).

Tris-HCl, pH 8.0, containing 0.1 M NaCl. The column is eluted with the same buffer at a flow rate of 8 ml/hr, and 2-ml fractions are collected. fX-Gla-EGF$_{NC}$ elutes after approximately 0.45 column volumes.

Characterization of fX-Gla-EGF$_{NC}$. The isolation procedure yields a fragment which is homogeneous on SDS-PAGE, and has two amino-terminal sequences in equimolar amounts that correspond to residues 1–5 of the light chain (Ala-Asn-Ser-Phe-Leu) and residues 154–158 of the heavy chain of factor X (Asn-Val-Ala-Pro-Ala). The Gla and Hya values obtained, 12.6 and 0.96 mol per mole of protein, respectively, are consistent with the amino acid sequence. The absorption coefficient ($A_{1\,cm}^{1\%}$ at 280nm) is 9.1 based on a molecular weight of 19,429. The recovery of fX-Gla-EGF$_{NC}$ is 50–55%.

The Gla-containing fragment and the heavy-chain peptide are separated by HPLC after reduction of the disulfide bonds. Reduction of fX-Gla-EGF$_{NC}$ is carried out by incubation of 60 μg of fragment for 1 hr at 37° in 50 μl of 0.5 M Tris-HCl, pH 8.5, containing 6 M guanidine hydrochloride, 5 mM EDTA, and 25 mM dithiothreitol (DTT). After reduction, 50 μl of 2.5 M Tris-HCl, pH 8.5, containing 6 M guanidine hydrochloride, 5 mM EDTA, and 60 mM iodoacetic acid is added, and the incubation continued for 1 hr in the dark. The sample is then subjected to HPLC on a C$_4$ column (2.1 × 30 mm, BU-300 Aquapore Butyl 7 μm, Brownlee Labs, Santa Clara, CA) equilibrated at 45° with 50 mM phosphate buffer, pH 6.8, and eluted with a linear gradient of acetonitrile [0 to 50% (v/v) over 60 min] in the same buffer at a flow rate of 0.1 ml/min. The Gla-containing fragment and the heavy-chain peptide of fX-Gla-EGF$_{NC}$ elute with 20 and 30% acetonitrile, respectively. HPLC in trifluoroacetic acid is not possible, as the Gla-containing fragments precipitate at low pH.

Amino acid analysis indicates that the Gla-containing fragment consists of residues 1–140 of factor X, i.e., the intact light chain, and a disulfide-linked peptide derived from the heavy chain (residues 154–183; Table I).[25]

Isolation of Two EGF Modules (fX-EGF$_{NC}$)

Isolation of fX-EGF$_{NC}$ is achieved by chymotryptic digestion of fX-Gla-EGF$_{NC}$ (from which lysine blocking groups have been removed).

Chymotryptic Digestion. fX-Gla-EGF$_{NC}$ (1.0–2.5 mg/ml) in 50 mM Tris-HCl, pH 8.0, containing 0.1 M NaCl and 2 mM EDTA, is digested with 5 μg of α-chymotrypsin (EC 3.4.21.1, type II, Sigma) per milligram of fX-Gla-EGF$_{NC}$ for 20 min. Digestion is terminated by adding DFP and phenylmethylsulfonyl fluoride (PMSF) to final concentrations of 2 mM. The buffer contains EDTA because chymotryptic cleavage between Tyr-44

TABLE I

AMINO ACID COMPOSITION OF f X-Gla-EGF$_{NC}$, f X-EGF$_{NC}$, f X-EGF$_C$, f X-Gla-EGF$_N$, f X-EGF$_N$, AND f X-helix-EGF$_N$

Amino acid	Gla-EGF$_{NC}$ (1–140 + 154–183)	EGF$_{NC}$ (45–140 + 154–171/183)	EGF$_C$ (88–128)	Gla-EGF$_N$ (1–86)	EGF$_N$ (45–86)	Helix-EGF$_N$ (29–86)
Asp	17.8 (16)[a]	14.5 (12/12)	4.9 (5)	9.2 (9)	5.1 (5)	7.0 (7)
Thr[b]	7.5 (9)	6.7 (6/8)	1.1 (1)	3.9 (4)	2.9 (3)	3.6 (4)
Ser[b]	11.9 (12)	8.4 (8/9)	5.6 (6)	4.6 (4)	1.0 (1)	2.0 (2)
Glu	33.8 (30)	17.4 (15/16)	5.0 (5)	20.6 (20)	6.3 (6)	11.3 (11)
Pro	4.8 (4)	5.6 (4/4)	ND[c]	1.0 (1)	1.0 (1)	ND
Gly	16.0 (17)	16.2 (13/16)	4.2 (4)	8.5 (8)	7.0 (7)	9.4 (7)
Ala	10.5 (10)	6.6 (6/6)	1.2 (1)	5.1 (5)	1.0 (1)	2.2 (2)
Cys[d]	15.6 (16)	13.2 (14/14)	4.6 (6)	6.8 (8)	5.2 (6)	ND
Val[e]	6.6 (7)	5.1 (4/5)	2.9 (3)	2.5 (2)	<0.1 (0)	1.4 (1)
Met	0.5 (1)	0.8 (1/1)	<0.1 (0)	<0.1 (0)	<0.1 (0)	<0.1 (0)
Ile[e]	2.1 (3)	2.5 (2/3)	0.9 (1)	1.0 (1)	1.0 (1)	1.0 (1)
Leu	9.6 (9)	6.3 (5/5)	1.9 (2)	5.1 (5)	1.0 (1)	1.1 (1)
Tyr	3.0 (3)	2.6 (2/2)	0.9 (1)	1.7 (2)	1.0 (1)	1.8 (2)
Phe	8.4 (9)	6.6 (5/6)	1.1 (1)	4.8 (5)	2.0 (2)	3.8 (4)
His	3.2 (3)	3.8 (3/3)	1.2 (1)	2.3 (2)	2.0 (2)	1.8 (2)
Lys	8.9 (9)	7.4 (6/7)	1.0 (1)	5.1 (5)	3.2 (3)	3.2 (4)
Arg	8.6 (9)	6.5 (6/7)	2.9 (3)	2.8 (3)	1.0 (1)	1.0 (1)
Trp	ND	ND	ND	ND	ND	ND
Hya	0.96 (1)	1.02 (1/1)	<0.1 (0)	0.77 (1)	1.05 (1)	1.26 (1)
Gla	12.6 (12)	<0.1 (0/0)	<0.1 (0)	11.4 (12)	<0.1 (0)	0.24 (4)

[a] The calculated numbers of residues according to sequence data are shown in parentheses.
[b] Values obtained by extrapolation to zero hydrolysis time.
[c] ND, Not determined.
[d] Determined as cystine.
[e] Determined after 72-hr hydrolysis.

and Lys-45 (i.e., between the Gla module and the first EGF module) is very slow in the presence of calcium ions.[25]

Ion-Exchange Chromatography. The chymotryptic digest is applied to a Q-Sepharose Fast Flow column (1.6 × 13 cm) equilibrated with 50 mM Tris-HCl, pH 7.5. Elution is accomplished with a 500-ml linear NaCl gradient (0–0.8 M) in the same buffer at a flow rate of 50 ml/hr. Fractions of 10 ml are collected and the Hya content is measured. The Hya-containing material elutes as a single peak at ≈0.2 M NaCl, and the Gla module elutes at ≈0.5 M NaCl.

Characterization of f X-EGF$_{NC}$. The Hya-containing fragment has two amino-terminal sequences in equimolar amounts that correspond to residues 45–49 of the light chain of factor X (Lys-Asp-Gly-Asp-Gln) and to residues 154–158 of the heavy chain (Asn-Val-Ala-Pro-Ala), respectively. The absorption coefficient ($A_{1\,cm}^{1\%}$ at 280 nm) of fX-EGF$_{NC}$, using a molecular weight of 13,699, is 7.4. The overall recovery of this fragment is about 35%, and the recovery from fX-Gla-EGF$_{NC}$ about 65%.

The peptide chains can be separated by HPLC after reduction of the intact fragment following the procedure described for fX-Gla-EGF$_{NC}$. However, in this case the column is equilibrated with 0.1% (v/v) trifluoroacetic acid, and elution is accomplished with a linear acetonitrile gradient (0–35% over 60 min) in 0.1% trifluoroacetic acid. The heavy-chain peptide and the EGF region elute with 27 and 30% acetonitrile, respectively. The amino acid analysis of the intact fragment (Table I) and the isolated peptide chains[25] indicates that the portion from the light chain consists of residues 45–140 (the two EGF modules), and that the material derived from the heavy chain is a mixture of two peptides with the same aminoterminus (Asn-154) but different carboxy termini, either Met-171 or Arg-183. Gla is not detected, and the Hya content is 1.0 mol per mole of fragment.

Isolation of Carboxy-Terminal EGF Module (fX-EGF$_C$)

This fragment can be isolated by digestion of fX-Gla-EGF$_{NC}$ with *Staphylococcus aureus* V8 protease.

Cleavage with Staphylococcus aureus V8 Protease. fX-Gla-EGF$_{NC}$ (≈ 1 mg/ml), in 50 mM Tris-HCl, 0.1 M NaCl, and 2 mM EDTA at pH 8.0, is digested with V8 protease (protease from *S. aureus* strain V8, type XVII, Sigma; 50 μg per milligram of fX-Gla-EGF$_{NC}$) for 90 min at 37°. The reaction is terminated by the addition of DFP to a final concentration of 2 mM.

Gel Filtration. The digest is concentrated to 1–2 ml using a Filtron Novacell (cutoff level 1 kDa) and chromatographed on a Sephadex G-50 Superfine column (1.6 × 90 cm) equilibrated with the above buffer and eluted at a flow rate of 12 ml/hr. The peak containing fX-EGF$_C$, identified by amino-terminal sequence analysis, elutes after approximately 0.65 column volumes.

Characterization of fX-EGF$_C$. The isolated fragment has the aminoterminal sequence Ile-Xxx-Ser-Leu (where Xxx is a Cys residue), corresponding to residues 88–91 of the light chain of bovine factor X. The amino acid composition (Table I) indicates the fragment to consist of residues 88 to 128 of the light chain of factor X. Sequence analysis gives no evidence of internal cleavages. The absorption coefficient ($A_{1\,cm}^{1\%}$ at 280 nm) is 4.0, assuming a molecular weight of 4460. The recovery of fX-EGF$_C$ from fX-Gla-EGF$_{NC}$ is 45–50%.

Isolation of Amino-Terminal EGF Module Linked to Gla Module (fX-Gla-EGF_N)

The fragment fX-Gla-EGF$_N$ is isolated after proteolytic cleavage between the two EGF modules, which requires prolonged digestion with trypsin at high concentration. Therefore, prior reversible modification of the lysine residues with citraconic anhydride is crucial for the isolation of an intact fragment.

Tryptic Digestion. Citraconylated[35] factor X (0.5–1.0 mg/ml) in 50 mM Tris-HCl, pH 8.0, containing 0.1 M NaCl and 2 mM EDTA, is digested with trypsin (60 μg per milligram of factor X) for 4 hr. DFP, to a final concentration of 2 mM, is added to inactivate the trypsin.

Ion-Exchange Chromatography. The tryptic digest is applied to a Q-Sepharose Fast Flow column and chromatographed as described for fX-Gla-EGF$_{NC}$, except that the NaCl gradient is from 0.1 to 0.8 M. Under these conditions, fX-Gla-EGF$_N$ elutes with approximately 0.6 M NaCl. The fragment is identified by Gla and Hya measurements after alkaline and acid hydrolysis, respectively.[36,37] It is nearly homogeneous on SDS-PAGE.

Gel Filtration. Fractions containing fX-Gla-EGF$_N$ are pooled, and the lysine blocking groups removed (see isolation of fX-Gla-EGF$_{NC}$). The pooled material is concentrated to 3 ml using a Filtron Novacell (cutoff level 3 kDa) and is chromatographed on a Sephadex G-50 Superfine column (1.5 × 97 cm) equilibrated with 50 mM Tris-HCl, pH 8.0, containing 0.1 M NaCl and 2 mM EDTA at a flow rate of 10 ml/hr. This step removes any remaining contaminating polypeptides and the blocking groups.

Characterization of fX-Gla-EGF_N. The isolated fragment has the amino-terminal sequence Ala-Asn-Ser-Phe-Leu, which corresponds to residues 1–5 in the light chain of bovine factor X. There is no evidence of internal cleavage. Results of the amino acid analysis indicate isolated fX-Gla-EGF$_N$ to consist of residues 1–86 of the light chain from bovine factor X (Table I). The Gla and Hya contents are 11.4 and 0.77 mol per mole of protein, respectively. Based on the Hya content and the absorbance at 280 nm the recovery of fX-Gla-EGF$_N$ is around 90%. The absorption coefficient of fX-Gla-EGF$_N$ ($A_{1cm}^{1\%}$ at 280 nm) is 9.9 using a molecular weight of 10,301.

Isolation of Amino-Terminal EGF Module (fX-EGF_N)

The amino-terminal EGF module of factor X is isolated by chymotryptic removal of the Gla module from fX-Gla-EGF$_N$, essentially as described for the isolation of fX-EGF$_{NC}$.

Chymotryptic Digestion. fX-Gla-EGF$_N$ (≈1 mg/ml) in 50 mM Tris-

HCl, pH 8.0, containing 0.1 M NaCl and 2 mM EDTA, is incubated for 40 min in the presence of 5 μg of α-chymotrypsin per milligram of fX-Gla-EGF$_N$. To terminate the cleavage, DFP and PMSF are added to 2 mM and 1 mM, respectively. The digestion yields two products, as judged by SDS-PAGE, that can be separated by gel filtration.

Gel Filtration. The volume of the chymotryptic digest is reduced to about 1 ml using a Speed Vac (Savant Instruments, Inc., Hicksville, NY) concentrator or a Filtron Novacell (cutoff level 1 kDa). The sample is applied to a Sephadex G-50 Superfine column (1.5 × 88 cm) equilibrated with 50 mM NH$_4$HCO$_3$. Elution is at 6 ml/hr and 2-ml fractions are collected. A volatile buffer is chosen to allow lyophilization of the purified fX-EGF$_N$. The chromatogram shows two peaks that are subjected to Hya and Gla analysis. The first peak contains the Gla module (12.9 mol Gla per mole of protein and no Hya), and the second peak contains the EGF module (no Gla and 1.05 mol Hya per mole of protein, Table I). It is noteworthy that the Gla module (44 amino acids) and the EGF$_N$ module (42 amino acids) are completely separated by gel filtration. This is probably to be attributed to the accumulation of negative charge on the Gla module, which leads to steric exclusion from the gel due to negative charges on the matrix.

Characterization of fX-EGF$_N$. The amino-terminal sequence of fX-EGF$_N$ is Lys-Asp-Gly-Asp-Gln (residues 45–49 of the light chain of factor X) due to chymotryptic cleavage carboxy-terminal of Tyr-44. Amino acid analysis shows Arg-86 to be the carboxy terminus of fX-EGF$_N$. The overall recovery (from factor X) of the fragment (based on the Hya content) is approximately 55% and that from fX-Gla-EGF$_N$ is 60%; the absorption coefficient ($A^{1\%}_{1cm}$ at 280nm) is 3.1, using a molecular weight of 4571. The structure of this fragment has been determined by two-dimensional nuclear magnetic resonance (2D NMR) spectroscopy.[38]

Isolation of Amino-Terminal EGF Module Linked to the Aromatic Cluster (fX-helixEGF$_N$)

The carboxy-terminal, presumably α-helical, part of the Gla module contains a cluster of aromatic amino acids (Phe-40, Trp-41, and Tyr-44). This region seems to be a nucleation site for the folding of the Gla module in the presence of calcium ions.[39] We have therefore devised a procedure to isolate fX-EGF$_N$ bound to the carboxy-terminal residues of the Gla module, fX-Gla-EGF$_N$ being used as starting material. Tryptic cleavage at

[38] M. Selander-Sunnerhagen, M. Ullner, E. Persson, O. Teleman, J. Stenflo, and T. Drakenberg, *J. Biol. Chem.* **267**, 19642 (1992).
[39] M. Soriano-Garcia, C. H. Park, A. Tulinsky, K. G. Ravichandran, and E. Skrzypczak-Jankun, *Biochemistry* **28**, 6805 (1989).

Arg-28 in the sequence Ala-Arg-Gla would yield the desired fragment. As trypsin does not cleave at an Arg residue that is adjacent to Gla, thermal decarboxylation of Gla is the first step. In the second step, the fragment is citraconylated to preclude tryptic cleavage at Lys.

Decarboxylation and Citraconylation. fX-Gla-EGF$_N$, dialyzed against 10% acetic acid and lyophilized, is subjected to thermal decarboxylation for 20 hr at 100° *in vacuo*, after which the Gla content is less than 0.2 mol per mole of protein.[26] Decarboxylated fX-Gla-EGF$_N$ is dissolved (2 mg/ml) by stirring for 2 hr in 0.1 M borate buffer, pH 8.2, and the Lys residues are modified with citraconic anhydride as described above. To remove excess citraconic anhydride, the sample is chromatographed on a Sephadex G-50 Superfine column (1.5 × 90 cm) equilibrated with 50 mM Tris-HCl, pH 8.0, containing 0.1 M NaCl and 2 mM EDTA at a flow rate of 8 ml/hr. The fractions containing the modified fragment are pooled and subjected to tryptic digestion.

Tryptic Digestion. Modified fX-Gla-EGF$_N$ ($\approx$ 5 mg/ml) is incubated at 37° for 30 min in the presence of 60 μg of trypsin per milligram of fX-Gla-EGF$_N$. The reaction is terminated by adding DFP to a final concentration of 5 mM. The lysine blocking groups are removed by incubating the sample at pH 3.0 overnight (see above). After deblocking, the pH is adjusted to 8.0 by addition of 2 M NaOH and DFP is added again to a final concentration of 5 mM.

Gel Filtration. The digest is applied to a Sephadex G-75 Superfine column (1.5 × 105 cm) equilibrated with 0.1 M NH$_4$HCO$_3$ at a flow rate of 8 ml/hr. The fragment elutes after about 0.62 column volumes.

Characterization of fX-helix-EGF$_N$. The isolated fragment has the amino-terminal sequence Glu-Val-Phe-Glu-Asp, corresponding to residues 29–33 of the light chain of bovine factor X. The amino acid composition and the Gla (0.24 mol per mole of protein) and Hya (1.0 mol per mole of protein) contents (Table I) are compatible with a fragment that contains residues 29–86, where the Gla residues in positions 29, 34, and 35 of the native fragment have been decarboxylated to Glu. The absorption coefficient is 8.2, assuming a molecular weight of 6166. The overall recovery of this fragment from factor X is 20%.

Isolation and Characterization of EGF Module-Containing Fragments from Bovine Factor IX

Isolation of Two EGF Modules Linked to Gla Module (fIX-Gla-EGF$_{NC}$)

A fragment containing the Gla module attached to the two EGF modules of bovine factor IX is isolated after limited chymotryptic digestion of bovine factor IX.

Chymotryptic Digestion. Purified bovine factor IX (0.5 mg/ml) is dialyzed against 50 mM Tris-HCl, pH 7.5, containing 0.1 M NaCl. After dialysis the protein is incubated with 5 μg α-chymotrypsin per milligram of factor IX for 30 min at 37° in the presence of 5 mM EDTA. The reaction is terminated by adding DFP and PMSF, each to a final concentration of 2 mM, and the fragment is isolated by ion-exchange chromatography and gel filtration (see Fig. 2).

Ion-Exchange Chromatography. The chymotryptic digest is chromatographed on a Q-Sepharose Fast Flow column (1.6 × 10 cm) equilibrated with 50 mM Tris-HCl, pH 7.5, containing 0.1 M NaCl. Elution is carried out with a linear NaCl gradient, from 0.1 to 0.8 M (2 × 200 ml), at a flow rate of 35 ml/hr and 4-ml fractions are collected. fIX-Gla-EGF$_{NC}$ is identified by SDS-PAGE and Gla and Hya measurements. It elutes with about 0.4 M NaCl.

Gel Filtration. The fIX-Gla-EGF$_{NC}$-containing pool is concentrated to 1 ml using a Filtron Novacell (cutoff level 3 kDa) and applied to a Sephadex G-75 Superfine column (1.6 × 80 cm) equilibrated with 0.1 M NH$_4$HCO$_3$ at a flow rate of 5 ml/hr. It elutes after about 0.5 column volumes.

Characterization of fIX-Gla-EGF$_{NC}$. The amino-terminal sequence analysis of the fragment indicates that it is at least 95% homogeneous. It has two sequences in equimolar amounts, Tyr-Asn-Ser-Gly-Lys-Leu, corresponding to residues 1–6 of bovine factor IX, and Val-Thr-Pro-Ile-Xxx-Ile, corresponding to residues 286–291 (where Xxx denotes a Cys residue). Amino acid analysis of the intact fragment (Table II) and the individual peptide chains[26] shows fIX-Gla-EGF$_{NC}$ to contain residues 1–144 and 286–296 of bovine factor IX. The fragment contains 11.8 mol of Gla and 1.1 mol of Hya per mole of protein. The absorption coefficient of fIX-Gla-EGF$_{NC}$ ($A^{1\%}_{1cm}$ at 280 nm) is 10.5, based on a molecular weight of the apoprotein of 18,170. The recovery from factor IX is 25–35%.

Isolation of Two EGF Modules (fIX-EGF$_{NC}$)

This fragment consists of the two EGF modules of bovine factor IX. The fragment is isolated after digestion of fIX-Gla-EGF$_{NC}$ with lysyl endopeptidase (*Achromobacterlyticus* protease I, EC 3.4.21.50, Wako Pure Chemical Industries, Japan).

Lysyl Endopeptidase Digestion. fIX-Gla-EGF$_{NC}$ (0.5 mg/ml) is dialyzed against 50 mM Tris-HCl, pH 7.5, containing 0.1 M NaCl. After dialysis, the fragment is incubated with lysyl endopeptidase (2 μg/mg fIX-Gla-EGF$_{NC}$) for 2 min at 37° in the presence of 5 mM EDTA. The buffer contains EDTA because cleavage at Lys-43 is abolished in the presence of

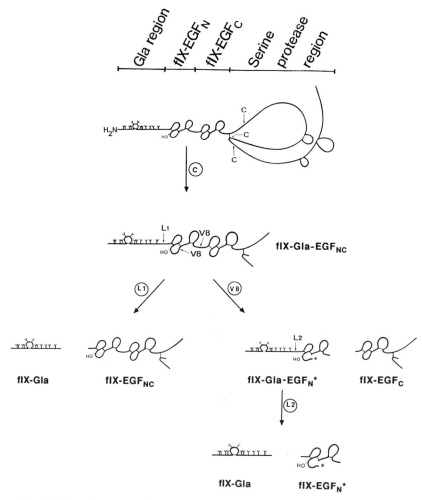

FIG. 2. Schematic representation of the procedure used to isolate fragments from bovine factor IX. Chymotryptic digestion (C) of intact bovine factor IX results in a fragment containing the Gla module and the two EGF modules, i.e., fIX-Gla-EGF$_{NC}$. The Gla module is cleaved from fIX-Gla-EGF$_{NC}$ by digestion with lysyl endopeptidase (L1) and fIX-EGF$_{NC}$ is obtained. Digestion of fIX-Gla-EGF$_{NC}$ with V8 protease (V8) results in two fragments, fIX-EGF$_C$ and fIX-Gla-EGF$_N^*$. fIX-EGF$_N^*$ can be isolated from fIX-Gla-EGF$_N^*$ by digestion with lysyl endopeptidase (L2). The asterisks denote that the fragments have an internal cleavage at Glu-70.

TABLE II

AMINO ACID COMPOSITION OF f IX-Gla-EGF$_{NC}$, f IX-EGF$_{NC}$, f IX-Gla-EGF$_N^*$, f IX-EGF$_N^*$, AND f IX-EGF$_C$

Amino acid	Gla-EGF$_{NC}$ (1–144 + 286–296)	EGF$_{NC}$ (44–144 + 286–296)	Gla-EGF$_N^*$ (1–83)	EGF$_N^*$ (44–83)	EGF$_C$ (84–144 + 286–296)
Asp	19.2 (19)[a]	15.7 (16)	9.7 (10)	7.4 (7)	9.2 (9)
Thr	8.1 (8)	5.2 (5)	3.9 (4)	1.2 (1)	4.1 (4)
Ser[b]	9.0 (10)	7.7 (8)	4.0 (4)	2.0 (2)	5.4 (6)
Glu	22.9 (23)	11.2 (11)	19.0 (19)	7.7 (7)	4.4 (4)
Pro	ND[c]	ND	ND	ND	ND
Gly	10.7 (10)	8.7 (8)	7.2 (7)	4.3 (5)	3.2 (3)
Ala	6.2 (6)	4.9 (5)	2.1 (2)	1.1 (1)	4.0 (4)
Cys	ND	ND	ND	ND	ND
Val[d]	8.9 (9)	6.7 (7)	3.3 (3)	1.3 (1)	5.3 (6)
Met	0.9 (1)	1.0 (1)	0.9 (1)	0.8 (1)	< 0.1 (0)
Ile[d]	5.2 (5)	5.0 (5)	1.1 (1)	1.0 (1)	3.9 (4)
Leu	6.1 (6)	3.5 (4)	3.1 (3)	1.0 (1)	2.8 (3)
Tyr	4.8 (5)	3.7 (4)	2.9 (3)	1.9 (2)	1.8 (2)
Phe	7.3 (7)	3.5 (3)	4.9 (5)	1.0 (1)	2.3 (2)
His	1.3 (1)	1.2 (1)	< 0.1 (0)	<0.1 (0)	1.1 (1)
Lys	11.9 (13)	7.7 (8)	6.2 (6)	0.8 (1)	5.5 (7)
Arg	7.1 (8)	4.7 (5)	3.0 (3)	< 0.1 (0)	4.2 (5)
Trp	ND	ND	ND	ND	ND
Hya	1.1 (1)	1.2 (1)	1.1 (1)	1.0 (1)	< 0.1 (0)
Gla	12.6 (12)	< 0.1 (0)	12.4 (12)	< 0.1 (0)	< 0.1 (0)

[a] The calculated numbers of residues according to sequence data are shown in parentheses.
[b] Values obtained by extrapolation to zero hydrolysis time.
[c] ND, Not determined.
[d] Determined after 72-hr hydrolysis.

divalent cations.[26] To terminate the reaction, DFP and PMSF are added to a final concentration of 2 mM each.

Ion-Exchange Chromatography. The digest is applied to a Q-Sepharose Fast Flow column (1.6 × 10 cm), equilibrated with 50 mM Tris-HCl, pH 7.5. Elution is accomplished with a 300-ml linear NaCl gradient (0–0.8 M) in the same buffer at a flow rate of 25 ml/hr, and 2-ml fractions are collected. Material containing Hya but no Gla elutes with 0.25 M NaCl. It is pooled and concentrated using a Filtron Novacell (cutoff level 3 kDa).

Gel Filtration. The fIX-EGF$_{NC}$ pool is chromatographed on a Sephadex G-75 column (1.6 × 80 cm), as described for the isolation of fIX-Gla-EGF$_{NC}$ (see above).

Characterization of fIX-EGF$_{NC}$. This isolation procedure yields a fragment with two amino-terminal sequences: Gln-Tyr-Val-Asp-Gly-Asp corresponding to residues 44–49 of bovine factor IX and Val-Thr-Pro-Ile-Xxx-Ile (where Xxx corresponds to Cys) from residues 286–291. An internal cleavage in the carboxy-terminal EGF module at Lys-96 is present in ≈ 10% of the material. Results of amino acid analysis (Table II) show the fragment to contain no Gla and 1.2 mol of Hya per mol of protein and are compatible with the composition of residues 44–144 and 286–296 of bovine factor IX. The two peptides are linked by a disulfide bond. The absorption coefficient ($A_{1\,cm}^{1\%}$ at 280 nm) of fIX-EGF$_{NC}$ is 9.5, assuming a molecular weight of the apoprotein of 12,390. The recovery relative to fIX-GlaEGF$_{NC}$ is 30–40%.

Isolation of Amino-Terminal EGF Module Linked to Gla Module (fIX-Gla-EGF$_N^*$) and Carboxy-Terminal EGF Module (fIX-EGF$_C$)

Both the amino-terminal EGF module attached to the Gla module and the carboxy-terminal EGF module of the bovine factor IX molecule can be isolated by proteolytic cleavage of fIX-Gla-EGF$_{NC}$ with protease from *S. aureus* strain V8.

V8 Protease Digestion. fIX-Gla-EGF$_{NC}$ (0.5 mg/ml), isolated as described in the previous section, is dialyzed against 50 m*M* Tris-HCl, pH 7.5, containing 0.1 *M* NaCl. After dialysis, the protein is incubated with V8 protease (100 μg per milligram of fIX-Gla-EGF$_{NC}$) for 3 hr at 37° in the presence of 5 m*M* EDTA. The digestion is terminated by adding DFP to a final concentration of 2 m*M*. SDS-PAGE of the digestion products reveals two fragments with apparent molecular weights of 14,000 and 16,000, respectively.

Gel Filtration. The cleavage products are concentrated to a volume of about 1 ml using a Filtron Novacell (cutoff level 3 kDa) and chromatographed on a column of Sephadex G-50 Superfine (1.6 × 80 cm) equilibrated with 50 m*M* Tris-HCl, pH 7.5, containing 0.1 *M* NaCl, eluted at 6 ml/hr. Fractions of 2 ml are collected. Two peaks are obtained. The material in each peak is characterized by amino-terminal sequence determination, amino acid analysis, and Gla and Hya measurements.

Characterization of fIX-Gla-EGF$_N^*$ and fIX-EGF$_C$. The material in the first peak corresponds to the higher molecular weight fragment. Two amino-terminal sequences in equimolar amounts are found in this fragment, Tyr-Asn-Ser-Gly-Lys-Leu corresponding to residues 1–6 of intact bovine factor IX, and Xxx-Trp-Xxx-Gln-Ala-Gly corresponding to residues 71–76 (Trp-72 has a disulfide-linked Cys on either side) indicating that the isolated fIX-Gla-EGF$_N^*$ has an internal cleavage at Glu-70. The amino acid analysis (Table II) shows a Gla and Hya content of 12.4 and

1.1 mol per mole of protein, respectively, and an amino acid composition that is compatible with a peptide containing residues 1–83 of bovine factor IX, i.e., the amino-terminal EGF module attached to the Gla module (fIX-Gla-EGF$^*_N$). The absorption coefficient of the fragment is 16.5, assuming a molecular weight of the apoprotein of 10,220, and the recovery relative to fIX-Gla-EGF$_{NC}$ is 30–40%.

The material in the second peak has two sequences in equimolar amounts: Leu-Asp-Ala-Thr and Val-Thr-Pro-Ile, corresponding to residues 84–87 and 286–289, respectively, of intact bovine factor IX. No Gla or Hya is found by amino acid analysis. The amino acid composition of the fragment (Table II) is consistent with a peptide consisting of residues 84–144 and residues 286–296, i.e., the carboxy-terminal EGF module linked to a peptide from the serine protease region by a disulfide bond (fIX-EGF$_C$). The absorption coefficient is 3.7, assuming a molecular weight of 7980. The recovery relative to fIX-Gla-EGF$_{NC}$ is 30–40%.

Isolation of Amino-Terminal EGF Module (fIX-EGF$^*_N$)

This fragment can be isolated after digestion of fIX-Gla-EGF$^*_N$ by lysyl endopeptidase using the conditions described for the isolation of fIX-EGF$_{NC}$ (see above).

Lysyl Endopeptidase Digestion. fIX-Gla-EGF$^*_N$ (0.5 mg/ml), isolated as described above, is digested with lysyl endopeptidase (2 μg per milligram of fIX-Gla-EGF$^*_N$) at 37° in a buffer containing 50 mM Tris-HCl, pH 7.5, 0.1 M NaCl, and 5 mM EDTA. After 2 min the reaction is terminated by addition of PMSF to a final concentration of 2 mM.

Gel Filtration. The proteolytic digest is concentrated using a Filtron Novacell (cutoff level 1 kDa) and chromatographed on a Sephadex G-50 Superfine column (1.6 × 80 cm) as described above for the isolation of fIX-Gla-EGF$^*_N$. Two peaks are obtained.

Characterization of fIX-EGF$^*_N$. The high-molecular-weight material contains Gla but not Hya, and is identified as the Gla module (residues 1–43). The material in the second peak contains 1.0 mol Hya per mole of protein but no Gla, and has an amino acid composition compatible with a peptide containing residues 44–83 of intact bovine factor IX (Table II). The amino-terminal sequence analysis reveals two sequences in equimolar amounts: Gln-Tyr-Val-Asp-Gly (residues 44–48 of bovine factor IX) and Xxx-Trp-Xxx-Gln-Ala-Gly corresponding to residues 71–76, which was already present in the starting material (see isolation of fIX-Gla-EGF$^*_N$). The absorption coefficient ($A^{1\%}_{1cm}$ at 280 nm) is 14.9, assuming a molecular weight of the apoprotein of 4440. The recovery relative to fIX-Gla-EGF$^*_N$ is 35–45%.

Isolation and Characterization of Protein C Fragments

Isolation of Two EGF Modules Linked to Gla Module (pC-Gla-EGF$_{NC}$)

A peptide containing the Gla module and the two EGF modules can be isolated from bovine protein C using a procedure that involves citraconylation of the lysine side chains of protein C and tryptic digestion of the citraconylated protein, a procedure similar to that described for the isolation of fX-Gla-EGF$_{NC}$.

Citraconylation. Bovine protein C is citraconylated[35] as described for bovine factor X earlier in this chapter.

Tryptic Digestion. Citraconylated bovine protein C (1.8–2.5 mg/ml) in 50 mM Tris-HCl, pH 7.5, containing 0.1 M NaCl and 5 mM EDTA, is digested with 20 μg of trypsin per milligram of protein C for 5 min at 37°. The reaction is terminated by the addition of DFP to a final concentration of 10 mM, and the sample is dialyzed against 50 mM Tris-HCl, pH 8.0, containing 1 mM EDTA. SDS-PAGE of the tryptic digest shows a main product with an apparent molecular weight of 28,000.

Ion-Exchange Chromatography. The digest is chromatographed on a Q-Sepharose Fast Flow column on top of which is a layer (1.75 ml) of Sepharose 4B with immobilized soybean trypsin inhibitor (1.5 mg of inhibitor/ml gel). The column is equilibrated with 50 mM Tris-HCl, pH 8.0, containing 1 mM EDTA, and eluted stepwise with the same buffer containing (1) 0.1 M NaCl, (2) 0.2 M NaCl, (3) 0.2 M NaCl containing 20 mM CaCl$_2$, and finally (4) 0.8 M NaCl. The flow rate is 50 ml/hr and 5-ml fractions are collected. The peak containing pC-Gla-EGF$_{NC}$ is eluted with 20 mM CaCl$_2$ and 0.2 M NaCl as judged by SDS-PAGE and the Gla and Hya measurements. The pC-Gla-EGF$_{NC}$-containing fractions are pooled and the pH is lowered to 3.0 to remove the lysine blocking groups (see isolation of fX-Gla-EGF$_{NC}$).

Gel Filtration. After deblocking and neutralization, the sample is concentrated and subjected to gel filtration to remove trace impurities and blocking groups. pC-Gla-EGF$_{NC}$ is applied on a Sephadex G-75 column (1.6 × 90 cm) equilibrated with 0.1 M NH$_4$HCO$_3$ at a flow rate of 8 ml/hr and 2-ml fractions are collected. The fragment elutes as a single peak.

Characterization of pC-Gla-EGF$_{NC}$. The isolated fragment is at least 95% homogeneous, and the amino-terminal sequence analysis gives two sequences in equimolar amounts: Ala-Asn-Ser-Phe-Leu-Xxx-Xxx-Leu corresponding to residues 1–8 of the light chain of bovine protein C, and Leu-Ala-Lys-Pro-Ala-Thr-Leu-Ser corresponding to residues 108–115 of the heavy chain. The Gla residues in positions 6 and 7 are not detected due to poor extraction from the protein sequencer.[40] Results of amino acid

---

[40] P. Fernlund and J. Stenflo, *J. Biol. Chem.* **257**, 12170 (1982).

TABLE III
AMINO ACID COMPOSITION OF pC-Gla-EGF$_{NC}$ AND pC-EGF$_{NC}$

Amino acid	Gla-EGF$_{NC}$ (1–143 + 108–131)	EGF$_{NC}$ (42–143 + 108–131)
Asp	13.2 (14)[a]	10.1 (10)
Thr	4.8 (5)	3.3 (3)
Ser[b]	10.8 (12)	8.6 (10)
Glu	27.4 (25)	13.9 (13)
Pro	9.2 (9)	8.4 (8)
Gly	16.8 (16)	15.0 (15)
Ala	9.2 (9)	5.7 (6)
Cys	ND[c]	ND
Val[d]	6.3 (6)	4.3 (4)
Met	1.9 (2)	1.2 (1)
Ile[d]	3.8 (4)	2.4 (3)
Leu	11.6 (11)	8.1 (9)
Tyr	3.5 (3)	3.8 (3)
Phe	8.4 (8)	3.8 (4)
His	4.2 (4)	3.9 (4)
Lys	4.4 (4)	3.4 (4)
Arg	14.0 (13)	9.8 (10)
Trp	ND	ND
Hya	1.1 (1)	1.0 (1)
Gla	12.7 (11)	<0.1 (0)

[a] The calculated numbers of residues according to sequence data are shown in parentheses.
[b] Values obtained by extrapolation to zero hydrolysis time.
[c] ND, Not determined.
[d] Determined after 72-hr hydrolysis.

composition analysis of the intact fragment (Table III) and of the two isolated peptide chains[30] are compatible with a peptide containing residues 1–143 of the light chain of bovine protein C that is disulfide linked to a peptide derived from the heavy chain, corresponding to residues 108–131. The Gla and Hya contents are 12.7 and 1.1 mol per mole of protein, respectively. The absorption coefficient ($A_{1\,cm}^{1\%}$ at 280 nm) is 6.9. The overall recovery of the fragment is 15–25%.

Isolation of Two EGF Modules (pC-EGF$_{NC}$)

This fragment consists of the two EGF modules of bovine protein C and is isolated from Gla-domainless bovine protein C by limited tryptic digestion. Gla-domainless bovine protein C rather than intact protein C is used as starting material to obviate the risk of contamination with the Gla module.

Tryptic Digestion. Gla-domainless bovine protein C (3.6–14.4 mg/ml)[22] is incubated at 37° for 30 min in a digestion mixture containing 50 mM Tris-HCl, pH 7.5, 0.1 M NaCl, 2 mM EDTA, and 20 μg of trypsin per milligram of Gla-domainless protein C. The hydrolysis is terminated by adding DFP to a final concentration of 10 mM, and the digestion products are separated by gel filtration.

Gel Filtration. The tryptic digest is chromatographed on an AcA 44 column (1.6 × 94 cm) equilibrated with 0.1 M NaHCO$_3$ and eluted at a flow rate of 7 ml/hr. The peak containing pC-EGF$_{NC}$, identified by SDS-PAGE, Hya measurement, and amino-terminal sequence analysis, elutes after approximately 0.45 column volumes. An identical result is obtained if intact protein C is used as starting material.

Characterization of pC-EGF$_{NC}$. This isolation procedure results in a fragment that appears homogeneous on SDS-PAGE, with an apparent molecular weight of 22,000, and gives two amino-terminal sequences in equimolar amounts: Ser-Lys-Tyr-Ser-Asp-Gly corresponding to residues 42–47 of the light chain of bovine protein C and Leu-Ala-Lys-Pro-Ala-Thr corresponding to residues 108–113 of the heavy chain. There is no evidence of internal cleavages, and the amino acid analysis shows the fragment to contain 1.0 mol Hya per mole of protein and less than 0.06 mol Gla per mole of protein (Table III). The amino acid composition of the intact fragment (Table III) and the isolated peptides[29] (after reduction, alkylation, and HPLC separation; see above) is consistent with a peptide containing residues 42–143 of the light chain of bovine protein C, disulfide linked to a peptide derived from the heavy chain, corresponding to residues 108–131. The absorption coefficient of the fragment is 5.8. The recovery from Gla-domainless protein C is 50–60%.

Concluding Remarks

The procedures described in this chapter have been devised to isolate EGF modules from the noncatalytic parts of factors IX and X and protein C, either alone or linked to an intact Gla module. As bovine coagulation factors are used, tens of milligrams of several of the fragments can be isolated with moderate effort. To isolate certain of the fragments, reversible modification of Lys residues by citraconylation is required. The EGF and Gla modules are stable at room temperature in the absence of reducing agents, and the removal of the Lys blocking groups at low pH has no detrimental effect on the fragments. At present, more than 20 mg of the fragment fX-EGF$_N$ has been isolated from citraconylated factor X. The structure of this fragment has been determined with 2D NMR spectroscopy.[38]

Available evidence indicates that the Gla modules retain their native conformations when linked to the EGF modules. Limited proteolysis has been used to separate the Gla module from the EGF modules in fIX-Gla-EGF$_{NC}$ and fX-Gla-EGF$_{NC}$. The Ca^{2+} dependence of the proteolysis was found to be identical to that observed when the Gla module is removed from the intact coagulation factors.[21] To monitor Ca^{2+}-induced conformational changes, the intrinsic protein fluorescence was measured.[25] On titrations of the free Gla module, fluorescence was quenched at 2 to 5 mM Ca^{2+}. In the fragment in which the Gla module is linked to an EGF module the fluorescence quenching occurred at ≈ 1 mM Ca^{2+}, i.e., the same concentration as in the intact proteins. This indicates that the free Gla module, cleaved in the aromatic cluster region (residues 41–45 in factor IX), cannot fold to a native conformation even in the presence of calcium ions. In contrast, the Gla module apparently has normal calcium ion binding properties when it is linked to one or two EGF modules. Based on these experiments the EGF module-containing fragments are considered to retain their native conformations. Their use in the study of structure–function relationships in factors IX,[26,27,41] X,[24,25,38,42] and protein C[28–30] therefore appears justified.

[41] J. Astermark and J. Stenflo, *J. Biol. Chem.* **266,** 2438 (1991).
[42] E. Persson, C. Valcarce, and J. Stenflo, *J. Biol. Chem.* **266,** 2453 (1991).

[25] Role of Propeptide in Vitamin K-Dependent γ-Carboxylation

By KAREN J. KOTKOW, DAVID A. ROTH, THOMAS J. PORTER, BARBARA C. FURIE, and BRUCE FURIE

Introduction

The vitamin K-dependent proteins involved in hemostasis, prothrombin, factor IX, factor X, factor VII, protein C, and protein S, are synthesized in the liver. Several posttranslational modifications must occur before secretion to yield the biologically active forms of these proteins that circulate in blood. In addition to processing characteristic of most secreted proteins, including signal peptide cleavage, propeptide cleavage, disulfide bond formation, and glycosylation, these proteins undergo a unique amino acid modification in which glutamic acid residues are modified to γ-carboxyglutamic acid (Gla).[1,2] Reduced vitamin K is a cofactor in this reaction, which is catalyzed by a vitamin K-dependent carboxylase. The reaction also requires molecular oxygen, carbon dioxide, and a suitable protein substrate. In the absence of vitamin K or in the presence of the vitamin K antagonist warfarin, this modification cannot occur, resulting in the synthesis of biologically inactive protein.[3-5]

The vitamin K-dependent coagulation proteins are zymogens, which must be enzymatically cleaved to generate active serine proteases. Several structural domains distinct from the serine protease domain of these proteins are responsible for membrane interaction and protein complex formation. The amino-terminal domain of the vitamin K-dependent proteins is rich in γ-carboxyglutamic acid residues. This amino acid modification confers metal-binding properties, which induce a conformational transition in these proteins. The calcium-stabilized conformer of the vitamin K-dependent proteins binds membrane surfaces.[6-8] Significant enhancement of the activity of vitamin K-dependent blood clotting enzymes (factors IXa, Xa, and VIIa) is observed when these proteins are assembled on membrane surfaces in the presence of calcium, membrane-bound cofactors (factor Va and VIIIa), and vitamin K-dependent membrane-bound substrates (factor X and prothrombin) (for review, see Mann[9]).

The mature vitamin K-dependent proteins of blood and bone — prothrombin, factor IX, factor X, factor VII, protein S, protein C, bone Gla protein, and matrix Gla protein — share extensive sequence homology in their γ-carboxyglutamic acid-rich domains. However, the uncarboxylated forms of these mature proteins do not serve as good substrates for vitamin K-dependent carboxylation, suggesting that the substrate for the vitamin K-dependent carboxylase was not simply the uncarboxylated form of the secreted protein.[10] Examination of the cDNA sequences for these proteins revealed that they are synthesized in a precursor form with a signal peptide and a propeptide (for review, see Furie and Furie[11]). Although the

[1] J. Stenflo, P. Fernlund, W. Egan, and P. Roepstorff, *Proc. Natl. Acad. Sci. U.S.A.* **71**, 2730 (1974).
[2] G. L. Nelsestuen, T. H. Zytkovicz, and J. B. Howard, *J. Biol. Chem.* **249**, 6347 (1974).
[3] P. O. Ganrot and J. E. Nilehn, *Scand. J. Clin. Lab. Invest.* **22**, 23 (1968).
[4] G. L. Nelsestuen and J. W. Suttie, *J. Biol. Chem.* **247**, 8176 (1972).
[5] J. Stenflo and P. O. Ganrot, *J. Biol. Chem.* **247**, 8160 (1972).
[6] G. L. Nelsestuen, *J. Biol. Chem.* **251**, 5648 (1976).
[7] F. G. Prendergast and K. G. Mann, *J. Biol. Chem.* **252**, 840 (1977).
[8] M. Borowski, B. C. Furie, S. Bauminger, and B. Furie, *J. Biol. Chem.* **261**, 4969 (1986).
[9] K. G. Mann, *Trends Biochem. Sci.* **12**, 229 (1987).
[10] B. A. M. Soute, C. Vermeer, M. DeMetz, H. C. Hemker, and H. R. Lijnen, *Biochim. Biophy. Acta* **67**, 101 (1981).
[11] B. Furie, and B. C. Furie, *Cell (Cambridge, Mass.)* **53**, 505 (1988).

signal sequences demonstrate the variation in sequence, marked sequence homology was evident in the propeptides. Both phenylalanine at -16 and alanine at -10 are completely conserved. Well-conserved hydrophobic residues are found at positions -17, -7, and -6. Well-conserved basic amino acids are found at residues -4, -3, -2, and -1. The sequence homology displayed by the propeptides of the functionally diverse vitamin K-dependent proteins, including those of both plasma and bone, led to the hypothesis that the propeptide might serve to designate specific precursor polypeptides for vitamin K-dependent carboxylation.[12]

To test this hypothesis and to study the role of the propeptide in carboxylation, modifications were made in the propeptide-encoding region of the factor IX cDNA by site-specific mutagenesis.[13] The cDNA constructions were expressed in Chinese hamster ovary (CHO) cells and the amount of carboxylated recombinant factor IX species was determined. The sequence encoding the propeptide was completely deleted from the factor IX cDNA. Additionally, two point mutations were made involving amino acids that are completely conserved among the vitamin K-dependent proteins; phenylalanine at -16 was changed to alanine, and alanine at -10 was changed to glutamic acid.

Recombinant wild-type factor IX was carboxylated when expressed in this cell system. However, deletion of the propeptide completely abolished carboxylation of factor IX. Amino acid substitutions of alanine for phenylalanine at -16 and glutamic acid for alanine at -10 in the propeptide resulted in severely impaired carboxylation. In addition, no effect on signal peptide cleavage, propeptide cleavage, or β-hydroxylation was observed in these recombinant factor IX species.[14] The results of these studies demonstrate that the propeptide functions as a carboxylation recognition site designating the adjacent glutamic acid residues for vitamin K-dependent carboxylation.

Similar studies were performed to define the carboxylation recognition site within the propeptide of prothrombin.[15] Partial inhibition of carboxylation was observed with mutations of histidine to glycine at residue -18, valine to serine at -17, leucine to glycine or aspartic acid at -15, and alanine to aspartic acid at -10. In contrast, mutations of alanine to serine at residue -14 and serine to valine at -8 had no effect on carboxylation. When considered with earlier data these results localize the carboxylation

[12] L. C. Pan and P. A. Price, *Proc. Natl. Acad. Sci. U.S.A.* **82**, 6109 (1985).
[13] M. J. Jorgensen, A. B. Cantor, B. C. Furie, C. L. Brown, C. B. Shoemaker, and B. Furie, *Cell (Cambridge, Mass.)* **48**, 185 (1987).
[14] M. J. Rabiet, M. J. Jorgensen, B. Furie, and B. C. Furie, *J. Biol. Chem.* **262**, 14898 (1987).
[15] P. Huber, T. Schmitz, J. Griffin, M. Jacobs, C. Walsh, B. Furie, and B. C. Furie, *J. Biol. Chem.* **265**, 12467 (1990).

FIG. 1. Residues involved in the γ-carboxylation recognition site. The sequence of the propeptide of prothrombin is shown. Components of the γ-carboxylation recognition site whose mutation inhibits carboxylation are black. Residues that are not part of the γ-carboxylation recognition site are stippled. Residues that have not been evaluated are white. Reproduced with permission from Huber *et al.*[15]

recognition site to residues -18, -17, -16, -15, and -10 (Fig. 1). Mutations at these residues clearly affected carboxylation. Mutations at residues that are not part of the carboxylation recognition site, including alanine -14, serine -8, arginine -4 and arginine -1, did not affect the extent of carboxylation.

In addition, a different approach toward the understanding of the nature of the substrate requirements of the carboxylase was developed. We reasoned that if the propeptide contains a critical carboxylation signal in the naturally occurring substrate, it would be expected that a synthetic peptide substrate incorporating the propeptide would bind with higher affinity to the carboxylase than the pentapeptide, FLEEL (Phe-Leu-Glu-Glu-Leu). FLEEL is a peptide modeled on a sequence in bovine prothrombin which is carboxylated *in vitro*.[16,17] With this approach, the ability of various peptide substrates to be carboxylated by a crude bovine liver microsomal preparation containing the carboxylase was examined.[18] ProPT28 is a synthetic peptide based on amino acids -18 to $+10$ of prothrombin, in which -18 to -1 is the propeptide and $+1$ to $+10$ is the first 10 residues of the mature prothrombin. It is carboxylated with a K_m of 3.6 μM. In contrast, the pentapeptide FLEEL was carboxylated with a K_m of 2200 μM in this system. A 10-residue peptide based on residues $+1$ to $+10$ of prothrombin and a 20-residue peptide based on residues -10 to $+10$ of prothrombin both lack an intact propeptide, and both peptides were poor substrates for the carboxylase, displaying K_m values of 1000 and 850 μM, respectively. A derivative of ProPT28 which contained the substitution of alanine for phenylalanine at -16 is carboxylated with a K_m of 200 μM. This is significantly different than the K_m of 3.6 μM obtained for ProPT28. These results confirm that an intact carboxylation recognition site is required for efficient carboxylation *in vitro* and that this site is essential for high-affinity binding of the vitamin K-dependent carboxylase.

[16] J. W. Suttie, J. M. Hageman, S. R. Lehrman, and D. H. Rich, *J. Biol. Chem.* **251**, 5827 (1976).
[17] B. A. M. Soute, M. M. W. Ulrich, and C. Vermeer, *Thromb. Haemostasis* **57**, 77 (1987).
[18] M. M. W. Ulrich, B. Furie, M. Jacobs, C. Vermeer, and B. C. Furie, *J. Biol. Chem.* **263**, 9697 (1988).

Determination of the secondary structure of the prothrombin propeptide by two-dimensional NMR has revealed the context of the carboxylation recognition site. The propeptide includes an α helix from residues -13 to -3. The bulk of the carboxylation recognition site is N-terminal to the helix, but a component of the site resides within the helix.[19] It would appear that the carboxylation recognition site, mainly defined by flexible random coil secondary structure, is extended by the helix linking the carboxylation recognition site to the γ-carboxyglutamic acid-rich region.

Methods

Expression and Isolation of Recombinant Vitamin K-Dependent Proteins

Cell Culture

Chinese Hamster Ovary Cells. The dihydrofolate reductase-deficient Chinese hamster ovary cell line, CHO DUKX-B11, was employed for the development of stable transfected cell lines expressing recombinant vitamin K-dependent proteins.[20] These cells, after transfection with the cDNA in the expression plasmid pMT2, are able to process and secrete carboxylated vitamin K-dependent proteins when grown in cell culture media supplemented with vitamin K_1.[21] Typically, we obtain $1-2$ μg/ml of fully carboxylated prothrombin using this expression system. Expression of factor IX cDNA yields between 60 and 90% carboxylated recombinant factor IX with levels reaching $1-2$ μg/ml. Recombinant factor X is efficiently processed to a two-chain molecule in this system; only $5-15\%$ of the recombinant factor X is a single-chain species. Carboxylation levels of recombinant factor X vary from 65 to 80%, but it is possible to obtain a population of factor X with carboxylation levels of $80-95\%$ by immunoaffinity chromatography using conformation-specific antibodies directed against a metal ion-stabilized conformer of factor X. Typical yields of recombinant factor X are $0.5-0.7$ μg/ml. The propeptide of these recombinant vitamin K-dependent proteins is efficiently cleaved at these expression levels. Amplification of vitamin K-dependent-expressing CHO cells with methotrexate results in lower efficiency of γ-carboxylation and propeptide cleavage.

[19] D. G. Sanford, C. Kanagy, J. L. Sudmeier, B. C. Furie, B. Furie, and W. W. Bachovchin, *Biochemistry* **30,** 9835 (1991).
[20] G. Chasin and L. A. Urlaub, *Proc. Natl. Acad. Sci. U.S.A.* **77,** 4216 (1980).
[21] R. J. Kaufman, L. C. Wasley, B. C. Furie, B. Furie, and C. B. Shoemaker, *J. Biol. Chem.* **261,** 9622 (1986).

Untransfected CHO cells are grown in α-modified Eagle's medium lacking ribonucleosides and deoxyribonucleosides (GIBCO, Grand Island, NY) containing 10% heat-inactivated fetal calf serum; 5 μg/ml vitamin K_1 (Aquamephyton, Merck Sharp and Dohme); 10 mM HEPES, pH 7.2; thymidine, adenosine, and deoxyadenosine (10 μg/ml each); and penicillin and streptomycin (100 μg/ml each). This cell line is passaged by trypsinization of confluent monolayers with a balanced salt solution containing 0.05% trypsin, 0.53 mM EDTA (GIBCO).

The CHO DUKX-B11 cell line can be successfully transfected by electroporation,[15] calcium phosphate coprecipitation,[22] and lipofectin[23] (BRL, Gaithersburg, MD) according to standard procedures. These cells were transfected with 20 μg of a single expression plasmid containing both the specific vitamin K-dependent protein cDNA and the dihydrofolate reductase cDNA. Cells transfected by electroporation or calcium phosphate coprecipitation were cultured in the medium described above. Cells transfected by lipofectin were also cultured in this medium except serum was omitted. Transfected cells were selected for their ability to grow in nucleoside-depleted culture medium (i.e., to express dihydrofolate reductase activity). Two days after transfection the cells were subcultured into the same medium except that the nucleosides were omitted and dialyzed heat-inactivated serum was used ("selective medium"). After 1 week of growth in selective medium, colonies of transfected cells were visible. The colonies were screened using a filter-binding immunoassay to identify those colonies secreting the highest level of recombinant protein.[24] The positive colonies were then picked by isolating each colony with a cloning ring (Bellco, Vineland, NJ) and trypsinizing the cells contained within the cloning ring. The cells from individual colonies were transferred to 16-mm culture wells where they were propagated in selective medium as a separate cell line. We have easily obtained large volumes of conditioned media from transfected CHO cell lines expressing vitamin K-dependent proteins by seeding them into a Cell Factory (Nunc, Naperville, IL) containing 2 liters of selective medium and then replacing the medium in the Cell Factory each week. We have carried the cells in a Cell Factory for up to 6 weeks with continued production of high-quality protein.

293 Cells. An adenovirus-transformed human embryonic kidney cell line, 293 (American Tissue Culture Collection, Rockville, MD; CRL 1573), has also been used to develop stable cell lines expressing recombi-

---

[22] R. J. Kaufman and P. A. Sharp, *Mol. Biol.* **159,** 601 (1982).
[23] P. L. Feigner, T. R. Gadek, M. Holm, R. Roman, H. W. Chan, M. Wenze, J. P. Northrop, G. M. Ringold, and M. Danielsen, *Proc. Natl. Acad. Sci. U.S.A.* **84,** 7413 (1987).
[24] A. A. McCracken and J. L. Brown, *BioTechniques* **2,** 82 (1984).

nant vitamin K-dependent proteins.[25] The 293 cells appear to be able to produce and carboxylate some vitamin K-dependent protein substrates more efficiently than the CHO DUKX-B11 cell line.[26] In addition, they may be transfected with a variety of expression plasmids because identification of positive transfectants depends on the expression of a cotransfected plasmid containing a selectable drug resistance marker. The 293 cells grow slower than CHO cells and do not adhere as well to culture plates, making the screening of transfected colonies by the filter immunoassay technique difficult.

Untransfected 293 cells are grown and maintained in Dulbecco's modified Eagle's medium (GIBCO) containing 10% fetal calf serum, 5 μg/ml vitamin K_1, 10 mM HEPES, pH 7.2, and 100 μg/ml each penicillin and streptomycin. They are passaged using Versene (8.0 g/liter NaCl, 0.4 g/liter KCl, 0.2 g/liter EDTA, 5 mg/liter phenol red). The 293 cells are transfected by the calcium phosphate coprecipitation technique using 20 μg of expression plasmid and 2 μg of Hsp Neo.[27] Two days after transfection the cells are fed with selective media containing 0.1% Geneticin (G418; GIBCO). After 7–10 days in selective medium, colonies are visible. These colonies are picked and grown as separate transfected cell lines. Large volumes of conditioned media are also obtained from transfected 293 cell lines by growing and refeeding them in the Cell Factory (Nunc) apparatus.

Construction of Expression Plasmids

The cDNA encoding a specific vitamin K-dependent protein is cloned into the expression vector pMT2.[21] The pMT2 expression plasmid, a modified form of the expression vector p91023, was the generous gift of Dr. R. Kaufman (Genetics Institute, Cambridge, MA). In this vector the dihydrofolate reductase cDNA resides downstream of the insert cloning site.

Purification of Recombinant Proteins

Recombinant vitamin K-dependent proteins are purified from cell culture supernatant by immunoaffinity chromatography. To obtain fully carboxylated recombinant prothrombin cell culture supernatant is concentrated 10-fold using an Amicon (Danvers, MA) RA2000 concentrator, made 3 mM in calcium chloride, and 0.1% in Tween 20, and then loaded

[25] F. L. Graham, J. Smiley, W. C. Russell, and R. Nairn, J. Gen. Virol. 36, 59 (1977).
[26] B. W. Grinnell, J. D. Walls, C. Marks, A. L. Glasebrook, D. T. Berg, S. B. Yan, and N. U. Bang, Blood 76, 2546 (1990).
[27] G. T. Williams, T. K. McClanahan, and R. I. Morimoto, Mol. Cell. Biol. 9, 2574 (1989).

onto a Sepharose 4B affinity matrix coupled with conformation-specific antiprothrombin:Ca(II) antibodies (for a full description of antibody populations used in these purification methods, see below). The column is washed with 20 mM Tris-HCl, 1.0 M NaCl, 1 mM benzamidine, 3 mM CaCl$_2$, 0.1% Tween 20, pH 8.1. The recombinant prothrombin is eluted with 20 mM Tris-HCl, 0.15 M NaCl, 5 mM EDTA. To isolate all recombinant prothrombin species regardless of carboxylation state, Tween 20 is added to concentrated cell culture supernatant to 0.1% and the supernatant loaded onto a Sepharose 4B affinity matrix coupled with antiprothrombin:total antibodies. The colunmn is washed with 1.5 liters of 20 mM Tris-HCl, 1.0 M NaCl, 1 mM benzamidine, 3 mM CaCl$_2$, 0.1% Tween 20, pH 8.1, before elution with 4 M GuHCl. All purified recombinant prothrombin samples are extensively dialyzed against 20 mM Tris-HCl, 0.15 M NaCl, pH 8.1, at 4°. This procedure has been successfully employed with other vitamin K-dependent proteins.

Direct γ-Carboxyglutamic Acid Analysis of Vitamin K-Dependent Proteins

The method of determining the γ-carboxyglutamic acid content of vitamin K-dependent protein samples described here is a modification of the method developed by Kuwada and Katayama.[28] This technique is simple and sensitive, allowing determination of the γ-carboxyglutamic acid content of picomoles of sample. Each sample is purified by reversed-phase high-performance liquid chromatography (HPLC). The sample is hydrolyzed into its constituent amino acids under alkaline conditions in order to preserve the γ-carboxyglutamic acid.[29] The alkaline hydrolyzate is then derivatized with a reagent containing o-phthalaldehyde (OPA) and ethanethiol (ET). The derivatization reaction results in an amino acid/OPA/ET fluorescent complex in which the OPA/ET reagent is covalently bound to the primary amine of the amino acid residue (Fig. 2). Anion-exchange HPLC and fluorescence spectroscopy is employed to separate and detect the labeled amino acids. Under the chromatography conditions employed in this method, the separation of glutamic acid, aspartic acid, and γ-carboxyglutamic acid is optimized. The Gla content of the sample is calculated from the ratios of the peak areas of the three amino acids using a standard such as plasma-derived prothrombin for comparison. Because the Gla content of the protein sample is determined only as a relative value, it is extremely important that only pure protein samples are used for the analysis.

[28] M. Kuwada and K. Katayama, *Anal. Biochem.* **131**, 173 (1983).
[29] P. V. Hauschka, *Anal. Biochem.* **80**, 212 (1977).

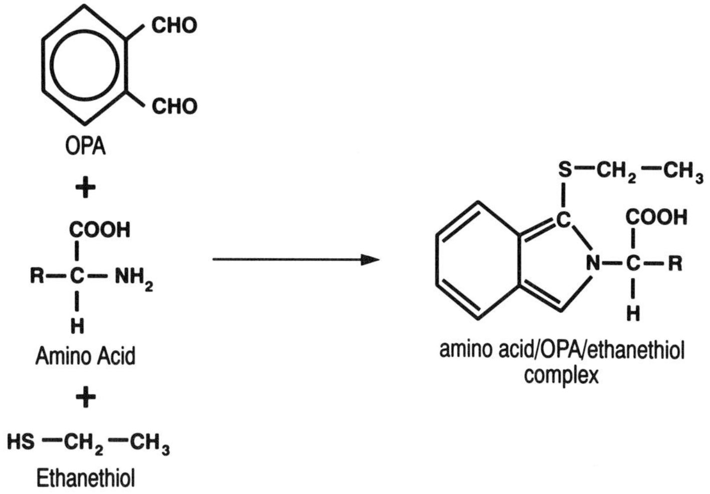

FIG. 2. *o*-Phthalaldehyde reaction with an amino acid in the presence of ethanethiol. The reaction forms an amino acid/OPA/ethanethiol fluorescent complex.

Reversed-Phase HPLC Purification of Samples. Vitamin K-dependent protein samples are purified by HPLC using a C_4 reversed-phase cartridge and a Beckman HPLC system equipped with System Gold software (version 310, Fullerton, CA). Each sample (300–500 pmol) is purified on an Aquapore C4 300 A cartridge (Brownlee Laboratories, Applied Biosystems, Foster City, CA) (3 cm × 4.6 mm) with a flow rate of 1 ml/min [solvent A, 0.1% (v/v) trifluoroacetic acid in water; solvent B, 0.1% (v/v) trifluoroacetic acid in acetonitrile]. A linear gradient of 15–50% solvent B in 35 min (1.0% B/min) is employed. Ultraviolet absorbance is monitored at 215 and 280 nm. Protein fractions are collected into 1.5-ml microfuge tubes, immediately frozen in a bath of dry ice and ethanol, and lyophilized.

Alkaline Hydrolysis and Neutralization. Each lyophilized protein sample is resuspended in 50 μl of 2.5 *M* KOH and transferred to a small acid-washed, pyrolyzed glass tube (WISP tubes, Waters, Milford, MA). The tubes containing the samples are placed in a glass screw-cap reagent chamber (Pierce, Rockford, IL) fitted with a Mininert valve closure (Pierce). One ml of 2.5 *M* KOH is added to the bottom of the hydrolysis chamber to maintain sample volume. The Pierce vial containing the samples is placed alternatively under vacuum for 1 min and then flushed with N_2 for 1 min, for three cycles. Finally, the vial is sealed under vacuum. The hydrolysis chamber containing the samples is incubated in a sand bath at 100 to 110° for 16 hr. After the alkaline hydrolysis is complete the samples are removed from the hydrolysis chamber and neutralized by the addition of small aliquots of 70 and 7% perchloric acid until the pH of each hydrolyzate is approximately 4. The insoluble potassium perchlorate is pelleted by spinning the sample in a microfuge at 14,000 g for 10 min at 22°. The supernatant is removed to an acid-washed, pyrolyzed glass tube.

Preparation of OPA/ET Reagent. The OPA/ET reagent is made by dissolving 10 mg of OPA (Sigma, St. Louis, MO) into 0.5 ml HPLC-grade methanol. Ethanethiol (5 μl) (Fluka, Ronkonkoma, NY) is added to the OPA and vortexed. Next, 1.0 ml of 0.15 M sodium borate buffer, 0.2% Brij 35, pH 10.5, is added. The solution is vortexed and stored under N_2 at room temperature. The OPA/ET reagent is kept overnight before use. Although it is preferable to make the OPA/ET reagent the day before each analysis, 4–5 μl of ethanethiol may be added after a few days to extend the life of the reagent.

Derivatization with OPA/ET Reagent. The alkaline hydrolyzates are derivatized by mixing an aliquot of the hydrolyzate (50 μl) with 50 μl of the OPA/ET reagent. The sample is incubated for 2 min at room temperature and then 100 μl of 0.1 M KH_2PO_4 in 66% acetonitrile is added. The sample is vortexed and protected from light until it is ready for injection. Because the fluorescence yield of the derivatized amino acids in the sample decreases with time, samples are not derivatized more than 1 hr before they are subjected to amino acid analysis.

Anion-Exchange HPLC. Our HPLC system consists of Beckman 126 pumps with System Gold software (version 310) coupled to an Applied Biosystems (Foster City, CA) Model 980 fluorescence detector. The fluorescence detector is equipped with a 418-nm bandpass filter. The System Gold HPLC unit is connected to a WISP autosampler (Waters). A Nucleosil 5SB anion-exchange column (4.6 × 250 mm; Macherey-Nagel, Germany) equipped with a guard cartridge is used to perform the amino acid analysis. The column is run at room temperature under isocratic conditions at a flow rate of 1 ml/min; the mobile phase consists of equal volumes of 0.1 M sodium citrate buffer, pH 5.28, and acetonitrile. These conditions, a modification of the published procedure of Katayama and Kuwada, result in improved resolution of the γ-carboxyglutamic acid peak. The excitation monochromator of the fluorescence detector is set at 240 nm. For each analysis, 25–200 μl of derivatized sample is injected.

Calculation of Data to Determine Gla Content of Samples. The first clearly resolved peak with a retention time of about 16.7 to 19 min represents derivatized glutamic acid (Fig. 3). The second peak with a retention time of about 19.6 to 23 min represents derivatized aspartic acid. The third peak with a retention time of about 24.7 to 30 min represents derivatized γ-carboxyglutamic acid. Glutamine and asparagine are converted to glutamic acid and aspartic acid, respectively, during the alkaline

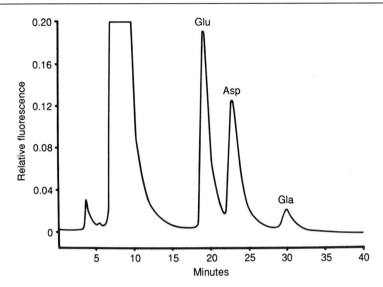

FIG. 3. HPLC chromatograph of a plasma-derived human prothrombin sample. The sample has been reversed-phase purified, base hydrolyzed, derivatized with OPA/ET reagent, and chromatographed as described.

hydrolysis. The areas of the peaks can be used to calculate the molar ratios of Glu:Gla, Asp:Gla, and Glu:Asp. The molar ratios for the samples are compared with the molar ratios obtained from a known protein standard, and the theoretical molar ratios of that standard protein. A correction factor is determined for the difference in the fluorescence yield of Gla compared with that of Glu + Gln or Asp + Asn in the protein standard. This correction factor is used to correct the values obtained for the unknown sample.

Determination of γ-Carboxyglutamic Acid in Vitamin K-Dependent Proteins by Immunologic Methods

The extent to which vitamin K-dependent proteins are γ-carboxylated has also been determined by a radioimmunoassay method using conformation-specific antibodies. In this method two immunoassays are performed, each using different fractions of a polyclonal antibody population raised against the vitamin K-dependent protein of interest. One population of polyclonal antibodies reacts exclusively with antigen epitopes which are stabilized by divalent metal ions. Because only carboxylated vitamin K-dependent proteins are able to assume this metal-stabilized conformer, this reagent is a highly sensitive probe for reporting the carboxylation state of the antigen [e.g., antiprothrombin:Ca(II) antibodies]. The other population of antibodies is directed against antigen epitopes which are not stabilized by metal ions. When employed in an immunoassay this antibody population reports the total amount of antigen present in the sample regardless of the extent to which it is carboxylated (e.g., anti-prothrombin:total antibodies). From the results of both radioimmunoassays the carboxylation state of the sample can be determined.

Preparation of Polyclonal Antibody Populations. A thorough discussion of conformation-specific antibodies is available (see Furie *et al.*[30]). This reference contains detailed technical procedures for obtaining conformation-specific antibodies, purifying antibody subpopulations, preparing affinity chromatography matrices, and developing radioimmunoassays. Therefore, only the specific procedures required to purify the antibody populations and perform the radioimmunoassays for this method of Gla analysis will be considered here.

Conformation-specific rabbit antiprothrombin antibodies are purified from serum obtained from a rabbit immunized with plasma-derived human prothrombin. The antisera is made 3 mM in calcium chloride and bound to a prothrombin-Sepharose 4B affinity column which has been equilibrated in 40 mM Tris-HCl, 150 mM sodium chloride, 3 mM calcium chloride, pH 8.1. The column is washed with 40 mM Tris-HCl, 1.0 M sodium chloride, 3 mM calcium chloride, pH 8.1. Conformation-specific antiprothrombin antibodies are eluted from the column with 40 mM Tris-HCl, 5 mM EDTA, pH 8.1. Antibodies remaining on the prothrombin-Sepharose column are then eluted with 4 M guanidine hydrochloride (antiprothrombin:total antibodies). The purified antibody populations are dialyzed into 40 mM tris-HCl, 150 mM sodium chloride, pH 8.1, at 4°. Conformation-specific antibodies and total antibody populations have been isolated from antisera raised against other vitamin K-dependent proteins.

Competition Radioimmunoassays. The displacement of ^{125}I-labeled prothrombin from antiprothrombin antibodies is measured using a solution-phase competition radioimmunoassay. Plasma-derived human prothrombin is labeled with ^{125}I by the lactoperoxidase method using Enyzmobeads (Bio-Rad, Richmond, CA).[31,32] Antiprothrombin:total antibodies (1.1×10^{-9} M) are added to a reaction mixture which includes ^{125}I-labeled prothrombin (1.7×10^{-10} M) and varying concentrations of competitor. All components of the assay are diluted into Tris-buffered saline containing 1 mM benzamidine, 3 mM EDTA, and carrier rabbit γ-globulin, unless

[30] B. Furie, R. A. Blanchard, D. J. Robison, M. M. Tai, and B. C. Furie, this series, Vol. 84, p. 60.
[31] J. J. Marcholonis, *Biochem. J.* **113**, 299 (1969).
[32] M. Morrison and G. S. Bayse, *Biochemistry* **9**, 2995 (1970).

the unknown competitors are obtained by a recombinant expression system and not isolated from cell culture supernatant. In that case all components of the assay are diluted into tissue culture media containing 1 mM benzamidine, 3 mM EDTA, and carrier rabbit γ-globulin. In a separate assay, conformation-specific antiprothrombin:Ca(II) antibodies (3.4 × 10^{-10} M) are added to the same reaction mixture except that 3 mM calcium chloride replaces EDTA. After overnight incubation at 4°, the bound ^{125}I-labeled prothrombin is precipitated by the addition of goat antirabbit immunoglobulin. After centrifugation the supernatant is removed and the precipitate is assayed for ^{125}I in a Beckman Gamma 8000 gamma counter. A standard curve is prepared with known concentrations (1.7 × 10^{-8} to 3.3 × 10^{-11} M) of plasma-derived human prothrombin to determine unknown antigen concentrations. The percentage of carboxylation of the sample is determined by dividing the concentration of the carboxylated antigen by the concentration of the total antigen and multiplying by 100. Analogous radioimmunoassays have been employed to determine the carboxylation state of other vitamin K-dependent protein samples.

In Vitro Carboxylation Assay

Preparation of Crude Vitamin K-Dependent Carboxylase

Investigation of the kinetics of enzyme:substrate interactions have been accomplished using a partially purified carboxylase preparation. Our method of preparation, a variation of several previously published purification strategies,[17,33] entails preparing microsomes from bovine liver and then solubilizing the carboxylase from these microsomes with the nondenaturing zwitterionic detergent 3-[(3-cholamidopropyl)dimethylammonio]-1-propane sulfonate (CHAPS) in the presence of high salt concentrations.[34]

Preparation of Bovine Liver Microsomes. Fresh bovine liver is obtained at a slaughterhouse. It is sliced immediately into 2-cm slices and submerged in ice-cold (4°) buffer [150 mM sodium chloride, 50 mM Tris-HCl, 5% (v/v) glycerol, pH 7.5]. All subsequent steps are done at 4° unless otherwise stated. The liver slices are cut into 2-cm cubes and homogenized in an equal volume of homogenization buffer [150 mM sodium chloride, 50 mM Tris-HCl, 5% (v/v) glycerol, 1 mM EDTA, 1 mM benzamidine, 1 mM phenylmethylsulfonyl fluoride, pH 7.5] using a Polytron homogenizer Model PT-MR 3000 (Kinematica, Littau, Switzerland) equipped with a large probe at a setting of 18 for 1 to 5 min. During this time care must be taken to avoid frothing of the homogenate. The homogenate is filtered

[33] J.-M. Girardot, *J. Biol. Chem.* **265,** 15008 (1982).
[34] J.-M. Girardot and B. C. Johnson, *Anal. Biochem.* **121,** 315 (1982).

through gauze and additional tissue debris is sedimented at 10,000 g at 4° for 10 min. The supernatant is filtered through gauze, and microsomes from this supernatant are sedimented in an ultracentrifuge at 130,000 g at 4° for 60 min. The supernatant from this centrifugation is discarded and the soft microsomal pellet, which represents approximately 15% of the starting volume, is transferred to a beaker and homogenized with the Polytron for 30 sec. The microsomes are then quickly frozen and stored at −80°. The protein concentration of the bovine liver microsomal preparation is approximately 75 mg/ml, as determined by deoxycholate–trichloroacetic acid precipitation followed by quantitation using the Folin phenol method.[35]

Solubilization of Vitamin K-Dependent Carboxylase from Bovine Liver Microsomes. Bovine liver microsomes are thawed at 4°. An equal volume of solution containing 2% CHAPS, 2 M sodium chloride, 5 mM DTT, 2 mM benzamidine, 0.2 mg/ml soybean trypsin inhibitor is added to the thawed microsomes. The mixture is stirred for 30 min in an ice-water bath to solubilize carboxylase activity from the microsomes. Microsomal debris is sedimented in an ultracentrifuge as previously described at 130,000 g at 4° for 60 min. The supernatant is decanted and ammonium sulfate (Sigma, grade III) to a final concentration of 55% saturation is slowly added while stirring the supernatant at 4°. The pH is maintained at pH 7.5 with 1 M Tris-HCl, pH 7.5. The solution is stirred at 4° for 20 min. The precipitated protein is collected by centrifugation at 10,000 g for 10 min at 4°. The floating pastelike pellet that contains carboxylase is separated from the fluid phase. The pellet is slightly diluted with homogenization buffer to decrease its viscosity and then homogenized with the Polytron for 30 sec as described above. The CHAPS/sodium chloride-solubilized ammonium sulfate-precipitated bovine liver microsomal protein is aliquoted, quick frozen, and stored at −80° for several months without loss of activity. The resultant microsomal protein concentration is approximately 75 to 100 mg/ml and has a specific carboxylase activity of approximately 4 × 10⁵ cpm/mg/hr when $^{14}CO_2$ and the synthetic pentapeptide FLEEL are used as a substrate in the *in vitro* carboxylation assay described below.

In Vitro Carboxylation Assay

This assay, which is a modification of the method of Vermeer and Soute,[36] measures the amount of $^{14}CO_2$ incorporated into a substrate by a crude preparation of bovine liver vitamin K-dependent carboxylase. The assay allows for evaluation of various synthetic peptides to serve as substrates for the vitamin K-dependent carboxylase.

---

[35] G. L. Peterson, this series, Vol. 91, p. 95.

[36] C. Vermeer and B. A. M. Soute, *in* "Current Advances in Vitamin K Research" (J. W. Suttie, ed.), p. 25. Elsevier, New York, 1988.

Preparation of Reduced Vitamin KH$_2$. One hour prior to beginning the carboxylase assay, vitamin K is reduced as follows: 1 ml of vitamin K$_1$ (phytonadione injection, 10 mg/ml, USP, Abbott Laboratories, Chicago, IL) is transferred to a foil-wrapped capped tube. 2-Mercaptoethanol (5 μl) and 1 mg solid sodium borohydride are added. The solution is vortexed. Additions of 1 mg of sodium borohydride are added until the solution is clear. The reduction is allowed to proceed at room temperature for at least 1 hr in a foil-wrapped capped tube.

Preparation of FLEEL Peptide Stock Solution. One ml of water is added to 100 mg lyophilized synthetic peptide FLEEL (Sigma). Then 200 μl of 1 N NaOH is added, and the solution is vortexed until the peptide dissolves. The pH of the solution should be between 7 and 8. The peptide solution is transferred to a microfuge tube and the volume is brought to 1.5 ml with water. The final concentration of this peptide stock is 100 mM. Other synthetic peptide substrates are synthesized and purified according to standard procedures.[18]

Preparation of Reaction Samples. The amount of $^{14}CO_2$ incorporated into peptide substrates is measured in reaction mixtures of 125 μl final volume. Each 125-μl reaction contains 10 μCi NaH$^{14}CO_3$ (2 mCi/ml, 56.1 mCi/mmol; Amersham, Amersham, UK), 5 μl vitamin KH$_2$ prepared as described, 12.5 μl 100 mM FLEEL (or other synthetic peptide substrates at appropriate concentration), 1 μl 1 M dithiothreitol in phosphate-buffered saline, pH 7.4, and 0.1% (w/v) CHAPS. The 100-μl reaction cocktail is maintained on ice until initiation of the reaction by addition of 25 μl of partially purified carboxylase (approximately 1 mg of protein with a specific activity of 4 $\times$ 10^5 cpm/mg/hr). All work is done in a fume hood. Tubes are capped immediately after initiation of the reaction and incubated for 30 min at 25°. Longer incubations maybe required for less active carboxylase preparations. The reaction tubes are uncapped and transferred in a chemical fume hood to a vacuum bell jar sealed with a silicon membrane. The vacuum bell jar is fitted to a vacuum pump protected by a 1 N NaOH trap to capture free $^{14}CO_2$. The reaction is quenched by the addition of 1 ml 10% trichloroacetic acid to each tube using a syringe and needle to make additions through the silicon membrane. The samples are kept under vacuum for 5 min. To remove traces of remaining unincorporated $^{14}CO_2$ and reduce the volume for scintillation counting the samples are transferred to glass scintillation vials containing boiling chips and boiled on a heating plate until about 0.3 to 0.5 ml of solution remains in each vial. After cooling, 5 ml of Atomlight scintillation fluid (DuPont-NEN, Boston, MA) is added to each vial. The samples are counted in a Beckman LS1801 liquid scintillation counter. A background value, obtained by carrying out the reactions in the absence of vitamin K, is subtracted from the counts recorded for each sample.

[26] Expression of Recombinant Vitamin K-Dependent Proteins in Mammalian Cells: Factors IX and VII

By KATHLEEN L. BERKNER

Introduction

The ability to generate cell lines expressing recombinant vitamin K-dependent proteins provides a powerful system for their analysis. Because only a single recombinant protein is expressed, isolation of the protein does not have the problem of trace contamination with other blood proteins. This feature provides the additional benefit that the recombinant protein undergoes substantially less exposure to serine proteases during purification, and thus can be isolated as a homogeneous preparation of the zymogen form. The vitamin K-dependent proteins undergo several posttranslational modifications, which in many cases include propeptide processing, γ-carboxylation, aspartyl β-hydroxylation, carbohydrate addition (both N-linked and O-linked), and proteolytic cleavage to a two-chain form. Cell lines expressing these recombinant proteins can be manipulated to block many of these modifications, making it possible to analyze the effect of these modifications on activity. For example, culturing the cells in tunicamycin eliminates N-linked glycosylation, or depleting the vitamin K in the media can be used to abolish carboxylation.[1,2] Inhibition of aspartyl β-hydroxylation and its effect on factor IX activity have also been examined.[3] Another useful manipulation is the ability to label these recombinant protein-producing cell lines *in vivo,* which makes it possible to study intracellular processing events.

Mutational analysis of recombinant vitamin K-dependent proteins has also provided an important approach for analyzing structure–function relationships of these proteins. For example, mutations in the propeptide,[4-7] in the aspartyl residue that undergoes β-hydroxylation,[8,9] in the Gla

[1] K. Berkner, S. Busby, E. Davie, C. Hart, M. Insley, W. Kisiel, A. Kumar, M. Murray, P. O'Hara, R. Woodbury, and F. Hagen, *Cold Spring Harbor Symp. Quant. Biol.* **51,** 531 (1986).

[2] S. Busby, A. Kumar, M. Joseph, L. Halfpap, M. Insley, K. Berkner, K. Kurachi, and R. Woodbury, *Nature (London)* **316,** 271 (1985).

[3] C. K. Derian, W. VanDusen, C. T. Przysiecki, P. N. Walsh, K. L. Berkner, R. J. Kaufman, and P. A. Friedman, *J. Biol. Chem.* **264,** 6615 (1989).

[4] D. C. Foster, M. S. Rudinski, B. G. Schach, K. L. Berkner, A. A. Kumar, F. S. Hagen, C. A. Sprecher, M. Y. Insley, and E. W. Davie, *Biochemistry* **26,** 7003 (1987).

[5] M. J. Rabiet, M. J. Jorgensen, B. Furie, and B. C. Furie, *J. Biol. Chem.* **262,** 14895 (1987).

domain,[10] and in the endoproteolytic processing site[11,12] have all provided functional information on factor IX, factor VII, and protein C. Fusion proteins between the vitamin K-dependent proteins can also be generated, and were used to show, for example, that the Gla domain of one protein (factor IX) can substitute for another (factor VII) in generating a protein with coagulant activity.[1,13]

Biologically active vitamin K-dependent proteins have been expressed in a variety of cell lines.[1,2,14-19] The extensive posttranslational modifications that these proteins undergo present a unique set of challenges not found for other recombinant proteins. Production of an accurately processed recombinant protein is not always routine, and more extensive characterization of the recombinant product is required. This chapter uses factor IX and factor VII as models for the generation of cell lines that express vitamin K-dependent proteins and for the characterization of these recombinant products. Some of the limitations that have been observed in various posttranslational modifications will be discussed. Two factors, the production level and the choice of cell line, impact significantly in producing a recombinant protein as qualitatively identical to the cognate plasma protein as possible. The effect that these two parameters have on the expression of factor IX will be described in detail. Not all vitamin K-

[6] M. J. Jorgensen, A. B. Cantor, B. C. Furie, C. L. Brown, C. B. Shoemaker, and B. Furie, Cell (Cambridge, Mass.) 48, 185 (1987).
[7] S. J. Busby, K. L. Berkner, L. M. Halfpap, J. E. Gambee, and A. A. Kumar, in "Current Advances in Vitamin K Research: A Steenbock Symposium" (J. W. Suttie, ed.), p. 173. Elsevier, New York, 1988.
[8] A.-K. Ohlin, G. Landes, P. Bourdon, C. Oppenheimer, R. Wydro, and J. Stenflo, J. Biol. Chem. 263, 19240 (1988).
[9] D. J. G. Rees, I. M. Jones, P. A. Handford, S. J. Walter, M. P. Esnouf, K. J. Smith, and G. G. Brownlee, EMBO J. 7, 2053 (1988).
[10] L. Zhang and F. J. Castellino, Biochemistry 29, 10828 (1990).
[11] D. C. Foster, C. A. Sprecher, R. D. Holly, J. E. Gambee, K. M. Walker, and A. A. Kumar, Biochemistry 29, 347 (1990).
[12] P. Wildgoose, K. L. Berkner, and W. Kisiel, Biochemistry 29, 3413 (1990).
[13] K. L. Berkner, D. E. Prunkard, J. E. Gambee, K. M. Walker, L. M. Halfpap, S. J. Busby, and A. A. Kumar, in "Current Advances in Vitamin K Research: A Steenbock Symposium" (J. W. Suttie, ed.), p. 199. Elsevier, New York, 1988.
[14] D. S. Anson, D. E. G. Austen, and G. G. Brownlee, Nature (London) 315, 683 (1985).
[15] H. de la Salle, W. Altenburger, R. Elkaim, K. Dott, A. Dieterlé, R. Drillien, J. P. Cazenave, P. Tolstoshev, and J. P. Lecocq, Nature (London) 316, 268 (1985).
[16] B. W. Grinnell, D. T. Berg, J. Walls, and S. B. Yan, Biotechnology 5, 1189 (1987).
[17] R. J. Kaufman, L. C. Wasley, B. C. Furie, B. Furie, and C. B. Shoemaker, J. Biol. Chem. 261, 9622 (1986).
[18] C. Oppenheimer and R. Wydro, in "Current Advances in Vitamin K Research: A Steenbock Symposium" (J. Suttie, ed.), p. 165. Elsevier, New York, 1988.
[19] J. W. Suttie, Thromb. Res. 44, 129 (1986).

dependent proteins respond identically, however, in the same expression system. Factor VII and factor IX, for example, differ in their efficiency of carboxylation and both of these proteins are secreted as single chains, in contrast to protein C, which undergoes cleavage to a two-chain form during secretion. Relevant observations about the expression of factor VII and, to a limited extent, protein C, will therefore be included as well.

Generation of Cell Lines Producing Recombinant Vitamin K-Dependent Proteins

Because the vitamin K-dependent proteins require so many different modifications for activity, the main emphasis for their expression has been to find an appropriate cell line (as described below). Carboxylation, in particular, is limiting in many cell lines, and has therefore been a focal point in choosing an optimal one. Several different vectors and cell lines have been analyzed for efficacy of expression of biologically active factor VII. The main concern with vector usage is that it function efficiently in the desired cell type. Promoters that have been tested include the SV40 early promoter, the Ad2 major late promoter, and the metallothionein promoter. Vectors containing these promoters were tested for transient activity in a wide variety of cell lines using a reporter gene encoding chloramphenicol acetyltransferase, and many of these lines were efficient for expression.

This type of testing led to the development of an Ad2 major late promoter-based vector (Fig. 1) that has been versatile in a number of cell lines of interest. cDNAs encoding human factor VII, factor IX, and protein C have all been expressed in the vector. In some cases the 5' untranslated sequences have been modified, e.g., to remove a long GC tract in a factor IX cDNA or to delete an upstream, out of frame ATG that decreases translation efficiency of the initiation methionine. The 3' untranslated sequences do not appear to affect expression efficiency. Factor VII, for example, contains a 1-kb 3' untranslated region, and expression results using a full-length or 3' untranslated–deleted cDNA are indistinguishable. The cDNAs encoding the vitamin K-dependent proteins are inserted into the vector downstream of a set of splice signals that ultimately adjoin an Ad2 tripartite leader sequence to the mRNAs being expressed. This sequence has been shown to increase translation efficiency in heterologous gene expression.[20-24] The last important element in the vector is the SV40

[20] J. Logan and T. Shenk, *Proc. Natl. Acad. Sci. U.S.A.* **81**, 3655 (1984).
[21] K. L. Berkner and P. A. Sharp, *Nucleic Acids Res.* **13**, 841 (1985).
[22] A. R. Davis, B. Kostek, B. B. Mason, C. L. Hsiao, J. Morin, S. K. Dheer, and P. P. Hung, *Proc. Natl. Acad. Sci. U.S.A.* **82**, 7560 (1985).

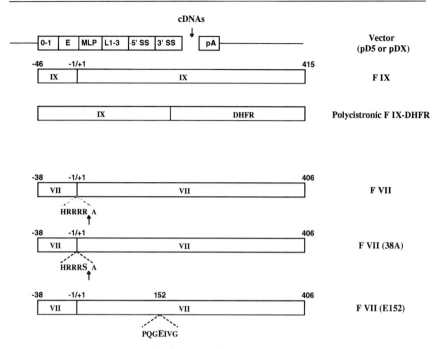

FIG. 1. Constructions encoding factor IX, factor VII, and the pD5/pDX mammalian cell expression vector. The numbers above each construction correspond to the amino acid number for the leader (negative numbers, composed of the signal sequence and the propeptide) and mature portion of each protein. In all cases but one, the leader and mature part of the protein are each indicated by an open box. The exception is the polycistronic construct, which comprises the entire open reading frame for factor IX and dihydrofolate reductase (DHFR) (each indicated by a boxed segment). The mutations for factor VII are highlighted by larger lettering. The expression vector contains a unique *Bam*HI (pD5) or *Eco*RI (pDX) site for cDNA insertion. The vectors contain the leftmost Ad5 terminal sequences (0–1), the SV40 enhancer (E), the Ad2 major late promoter (MLP), a cDNA encoding the Ad2 tripartite leader (L1–3), splice signals (5′ SS, 3′ SS), and an SV40 polyadenylation signal (pA).

enhancer (E in Fig. 1), which significantly increases expression in a number of cell lines. In 293 cells, for example, the levels of chloramphenicol acetyltransferase and factor VII expression are increased approximately 10-fold by the inclusion of the enhancer in the vector.

To determine the best cell lines for the expression of factor VII, factor VII/pDX (Fig. 1) was first transiently transfected into a wide variety of cell

23 R. J. Kaufman, *Proc. Natl. Acad. Sci. U.S.A.* **82,** 689 (1985).
24 K. L. Berkner, B. S. Schaffhausen, T. M. Roberts, and P. A. Sharp, *J. Virol.* **61,** 1213 (1987).

lines. Secreted levels of biologically active factor VII ranged from 10 to 200 ng/ml/day, and those lines that exhibited factor VII activity were then tested for ease and efficacy of isolating stable transfectants. Hygromycin, G418, and methotrexate were tested as selective agents. Almost all of the cell lines (listed below) generated stable transfectants using the phosphotransferase gene and G418. Methotrexate and the dihydrofolate reductase (DHFR) gene were useful only for CHO cells and BHK cells.

Three permutations have been used to express the cDNAs encoding human factor VII, factor IX, or protein C. These include placing the selectable gene and the gene of interest on the same plasmid or on different plasmids. Containment on the same plasmid is useful for cell lines [e.g., Hep G2 cells or human hepatoma ATCC (Rockville, MD) HTB-52 cells] that are inefficiently transfected. With other cell lines (e.g., CHO cells, BHK cells, or 293 cells) that generate transfectants more readily, however, there is no great advantage in having both genes on the same plasmid. With BHK cells transfected with three separate plasmids encoding factor VII, factor IX, and dihydrofolate reductase, for example, about half of the transfectants expressed all three genes. The third alternative is to use polycistronic-encoding constructs: the cDNAs encoding factor IX or factor VII and DHFR or phosphotransferase are both positioned downstream of the same promoter (Fig. 1).[25] Because the translational efficiency of a cDNA decreases when it is in the second cistron position, the order of the gene of interest and the selectable gene is important.[25]

In general, the considerations in choosing a suitable cell line for expressing a recombinant protein include (1) ease of isolating transfectants, (2) growth rate of the cell line, (3) ability of the cell line to grow in serum-free media or (4) in suspension, (5) ability to amplify the gene product, (6) stability of the cell line in the absence of selection, (7) production levels of the recombinant protein, and (8) scale-up capabilities. In addition, the vitamin K-dependent proteins require that a number of different posttranslational modifications be performed efficiently. Our studies of factor VII and factor IX production have included two rat hepatoma lines, two human hepatoma lines (HuHep ATCC HTB-52 and Hep G2 cells), BHK cells, CHO cells, and 293 cells. Factor IX expression in rat and human hepatoma lines, in CHO cells, and in mouse fibroblasts has also been reported by others.[14,15,17] CHO cells have provided several advantages for the production of noncarboxylated proteins, such as the potential for gene amplification or growth in suspension and in serum-free media. However, they are inefficient in both γ-carboxylation and propeptide processing of factor IX and factor VII, making them a less attractive

[25] K. L. Berkner, E. Boel, and D. Prunkard, in "Viral Vectors" (Y. Gluzman and S. H. Hughes, eds.), p. 56. Cold Spring Harbor Lab., Cold Spring Harbor, New York, 1988.

candidate cell source. The hepatoma lines were tested because liver is the source of several known vitamin K-dependent proteins. Colonies derived from the human hepatoma line (ATCC HTB-52) are efficient in the secretion of both factor IX (described below) and factor VII. However, generating these colonies is a lengthy project, and offers no real advantage over 293 cells, which accurately process factor VII and factor IX and which are easier to work with. Hep G2 cells naturally secrete several vitamin K-dependent proteins[26,27]; however, production of colonies expressing recombinant factor VII takes months and the expression levels are poor (e.g., 20–100 ng/ml/day).[28]

By far the most useful cell lines for expression have been BHK cells and 293 cells. Establishing a factor IX, protein C, or factor VII-producing cell line is easier in BHK cells, as will be detailed below, and BHK cells can amplify exogenously added DHFR, along with a cotransfected gene of interest. BHK cells have also been readily amenable to scale-up for isolating large amounts of protein. However, some of the posttranslational modifications are limiting in BHK cells with increased levels of production, resulting in the secretion of incompletely modified protein (described below). It is therefore critical to isolate cell lines expressing below these limiting levels (e.g., ≤ 5 μg/ml/day for factor VII or ≤ 0.5 μg/ml/day for factor IX). With 293 cells the limiting levels are substantially higher (e.g., at least 20-fold more with factor IX), making it the best choice to obtain a cell line that expresses higher levels of vitamin K-dependent proteins. With BHK cells and 293 cells, expression of endogenous vitamin K-dependent proteins has not been detected.[29]

Generation of Factor VII and Factor IX-Producing Cell Lines

Reagents

DNAs: Plasmid DNAs and salmon sperm DNA are stored at 4° in sterile 10 mM Tris, pH 8, 1 mM EDTA; CsCl-banded plasmid DNA or DNA isolated from rapid preparations[30] can both be used, although lower transfection efficiencies are obtained with the latter

Trypsin, 0.25% (JRH Biosciences, Lenexa, KS)

Versene, 0.2% EDTA in phosphate-buffered saline (PBS) (GIBCO, Grand Island, NY)

[26] D. S. Fair and B. R. Bahnak, *Blood* **64**, 194 (1984).
[27] D. S. Fair and R. A. Marlar, *Blood* **67**, 64 (1986).
[28] K. L. Berkner, data not shown (1985).
[29] K. L. Berkner, data not shown (1989).
[30] J. Sambrook, E. F. Fritsch, and T. Maniatis, "Molecular Cloning: A Laboratory Manual." Cold Spring Harbor Lab., Cold Spring Harbor, New York, 1989.

Media: Dulbecco's modified Eagle's medium (MEM) (GIBCO), 0.01%
(w/v) penicillin – streptomycin – neomycin (GIBCO), 0.1 mg/ml pyruvate (Irvine, Santa Ana, CA), 10% dialyzed fetal bovine serum (JRH
Biosciences), glutamine (2 mM; JRH Biosciences). Serum-less media
is identical except that no serum is added.

Phosphate-buffered saline (GIBCO)

Methotrexate (Sigma, St. Louis, MO): A 10^{-1} M stock is prepared in
dimethyl sulfoxide (DMSO) and then diluted into PBS

G418 (GIBCO): Prepared in PBS as a 50 mg/ml stock

Nitrocellulose filters (Millipore, Bedford, MA): HATF 08225 or HATF
13750 are the only ones used because they are manufactured without
cytotoxic surfactants;[31] autoclaved 15 min on liquid cycle in aluminum foil with the blue interleaves in place

Teflon mesh (Spectrum Medical Industries): Mesh pieces are pretreated
in 0.1 N HCl for 1 hr and rinsed in water prior to autoclaving. After
use, the meshes can be recycled by incubating in trypsin for several
hours and then rinsing with water

Western A buffer: 50 mM Tris, pH 7.4, 5 mM EDTA, 0.05% Nonidet
P-40 (NP-40), 150 mM NaCl, and 0.25% gelatin

Primary antibody: Rabbit polyclonal α-factor VII or α-factor IX antisera
diluted 1:200 in Western A buffer. These dilutions can be reused
several times and are stored at 4° in 0.02% (w/v) sodium azide between uses

Secondary antibody. Goat α-rabbit conjugated to horseradish peroxidase
(Cappel, Westchester, PA), diluted 1:1000 in Western A buffer

Western B buffer: 50 mM Tris, pH 7.4, 5 mM EDTA, 0.05% (v/v)
NP-40, 1 M NaCl, 0.4% sarcosyl, and 0.25% gelatin (w/v)

Color reagent: Horseradish peroxidase color development reagent
(60 mg, Bio-Rad, Richmond, CA) is dissolved in 20 ml methanol and
then added to 100 ml 50 mM Tris, pH 7.4, 150 mM NaCl, and 100 μl
H_2O_2 (30% solution)

BHK cells and human hepatoma cells (ATCC HTB52) are seeded 1 day
before transfection so that the confluency of a 10-cm dish is 20–30%
during the transfection. The 293 cells are split 2 days prior to transfection
to give a 50% confluency at the time of transfection. If the 293 cells are too
confluent or are not given sufficient time to recover between seeding and
transfection, the cells shed badly during the transfection. Plasmids encoding the expression cassettes [10 μg factor IX/pD5 (Fig. 1) or factor VII/
pDX[12]] are cotransfected with 10 μg of sonicated salmon sperm DNA
and 1 μg of plasmid DNA containing a G418 resistance (for 293 and

---

[31] A. A. McCracken and J. L. Brown, *BioTechniques,* Mar./Apr., p. 82 (1984).

human hepatoma cells) or DHFR (for BHK cells) encoding sequence.[32,33] DNA is added to the cells using calcium phosphate precipitation,[34,35] including a glycerol shock 5–7 hr after the addition of $CaCl_2$. A mock transfection, using only salmon sperm DNA, is performed in parallel to monitor the selection process. After 2 days, at which time the plates should be 90–100% confluent, the cells are removed with trypsin (BHK or human hepatoma cells) or Versene (293 cells) and distributed into duplicate 150-mm dishes at 1:10, 1:50, 1:100, and 1:500 dilutions in media containing 150 nM methotrexate (BHK cells) or 0.1% G418 (293 and human hepatoma cells). Media are changed weekly, and after 10–14 days (BHK cells), 3–4 weeks (293 cells), or 4–6 weeks (human hepatoma cells) colonies are visible, even to the naked eye. Mock transfected plates should have no colonies and few viable cells remaining. Mock transfected 293 cells, however, occasionally exhibit a high background.

The decision by which individual colonies are chosen depends on the subsequent use of the protein being isolated from that cell line. If a high-level production cell line is desired, for example, an immunofilter assay (described below) is used to screen the colonies. However, in many cases what is desired is an accurately posttranslationally modified protein, in which case several dozen random colonies are isolated; these will go through some characterization before a particular colony is chosen for production. At a minimum, this analysis would include levels of secretion using an ELISA and *in vivo* labeling plus gel analysis to determine the molecular weight form(s) of the secreted protein. Biological activity measurements and Gla analysis might also be used in the initial screening. Other considerations in choosing a final cell line would be the growth rate, ability to grow in media reduced or depleted in serum, and the stability of recombinant protein production in the absence of selection. Characterization of multiple colony isolates for any given construction is critical to avoid artifacts. The proteins undergo several posttranslational modifications, and different cell lines vary in their capacity for effecting these modifications.

The immunofilter assay, adapted from McCracken and Brown,[31] is a reliable and useful technique for the isolation of cell lines of interest, especially where large numbers of colonies need to be screened. In this protocol, colonies generated by transfection are overlayed with a Teflon mesh and then a nitrocellulose filter, and secreted protein absorbed to the

[32] P. Southern and P. Berg, *J. Mol. Appl. Genet.* **1**, 327 (1982).
[33] C. C. Simonsen and A. D. Levinson, *Proc. Natl. Acad. Sci. U.S.A.* **80**, 2495 (1983).
[34] M. Wigler, S. Silverstein, L. S. Lee, A. Pellicer, Y. Cheng, and R. Axel, *Cell (Cambridge, Mass.)* **11**, 223 (1977).
[35] F. L. Graham and A. J. van der Eb, *Virology* **52**, 456 (1973).

nitrocellulose is detected immunologically. The colony size and, to a very rough approximation, the level of expression are reflected in the size and intensity of the spots that appear on the nitrocellulose. The immunofilter assay is particularly useful with 293 cells. High background (i.e., nonexpressing colonies) is often observed after transfection of this cell line. The percent of positive colonies isolated per transfection varies between 5 and 50%. Moreover, it is desirable to isolate several colonies expressing any particular protein. The immunofilter assay is also useful in transfection experiments where more than one cDNA of interest is added to the same cell, because the frequency of obtaining a cell line expressing both DNAs is lower.

Plates to be assayed are rinsed with PBS and 2.5 ml (for a 100-mm dish) of serum-less medium is added. Both solutions are prewarmed to 37° before use. Using two pairs of sterile forceps (i.e., dipped in ethanol and flamed), the Teflon mesh is prewet in PBS and then gently laid on the cells, being especially careful to avoid trapped air bubbles. The nitrocellulose filter is then laid over the Teflon mesh and the filter orientation is marked on the bottom of the plate to correspond to marks placed on the nitrocellulose filter before autoclaving. After a 1- to 6-hr incubation at 37° the nitrocellulose filter is removed and placed in Western A buffer. If cells are being probed for expression of two different proteins, two separate nitrocellulose filters are used and each is incubated with the cells for 1–3 hr. The cells are monitored for viability using a microscope and then fluid changed to their usual media. Recovery of BHK cells is routine, but with 293 cells occasionally the colonies lift off the dish. Reserve transfected plates are clearly desirable.

Nitrocellulose filters are incubated in Western A buffer at room temperature for 1 hr or at 4° for 16 hr. Approximately 10 ml of solution per filter is used throughout this procedure. Primary antibody in fresh Western A buffer is added and after 1 hr at room temperature the filters are washed in Western A buffer. Secondary antibody is incubated with the filters at room temperature for 1 hr; the filters are then washed in Western B buffer, rinsed in water, and color reagent is added. The plates are rotated on a shaker until color develops (usually 30 sec to 5 min) and the reaction is stopped by transferring the filter to H_2O.

Individual colonies are then selected by trypsinization in cloning cylinders and expanded in 35-mm plates to generate enough cells so that an enzyme-linked immunosorbent assay (ELISA)[36] can be performed. The ELISA is generally performed on medium from a nearly confluent dish

---

[36] E. Harlow and D. Lane, "Antibodies: A Laboratory Manual." Cold Spring Harbor Lab., Cold Spring Harbor, New York, 1988.

(approximately 10^5 cells). Subsequent characterizations (e.g., clotting assays or *in vivo* labeling) are carried out when roughly 10^6 cells are obtained.

Both the transfection and passaging of cell lines are performed using vitamin K-deficient media. Carrying the cells in vitamin K-deficient media allows subsequent comparisons to be made on both carboxylated and noncarboxylated proteins. To measure, e.g., biological activity or other characteristics dependent on carboxylation, cells are fluid changed into vitamin K-containing medium (described below) 1 day before use in an experiment. Vitamin K manifests its effects almost immediately after addition to the medium and any protein originally synthesized in the vitamin K-depleted cells should be secreted from the cell by 24 hr. Comparing protein secreted from vitamin K-depleted or -containing cells cannot be accomplished by simply fluid-changing a vitamin K-containing cell line into medium lacking vitamin K. The vitamin K appears to be stored by the cells and it can take several weeks to deplete intracellular stores.[28]

Characterization of Cell Lines

Biological Activity Measurement

Reagents

Serum-free medium: A 1:1 mixture of Dulbecco's MEM (JRH Biosciences) and Ham's F12 (JRH Biosciences) supplemented with 0.1 mg/ml pyruvate (Irvine), 2 mM glutamine (JRH Biosciences), 0.01% penicillin–streptomycin–neomycin (GIBCO), insulin (5 mg/liter; GIBCO), selenium (3 μg/liter; Aldrich, Milwaukee, WI), transferrin (20 mg/liter; JRH Biosciences), fetuin (10 mg/liter; Sigma), HEPES, pH 7.2 (25 mM; JRH Biosciences)
Vitamin K (phytonadione; Merck, Sharp and Dohme, Westpoint, PA)
Factor IX-deficient plasma (George King, Overland Parks, KS)
Normal plasma (George King)
Phosphate-buffered saline (GIBCO)
APTT reagent (Actin FS; American Dade, Miami, FL)
25 mM CaCl$_2$
Factor VII-deficient plasma (George King)
Thromboplastin C (American Dade; contains CaCl$_2$)

Factor VII activity is measured using a one-step prothrombin time assay.[37] To construct the standard curve, normal plasma is serially diluted

[37] G. J. Broze, Jr. and P. W. Majerus, this series, Vol. 80, p. 228.

into PBS (from 1 : 5 to 1 : 640). Dilutions (100 μl) are mixed with 100 μl of factor VII-deficient plasma and 200 μl of thromboplastin C, and the clotting time is measured on an MLA Electra 800 automatic coagulation timer (Medical Laboratory Automation, Inc., Pleasantville, NY). The standard curve is based upon a value of 400 ng/ml for factor VII in normal plasma.[38] Media samples containing factor VII are diluted into PBS so that the clotting time, when measured, falls along the standard curve. The minimal dilution is twofold and 100 μl of diluent is used in the assay.

Factor IX activity is determined using an activated partial thromboplastin time (APTT)-based assay.[39] Serial dilutions of normal plasma (from 1 : 10 to 1 : 1280) in PBS are used to generate a standard curve. Dilutions (100 μl) are mixed with 100 μl of factor IX-deficient plasma and 100 μl of APTT reagent. After a 5 min incubation at 37°, 25 mM CaCl$_2$ (100 μl) is added and the clotting time is measured. Media samples diluted at least 1 : 2 into PBS (100 μl) are treated identically to measure the level of recombinant factor IX. The standard curve is based on a value of 3 μg/ml for factor IX in normal plasma,[40] which agreed with values obtained using an ELISA.

To prepare media samples for assay, cells (usually 100-mm dishes with cells near or at confluency, ~2–5 × 10^6 cells) are rinsed with PBS and then incubated in 5 ml of serum-free media for 1–2 days. Reliable activity measurements can also be obtained with media samples containing up to 1% dialyzed fetal calf serum, and so for cell lines that do not survive well in serum-free media, serum can be included. With activity measurements of protein C, reliable activity determinations cannot be obtained even in serum-free media, necessitating purification of the protein before measuring its activity. When media samples containing factor IX or factor VII are assayed, two controls are run in parallel: untransfected cells corresponding to the recombinant-expressing cell lines being analyzed and the recombinant cell line cultured in medium lacking vitamin K.

We have examined the effect of cell confluency on activity of factor VII and factor IX with two different cell lines (BHK and 293 cells) and observed no effect. For obtaining activity measurements, then, a near confluent plate is used to generate a more concentrated protein sample. Clotting values are compared to factor IX or factor VII protein concentrations determined using an ELISA to determine the specific activity of the protein.

Table I shows that it is possible to isolate cell lines that express fully active factor VII or factor IX. Activity is dependent on inclusion of vitamin

[38] D. S. Fair, *Blood* **62**, 784 (1983).
[39] J. P. Miletich, C. M. Jackson, and P. W. Majerus, *J. Biol. Chem.* **253**, 6908 (1978).
[40] J. P. Miletich, G. J. Broze, Jr., and P. W. Majerus, this series, Vol. 80, p. 221.

TABLE I
BIOLOGICAL ACTIVITY OF RECOMBINANT FACTOR IX AND FACTOR VII

Protein	Cell line[a]	ELISA (μg/ml/day)	Activity (μg/ml/day)[b]	Active (%)
FVII	BHK	4.8	4.9	102
FVII	Human hepatoma	2.2	2.5	114
FIX	BHK (IXE)	0.4	0.4	100
FIX	BHK (D41-11)	15.4	0.6	4
FIX	293 (D30-1)	16.2	15.7	97
FIX	Human hepatoma	5.2	4.9	94

[a] Cell lines were all cultured in vitamin K-containing media. The cell line names in parentheses designate individual factor IX-producing lines.
[b] As determined in prothrombin time or APTT-based assays using a normal plasma standard curve (see text).

K in the medium.[1,2] In initial experiments, the production of active factor VII from BHK cells grown in media containing vitamin K concentrations over a range of 50 ng/ml to 50 μg/ml was measured.[28] Activity was observed even at the lowest concentration of vitamin K used. At 50 μg/ml cytotoxicity was observed. Concentrations of 5 μg/ml are routinely used in media containing serum; concentrations of 0.5 μg/ml are used in serum-free media. At these concentrations there is no observed effect on cell growth and the vitamin K concentration is in sufficient excess so that there is no concern over the long-term stability of vitamin K.

Although it is possible to isolate cell lines expressing fully active factor IX or factor VII, individual cell lines vary widely in their production levels and this variation directly affects the ultimate activity of the vitamin K-dependent protein being expressed. An example of this observation with factor IX in BHK cells is shown in Table I. A BHK cell line (IXE) expressing factor IX at 0.4 μg/ml/day produces fully active material, while a cell line (D41-11) secreting levels of 15.4 μg/ml/day expresses a protein with very little activity. Impaired posttranslational modifications (i.e., pro-peptide processing, glycosylation, and carboxylation) are observed with higher factor IX production levels in BHK cells (discussed below). Under-carboxylated factor IX with very little biological activity has also been reported in CHO cells expressing factor IX at high levels.[17] With both the CHO and BHK cells, a subpopulation of fully active factor IX can be isolated using Ca^{2+}-dependent α-factor IX antibodies.[17,29] In contrast to BHK and CHO cells, both human hepatoma cells and 293 cells can produce factor IX at higher levels, with full biological activity (Table I), demonstrating the greater capacity of these two cell lines for effecting the modifications required for activity.[1]

Impaired modification and decreased activity are also observed with BHK cell lines expressing factor IX at 5 µg/ml/day. This observation is of interest because factor VII produced at comparable levels (Table I) is fully active and accurately modified posttranslationally.[1,41] Thus, these two closely related proteins are not processed with identical efficiency during secretion. This highlights the fact that not all of the recombinant vitamin K-dependent proteins behave identically in a particular expression system.

In Vivo Labeling and Gel Analysis

Reagents

Labeling medium: Dulbecco's -Cys, -Met MEM (JRH Biosciences) supplemented with 0.01% penicillin–streptomycin–neomycin (GIBCO), 1% dialyzed fetal bovine serum (JRH Biosciences), 0.1 mg/ml pyruvate (Irvine), 2 mM glutamine (JRH Biosciences)

Phytonadione (vitamin K; Merck, Sharp and Dohme): final concentration of 1 µg/ml

Express (NEN) protein ^{35}S-Labeling mix: This is a mixture of [^{35}S]Cys and [^{35}S]Met (at an approximate ratio of 1:4); it is used at a final concentration of 20 µCi/ml

Phosphate-buffered saline (PBS)

Penman lysis buffer:[42] 10 mM HEPES, pH 7.4, 50 mM NaCl, 2.5 mM $MgCl_2$, 0.3 M sucrose, and 0.5% Triton X-100, containing 2 mM phenylmethylsulfonyl fluoride (PMSF; Sigma)

RIP-A lysis buffer: 10 mM Tris, pH 7.5, 1% deoxycholic acid, 1% Triton X-100, 0.1% SDS, 5 mM EDTA, and 0.15 M NaCl, containing 2 mM PMSF.

Two days prior to labeling, cells are seeded in 100-mm dishes so that the confluency is 50% (BHK or human hepatoma cells) or 70% (293 cells) at the time of labeling. One day before labeling, any cells to be pulsed in vitamin K are fluid changed into medium containing phytonadione. If a comparison of a cell line is being made in the presence or absence of vitamin K, the companion plate lacking vitamin K is also fluid changed, because adding fresh medium to the cells can stimulate growth and affect the ultimate number of cells being labeled. For cell lines expressing factor IX, factor VII, or protein C, a control plate, containing the untransfected progenitor cell line, is processed in parallel.

[41] L. Thim, S. Bjoern, M. Christensen, E. M. Nicolaisen, T. Lund-Hansen, A. H. Pedersen, and U. Hedner, *Biochemistry* 27, 7785 (1988).
[42] A. Ben-Ze'ev, A. Duerr, F. Solomon, and S. Penman, *Cell (Cambridge, Mass.)* 17, 859 (1979).

To label cells, medium is aspirated from each dish, which is then rinsed with 2 ml of sterile, prewarmed PBS. Then 2 ml of labeling medium containing radioisotope and, as appropriate, vitamin K is added. Cells are incubated between 1 and 10 hr and it has been determined that even after 10 hr there is sufficient Cys and Met available to support protein synthesis. Because the ^{35}S radiolabel generates volatile by-products, the plates are stored in a CO_2 incubator containing a charcoal tray. To harvest the secreted protein, medium is adjusted to 1 mM PMSF, then centrifuged at 1000 g for 5 min at 4° to remove any cellular debris, and stored at $-20°$ until use. Intracellular protein is isolated by rinsing the cells, while on the plate, with 2 ml of cold PBS and then lysing them in 2 ml of Penman lysis buffer[42] for 5 min at 4°. Lysis is obvious by visualization under a microscope: only the framework of the cell, including the nuclei, appears to be left intact. The lysate is spun for 1000 g for 5 min at 4° to remove particulates, then stored at $-20°$ until use. An Alternative extraction method is performed as above, but uses RIP-A buffer instead. Both extractions have been shown to be effective in recovering all of the native intracellular factor IX, factor VII, or protein C. However, with occasional mutant forms of these proteins and with some other recombinant proteins [e.g., tissue plasminogen activator (TPA)], less protein is extracted using the Penman lysis buffer than is obtained using the RIP-A lysis buffer (which contains SDS). As a practical consideration, RIP-A buffer lyses the nuclei, generating a highly viscous solution which makes quantitative manipulations during immunoprecipitation difficult. So if the protein of interest can be shown to be fully extractable in the Penman lysis buffer (which is ascertained for each new protein by comparing the results using each of the two lysis buffers), then it is the extraction method of choice.

Cellular lysates and media are quantitatively immunoprecipitated with antibodies.[36] Monoclonal antibodies are usually used because they provide a lower background than polyclonal antibodies, especially in the analysis of intracellular protein. An important consideration in monoclonal antibody choice is the lack of sensitivity to posttranslational modifications, which can vary depending on the level of protein production in a given cell line (see below) and which may not be known for intracellular material. A number of monoclonal antibodies are therefore tested and compared with results obtained using polyclonal antibodies.

In vivo labeling and gel analysis can provide a significant amount of information, including the molecular weight(s) of both secreted and intracellular recombinant proteins and the efficiency of secretion of the protein. The effect of manipulations of culture conditions (e.g., blocking carboxylation by depleting vitamin K or glycosylation by including tunicamycin) on the protein can also readily be examined. Finally, the analysis is amen-

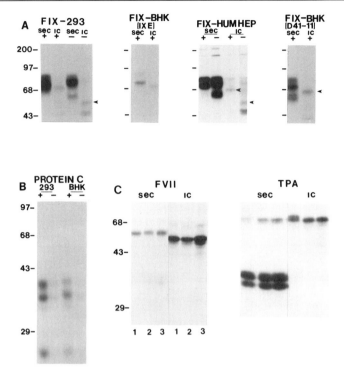

FIG. 2. *In vivo* labeling of factor IX-, factor VII-, and protein C-producing cell lines. Labeling of BHK cells, 293 cells, and human hepatoma (hum hep) cells is performed for 6 hr (A and B) or 2 hr (C) as described in the text. The "+" or "−" above individual lanes refers to whether vitamin K was present in the media where the cells were cultured. Where this is not indicated, the media contained vitamin K. The secreted (sec) and intracellular (IC) proteins are immunoprecipitated and analyzed on gels. (A) The polyclonal α-factor IX antibody cross-reacts with several intracellular proteins. The band specific to factor IX-producing cells, determined by comparisons with control untransfected cells, is indicated by an arrowhead. The horizontal bars indicate the molecular weight markers, given at the far left of each picture. The factor IX-293 cell line is D30-1, the same cell line described in Table I. (C) Protein from independent BHK colonies (1–3), which express both TPA and factor VII, was immunoprecipitated with α-factor VII (left) or α-TPA (right) monoclonal antibodies.

able to processing many samples, and the screening of multiple isolates helps to avoid studying potential artifacts that arise from having picked a clone that produces an incorrectly processed protein.

An example of the analysis of factor IX secreted from three different cell lines is shown in Fig. 2A. Three of these cell lines, factor IX-293 (D30-1), factor IX-hum hep, and factor IX-BHK (IXE), produce fully active factor IX (Table I). All three lines secrete factor IXs with similar molecular weights when the cells are grown in vitamin K-containing media. In each case a doublet of factor IX protein is observed, and the lower molecular weight form has been shown to comigrate with plasma factor IX (as described below). The heterogeneity in molecular weights is probably due to glycosylation differences because tunicamycin treatment generates a single band, corresponding to the lower molecular weight form.[29]

The efficiency of secretion of active factor IX in 293 cells and human hepatoma cells is substantially higher than in BHK cells. When different cell lines are labeled using identical conditions and are then immunoprecipitated in antibody excess, the band intensity after gel electrophoresis and autoradiography quantitatively reflects the level of factor IX secreted. For example, in Fig. 2A the relative band intensities of secreted protein correlate with the ELISA values (Table I). The secretion of active factor IX from IXE cells, then, is considerably lower than from human hepatoma or 293 cells (Fig. 2A, Table I). [It is possible to isolate BHK cell lines secreting more factor IX (e.g., D41-11, Fig. 2A), but most of the protein is inactive (Table I).] In comparing the percent secreted or intracellular factor IX in each of the three cell lines producing active factor IX during a 6-hr pulse, one observes that >99% of the factor IX is secreted from 293 cells and human hepatoma cells while only about half of the factor IX has exited from the IXE cells. Inefficient secretion of factor VII[13] (Fig. 2C) and of protein C in BHK cells has also been observed. This does not represent a generalized block to secretion in this cell line. Other proteins, such as TPA, even when cotransfected into the same BHK cell line with a vitamin K-dependent protein (i.e., factor VII), are efficiently secreted (Fig. 2C).

In both BHK and 293 cells, quantitative differences in the secretion of several vitamin K-dependent proteins are observed when the cell lines expressing those proteins are labeled with or without vitamin K in the medium. Approximately two- to fivefold more protein C (Fig. 2B), factor VII,[1] or factor IX[13] is secreted in the presence of vitamin K. Qualitative differences in protein C and in factor IX expressed in the presence or absence of vitamin K are observed, as well. Figure 2A shows that human hepatoma cells and 293 cells cultured in vitamin K-depleted media secrete a heterogeneous population of factor IX molecules, and this is observed with BHK cells as well[13] (Fig. 3). Tunicamycin inhibition experiments indicate that the heterogeneity is due, at least in part, to underglycosylation of factor IX, suggesting that the extent of carboxylation ultimately affects the extent of glycosylation.

The heterogeneity in molecular weight forms observed for factor IX secreted from cell lines lacking vitamin K is also observed for BHK cells overproducing factor IX (i.e., over levels that are carboxylated). This can be seen with the D41-11 cell line in Fig. 2A. As mentioned, heterogeneity is

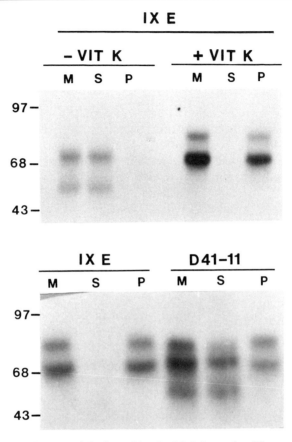

FIG. 3. Barium citrate precipitation of *in vivo*-labeled proteins. The medium starting sample (M), nonbarium precipitable supernatant (S), and solubilized barium citrate precipitate (P) were all immunoprecipitated and analyzed on gels. All starting media samples were media from cells labeled *in vivo* in the presence of vitamin K, except as noted. IXE and D41-11 are two BHK cell lines expressing factor IX (Table I).

due in part to glycosylation differences. In addition, amino acid sequence analysis of purified factor IX from D41-11 cells shows that only about half of the propeptide is processed.[28] Whether impaired processing is a direct consequence of poor carboxylation in D41-11 cells or whether the BHK cells are saturated for this processing capacity is not clear.

Factor VII and factor IX both circulate in blood in the single-chain form, and this is the only structure we have observed with the expression of these proteins in several different cell lines. In contrast, protein C and factor X, which circulate in blood as two-chain molecules, are cleaved during secretion from BHK or 293 cells (shown for protein C in Fig. 2B).

Barium Citrate Precipitation

Reagents

Phenylmethylsulfonyl fluoride (Sigma): 200 mM in 2-propanol
0.5 M sodium citrate
1 M $BaCl_2$
$BaCl_2$/NaCl wash: 0.1 M $BaCl_2$ and 0.1 M NaCl
Resolubilization buffer: 0.15 M sodium citrate and 0.1% bovine serum albumin in PBS

Proteins that are extensively γ-carboxylated can be adsorbed selectively to barium citrate. Thus, as a quick, qualitative assessment of the carboxylation of recombinant vitamin K-dependent proteins expressed in various cell lines, barium citrate precipitation[43] of radiolabeled proteins followed by immunoprecipitation can be useful. For example, when this procedure is used as a screen for analyzing many independent clones producing a particular recombinant protein, colonies expressing at levels that saturate carboxylation capacities are easily detected. Incorporating barium citrate precipitation analysis as a screen is convenient, because colonies are almost always labeled *in vivo* for immunoprecipitation and gel analysis, so the starting material for this procedure is already available.

Two aliquots of medium (400 μl) from *in vivo*-labeled cells are precleared with *Staphylococcus aureus*[36] and then adjusted to 1 mM PMSF. One of the samples is reserved on ice for subsequent immunoprecipitation. The other sample is citrated (18 μl 0.5 M sodium citrate) and kept on ice for 10 min. After the addition of 18 μl 1 M $BaCl_2$, the sample is vortexed, incubated on ice for 1 hr, and then spun at 10,000 g for 5 min at 4°. The nonbarium citrate precipitable supernatant is stored at 4° for subsequent immunoprecipitation and the pellet is resolubilized in 0.1 M $BaCl_2$/0.1 M NaCl (400 μl) by vigorous vortexing. After incubation for 1 hr on ice, the sample is respun, the supernatant is discarded, and the pellet is dissolved in 400 μl resolubilization buffer. This resolubilized pellet, the nonbarium citrate precipitable supernatant, and the initial medium sample are then immunoprecipitated.[36]

Two features are important to the success of this procedure. One is the long incubation times on ice to assure quantitative precipitation. The other is that the pellet be completely resolubilized during the vortexing after the $BaCl_2$/NaCl wash solution is added. Any residual barium citrate will produce a precipitate that results in a high background during the subsequent immunoprecipitation.

---

[43] O. P. Malhotra, *Thromb. Res.* **15**, 427 (1979).

Figure 3 illustrates how barium citrate precipitation analysis can be used to distinguish carboxylated and noncarboxylated protein. When a fully active factor IX produced at low levels in BHK cells (IXE, Fig. 2, Table I) is carried through this procedure, for example, the protein is fully barium citrate precipitable if the cells are cultured in vitamin K, and totally nonbarium citrate precipitable if the medium is depleted of vitamin K (Fig. 3). As mentioned above, this procedure easily detects cell lines producing undercarboxylated protein. Figure 3 shows the precipitation profile of protein from a BHK cell line producing factor IX at increased levels (D41-11, Table I). Much of the secreted factor IX is nonbarium citrate precipitable, indicating that the protein is poorly carboxylated. This prediction was borne out by purification of factor IX from D41-11 cells and Gla quantitation of the purified protein (described below). Thus, this procedure, when carefully followed, is a quick screen providing a rough indication of the extent of carboxylation, which can be useful in choosing candidate cell lines for further study.

Purification and Characterization of Recombinant Factor VII and Factor IX

Purification

Reagents

Serum-free medium and medium containing serum are described in previous sections. The latter is supplemented with either 1% or 10% fetal calf serum (JRH Biosciences). Serum-free medium and medium containing 1% serum have 1 μg/ml vitamin K (Merck, Sharp and Dohme) and medium containing 10% serum has 5 μg/ml vitamin K

Phosphate-buffered saline

100 mM CaCl$_2$ in PBS

0.25 M sodium citrate, pH 4

α-Factor VII antibody: A purified monoclonal antibody (1.5.4.1[1]) is coupled to CNBr-activated Sepharose (Pharmacia) at 2 mg/ml, according to the manufacturer's instructions

α-Factor IX antibody: ESN-4 (American Diagnostica, Greenwich, CT), a Ca^{2+}-dependent monoclonal antibody (that does not distinguish between carboxylated and noncarboxylated factor IX), is coupled to CNBr-activated Sepharose at 1 mg/ml

Because culture conditions and production levels of individual cell lines are variable, the method used to purify a recombinant protein varies accordingly. Two parameters, i.e., the ability of the cell line to survive in low or no serum and to survive in the absence of selection, are usually tested on multiple clones secreting the same recombinant protein, before scaling-up the culture. Other considerations are the amounts of recombinant protein required and whether one wants to isolate the total factor IX or the factor VII population secreted from a given cell line. Described below are two examples for the purification of factor IX or factor VII using monoclonal antibodies that do not distinguish between carboxylated and uncarboxylated protein. Monoclonal antibodies that recognize both conformational and metal ion-dependent epitopes can be used, as well, to select a subpopulation of carboxylated factor IX, factor VII, or protein C.[41,44–46]

Factor IX is purified from a BHK-producing cell line (IXE) by first seeding the cells into a large tissue culture chamber (1200 cm^2 unit; Nunc) in medium (500 ml) containing 10% fetal calf serum. When the cells reach confluency, they are changed into medium as above but with 1% serum. Medium is harvested at 3- to 4-day intervals, centrifuged to remove any cellular debris, and frozen. When approximately 10 liters is collected, the medium is concentrated 20-fold using an RA2000 (Micron, Westboro, MA) concentrator at room temperature. The sample is then applied to an α-factor IX-Sepharose column (ESN4, 20 ml). The resin is washed with 10 column volumes of PBS and the factor IX is eluted with PBS containing 100 mM CaCl$_2$ and quantitated using an ELISA.[36]

To purify recombinant native factor VII and a factor VII mutated at the activation site (amino acid 152,[12] Fig. 1), BHK cells expressing each protein (at levels of ∼2–5 μg/ml/day) are grown in 150-mm dishes (five each) containing medium (25 ml per plate) with 1% fetal calf serum and 150 nM methotrexate. Medium is collected every 3–4 days, centrifuged, and then frozen. Accumulated medium (250 ml to 1 liter) is passed over an α-factor VII-Sepharose column (10 ml). After washing with 10 column volumes of PBS, the factor VII is eluted with 0.25 M sodium citrate, pH 4, and then dialyzed into PBS. Both the factor IX and factor VII purifications are carried out at 4°.

The cell conditions for scaling-up protein production are made fairly flexible by the fact that it is possible to readily generate suitable monoclonal antibodies against human factor VII and human factor IX. Protein purification using a single immunoaffinity step is usually all that is required to achieve homogeneity. The levels of recombinant protein secreted are generally in the range of 0.2–20 mg/liter/day. Where less than 1 mg of protein is required, the medium is usually loaded directly onto the column.

[44] H. A. Leibman, S. A. Limentani, B. C. Furie, and B. Furie, *Proc. Natl. Acad. Sci. U.S.A.* **82,** 3879 (1985).

[45] W. R. Church, T. Messier, P. R. Howard, J. Amiral, D. Meyer, and K. G. Mann, *J. Biol. Chem.* **263,** 6259 (1988).

[46] T. Nakagaki, D. C. Foster, K. L. Berkner, and W. Kisiel, *Biochemistry* **30,** 10819 (1991).

When larger volumes of medium are used, a concentration step, as described above, is performed. Both BHK cells and 293 cells can be adapted to culturing in serum-free medium, but the cells survive for a longer time if 1% serum is included in the medium. Using low levels of serum does not compromise the purification of factor IX or factor VII and it prolongs the length of time over which protein can be isolated from the cells. With the IXE cell line described above, for example, media were harvested from the same Nunc chamber over a 2-year period. Secreted recombinant protein is quite stable in the medium, and is not affected by the cell confluency. The stability of the recombinant protein in a stationary culture can be tested. *In vivo*-labeled factor IX, for example, was placed into the medium of IXE (BHK) cells producing factor IX, and the culture was incubated for 1 week. The radiolabeled factor IX was then analyzed on an acrylamide gel following immunoprecipitation. No degradation of factor IX was observed, even with this very sensitive method of detection.[28]

An optional method for scaling-up cell density is to adapt the cells to growth in suspension. This has been done, for example, with the D30-1 cell line, a 293 cell line expressing factor IX (Fig. 2A). Adaptation to cell suspension is useful if the cell line will be used extensively. If only a few milligrams of protein are required, however, this approach is probably not worth the effort because the growth in suspension of many of the cell types is not routine.

Purification of factor VII and factor IX is monitored by gel electrophoresis and Coomassie blue and/or silver staining. With factor IX, multiple molecular weight forms are observed (e.g., Fig. 2) and so Western analysis is also performed to confirm the identity of the stained bands as factor IX. A gel analysis of purified plasma and recombinant factor IX and of two recombinant factor VII molecules is shown in Fig. 4. With both factor VII and factor IX it is possible to isolate protein that is virtually all in the single-chain form. One of the lanes in Fig. 4 was included, however, to show that occasionally during purification some cleavage of factor IX can be observed. Because factor IX is extremely stable in medium on the cells, this breakdown presumably occurs during purification. The protein may become more susceptible to proteolysis following medium concentration, and the fact that proteolysis occurs only occasionally is likely due to variability in cell lysis.

Gla Analysis

Reagents

0.1 *M* NH$_4$HCO$_3$ (Aldrich)
5 *M* KOH (Aldrich): Freshly made with HPLC-grade water (Baxter, Muskegon, MI) and bubbled for 30 sec with N$_2$

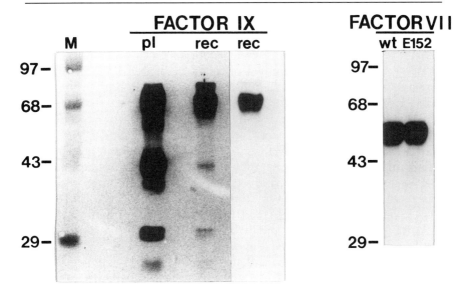

FIG. 4. Gel analysis of purified factor IX and factor VII. Plasma factor IX (pl) and recombinant factor IX (rec) isolated from the BHK cell line IXE were electrophoresed under reducing conditions along with molecular weight markers (M) and stained with Coomassie blue. The lower (i.e., <50,000) molecular weight forms migrate at the positions of the factor IX heavy and light chains. A native factor VII (wt) and a factor VII mutated from an Arg to a Glu at amino acid 152 (E152) are also shown.

Cation-exchange columns (Dowex 50W-X8 resin, 100–200 mesh, Bio-Rad): 3 ml, prewashed with 20 ml 1.0 M NH$_3$

1.0 M NH$_3$: 6 ml 17M NH$_3$ Suprapur (Merck) plus 94 ml HPLC-grade water

OPA/methanol/2-MSH: 50 mg o-phthalaldehyde (Pierce) in 1.3 ml methanol (Aldrich) with 50 μl 2-mercaptoethanol (2-MSH) (Pierce), made fresh daily and kept on ice, protected from light[47]

1 M Potassium borate buffer, pH 10.4 (Pierce)

1.0 M KH$_2$PO$_4$, pH 4.0

Amino acid standard H (Pierce): Contains 17 of the common amino acids, most at 2.5 μmol/ml in 0.1 N HCl

Individual amino acid stocks (Glu, Asp, Ala, Gla; Pierce, 2.5 μmol/ml): These are approximately 100× stocks and are quantitated by amino acid analysis

L-[U-^{14}C]Leucine (Amersham, ~300 mCi/mmol)

HPLC buffer: 5–50% acetonitrile (EM Science, Gibbstown, NJ) in 20 mM sodium citrate (Aldrich), 0.1% trifluoroacetic acid (v/v, Pierce), pH 5.5. The buffers are usually made fresh every few days

[47] Y. Haroon, *Anal. Biochem.* **140**, 343 (1984).

Because Gla quantitation is such a sensitive procedure, minimal cell culture scale-up is required to generate sufficient protein for analysis. Generally, one or two 150-mm plates cultured for 5–10 days will yield enough protein for study. Samples of factor IX or factor VII are affinity purified, dialyzed into ammonium bicarbonate and then either lyophilized and reconstituted in water (for dilute samples) or processed directly. Protein (usually 50–500 pmol in 100 μl) is mixed with 100 μl 5 M KOH in a 5-ml Teflon conical-bottom vial (Pierce, Tuftainer) fitted to a hydrolysis vial cap (Waters 07363)[48] and frozen. Alternatively, if several proteins are being processed at a time, the samples are hydrolyzed in polyallomer tubes (8 × 51-mm Quick Seal centrifuge tubes, Beckman). The tubes are heat sealed and then punctured several times with a needle, near the top of the tubes. The samples are frozen and then inserted into a hydrolysis vial (Waters). The Teflon vial or the hydrolysis vial is evacuated and flushed four times with argon on a Pico · Tag workstation (Millipore/Waters). After final evacuation, the tubes are incubated at 110° for 20 hr and then cooled to room temperature. Samples are loaded onto Dowex columns, followed by the addition of 2 ml of 1 M ammonium hydroxide. The first 0.5 ml is discarded and the next 1 ml is collected, frozen, and lyophilized on a Speed Vac concentrator (Savant, Farmingdale, NY). A radiolabeled base-stable amino {e.g., [^{14}C]leucine, 10^4 counts/min (cpm)} that elutes well away from the amino acids being quantitated (Fig. 5A) is usually added to the protein samples prior to base digestion, to monitor amino acid recoveries. The small amount of radiolabeled [^{14}C]Leu does not contribute significantly to the Leu signal, nor does it contain contaminants that otherwise interfere with the HPLC analysis. Amino acid recoveries from the Dowex columns are usually 75–85%.

Lyophilized amino acids are resuspended in 100 μl of HPLC-grade water and samples (5–20 μl) are mixed with 50 μl of OPA/methanol/2-MSH and 100 μl of 1 M sodium borate, pH 10.4. After 3 min, the samples are neutralized with 100 μl of 1.0 M KH$_2$PO$_4$, pH 4, vortexed, and filtered through a 0.45-μm Acrodisc (Gelman, Ann Arbor, MI) attached to a 3-ml syringe. Exactly 1 min postneutralization, 20 μl of sample is injected (Rheodyne, model 7120, equipped with a 10-μl sample loop) and analyzed by HPLC on a reversed-phase column (ODS hypersil 5 μM, 100 × 2.1 mm, Hewlett Packard, Wilmington, DE). A linear 5–23% acetonitrile gradient is applied (VISTA model 5500, VARIAN; flow rate 1 ml/min) for 11 min, followed by a linear 23–50% acetonitrile gradient for the next 4 min. OPA–amino acid fluorescence is measured on a fluorimeter (Hitachi F1050, λ_{ex} 323 nm, λ_{em} 430 nm, time constant 0.3, sensitivity 0.01)

[48] S. C. B. Yan, P. Razzano, Y. B. Chao, J. D. Walls, D. T. Berg, D. B. McClure, and B. W. Grinnell, *Biotechnology* 8, 655 (1990).

with a flow-through cell and quantitated using a Hewlett Packard 3396 Series II integrator. Separation of Gla, Glu, and Asp is reproducible (Fig. 5A) and there is little background around the Gla peak, as determined by analyzing factor IX purified from BHK cells cultured in the absence of vitamin K (Fig. 5B–C).

Because of the instability of the OPA–amino acid adduct, derivatization and subsequent injection are carefully timed. Under these conditions, reproducibility is excellent; e.g., with replicate samples of individual amino acids or base-hydrolyzed protein the range is only $\pm 2\%$.

For the Gla determination of an unknown protein sample, the following controls are processed in parallel: (a) a blank control (H_2O or ammonium bicarbonate) to check the purity of the solvents, (b) the plasma protein cognate of the recombinant protein analyzed, (c) an amino acid hydrolyzate (Fig. 5A) to monitor the integrity of the column, (d) occasionally the unknown protein sample (isolated from cells depleted in vitamin K), processed in parallel with the protein purified from vitamin K-containing medium, and (e) a mixture of Glu, Gla, and Asp, at several different concentrations, used to construct a standard curve. Asp is included in the quantitation to ensure that the Asp/Glu ratio is consistent with that predicted for the pure protein. Obviously a critical element to the success of this procedure is the purity of the protein. Aliquots of samples to be analyzed for Gla content are always electrophoresed in denaturing acrylamide gels and analyzed using both Coomassie blue and silver stains. However, contamination with a small amount of several different proteins could go undetected using this approach, and HPLC analysis of multiple amino acids affords an additional check on the homogeneity of a protein preparation.

When Gla quantitation is performed on recombinant factor IX or factor VII expressed in BHK or 293 cells, the results shown in Table II are obtained. BHK cells (i.e., IXE) producing factor IX at low levels secrete a factor IX that is extensively carboxylated, and this result is consistent with the activity measurement for this protein (Table I). Carboxylation, as expected, is dependent on the inclusion of vitamin K in the medium; BHK cells (i.e., IXE) cultured in the absence of vitamin K produce factor IX that is not detectably carboxylated (Fig. 5C, Table II). Factor IX carboxylation capacity in BHK cells is limited, however. With a cell line (D41-11) secreting increased levels of factor IX, the protein is poorly carboxylated (Table II). This result correlates with both the decreased carboxylation indicated by barium citrate precipitation analysis (Fig. 3) and with the poor activity observed for this protein (Table I).

Inefficient factor IX carboxylation in BHK cells contrasts with what is observed for factor VII.[1] At BHK production levels of 5 μg/ml/day, factor VII is almost completely carboxylated (Table II), whereas factor IX is

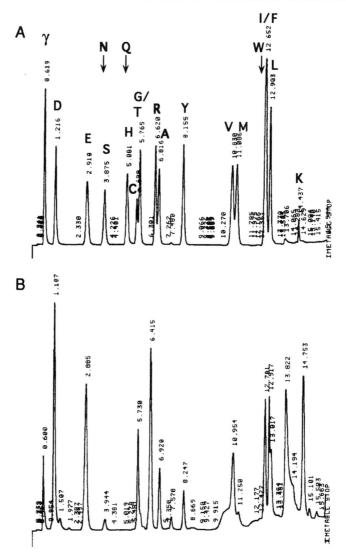

FIG. 5. HPLC analysis of base-hydrolyzed recombinant factor IX. (A) A reversed-phase separation (see text) of an amino acid hydrolyzate (Pierce) supplemented with Gla (γ). Positions of each amino acid are determined by chromatographing individual amino acids either alone or mixed with the amino acid hydrolyzate. The hydrolyzate does not contain Asn (N), Gln (Q), or Trp (W), whose positions are indicated by an arrow. Proline does not give a fluorescent adduct and Lys (K) fluorescence is less efficient [M. Roth, *Anal. Chem.* **43**, 880 (1971); J. R. Benson and P. E. Hare, *Proc. Natl. Acad. Sci. U.S.A.* **72**, 619 (1975)]. The separation profile for recombinant factor IX purified from BHK cells (IXE) cultured in the presence (B) or absence (C) of vitamin K is shown.

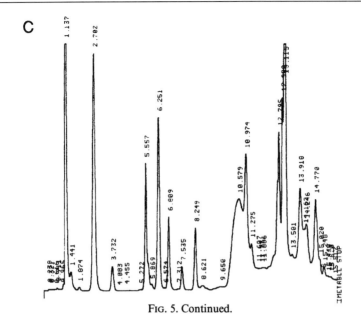

FIG. 5. Continued.

poorly carboxylated.[29] Thus, not all of the vitamin K-dependent proteins respond similarly in the same expression system. Inefficient carboxylation of factor IX in BHK cells also contrasts with the more efficient carboxylation of factor IX observed in 293 cells. In a 293 cell line (D30-1) expressing factor IX at levels of 16 μg/ml/day, the factor IX is extensively carboxylated (Table II). This result is consistent with the high activity observed for this protein (Table I), and demonstrates the higher capacity of 293 cells for carboxylation.

Propeptide Processing

Processing of the recombinant factor IX propeptide is not as efficient as for the factor VII propeptide in BHK cells, and is dependent, like carboxylation, on production levels. Factor IX secreted at low levels from BHK cells (IXE), when purified and subjected to Edman degradation, gives an amino acid sequence indistinguishable from that of plasma factor IX,[13] reflecting accurate propeptide processing. However, at elevated levels of expression a mixed sequence is obtained, at an approximate 1:1 ratio, resulting from the secretion of mature and propeptide-containing factor IX[29] (Table II). In contrast, the factor VII propeptide is completely pro-

TABLE II
GLA QUANTITATION AND PROPEPTIDE PROCESSING OF RECOMBINANT
FACTOR IX AND FACTOR VII

Protein source	Production level (μg/ml/day)	Gla (mol/mol)[b]	Propeptide processed (%)[c]
Plasma factor IX	—	11.8	N.D.
IXE (factor IX/BHK)	0.4	9.5	100
D41-11 (factor IX/BHK)	15.4	2.7	~50
D30-1 (factor IX/293)	16.2	9.2	N.D.
IXE (factor IX/BHK) ($-K$)[a]	0.2	0.2	N.D.
Factor VII/BHK	4.8	9.5	100

[a] Factor IX was purified, in this case, from IXE cells cultured in medium lacking vitamin K.

[b] Plasma factor IX contains 12 Gla residues [K. H. Choo, K. G. Gould, D. J. G. Rees, and G. G. Brownlee. *Nature* (*London*) **299**, 178 (1982); K. Kurachi and E. W. Davie, *Proc. Natl. Acad. Sci. U.S.A.* **79**, 6461 (1982)] and plasma factor VII contains 10 Gla residues [F. S. Hagen, C. L. Gray, P. O'Hara, F. J. Grant, G. C. Saari, R. G. Woodbury, C. E. Hart, M. Insley, W. Kisiel, K. Kurachi, and E. W. Davie, *Proc. Natl. Acad. Sci. U.S.A.* **83**, 2412 (1986)].

[c] Determined by the detection of propeptide sequence in the amino acid sequence analysis. N.D., Not determined.

cessed,[41] even at production levels (4.8 μg/ml/day) above those levels that saturate factor IX propeptide processing.[29] Thus BHK cells process (i.e., carboxylate and propeptide cleave) factor VII more efficiently than factor IX. There is a direct correlation between efficiency of carboxylation and efficiency of propeptide processing. However, whether carboxylation of factor IX or factor VII actually affects the efficiency of propeptide cleavage is not known. Inefficient propeptide cleavage of factor IX in CHO cells has also been reported,[49] and a mixture of mature and propeptide-containing protein C secreted from C127 cells has also been observed.[18]

Inhibition of propeptide cleavage by mutational manipulation of a recombinant vitamin K-dependent protein can be used to identify the propeptide cleavage site of that molecule. With factor VII, for example, no natural variants with impaired propeptide processing have been isolated. A recombinant factor VII variant [FVII (38A), Fig. 1] altered at the -1 amino acid (Arg $\rightarrow$ Ser) was generated[7] that mimicked a factor IX muta-

[49] A. Balland, T. Faure, D. Carvallo, P. Cordier, P. Ulrich, B. Fournet, H. de la Salle, and J. P. Lecocq, *Eur. J. Biochem.* **172**, 565 (1988).

TABLE III
N-TERMINAL AMINO ACID SEQUENCE OF WILD-
TYPE FACTOR VII AND FACTOR VII
PROPEPTIDE MUTANT 38A[a]

Wild-type factor VII (starts at +1)	38A (starts at −18)
A	A (491)[b]
N	V (244)
A	F (243)
F	V (221)
L	T (100)
γ	Q (179)
γ	E (106)
L	E (139)
R	A (116)
P	H (20)
G	G (74)
S	V (68)
L	L (57)
γ	H (9)
R	R (15)
γ	R (28)
C	R (9)
K	S (19)
γ	A (24)
γ	N (21)
Q	A (24)
C	F (16)
	L (16)

[a] Data and part of table were taken from "Current Advances in Vitamin K Research" (J. W. Suttie, ed.). Elsevier, New York, 1988.
[b] Numbers in parentheses refer to picomoles of amino acid.

tion which exhibited impaired propeptide processing.[50] When protein purified from BHK cells expressing this protein was subjected to Edman degradation, a single sequence starting at −18 in the propeptide sequence was observed (Table III).

Acknowledgments

The author thanks Susan Lingenfelter and Kurt Runge for valuable assistance in preparing this chapter.

[50] D. L. Diuguid, M. J. Rabiet, B. C. Furie, H. A. Leibman, and B. Furie, *Proc. Natl. Acad. Sci. U.S.A.* **83,** 5803 (1986).

[27] Thioester Peptide Chloromethyl Ketones: Reagents for Active Site-Selective Labeling of Serine Proteinases with Spectroscopic Probes

By PAUL E. BOCK

Introduction

Plasma proteolytic enzymes are regulated by specific macromolecular interactions that can be quantitated and characterized with fluorescence, absorbance, electron spin resonance, and nuclear magnetic resonance (NMR) techniques based on covalent labeling of the proteinases with extrinsic spectroscopic probes. A common initial goal of fluorescence studies of proteinase interactions is simply to prepare an enzyme derivative that provides a spectral change for observing the molecular event of interest. Achieving this goal, however, is often frustrated by the difficulties associated with the required protein chemical modification, coupled with a lack of ability to predict whether the derivatives that can be prepared will exhibit the necessary spectral properties. Although many group-selective labeling reagents are available,[1] representing a broad range of spectral properties, the absence of unique nucleophilic residues in the plasma proteinases typically restricts labeling with these reagents to heterogeneous modification of surface residues, which limits the utility of the derivatives. Because many regulatory interactions of the proteinases do not directly involve the catalytic site, active site-specific labeling methods have been preferred for incorporating probes in a well-defined location, where they are positioned to report functionally significant events. Analogs of irreversible inhibitors that react with the catalytic site serine residue[2-7] and derivatives of amino acid and peptide chloromethyl ketones that alkylate the histidine residue[8-18] have been used to incorporate fluorescence

[1] R. P. Haugland, "Molecular Probes Handbook of Fluorescent Probes and Research Chemicals." Molecular Probes, Eugene, OR, 1989.
[2] L. J. Berliner and S. S. Wong, *J. Biol. Chem.* **249** 1668 (1974).
[3] W. L. C. Vaz and G. Schoellmann, *Biochim. Biophys. Acta* **439**, 194 (1976).
[4] D. J. Epstein, H. A. Berman, and P. Taylor, *Biochemistry* **18**, 4749 (1979).
[5] J. C. Hsia, D. J. Kosman, and L. H. Piette, *Arch. Biochem. Biophys.* **149**, 441 (1972).
[6] R. P. Haugland and L. Stryer, *in* "Conformation of Biopolymers" (G. N. Ramachandran, ed.), Vol. 1, p. 321. Academic Press, New York, 1967.
[7] A. R. Moorman and R. H. Abeles, *J. Am. Chem. Soc.* **104**, 6785 (1982).
[8] G. Schoellmann, *Int. J. Pept. Protein Res.* **4**, 221 (1972).
[9] G. S. Penny and D. F. Dyckes, *Biochemistry* **19**, 2888 (1980).
[10] S. Krishnaswamy, E. B. Williams, and K. G. Mann, *J. Biol. Chem.* **261**, 9684 (1986).

METHODS IN ENZYMOLOGY, VOL. 222

probes[3,4,6–10,13–18] and other types of labels.[2,5,11,12] All of these reagents, however, are limited to incorporation of a single probe for each inhibitor analog synthesized and, prior to a recently described, extended series of fluorescent peptide chloromethyl ketone derivatives,[15] to a small number of fluorescence probes.

The active site-selective labeling approach was developed in an attempt to overcome the limitations of these methods. This new strategy combines the site specificity of inhibitor affinity labeling with the versatility of group-selective chemical modification to enable active site labeling of proteinases with a wide variety of probes by use of a single affinity-labeling reagent.[19–21] As illustrated in Scheme I,[22] irreversible inactivation of a serine protein-

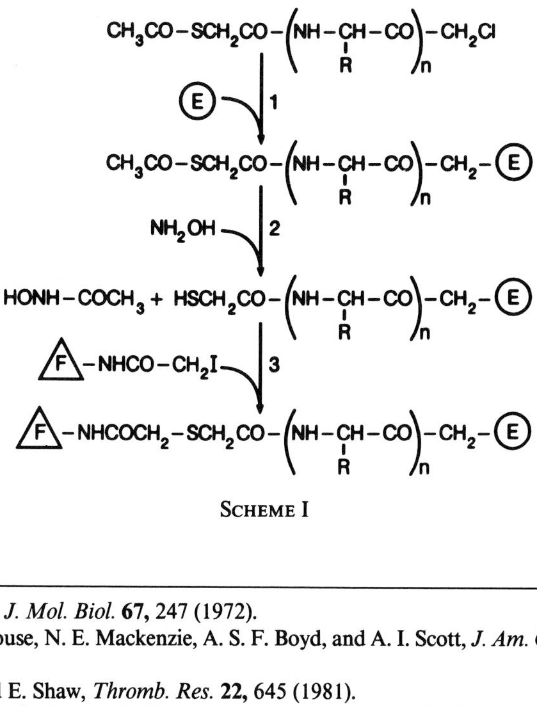

$$CH_3CO-SCH_2CO-\left(NH-\underset{R}{CH}-CO\right)_n-CH_2Cl$$

$$\textcircled{E} \searrow 1$$

$$CH_3CO-SCH_2CO-\left(NH-\underset{R}{CH}-CO\right)_n-CH_2-\textcircled{E}$$

$$NH_2OH \searrow 2$$

$$HONH-COCH_3 + HSCH_2CO-\left(NH-\underset{R}{CH}-CO\right)_n-CH_2-\textcircled{E}$$

$$\triangle F-NHCO-CH_2I \searrow 3$$

$$\triangle F-NHCOCH_2-SCH_2CO-\left(NH-\underset{R}{CH}-CO\right)_n-CH_2-\textcircled{E}$$

SCHEME I

[11] D. J. Kosman, *J. Mol. Biol.* **67,** 247 (1972).

[12] J. P. G. Malthouse, N. E. Mackenzie, A. S. F. Boyd, and A. I. Scott, *J. Am. Chem. Soc.* **105,** 1685 (1983).

[13] C. Kettner and E. Shaw, *Thromb. Res.* **22,** 645 (1981).

[14] M. E. Nesheim, C. Kettner, E. Shaw, and K. G. Mann, *J. Biol. Chem.* **256,** 6537 (1981).

[15] E. B. Williams, S. Krishnaswamy, and K. G. Mann, *J. Biol. Chem.* **264,** 7536 (1989).

[16] G. Schoellmann, G. Striker, and E. B. Ong, *Biochim. Biophys. Acta* **704,** 403 (1982).

[17] N. C. Genov and R. N. Boteva, *Biochem. J.* **238,** 923 (1986).

[18] K. Peters, O. Batz, W. E. Hohne, and S. Fittkau, *Biomed. Biochim. Acta* **43,** 909 (1984).

[19] P. E. Bock, *Biochemistry* **27,** 6633 (1988).

[20] P. E. Bock, *J. Biol. Chem.* **267,** 14963 (1992).

[21] P. E. Bock, *J. Biol. Chem.* **267,** 14974 (1992).

[22] Reprinted with permission from P. E. Bock, *Biochemistry* **27,** 6633 (1988). Copyright [1988] American Chemical Society.

ase with a peptide chloromethyl ketone derivative containing a thioester group at the amino terminus results in covalent incorporation of the thioester inhibitor by alkylation of the catalytic site histidine residue (reaction 1). Subsequent mild treatment of the enzyme–inhibitor complex with hydroxylamine generates a thiol group (reaction 2). Because few plasma proteinases contain natural thiols, this group can be used as a unique site for group-selective chemical modification with thiol-reactive probes, such as fluorescence probe iodoacetamides (reaction 3). The results of evaluating the specificity of the labeling reactions for several proteinases with two thioester peptide chloromethyl ketones suggest that the approach is sufficiently general to allow essentially any trypsin-like serine proteinase that lacks reactive thiols to be specifically labeled at the active site with many different probes by use of just one thioester peptide (arginine) chloromethyl ketone.[21] For fluorescence studies of proteinase interactions, the increased versatility of the approach overcomes the problem of not being able to predict which label will provide the most useful derivative by enabling a greater variety of them to be readily prepared and evaluated.

Accessible methods for synthesis, purification, and quantitative assay of the thioester peptide chloromethyl ketones, N^{α}-[(acetylthio)acetyl]-D-Phe-Pro-Arg-CH$_2$Cl (ATA-FPR-CH$_2$Cl) and N^{α}-[(acetylthio)acetyl]-D-Phe-Phe-Arg-CH$_2$Cl (ATA-FFR-CH$_2$Cl), are described here.[19,20] Procedures for active site-selective labeling of proteinases with fluorescence probes, quantitation of the reactions, and evaluation of competing side reactions are presented.[19–21] A main application of the approach to proteinase interactions is illustrated by the results obtained in screening an array of fluorescent human α-thrombin derivatives to identify those that report the binding of the fragment 2 domain of prothrombin.[21] Members of an array of 16 thrombin derivatives prepared with the two thioester peptide chloromethyl ketones and eight fluorescence probes signal a change in the environment of the active site accompanying fragment 2 binding by changes in probe fluorescence intensity, consistent with the previous observation of changes in the catalytic activity of bovine thrombin associated with the interaction.[23–25] The unique spectral response to fragment 2 binding shown by each thrombin derivative, depending on the structure of the probe and connecting peptide inhibitor, suggests that the opportunity to employ several different labels by use of the active site-selective labeling approach makes accessible a more detailed view of changes in the environment of the catalytic site associated with the binding of regulatory macromolecules.

[23] K. H. Myrmel, R. L. Lundblad, and K. G. Mann, *Biochemistry* **15**, 1767 (1976).
[24] H. V. Jakubowski, M. D. Kline, and W. G. Owen, *J. Biol. Chem.* **261**, 3876 (1986).
[25] F. J. Walker and C. T. Esmon, *J. Biol. Chem.* **254**, 5618 (1979).

Quantitative Assays for Thioester Peptide Chloromethyl Ketones: ATA-FPR-CH$_2$Cl and ATA-FFR-CH$_2$Cl

Assays that independently measure the thioester group of the compounds and the chloromethyl ketone group, based on its requirement for inhibitory function, are used for following purification of the inhibitors, characterization of the purified compounds, and quantitation of the labeling reactions.[19-21]

Thioester Kinetic Burst Assay

The concentrations of ATA-FPR-CH$_2$Cl and ATA-FFR-CH$_2$Cl can be determined on the basis of the thioester concentration by using the spectrophotometric thiol reagent, 5,5'-dithiobis(2-nitrobenzoic acid) (DTNB) to measure the amount of thiol generated when the thioester group reacts with hydroxylamine.[19,26] The assay is performed as a kinetic burst amplitude measurement, initiated by addition of NH$_2$OH to a mixture of inhibitor and a large molar excess of DTNB. The kinetic assay method is necessary for accurate determination of the thioester concentration because of the instability of the free-thiol form of the inhibitor and the need to correct the absorbance increase accompanying the DTNB–inhibitor thiol reaction for breakdown of DTNB by NH$_2$OH. The high concentration of DTNB in the assay ensures rapid and complete reaction of the thiol as it is produced, with less than 10% DTNB consumed over the course of the measurement. The kinetics of the thiol burst under these conditions reflect the rate of hydroxylaminolysis of the thioester and the reaction amplitude is equal to the thioester concentration.[27] The burst amplitudes are a linear function of inhibitor concentration and are independent of DTNB (0.1–1.0 mM) and NH$_2$OH (0.02–0.4 M) concentrations. The thiol burst is exponential, with a pseudo first-order rate constant of 0.01 sec^{-1} (~ 1 min half-time) at 0.1 M NH$_2$OH, independent of inhibitor and DTNB concentrations over the above ranges.[28]

Reagents

pH 7 Buffer: 0.1 M HEPES, 0.3 M NaCl, 1 mM EDTA, pH 7.0, Millipore (Bedford, MA) (0.45 μm) filtered

---

[26] P. C. Jocelyn, this series, Vol. 143, p. 44.

[27] Decreases in 412-nm absorbance from alkylation of the 2-nitro-5-thiobenzoic acid product by the chloromethyl ketone are negligible, owing to the slow rate of this reaction.[19]

[28] The observed first-order rate constant increases with a higher than first-power dependence on NH$_2$OH concentration, similar to the kinetics of NH$_2$OH reactions with some other thioesters.[29]

[29] T. C. Bruice and S. J. Benkovic, "Bioorganic Mechanisms," Vol. 1, p. 259. Benjamin, New York, 1966.

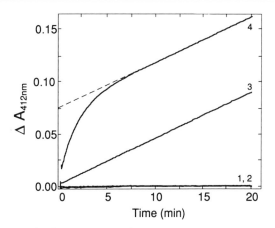

FIG. 1. Thioester kinetic burst assay of ATA-FPR-CH$_2$Cl. Tracings of the recorded changes in absorbance at 412 nm (ΔA_{412nm}), corrected for the background absorbance of DTNB, are shown for reactions containing DTNB alone (curve 1), ATA-FPR-CH$_2$Cl and DTNB (curve 2), NH$_2$OH added to DTNB at zero time (curve 3), and NH$_2$OH added to DTNB and ATA-FPR-CH$_2$Cl at zero time (curve 4). Final concentrations were 490 μM DTNB, 0.09 M NH$_2$OH, and 5.6 μM ATA-FPR-CH$_2$Cl. The dashed line represents the postburst linear rate extrapolated to zero time. Adapted with permission from Bock.[19] Copyright [1988] American Chemical Society

DTNB: 5 ± 0.5 mM solution in pH 7 buffer, Millipore filtered. This solution is stable for several months when stored frozen

Hydroxylamine: 1.0 M solution in pH 7 buffer, adjusted to pH 7.0 and filtered. This reagent is stable for at least 1 month at 4° but appears to become less effective over several months

Procedure. Reactions which measure the concentration of free thiol and the thioester concentration are performed in polystyrene semimicrocuvettes by recording the absorbance at 412 nm with time following 100-μl additions of DTNB and NH$_2$OH to a 900-μl solution of 3–30 μM inhibitor in pH 7 buffer, and a separate buffer control. Figure 1 shows the changes in absorbance observed for these reactions with ATA-FPR-CH$_2$Cl, performed individually and corrected for the background absorbance of about 0.1 due to DTNB. Compared to addition of DTNB to buffer alone, no significant difference in absorbance or change with time is seen on addition of DTNB to buffer containing the inhibitor (curves 1 and 2, Fig. 1), consistent with the absence of free thiols ($\pm 3\%$) in preparations of the purified compounds. Addition of NH$_2$OH to a DTNB–inhibitor mixture results in the exponential absorbance burst due to the thioester–NH$_2$OH reaction, followed by a continuing linear increase (curve 4, Fig. 1), which is accounted for by the breakdown of DTNB by NH_2OH, as seen in the control (curve 3, Fig. 1). The amplitude of the exponential phase is obtained by extrapolation of the linear postburst rate to the time of NH_2OH addition and correction for dilution. Similar analysis of the control reaction trace (curve 3, Fig. 1) typically yields a small dead-time absorbance increase accompanying NH_2OH addition, which is subtracted from the thioester burst amplitude. The thioester concentration is calculated from the corrected burst amplitude with an extinction coefficient of 14,150 M^{-1} cm^{-1} for the 2-nitro-5-thiobenzoic acid product.[30] The inhibitor and control reactions of the assay are usually either performed separately, with tandem DTNB and NH_2OH additions, or as a direct difference measurement in a dual-beam spectrophotometer. The recorded difference between an absorbance-balanced sample cuvette containing the inhibitor and a buffer blank of the same volume after sequential, equal volume additions of DTNB and NH_2OH to both cuvettes yields the free thiol and thioester reactions, respectively, corrected for the absorbance due to DTNB and the small change on addition of NH_2OH. Extrapolation of a residual positive or negative postburst rate, due to minor imbalance between the cuvettes, may still be required.

Stoichiometric Enzyme Inactivation Assay

The irreversibility of proteinase inactivation by the inhibitors and stability of the covalent complexes allow the concentration of functional inhibitor to be determined from the equivalent concentration of enzyme inactivated. This assay measures the concentration of competent chloromethyl ketone inhibitor and is independent of the measurement based on thioester groups. Following incubation of the enzyme with 0.2–0.8 equivalents of inhibitor for long enough to complete the inactivation reaction under conditions where titrations of the activity loss are linear, the inhibitor concentration is determined by multiplying the fraction of the initial activity lost times the known enzyme concentration. Enzyme concentrations are based on those established independently by active site titration.[31] For routine assay of ATA-FPR-CH_2Cl, 10 μl of 10 μM thrombin in 5 mM MES, 0.3 M NaCl, 1 mM EDTA, pH 6.0, is added to 90 μl of inhibitor diluted in pH 7 buffer containing 1 mg/ml polyethylene glycol (PEG) 8000. Incubations are carried out at 25° for 15–60 min before measurement of the residual thrombin activity from the initial rate of H-D-Phe-

[30] P. W. Riddles, R. L. Blakeley, and B. Zerner, *Anal. Biochem.* **94**, 75 (1979).
[31] T. Chase and E. Shaw, *Biochemistry* **8**, 2212 (1969).

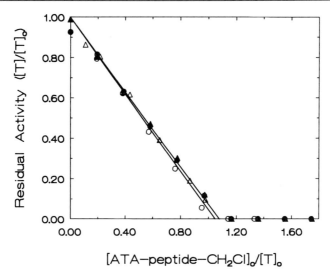

Fig. 2. Titrations of thrombin activity with ATA-FPR-CH₂Cl and ATA-FFR-CH₂Cl. The fraction of initial thrombin chromogenic substrate activity remaining ($[T]/[T]_0$) is plotted as a function of the ratio of initial inhibitor to enzyme concentration ($[\text{ATA-peptide-CH}_2\text{Cl}]_0/[T]_0$). Results are shown for incubations of 1.0 μM thrombin with ATA-FPR-CH₂Cl assayed after 1 hr (O) and 24 hr (●), and incubations of 9.4 μM thrombin with ATA-FFR-CH₂Cl assayed after 25 hr (△) and 116 hr (▲). Lines represent end points of 1.05 (ATA-FPR-CH₂Cl) and 1.08 (ATA-FFR-CH₂Cl).

Pip-Arg-*p*-nitroanilide (S2238, KabiVitrum) hydrolysis.[32] Titrations of thrombin with ATA-FPR-CH₂Cl under these conditions are linear and the residual activities are stable with time (Fig. 2). The sensitivity of the assay can be increased by lowering the enzyme concentration and the same procedure can also be used to assay other inhibitors, such as ATA-FFR-CH₂Cl, provided that conditions are used such that inactivation proceeds to completion without significant hydrolysis of the inhibitor occurring. First-order hydrolysis of peptide chloromethyl ketones at rates competitive

[32] Incubations of enzyme and inhibitor are carried out in PEG 20,000-coated polypropylene tubes and in buffers containing 1 mg/ml PEG 8000 to prevent enzyme adsorption artifacts.[33] Chromogenic substrate assays for thrombin are routinely performed with 100 μM S2238 in 0.1 M Tris-Cl, 0.1 M HEPES, 0.1 M NaCl, 1 mg/ml PEG, pH 7.8, at 25°, and monitored at 405 nm.[19,20,34,35] Thrombin active site titrations are performed with *p*-nitrophenyl-*p*-guanidinobenzoate.[19,20,31,36]

[33] Z. S. Latallo and J. A. Hall, *Thromb. Res.* **43**, 507 (1986).

[34] R. Lottenberg and C. M. Jackson, *Biochim. Biophys. Acta* **742**, 558 (1983).

[35] R. Lottenberg, J. A. Hall, M. Blinder, E. P. Binder, and C. M. Jackson, *Biochim. Biophys. Acta* **742**, 539 (1983).

[36] R. Lottenberg, J. A. Hall, J. W. Fenton, and C. M. Jackson, *Thromb. Res.* **28**, 313 (1982).

with parallel second-order enzyme inactivation results in titrations that are nonlinear at low enzyme and inhibitor concentrations.[20] Thus, thrombin titrations performed as described above with ATA-FFR-CH$_2$Cl, which inactivates thrombin with a 470-fold lower bimolecular rate constant compared to ATA-FPR-CH$_2$Cl, are linear only at enzyme concentrations > 1 μM (Fig. 2).[20]

Synthesis and Purification of ATA-FPR-CH$_2$Cl and ATA-FFR-CH$_2$Cl

The procedure currently used to prepare the N^α-[(acetylthio)acetyl] derivatives of both of the commercially available peptide chloromethyl ketones, FPR-CH$_2$Cl and FFR-CH$_2$Cl, is essentially the same as that originally described for ATA-FPR-CH$_2$Cl.[19,20] The compounds are synthesized by modification of the peptide α-amino group with succinimidyl (acetylthio)acetate (SATA)[37] under conditions which yield complete reaction, while limiting losses of inhibitory activity to less than 15%. Typically, SATA [Molecular Probes, Eugene, OR (preferred) or Calbiochem, La Jolla, CA] is dissolved in methanol at 45–50 mM, which is near the solubility limit, by mixing with a Pasteur pipette. Just before starting the reaction, 5–25 mg of the peptide chloromethyl ketone [Calbiochem (preferred) or Bachem (Torrance, CA)] is dissolved at 8–12 mM in 50 mM sodium phosphate buffer, pH 7.0. An equal volume of the SATA solution is added to the inhibitor and the mixture is incubated for 20 min at 20–25° before stopping the reaction by fivefold dilution with 25 mM sodium phosphate–H$_3$PO$_4$ buffer, pH 3.0, and immediately adjusting the pH to 3.0 with 1 N H$_3$PO$_4$.

The thioester peptide chloromethyl ketones are easily purified by conventional cation-exchange chromatography on sulfopropyl (SP)-Sephadex and adsorption chromatography on Sephadex G-10 at pH 3.0, where the compounds are stable.[19,20] Purification of the inhibitors is followed by measurements of the absorbance at 210 nm and by assays of thioester and inhibitor concentrations as described above. An approximate extinction coefficient of 20,000 M^{-1} cm^{-1} at 210 nm can be used to estimate the concentrations of the compounds for these assays. The synthetic reaction mixture adjusted to pH 3.0 is immediately applied onto a column of SP-Sephadex C-25 equilibrated at room temperature with 50 mM sodium phosphate–H$_3$PO$_4$, pH 3.0, and eluted with the same buffer. Similar results have been obtained with columns of 1.5 × 95–195 cm for 5- to 25-mg preparations, with the longer columns used for 25-mg preparations

[37] R. J. S. Duncan, P. D. Weston, and R. Wrigglesworth, *Anal. Biochem.* **132**, 68 (1983).

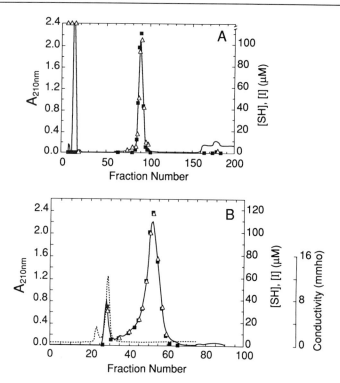

FIG. 3. Chromatographic purification of ATA-FPR-CH$_2$Cl. (A) A reaction mixture containing 5 mg of FPR-CH$_2$Cl and a fivefold excess of SATA was chromatographed on SP-Sephadex (1.5 × 95 cm) in 50 mM sodium phosphate–H$_3$PO$_4$, pH 3.0. The absorbance at 210 nm (A_{210nm}, —), thioester concentration ([SH], $\triangle$), and inhibitor concentration ([I], ■) were measured in 10-ml fractions. Residual FPR-CH$_2$Cl was eluted with buffer containing 0.5 M NaCl, starting at fraction 150. Fractions 87–94 were pooled. (B) Chromatography of the ATA-FPR-CH$_2$Cl recovered from the separation in (A) on Sephadex G-10 (1.5 × 97 cm) in 1 mM HCl. $A_{210\,nm}$ (—), conductivity (----), thioester concentration ($\triangle$), and inhibitor concentration (■) were measured in 3.6-ml fractions. Fractions 43–57 were pooled as the final product. Reprinted with permission from Bock.[19] Copyright [1988] American Chemical Society.

providing slightly improved resolution of minor side products. Figure 3A shows the results obtained for a representative preparation of ATA-FPR-CH$_2$Cl. The neutral and anionic reactants and products, including excess SATA and its hydrolysis products, elute at one column volume. The +1 charged ATA-FPR-CH$_2$Cl elutes with 5.5 column volumes of buffer as a peak of absorbance and coincident thioester and inhibitor concentration, preceded by one or more small absorbance peaks, which represent noninhibitory side products. Any residual +2 charged FPR-CH$_2$Cl remains bound to the column. Fractions containing ATA-FPR-CH$_2$Cl are pooled and lyophilized. The inhibitor is extracted from the residue with three to five ~5-ml aliquots of methanol and the combined extracts are dried at room temperature with a stream of N$_2$. The residue is dissolved in a small volume (2–6 ml) of 1 mM HCl and chromatographed on a column of Sephadex G-10 (1.5 × 95–120 cm) equilibrated with 1 mM HCl at room temperature. As shown in Fig. 3B, some ATA-FPR-CH$_2$Cl elutes with the salt peak while the majority elutes in a subsequent broad peak. Fractions containing the desalted inhibitor and exhibiting a constant (±10%) ratio of inhibitor to thioester concentration or absorbance at 210 nm are pooled and lyophilized. The final product, averaging a 35% yield, is dissolved in 1 mM HCl, typically at concentrations of 1–5 mM, centrifuged (Beckman Microfuge, 2 min), and stored in aliquots at −70°, conditions under which the inhibitor appears to be stable for years. An additional 10% yield of equally pure ATA-FPR-CH$_2$Cl can be obtained from a second run on Sephadex G-10 of the inhibitor recovered from the fractions that were not pooled from the first separation.

ATA-FFR-CH$_2$Cl is purified by the same procedure with an average yield of 40%.[20] This compound elutes from SP-Sephadex after 6.5 column volumes of pH 3.0 buffer and adsorbs more strongly to Sephadex G-10, eluting at ~1.5 column volumes as a single peak of desalted inhibitor with a long trailing edge. ATA-FFR-CH$_2$Cl is stored at concentrations up to 6 mM in 1 mM HCl.

Characterization of ATA-FPR-CH$_2$Cl and ATA-FFR-CH$_2$Cl

The purity of inhibitor preparations is routinely assessed by analytical reversed-phase HPLC on a C$_{18}$ silica column [MicroPak MCH-N-Cap-5, 4 × 150 mm (Varian, Sunnyvale, CA)] monitored by absorbance at 215 nm. The column is equilibrated at room temperature with 0.1% (v/v) trifluoroacetic acid in H$_2$O and developed with a linear, 0–100% gradient of 0.1% trifluoroacetic acid in CH$_3$CN at 1 ml/min over 60 min. As shown in Fig. 4, preparations of ATA-FPR-CH$_2$Cl elute as one major peak of inhibitor at 38% CH$_3$CN, representing 92 ± 5% of the integrated absorbance, with minor peaks often observed at 33 and 40% CH$_3$CN. The minor peaks appear to correspond to the noninhibitory degradation products and contaminant previously observed by HPLC under different solvent conditions.[19,38] ATA-FFR-CH$_2$Cl preparations elute as a major peak at 46% CH$_3$CN, representing 95 ± 7% of the integrated absorbance peaks, with minor species at 40 and 43% CH$_3$CN.[20]

---

[38] P. E. Bock, unpublished results, 1992.

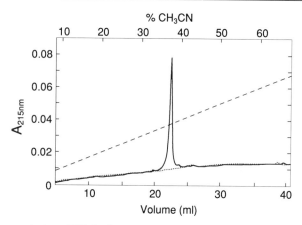

FIG. 4. Reversed-phase HPLC of ATA-FPR-CH$_2$Cl. Expanded portions of the A_{215nm} elution profiles obtained as described in the text are shown for 20-μl injections of 300 μM ATA-FPR-CH$_2$Cl (——) and a solvent blank ($\cdots$). No other peaks ascribable to the product are observed over the 0.1% trifluoroacetic acid, 0–100% CH$_3$CN linear gradient (----).

Preparations of the compounds are also characterized by the end points obtained in thrombin activity titrations based on inhibitor concentrations measured with the thioester assay, such as the titrations shown in Fig. 2. Representative preparations of ATA-FPR-CH$_2$Cl of 92 ± 5% purity by HPLC show titration end points of 1.04 ± 0.09 mol inhibitor thioester per mole of thrombin active sites, while preparations of ATA-FFR-CH$_2$Cl of 95 ± 7% purity by HPLC give end points of 1.08 ± 0.09 mol/mol.[20]

Active Site-Selective Labeling of Proteinases with Fluorescence Probes by Use of ATA-FPR-CH$_2$Cl or ATA-FFR-CH$_2$Cl

Successful site-specific chemical modification of proteins depends on the rates of the desired reactions being much faster than those of inevitable nonspecific reactions, as determined by both the intrinsic properties of the reagents and the reaction conditions. The specificity of active site-selective labeling by the use of ATA-FPR-CH$_2$Cl and ATA-FFR-CH$_2$Cl has been evaluated most extensively for the model system of human α-thrombin labeling with the fluorescence probe–iodoacetamide, 5-(iodoacetamido)fluorescein (5-IAF), by quantitating the specific and competing nonspecific reactions shown in Scheme II.[19,20] The same reagents and similar procedures have been used to label several other blood coagulation proteinases and related trypsin-like enzymes.[21] In the procedure most

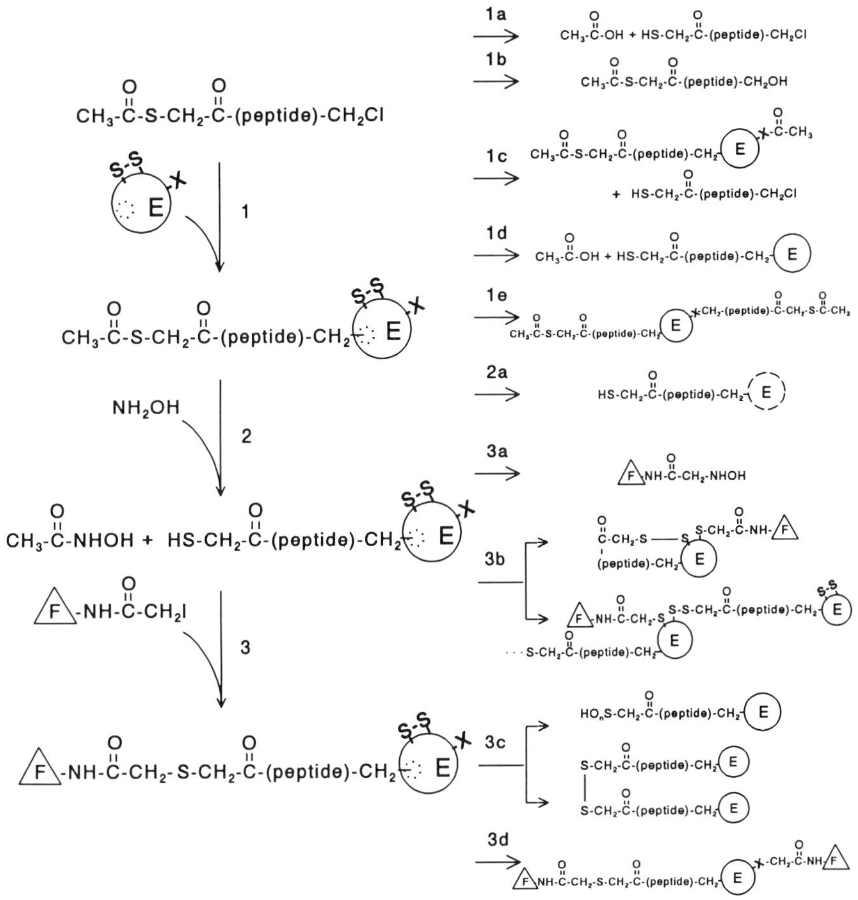

SCHEME II

extensively used, the three main reactions are reduced to two steps. In the first step, the enzyme is completely inactivated with a thioester peptide chloromethyl ketone (reaction 1, Scheme II). In the second step, incubation of the enzyme–inhibitor complex with NH_2OH in the presence of the fluorescence probe–iodoacetamide results in generation of the thiol group (reaction 2, Scheme II) together with its selective modification (reaction 3, Scheme II).

Table I summarizes the results obtained in evaluating the specificity of the first step of the procedure from measurements of the stoichiometry of inactivation, thioester inhibitor incorporation, and its dependence on prior blocking of the active site with an inhibitor such as $FPR-CH_2Cl$. Enzymes

TABLE I
ATA-Peptide-CH₂Cl/5-IAF Labeling of Various Proteinases[a]

Proteinase	ATA-Peptide-CH₂Cl	Activity titration end point (mol/mol)	Thioester incorporation (mol/mol)		5-AF incorporation (mol/mol)		
				Active site blocked	$-NH_2OH$	$+NH_2OH$	Active site blocked
α-Thrombin	ATA-FPR-CH₂Cl	1.04 ± 0.09	1.04 ± 0.06	0.03 ± 0.01	0.02 ± 0.01	0.97 ± 0.04	0.02 ± 0.01
	ATA-FFR-CH₂Cl	1.08	0.97	0.03	0.05	0.93	0.02
β/γ-Thrombin	ATA-FPR-CH₂Cl	1.05	1.03 ± 0.07	0.05	0.02 ± 0.02	0.99 ± 0.06	0.01 ± 0.01
	ATA-FFR-CH₂Cl	1.10	1.04	0.04	0.02	1.01	0.01
Meizothrombin 1[b]	ATA-FPR-CH₂Cl	—	1.01	—	0.03	0.96	—
Factor Xa	ATA-FPR-CH₂Cl	1.13	1.05	0.02	0.04	0.98	0.01
	ATA-FFR-CH₂Cl	1.02	1.01	0.02	0.03	0.97	0.01
Trypsin	ATA-FPR-CH₂Cl	1.13	1.03	0.07	0.03	0.83	0.02
	ATA-FFR-CH₂Cl	1.15	1.06	0.03	0.01	0.86	0.01
Factor IXa	ATA-FPR-CH₂Cl	—	0.97	0.04	0.04	0.94	0.03
	ATA-FFR-CH₂Cl	—	0.93	0.05	0.09	0.91	0.06
V-CP[c]	ATA-FPR-CH₂Cl	—	—	—	(−)	(+)	(−)
	ATA-FFR-CH₂Cl	—	—	—	(−)	(+)	(−)
Factor XIIa[c]	ATA-FPR-CH₂Cl	—	—	—	(+)	(+)	(−)
Kallikrein[c]	ATA-FPR-CH₂Cl	1.06[d]	—	—	(+)	(+)	(−)
	ATA-FFR-CH₂Cl	1.14	—	—	—	—	—

[a] Stoichiometries of the labeling reactions for several different proteinases are listed as mol/mol active sites. Activity titration end points, thioester incorporation, and 5-AF incorporation were determined as described in the text. All estimates of experimental error are ±2 SD, calculated from the averaged duplicate measurements of at least two experiments. The listed mean and error in the thrombin titration end point include data from titrations with four preparations of inhibitor. The mean and error listed for 5-AF incorporation include results from reactions with a two- to ninefold molar excess of 5-IAF. All other values are from reactions with a five- or sixfold excess. Results of qualitative labeling experiments are based on the appearance of fluorescein fluorescence associated with the proteins on SDS gels. Reprinted with permission from Bock.[21]

[b] Results for meizothrombin 1 are expressed as mol/mol protein instead of mol/mol active sites.

[c] Qualitative, SDS gel results.

[d] Estimated from the initial slope of a curved titration.

of different catalytic specificities show large, peptide sequence-dependent differences in their rates of inactivation by peptide chloromethyl ketones,[39,40] requiring some adjustment of the reaction conditions for inaction with the two thioester inhibitors. However, in all cases studied, the magnitudes of the inactivation rate constants, coupled with the high specificity of the active site-directed reaction and generally low nonspecific reactivity of the inhibitors, allow complete inactivation to be achieved under practical conditions. Inactivation is accompanied by incorporation of the thioester peptide chloromethyl ketones with stoichiometries of 0.98 ± 0.10 mol/mol active sites and with an active site specificity of $\geq 93\%$ estimated from the fraction prevented by blocking the active site.[20,21]

Reactions 2 and 3 (Scheme II) of the second step of the procedure have been similarly evaluated from the stoichiometry of 5-AF incorporation and its dependence on NH_2OH and on blocking the active site (Table I).[20,21] Unlike the dependence of the reaction rate on enzyme catalytic specificity and inhibitor structure seen for the first step, the rate of thiol generation (reaction 2, Scheme II) varied only about fivefold for the enzyme–inhibitor combinations studied, allowing the reactions to be performed under similar conditions in all cases. Averaging the results in Table I for all of the enzymes and inhibitors indicates that 0.92 ± 0.11 mol probe per mole of active sites is site-specifically incorporated, representing $\geq 93\%$ specific labeling. Similar results were obtained for three additional enzymes in experiments where labeling was judged qualitatively from the fluorescence associated with the proteins on SDS-polyacrylamide gels (Table I).

The procedures used for labeling the enzymes, quantitating the reactions, and assessing the side reactions listed in Scheme II are outlined below. Although the side reactions have not significantly compromised site-specific labeling of any of the enzymes studied thus far, they do occur and appear to account for the low level of nonspecific inhibitor and probe incorporation observed, as well as the atypical behavior of one enzyme, human plasma kallikrein, which is described further below. Methods for evaluating the side reactions are presented for use in characterizing the labeling of new enzymes and for consideration in designing reaction conditions to maximize the specificity of labeling. The methods are illustrated primarily with results obtained for human α-thrombin, ATA-FPR-CH$_2$Cl, and 5-IAF.[19,20]

[39] C. Kettner and E. Shaw, this series, Vol. 80, p. 826.

[40] J. C. Powers and J. W. Harper, in "Proteinase Inhibitors" (A. J. Barrett and G. Salvesen, eds.), p. 55. Elsevier, Amsterdam, 1986.

Step 1: Enzyme Inactivation (Scheme II, Reaction 1)

The ratio of the active site concentration of the stock enzyme solution to its absorbance at 280 nm is determined initially to enable later expression of inhibitor and probe incorporation in terms of the site concentration calculated from the protein absorbance. Absorbance measurements are made on mixtures of 200 μl of enzyme and 800 μl of 0.125 M Tris-Cl, 1.25 mM EDTA, 7.5 M guanidine, pH 8.5 (pH 8.5 Tris/guanidine buffer), against a reference with buffer substituted for enzyme.

All of the inactivation reactions for the enzymes studied thus far have been carried out at pH 7.0 and 25°, with other conditions ranging from 3 to 50 μM enzyme and a two- to fivefold molar excess of ATA-FPR-CH$_2$Cl or ATA-FFR-CH$_2$Cl in 0.05–0.10 M HEPES, 0.15–0.30 M NaCl, ±1 mM EDTA or 10–20 mM CaCl$_2$, and incubation for 1–23 hr, until the desired level of residual activity (usually < 1%) has been reached.[20,21] In experiments with thrombin (Table I), for example, a 1-hr incubation of ~ 30 μM enzyme with a twofold excess of ATA-FPR-CH$_2$Cl in pH 7 buffer was typically used to achieve <0.1% residual activity. Because the inactivation rate varies for different enzymes and inhibitors, the inhibitor concentration and incubation time should be adjusted and completion of the reaction monitored by the loss of activity. The results listed in Table I for thrombin and factor IXa show the effect of the extremes of these conditions that have been used. Incubation of factor IXa with an approximately threefold higher inhibitor concentration and for 20 times longer than thrombin resulted in equivalent stoichiometries of inhibitor incorporation and subsequent 5-AF labeling, with evidence of only slightly higher levels of nonspecific reactions.[21] In considering other possible buffer conditions for these reactions, it is essential to note that nucleophiles, including Tris, other primary amines, and azide, which react with thioesters, should be rigorously excluded. This applies to all of the preparative steps involving the inhibitors and thioester-enzyme derivatives as well as the analytical procedures, with the few exceptions indicated. Although HEPES buffers have been most extensively used, similar results have been obtained with phosphate buffers in preliminary studies.[38] Essentially no information is available regarding the effects of pH higher than 7.0 or temperatures above 25° on the thioester peptide chloromethyl ketones or their reactions.

Once inactivated, the enzyme is dialyzed against 50 mM HEPES, 0.3 M NaCl, 1 mM EDTA, pH 7.0, at 4° to remove the excess inhibitor. Preparations of ATA-FPR–thrombin or ATA-FFR–thrombin that have been quick-frozen in a dry ice/2-propanol bath and stored at −70° appear to be stable for months or years.

Measurement of Inhibitor Incorporation. The thioester burst assay described earlier is also used to quantitate incorporation of the inhibitor

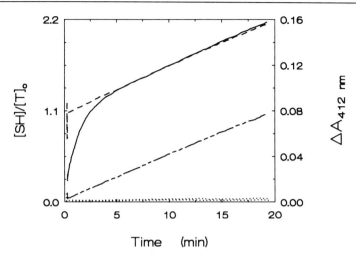

FIG. 5. Measurement of ATA-FPR-CH$_2$Cl thioester group incorporation into thrombin. Changes in absorbance (ΔA_{412nm}) recorded with time are shown for reactions of 5.2 μM ATA-FPR–thrombin, 480 μM DTNB, and 0.09 M NH$_2$OH in pH 7 buffer at 25°, corrected for the background due to DTNB. On the left axis ΔA_{412nm} is converted to the ratio of the concentration of thiol generated to the concentration of thrombin active sites ([SH]/[T]$_0$). Reaction mixtures: DTNB alone or ATA-FPR–thrombin mixed with DTNB at zero time (· · · ·); NH$_2$OH added to DTNB at zero time (– – –); NH$_2$OH added to a mixture of ATA-FPR–thrombin and DTNB at zero time (——); DTNB added at zero-time to ATA-FPR–thrombin that had been preincubated in 0.1 M NH$_2$OH for 30 min at pH 7.0 and 25° (- - - -). Reprinted with permission from Bock.[20]

accompanying reaction 1 (Scheme II).[20] As shown in Fig. 5, a mixture of 480 μM DTNB and 5.2 μM ATA-FPR–thrombin in pH 7.0 buffer shows no significant difference in absorbance at 412 nm compared to DTNB alone, indicating the absence of the free-thiol form of the inactivated enzyme. Addition of 0.09 M NH$_2$OH to ATA-FPR–thrombin and DTNB results in an exponential thiol burst and further linear increase due to breakdown of DTNB. Preincubation of ATA-FPR–thrombin with NH$_2$OH before addition of DTNB yields a dead-time burst, which represents the free-thiol enzyme concentration, demonstrating that the exponential process corresponds to reaction of NH$_2$OH with the thioester (Scheme II, reaction 2), independent of the rate of the succeeding DTNB–thiol reaction. Calculation of the exponential burst amplitude as described earlier and of the active site concentration from the 280-nm absorbance of the inactivated enzyme solution yields the stoichiometry of inhibitor incorporation, which for the assay of Fig. 5 was 1.04 mol inhibitor thioester groups per mole of active sites. As before, the assay may be performed as separate reactions and controls or as a difference measurement.[41] Stoichiometries of inhibitor incorporation obtained for thrombin and other enzymes by this procedure are listed in Table I.

Side Reactions (Scheme II; All Results Obtained in pH 7 Buffer at 25°[19-21]). Reaction 1a. Hydrolysis of the inhibitor thioester group, which can be measured by the appearance of thiols in an inhibitor–DTNB mixture, is slow, with ≤2% hydrolyzed over 20–27 hr.

Reaction 1b. Hydrolysis of the chloromethyl ketone is a first-order process, occurring with indistinguishable half-times of 41 and 36 hr for ATA-FPR-CH$_2$Cl and ATA-FFR-CH$_2$Cl, respectively.[20] This has generally been too slow to complicate achieving complete enzyme inactivation.

Reaction 1c. The extent of nonspecific acetylation of enzyme nucleophilic residues (X in Scheme II) by the inhibitors can be determined from the equivalent amount of thiol produced in a mixture of inhibitor, DTNB, and enzyme, relative to a DTNB–inhibitor control.[43] Results for thrombin and the inactive zymogen forms, prethrombin 1 and prothrombin, indicate ≤0.1 mol residues acetylated per mole of protein at 100 μM inhibitor over 20–27 hr.

Reaction 1d. Hydrolysis of the thioester group in the enzyme–inhibitor complex during the inactivation reaction and overnight dialysis steps does not occur to a significant extent for all but one of the enzymes studied. In the case of human plasma kallikrein, the thioester group is unstable once incorporated, spontaneously deacylating with a half-time of about 1 hr.[21] Consequently, labeling of this enzyme is active site-specific but does not require NH$_2$OH. This is atypical of the behavior observed for the remainder of the enzymes studied, where this reaction may account for the small difference between 5-AF incorporation in controls lacking NH$_2$OH and the active site-blocked enzymes, representing ≤0.04 mol/mol (Table I).

Reaction 1e. Nonspecific alkylation of the enzymes can be measured by the incorporation of thioester groups into the active site-blocked enzyme

---

[41] Indistinguishable assay results have been obtained with higher sensitivity by substitution of 4,4'-dipyridyl disulfide (~200 μM) for DTNB, and calculation of the thiol concentrations with an extinction coefficient of 19,800 M^{-1} cm^{-1} for the reactions monitored at 324 nm.[26,42]

[42] D. R. Grassetti and J. F. Murray, *Arch. Biochem. Biophys.* **119,** 41 (1967).

[43] It is best to use the active site-blocked enzyme or zymogen form of the proteinase for this measurement to avoid the possible contribution of hydrolysis of the thioester group of the incorporated inhibitor (Scheme II, Reaction 1d). Nonspecific acetylation measured in this way presumably includes O-acetylation of tyrosine residues, which may be partially reversed by the subsequent hydroxylamine treatment.[44]

[44] G. E. Means and R. E. Feeney, "Chemical Modification of Proteins," p. 71. Holden-Day, San Francisco, 1971.

(Table I) or the zymogen. Results for thrombin, prethrombin 1, and prothrombin indicate that this reaction results in ≤ 0.1 mol residues alkylated per mole of protein at 100 μM inhibitor over 20–27 hr.

Step 2: Thiol Generation and Fluorescence Probe Labeling (Scheme II, Reactions 2 and 3)

The much higher reactivity of thiols with iodoacetamides relative to amines, including NH_2OH, allows the thiol group to be generated and labeled in one reaction step, initiated by addition of NH_2OH to a mixture of enzyme–inhibitor complex and an excess of the probe. To prevent photochemical side reactions (not shown in Scheme II), all steps involving the probe should be performed under low light and solutions should be protected by covering them with aluminum foil. Solutions of the probes (Molecular Probes) are freshly prepared in buffer or solvents such as dimethyl sulfoxide or dimethylformamide that have been dried with Molecular Sieves (J. T. Baker, Phillipsburg, NJ) and probe concentrations are determined by absorbance or weight. Probe solutions in organic solvents should be prepared so that final solvent concentrations in the labeling reactions will be as low as possible (e.g., 1%). In studies with 5-IAF, solutions ≤ 200 μM have been made by suspending a small amount of the solid in pH 7.0 buffer, removing the excess by centrifugation (Microfuge, 2–4 min), and determining concentration from the absorbance at 498 nm in pH 8.5 Tris/guanidine buffer with an extinction coefficient of 84,000 M^{-1} cm^{-1}.[19-21]

Buffers and the 1.0 M NH_2OH to be used for the labeling reaction should be warmed to room temperature and degassed and/or purged with N_2 to minimize oxidative side reactions. In a typical labeling reaction, the ATA-FPR–enzyme or ATA-FFR–enzyme is mixed with pH 7.0 buffer and brought to 25° before addition of a fivefold molar excess of probe–iodoacetamide over 8–20 μM enzyme and initiation of the reaction by addition of 0.1 M NH_2OH. Incubation in the dark at 25° for 60–90 min has been sufficient to complete the reaction for all of the enzymes studied thus far. Completion of the reaction can be verified by the labeling stoichiometry and its independence on incubation time or probe concentration, and these conditions can be adjusted if necessary. Although it has not been necessary to compensate for the relatively small variation in the rates of thiol generation observed for different enzymes and inhibitors, this rate, as observed in the thioester assay (Fig. 5), may also be adjusted by changing the NH_2OH concentration.[38]

The labeling reaction mixture is chromatographed under low light on a column of Sephadex G-25 superfine equilibrated at room temperature with

10 mM HEPES, 0.3 M NaCl, 1 mM EDTA, pH 7.0, to remove the bulk of the excess dye. Strong adsorption of most of the free fluorescence probes to Sephadex results in efficient removal of the excess dye at column/sample volumes of ≥ 4 (ratios of 4–10 are routinely used), and this procedure in most cases is much more effective than dialysis.[38] Peak column fractions containing the labeled enzyme are identified by absorbance or visual inspection, using a hand-held UV lamp when necessary (366 nm, Blak-Ray Minearlight, UVP, San Gabriel, CA) and being careful to minimize the duration of the exposure. Residual free dye and NH_2OH are removed by dialysis against the desired buffer, the labeled enzyme is microfuged for 2 min, and quick-frozen aliquots are stored at $-70°$.

Measurement of Probe Incorporation. The stoichiometry of probe labeling is determined from absorbance measurements at 280 nm and at a well-separated wavelength characteristic of a probe model compound. Denaturing conditions are required for accurate quantitation of thrombin labeling, based on the requirement of these conditions to normalize the absorbance spectra of 5-AF–thrombin and that of 2-mercaptoethanol–5-AF used as a model compound.[19-20] To measure 5-AF incorporation, a 200-μl aliquot of labeled enzyme is mixed with 800 μl of pH 8.5 Tris/guanidine buffer and the absorbance at 280 nm is read against a blank with buffer substituted for enzyme. Then 10 μl of 100 mM dithiothreitol is added to both cuvettes and the 5-AF concentration is calculated with an extinction coefficient of 84,000 M^{-1} cm^{-1} from the absorbance at 498 nm, recorded after incubation for ≥ 10 min. The 280-nm absorbance is corrected for the contribution from the dye with the ratio, $\epsilon_{280nm}/\epsilon_{498nm} = 0.19$, determined from the spectrum of the model compound, and the active site concentration is calculated from the residual absorbance due to the protein. This procedure has also been used to quantitate labeling with other probes, such as 2-[(4'-iodoacetamido)anilino]naphthalene-6-sulfonic acid (IAANS), with an extinction coefficient for 2-mercaptoethanol–AANS of 26,600 M^{-1} cm^{-1} at 328 nm and $\epsilon_{280nm}/\epsilon_{328nm} = 0.78$.[21]

Figure 6 shows the results of quantitating the second step of the labeling procedure for ATA-FPR–thrombin as a function of 5-IAF concentration using the above procedures. A constant and maximal incorporation of 0.97 ± 0.04 mol 5-AF per mole of active sites was observed at 5-IAF concentrations above a twofold molar excess, in reactions accompanied by loss of the thiol group. Labeling was prevented by blocking the active site or omitting NH_2OH, with residual nonspecific incorporation as detailed in Table I, along with results obtained for other enzymes.

Examination of the fluorescence and protein-stained bands visible on SDS-polyacrylamide gels in samples from such labeling reactions, after removal of the free dye, can be used to assess the specificity of the reactions

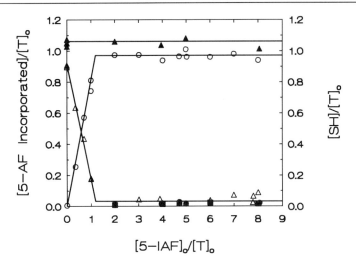

FIG. 6. Quantitation of 5-IAF labeling reactions of ATA-FPR–thrombin. Stoichiometries of 5-AF incorporation ([5-AF incorporated]/$[T]_0$) for reactions of ATA-FPR–α-thrombin in the presence (O) or absence (●) of NH_2OH and for active site-blocked thrombin (■) are plotted as a function of the molar ratio of 5-IAF to thrombin ($[5\text{-}IAF]_0/[T]_0$) in the pH 7 buffer, 25° incubations. Residual thioester content of ATA-FPR–thrombin samples ([SH]/$[T]_0$) after incubation with 5-IAF in the absence of NH_2OH (▲) and residual free thiols for reactions in the presence of NH_2OH (△), determined by DTNB assay. Reprinted with permission from Bock.[20]

for a new enzyme qualitatively, particularly if only small quantities are available (Table I). As shown for thrombin in Fig. 7, covalent labeling of the catalytic site-containing B chain is seen only for the sample of enzyme that was inactivated with the thioester peptide chloromethyl ketone and subsequently incubated with 5-IAF in the presence of NH_2OH. In SDS-gel electrophoresis experiments with the thioester-labeled enzymes, it is important to include freshly prepared iodoacetamide at a final concentration of 1–2 mM in the SDS sample preparation mixture for nonreduced samples, before addition of the proteins. The alkylating agent blocks the free thiol generated during sample preparation at 100° in SDS, which otherwise results in the appearance of numerous protein bands resulting from random thiol–disulfide exchange.

  Side Reactions (Scheme II; All Results Obtained at pH 7.0 and 25°[20,21]). Reaction 2a. The NH_2OH treatment is mild in comparison to that which cleaves susceptible peptide bonds,[45] and no cleavage has been observed by SDS-gel electrophoresis for any of the enzymes examined.

---

[45] P. Bornstein and G. Balian, this series, Vol. 47, p. 132.

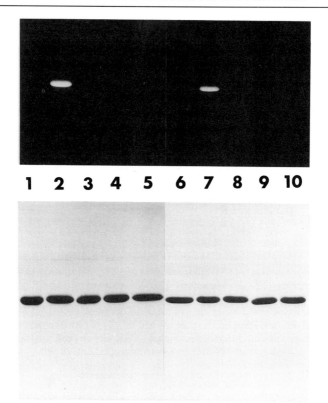

Fig. 7. SDS-gel electrophoresis of samples from 5-IAF labeling reactions of ATA-FPR–thrombin. The fluorescein fluorescence (top) and protein-stained bands (bottom) on 14% SDS gels of nonreduced (NR) and reduced (R) thrombin samples (8 μg) are shown. Lanes 1 (NR) and 6 (R), ATA-FPR-α-thrombin; lanes 2 (NR) and 7 (R), ATA-FPR–thrombin incubated with 5-IAF in the presence of NH$_2$OH; lanes 3 (NR) and 8 (R), ATA-FPR–thrombin incubated with 5-IAF in the absence of NH$_2$OH; lanes 4 (NR) and 9 (R), active site-blocked thrombin incubated with ATA-FPR-CH$_2$Cl, dialyzed and incubated with 5-IAF in the presence of NH$_2$OH; lanes 5 (NR) and 10 (R), thrombin denatured in SDS before treatment with ATA-FPR-CH$_2$Cl and incubation with 5-IAF in the presence of NH$_2$OH. Adapted with permission from Bock.[19] Copyright [1988] American Chemical Society.

Reaction 3a. Consumption of the fluorescence probe–iodoacetamide by reaction with NH$_2$OH does not appear to be significant, as indicated by the essentially stoichiometric 5-IAF titration of ATA-FPR–thrombin (Fig. 6). Although several other probe–iodoacetamides have also been used (see below), preliminary experiments with some reagents have been unsuccessful,[38] presumably due to their reaction with NH$_2$OH. However, it should be feasible to employ these and other thiol-reactive probes that may react at significant rates with NH_2OH by labeling the isolated free-thiol form of the enzyme–inhibitor complex. The free thiol form of ATA-FPR–thrombin, for example, can be isolated in $95 \pm 7\%$ yield by preincubation of ATA-FPR–thrombin with NH_2OH, followed by rapidly lowering the pH and removal of NH_2OH by dialysis.[20]

Reaction 3b. Introduction of the thiol group into proteins that contain disulfide bonds allows for the possibility of thiol–disulfide exchange, which could generate monomeric or polymeric side products labeled at enzyme sulfhydral groups. These extent of these side reactions can be assessed by determining the homogeneity of the location of the incorporated label in peptide mapping experiments. Amino acid analysis of the enzyme labeled with iodoacetic acid in place of the probe may also be used, because modification of the inhibitor thiol yields carboxymethylmercaptoacetate after acid hydrolysis, which is not detected, whereas the carboxymethylcysteine produced from labeling of the disulfide exchange product can be quantitated.[46] These reactions appear to occur only to a small extent for the free thiol forms of ATA-FPR–thrombin and ATA-FFR–thrombin, with ≤ 0.14 mol carboxymethylcysteine per mole of thrombin detected after incubation for 24-hr, much longer than required for probe labeling. Generation of the thiol together with its modification in one reaction step decreases the opportunity for these slower, competing reactions by limiting the lifetime of the free thiol group.

Reaction 3c. The thiol group can oxidize to unlabeled monomeric or disulfide-bonded dimeric products, causing lower than stoichiometric probe labeling and the appearance of an enzyme dimer on SDS gels. The extent of these reactions can be determined for the free thiol enzyme by measuring the loss of NH_2OH-generated thiol with DTNB. Losses of $50 \pm 20\%$ and similar decreases in 5-AF incorporation were observed for thrombin over 20 hr, as well as the appearance of an enzyme dimer. These oxidation products are minimized in the labeling procedure by the presence of EDTA, by reducing the level of dissolved oxygen, and by effective competition between the rate of thiol modification by the probe and the slower side reactions.

Reaction 3d. Nonspecific alkylation of enzyme nucleophilic residues (X in Scheme II) by the fluorescence probe–iodoacetamides is generally low under the pH 7.0 conditions used. The extent of this reaction estimated from 5-AF incorporation into untreated, native thrombin (~ 0.02 mol/mol)[20] is indistinguishable from the values of 0.02 ± 0.01 mol/mol

---

[46] Although it has not been tested, the same products are predicted after acid hydrolysis of the enzymes labeled with some of the fluorescence probes, which should allow similar quantitation of these side reactions by amino acid analysis of the fluorescent enzyme derivatives.

(Table I) obtained for the active site-blocked control, which reflects the sum of side reactions 1e and 3d (Scheme II), and the control lacking NH_2OH, which measures reaction 3d (Scheme II) plus any free thiol enzyme present from reaction 1d (Scheme II). This comparison suggests that direct alkylation by the probe–iodoacetamide accounts for a significant fraction of the low level of nonspecific labeling observed.

Screening Fluorescent Proteinase Derivatives for Reporters of Interactions

A main advantage of the ability to prepare a large number of enzyme derivatives that are active site-specifically labeled with different fluorescence probes is in identifying derivatives with optimum properties as reporters of proteinase interactions. In an example of this application, an array of 16 human α-thrombin derivatives was prepared with each of eight fluorescence probes connected to the active site by either ATA-FPR-CH_2Cl or ATA-FFR-CH_2Cl. These were then screened for changes in probe fluorescence intensity signaling the binding of the fragment 2 domain of prothrombin.[21] The maximum changes in fluorescence observed, shown in Fig. 8, provide some initial information about the behavior that can be expected of the derivatives in reporting interactions. Only about half of the derivatives exhibited changes of $> 10\%$ for investigating the interaction, confirming the likely need to test several in order to identify one that is truly useful for quantitative binding studies. Each of the derivatives appeared to show a unique spectral response to fragment 2 binding, with both the sign and magnitude of the fluorescence changes determined by the structure of the probe and the sequence of the connecting peptide. These results illustrate the lack of predictability in the behavior of the fluorescent derivatives in reporting changes in the environment of the active site resulting from interactions, and the advantage of screening a large number of them to select those with the most desirable properties. The behavior of members of the same array of derivatives in reporting other interactions indicates that the pattern of spectral responses may be characteristic for different macromolecular interactions, allowing derivatives that selectively signal particular interactions to be identified.[38,47]

Conclusions and Future Prospects

The active site-selective labeling approach was developed to expand the opportunities for applying spectroscopic techniques based on covalent

---

[47] P. J. Hogg, P. E. Bock, J. K. Labanowski, and C. M. Jackson, unpublished results, 1992.

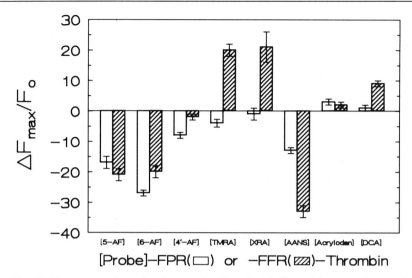

FIG. 8. Fluorescence intensity changes signaling the binding of fragment 2 to active site-labeled thrombin derivatives. The maximum fractional changes in probe fluorescence intensity ($\Delta F_{max}/F_0$), determined by analysis of fragment 2 titrations and expressed as percents, are plotted for various [probe]–FPR–thrombin (open bars) and [probe]–FFR–thrombin (shaded bars) derivatives. Error bars represent ±2 SD. The thiol-reactive probes represented are 5-(iodoacetamido)fluorescein (5-AF), 6-(iodoacetamido)fluorescein (6-AF), 4'-([[(iodoacetyl)amino]methyl]fluorescein (4'-AF), tetramethylrhodamine-5-(and-6)-iodo-acetamide (TMRA), rhodamine X iodoacetamide (XRA), 2-[(4'-iodoacetamido)anilino] naphthalene-6-sulfonic acid (AANS), 6-acryloyl-2-dimethylaminonaphthalene (acrylodan), and 7-diethylamino-3-[(4'-iodoacetylamino)phenyl]-4-methylcoumarin (DCA). Adapted with permission from Bock.[21]

probes to studies of the mechanisms and regulation of plasma proteinase reactions. The approach represents a new strategy for site-specific labeling in that it employs a single affinity label to incorporate a uniquely reactive group for subsequent selective modification by members of probe families with different structures and spectral properties, or probes of different types. The new combination of previous strategies employing group-selective reagents to incorporate thiols into proteins lacking them[37,48-51] and active site affinity labeling with peptide chloromethyl ketones[8-18] allows both site-specific labeling and an increased choice of probes from the many available as thiol reagents. Quantitation of the labeling reactions indicates

[48] I. M. Klotz and R. E. Heiney, Arch. Biochem. Biophys. 96, 605 (1962).
[49] J. Carlsson, H. Drevin, and R. Axén, Biochem. J. 173, 723 (1978).
[50] R. Benesch and R. E. Benesch, J. Am. Chem. Soc. 78, 1597 (1956).
[51] R. N. Perham and J. O. Thomas, J. Mol. Biol. 62, 415 (1971).

that the approach represents a practical alternative to the use of dedicated analogs of peptide chloromethyl ketones for active-site labeling, exhibiting a site specificity of $\geq 93\%$ and yielding equivalent stoichiometries of probe incorporation in those cases that can be compared.[15,20,21] Despite the potential for complications arising from side reactions, these reactions do not appear to present a more significant problem than is typically encountered in labeling natural protein thiols. Assessment of the sources of nonspecific labeling and comparison of the results obtained for different enzymes and inhibitors suggest that the specificity of labeling may be increased by refining the conditions of the inactivation and probe-labeling steps to better suit individual cases.

For fluorescence studies of proteinase interactions, the ability to prepare derivatives with many different probes overcomes the obstacle of not being able to predict which label will provide the spectral change needed to observe an interaction, and should also allow matching of probe spectral properties to the requirements of fluorescence anisotropy and resonance energy transfer studies. Results of the studies performed thus far indicate that the way in which changes in the proteinase active site environment associated with binding of regulatory macromolecules are reported as perturbations of the probes is a complex function of the structure as well as the spectral properties of the probe–peptide labels. The results for the thrombin–fragment 2 interaction,[21] along with those of a recent study of the thrombin–thrombomodulin interaction,[52] and results for other thrombin interactions[38,47] suggest that each peptide–probe label may provide a unique view of changes in vicinity of the active site, which may be distinctive for different macromolecular interactions. It is likely that the same derivatives will also differ individually in the fluorescence anisotropy changes that they display in response to interactions. On the basis of these observations, application of the active site-selective approach with one thioester peptide chloromethyl ketone can be estimated to provide roughly 50 experimental opportunities to achieve the initial goal of observing a fluorescence intensity or anisotropy signal for investigating each proteinase interaction of interest. The finding that proteolytic enzymes of diverse catalytic specificities can be active site-selectively labeled with the same thioester peptide chloromethyl ketones indicates that the approach can likely be applied to fluorescence studies of the proteinases of many systems, including the blood coagulation, contact activation, complement, protein C, and fibrinolysis pathways. Finally, the thioester peptide chloromethyl ketones should similarly enable investigation of these enzymes with

[52] J. Ye, N. L. Esmon, C. T. Esmon, and A. E. Johnson, *J. Biol. Chem.* **266**, 23016 (1991).

techniques based on incorporation of other types of labels, including chromophores, electron spin resonance probes, and NMR probes.

Acknowledgments

The majority of this work was performed at the American Red Cross Blood Services, Southeastern Michigan Region, Research Laboratory, Detroit, MI, and was supported by NIH Grant HL 38779 and in part by American Red Cross funds.

[28] Peptide Chloromethyl Ketones as Labeling Reagents

By E. Brady Williams and Kenneth G. Mann

Introduction

The utility of α-halo ketone derivatives of amino acids as covalent inhibitors of serine proteases has been recognized since their development in the 1960s by Shaw *et al.*[1,2] Much of the early work on these inhibitors has been previously reviewed in this series.[3] The earliest chloromethyl ketone derivatives were blocked, derivatized amino acids such as tosylphenylalanine chloromethyl ketone (TPCK) and tosyllysine chloromethyl ketone (TLCK). These reagents exhibited broad specificities toward serine proteases. Work in our laboratories and elsewhere has since focused on molecules with greater degrees of specificity toward specific proteases. This has largely been accomplished by using small peptides rather than amino acids as the base structure for the α-halo ketone. Such efforts have been particularly important in studies of the proteases of the coagulation and fibrinolytic pathways. A second major advance has been in the use of the covalent linking properties of chloromethyl ketones to carry marker groups into the enzymes. These groups then can be used in assays or to follow the proteins during studies of their interactions with each other and with other components of physiological systems. In this chapter, we will describe some of the methods we and others have found useful in preparing and using these reagents.

The synthesis of the chloromethyl ketone group generally involves reaction of a blocked amino acid with a chloroformate to produce a mixed

---

[1] G. Schoellmann and E. Shaw, *Biochem. Biophys. Res. Commun.* **7**, 31 (1962).
[2] E. Shaw, *Enzymes* **1**, 91 (1970).
[3] C. Kettner and E. Shaw, this series, Vol. 80, p. 826.

anhydride. This is then reacted with diazomethane to produce a diazomethyl ketone. The diazomethyl ketone reacts rapidly with ethanolic HCl to produce the chloromethyl ketone group. The chemistry of these reactions is adequately described elsewhere.[3] Many workers in the field have avoided these preparations due to the dangers associated with diazomethane. While these problems should not be minimized, use of proper glassware, a hood with good ventilation, gloves, and a safety shield is sufficient to make these preparations reasonable for well-equipped laboratories and trained personnel. To minimize dangers from explosion, the scale of the diazomethane reaction should be kept to around 50 mM or less. Clearseal nonground glassware is available from Aldrich (Milwaukee, WI). The kit includes instructions for safe preparation of diazomethane. A significant drawback in these preparations is the relatively low yields obtained. This has been especially true for the arginine chloromethyl ketone derivatives, where yields are typically in the 30–50% range. Attempts to improve the yields, primarily through the use of more soluble arginine derivatives, have not met with great success.[4]

Most preparations of peptide chloromethyl ketones have been accomplished by attaching additional blocked amino acids or peptides to the previously prepared amino acid chloromethyl ketone derivative. In these reactions, the principal difficulties are avoiding hydrolysis of the α-chloro ketone and cyclization reactions of the peptide derivatives. Hydrolysis of the chloromethyl ketone to the hydroxymethyl derivative can occur rapidly in basic aqueous media, and nucleophilic reagents such as amines and thiols can react rapidly in nonaqueous solvents. For these reasons, a major factor in designing synthesis, purification, and storage strategies is the minimization of times when the derivatives are subject to basic conditions. The most commonly used method for attaching peptides to the chloromethyl ketone-derived amino acid is mixed anhydride coupling. The rapid reaction rates, nonaqueous solvents, and minimum amounts of base required are favorable. A typical preparation of a tripeptide chloromethyl ketone is described below.

Preparation of L-Tyrosylglycylarginine Chloromethyl Ketone Dihydrochloride (YGR-ck · 2HCl)

N-*tert*-Butyloxycarbonyl-O-benzyltyrosylglycine (428 mg, 1.00 mmol) is dissolved in 5 ml of tetrahydrofuran at $-20°$ and treated with 1 equivalent (0.12 ml) of N-methylmorpholine followed by 1 equivalent (0.13 ml)

[4] R. T. Alpin, J. Christiansen, and G. T. Yang, *Int. J. Pept. Protein Res.* **21**, 555 (1983).

of isobutyl chloroformate. The mixed anhydride is allowed to form for 20 min at $-20°$. In a second flask, ε-nitro-L-arginine chloromethyl ketone hydrochloride (288 mg, 1.00 mmol) is dissolved in 2.0 ml of dimethylformamide (DMF) and cooled to $-20°$. The chloromethyl ketone solution is added to the mixed anhydride along with 1 equivalent (0.12 ml) of N-methylmorpholine. The reaction is stirred for 1 hr, slowly warming to $0°$.

The reaction is partitioned between cold 1 N sulfuric acid (25 ml) and ethyl acetate (50 ml) and the separated ethyl acetate layer is washed quickly with 20-ml portions of 1 N sulfuric acid, water, saturated sodium bicarbonate, water, and saturated sodium chloride. The remaining organic layer is dried over anhydrous magnesium sulfate and evaporated to an oil, which can be recrystalized from ethyl acetate/hexane to give a colorless solid. Yield is 85–90%, and the product shows little or no contamination with hydroxymethyl ketone, as determined by thin-layer chromatography (8 : 2 chloroform : methanol on silical gel) or by HPLC.

The free peptide chloromethyl ketone is obtained by removal of the protective groups using liquified hydrogen fluoride. The peptide is placed in a Teflon reaction vessel along with approximately 1 ml of anisole per gram of peptide. The anisole acts as a cation trap during the HF cleavage. The reaction vessel is cooled in a dry ice/acetone bath to $-60°$ and hydrogen fluoride is distilled into the container from a lecture bottle, using an all-Teflon system. The volume of HF should be 5–10 ml per gram of blocked peptide. The reaction is stirred and allowed to warm to room temperature for 1 hr, then cooled again to $-60°$. An aspirator is used to remove the hydrogen fluoride, taking care to avoid loss of product due to excessive foaming. Once vacuum is established, the vessel is allowed to warm; 1 hr or more may be required to remove all HF and reach room temperature. When all HF is removed, the resulting thick solution will generally be a dark red or violet, due to substitution products of the anisole. The oil is triturated with anhydrous ether two or three times, decanting and discarding the ether each time, then taken up in 1.0 M HCl in anhydrous ethanol (prepared by bubbling HCl gas into 100% ethanol). The resulting solution is evaporated to dryness and further dried in a desiccator over KOH, then dissolved in water and lyophilized. The peptide can then be chromatographed on a 1 × 25-cm column of Sephadex LH-20, swollen in methanol for 12 hr. The crude peptide is applied in methanol solution and chromatographed using methanol as the elutant. The elution can be followed using UV absorbance. The product, obtained by evaporation of the methanol and relyophilization of an aqueous solution of the resulting oil, typically shows small amounts of hyroxymethyl ketone on HPLC analysis.

TABLE I
FLUORESCENT CHLOROMETHYL KETONES

Product	Method	Yield (%)	Ex/Em[a] λ_{max} (nm)	Extinction (M^{-1} cm^{-1})
1,5-Dansyl-EGR-ck	SO$_2$Cl	52	320/520	4200
2,5-Dansyl-EGR-ck	SO$_2$Cl	32	378/475	3000
2,6-Dansyl-EGR-ck	SO$_2$Cl	45	359/435	5700
Fluoresceinyl-EGR-ck	ONSu	36	490/525	75,000
Rhodamine-X-EGR-ck	ONSu	25	560/590	85,000
Lissamine–rhodamine-EGR-ck	SO$_2$Cl	24	570/590	71,000
1-Pyrene-EGR-ck	SO$_2$Cl	52	349/398	30,000
1,5-Dansyl-FPR-ck	SO$_2$Cl	62	320/520	4200
2,6-Dansyl-FPR-ck	SO$_2$Cl	40	359/435	5700
Fluoresceinyl-FPR-ck	ONSu	32	490/525	75,000
Lissamine–rhodamine-FPR-ck	SO$_2$Cl	50	570/590	71,000
1,5-Dansyl-YGR-ck	SO$_2$Cl	40	320/530	4200

[a] Ex and Em are the excitation and emission maxima of the fluorescent group. Extinction coefficients are for the wavelength noted as the excitation maximum.

An alternative approach that has been reported for preparation of chloromethylated peptides[5,6] is the attachment of the chloromethyl ketone group to a peptide, rather than to the amino acid. In these preparations, the free acid, N-blocked peptide is reacted with a chloroformate to prepare the mixed anhydride, which can be converted to the diazomethyl ketone by reaction with diazomethane. This product is then converted to chloromethyl ketone using HCl in ethanol. Use of this route avoids the potential for hydrolysis of the chloromethyl ketone group during peptide coupling or workup. However, the requirement that amine and acid functional groups be blocked generally precludes the use of peptides produced by solid-phase (Merrifield) synthesis.

Derivatization of Peptide Chloromethyl Ketones

The covalent attachment of chloromethyl ketones to serine proteases makes them excellent carriers for labeling groups. The most widely used label has been the 5-dimethylamino-1-naphthalenesulfonamide group (dansyl). We have reported preparation of a variety of chloromethylated C-terminal arginine peptides with different fluorescent labeling groups (see Table I).[7] In each case, these are prepared by attachment of the fluorescent

[5] M. Szucs, S. Benghe, A. Borsadi, M. Wollimann, and G. Jareso, *Life Sci.* **32**, 2777 (1983).
[6] R. F. Venn and E. A. Barnard, *J. Biol. Chem.* **256**, 1529 (1981).
[7] E. B. Williams, S. Krishnaswamy, and K. G. Mann, *J. Biol. Chem.* **264**, 7536 (1989).

group to the unblocked peptide. The methods of attachment used depend on the kind of activated groups available on the fluorescent label. Most common are the sulfonyl chloride and N-hydroxysuccinimide ester groups. Typical preparations are described below.

Preparation of 2,6-Dansyl L-glutamylglycyl-L-arginine Chloromethyl Ketone Hydrochloride. A solution of Glu-Gly-Arg-ck · 2HCl (63 mg, 0.2 mmol) in 2.0 ml of methanol is cooled to 0° and treated with N-methylmorpholine (0.30 ml of 1.0 M solution in methanol). Dansyl chloride (56 mg, 0.20 mmol) is added, and the reaction is stirred for 10 min, warming to room temperature. The solution is acidified with 1.0 ml of 1.2 M HCl/ethanol and evaporated *in vacuo* to an oil. The oil is taken up in 0.1 M aqueous HCl and eluted from a Sephadex LH-20 column with methanol. The elution of the desired product is followed using fluorescence and Sakaguchi tests.[8] Removal of the methanol under reduced pressure gives an oil, which is dissolved in 0.01 M HCl and lyophilized to an off-white, hydroscopic solid. The yield is 72 mg (55%).

Preparation of 5- (and 6-)Carboxylfluoresceinyl-L-glutamylglycyl-L-arginine Chloromethyl Ketone Hydrochloride. A solution of Glu-Gly-Arg-ck · 2HCl (13.0 mg, 28 μmol) in water (150 μl) is treated in rapid succession with 5- (and 6-)carboxylfluorescein N-hydroxysuccinimide ester (15.0 mg, 32 μmol) in dimethylformamide (300 μl) followed by 1.0 M KHCO$_3$ (50 μl, 50 μmol). After 10 min at room temperature (22°), the reaction is stopped with 0.10 M HCl (400 μl) and the entire reaction mixture is applied to a 1 × 30-cm Sephadex LH-20 column (in methanol). Elution with methanol is followed using the Sakaguchi test and fluorescence. The elution gives a single band, which is positive to both tests. The product is pooled and the solvents are removed *in vacuo.* The resulting oil is taken up in 0.10 M HCl and lyophilized to an oil, which is taken up in 0.010 M HCl and stored frozen. Typical yield is 10.0 mg (36%). HPLC analysis indicates small amounts of hydroxymethyl ketone contaminants. Preparative HPLC can be used to obtain products essentially free of hydrolysis products. For this, a preparative scale C$_{18}$ column is used with a 20-min gradient of 0–70% acetonitrile/0.5% trifluoroacetic acid in water/0.5% trifluoroacetic acid.

With lysine peptides, a somewhat different approach is required. Although the difference in pK_a between the N-terminal amine and the guanidyl group make possible attachment of the fluorescent group directly to the free arginine derivative, lysine peptides must be labeled prior to removal of the ε-amino group protection. This then requires that the labeling group be

---

[8] J. M. Stewart and J. C. Young, "Solid Phase Peptide Synthesis." Freeman, San Francisco, 1969.

stable under the conditions required for blocking-group removal. We have prepared 1,5-dansyl-Tyr-Gly-Lys-ck from BOC-(*O*-Bz)-Tyr-Gly-(ε-Z)-Lys-ck as described below.

Preparation of 1,5-Dansyl-L-tyrosylglycyl-L-lysine Chloromethyl Ketone Dihydrochloride. BOC-(ε-Z)-Lys-ck was prepared by the method of Coggins *et al.*[9] and the BOC group is cleaved using 2.0 *M* HCl/ethanol. BOC-(*O*-Bz)-Tyr-Gly-COOH is coupled to the chloroketone using the mixed anhydride method as described above, to yield the fully blocked tripeptide chloromethyl ketone BOC-(*O*-Bz)-Tyr-Gly-(ε-Z)-Lys-ck.

The BOC group is removed using 2 *M* HCl/ethanol. The fully blocked peptide (72 mg, 0.10 mmol) is dissolved in 5 ml of 2 *M* HCl/ethanol and allowed to react for 5 min at 20°. The solvent is removed *in vacuo* and the resulting oil is triturated briefly in ether, then dried over KOH *in vacuo*. The resulting solid is taken up in methanol (2 ml) and treated with 1,5-dansyl chloride (24 mg, 0.10 mmol) and *N*-methylmorpholine in methanol (1 ml, 0.10 *M*). After 1 hr at 20°, the reaction mixture is acidified with 1 *M* HCl/ethanol and evaporated to an oil. The oil is taken up in methanol and chromatographed on Sephadex LH-20 (1 × 30 cm, in methanol), following the elution by UV absorbance and fluorescence. The product, 1,5-dansyl-(*O*-Bz)-Tyr-Gly-(ε-Z)-Lys-ck, is obtained by evaporation of the methanol and recrystallization from chloroform/cyclohexane. The yield is 76 mg (91%).

The dansylated product is then deprotected by HBr/acetic acid. The blocked, dansylated peptide is dissolved in 5 ml of 2 *M* HBr/acetic acid and stirred 30 min. The resulting solution is evaporated to an oil, dissolved in 1 ml of 1 *M* HCl/ethanol, reevaporated, taken up in water, and lyophilized. The oily residue is chromatographed on Sephadex LH-20 in methanol. The pooled product, identified by fluorescence and Pauly stain,[10] is collected by evaporation of the methanol and lyophilization from 0.01 *M* HCl. The resulting product, 1,5-dansyl-Tyr-Gly-Lys-ck · 2HCl is obtained in 32 mg (48%) yield. It is contaminated by traces of the hydroxymethyl ketone (by HPLC).

The shelf stability of these preparations varies considerably. The dansylated tripeptides are generally stable for at least 1 year if stored in the solid form at −70°, and solutions of the products in 10 m*M* HCl retain full activity for similar periods. Solid preparations of the free peptide chloromethyl ketones appear to decompose somewhat quicker, losing activity as inhibitors and producing significant levels of impurity when stored for 1 year. Similarly, rhodamine and fluorescene derivatives have a shorter shelf

[9] J. R. Coggins, W. Kray, and E. Shaw, *Biochem. J.* **137**, 579 (1974).
[10] Z. Pauly, *Hoppe-Seyler's Z. Physiol. Chem.* **42**, 508 (1904).

life. After 6 months, a rhodamine X derivative of glutamylglycylarginine chloromethyl ketone gave several fluorescent spots on thin-layer chromatographs, and had lost about 25% of its activity toward protease. In all cases, exposure to elevated temperatures for relatively brief times leads to decomposition, as does repeated freeze–thaw cycling. For these reasons, we generally aliquot new preparations into small vials that can be used individually.

Biotinylated Chloromethyl Ketones

The derivatization of peptide chloromethyl ketones with chemically or biochemically reactive groups offers another means of labeling proteases. We have utilized the strong interaction between biotin and avidin to both detect and remove active proteases in complex mixtures. The synthesis of the biotinylated inhibitors follows the protocol for N-hydroxysuccinimide ester reactions. A spacer of one or two 6-aminocaproyl residues must generally be added to obtain sufficient separation of the biotin from the labeled protein. Biotin-N-hydroxysuccinimide ester (biotin-ONSu) and derivatives with one and two 6-aminocaproyl spacers (biotin-CAP-ONSu, biotin-CAP-CAP-ONSu) are available commercially. A typical synthesis is described below.[7]

Preparation of Biotinyl-6-aminocaproyl-Tyr-Gly-Arg-ck. tBOC-(-O-Benzyl)-Tyr-Gly-(ε-nitro)-arg-ck (132 mg, 0.20 mmol) is dissolved in 2 M HCl/methanol (10 ml) for 10 min. The solution is evaporated to an oil and treated with a second aliquot of HCl/methanol for 20 min. The solvent is again removed *in vacuo* and ether (50 ml) is added. The solvent is decanted from the precipitated product and the residue is dried overnight over KOH at reduced pressure. The remaining solid is taken up in DMF (1 ml) and added to a solution of biotinyl-6-aminocaproic acid N-hydroxysuccinimide ester (91 mg, 20 mmol) in 2 ml DMF. To this is added a solution of KHCO$_3$ (40 mg, 40 mmol) in water (0.5 ml). The reaction is stirred for 20 min, then treated with HCl/methanol (5 ml) and ether (50 ml). The mixture separated into two phases. The upper layer (ether) is discarded and the lower layer is applied to a column of Sephadex LH-20 in methanol (1 × 20 cm). Elution with methanol gives a single peak (as detected by A_{280}). This is collected by precipitation with ether and filtration. The resulting solid gives a single band on thin-layer chromatography ($R_f = 0.21$, 5 : 1 chloroform : methanol on silica gel). The yield is 69 mg (35%).

The above blocked peptide (65 mg, 66 μmol) is treated with 1 ml of anisole and deprotected with liquid HF (5 ml). After stirring for 30 min, the HF is evaporated *in vacuo* and the resulting oil is stirred with ether (20 ml). After decanting the ether, the product is dried over KOH overnight.

The residue is taken up in 1.0 mM HCl and filtered to remove insoluble residues. The filtrate is lyophilized to an off-white solid, which is dissolved in methanol and chromatographed on Sephadex LH-20 (1 × 20.cm) in methanol. A single UV-positive, Sakaguchi-positive band is eluted. This is collected and the solvent is evaporated *in vacuo*. The residue is taken up in 1.0 mM HCl and lyophilized to yield 52 mg (87%). Samples are tested for their ability to inhibit the reaction of factor Xa with synthetic substrate (S2222) and to enhance the fluorescence of fluoresceinylavidin. The product is aliquoted and stored in frozen solution in 10 mM HCl.

In many applications, a second 6-aminocaproyl spacer is needed to ensure full biotin–avidin interactions. The preparation of biotinylcaproyl-caproyl-FPR-ck is analogous to the above method. The dicaproyl N-hydroxysuccinimide ester is commercially available.

Derivatization of Proteases

We have found that these peptide derivatives react with trypsin-type proteases generally, although considerable specificity can be observed in the rates of reaction. The level of specificity does not appear as high as reported by Kettner and Shaw[3] for the "unlabeled" peptide chloromethyl ketones. The specificity for protease versus zymogen is, however, total, allowing us to use these preparations to determine levels of activation in preparations of coagulation proteins or in plasma samples. Proteases labeled with fluorescent groups have also proved useful in studies of protein–protein and protein–phospholipid interaction, using both fluorescence polarization and fluorescence energy transfer techniques.[11] A typical protocol for labeling a protease is described below.

Preparation of 2,6-Dansyl-EGR-plasmin. Human plasmin (2.0 mg, 29 nmol) in HEPES/saline (4.0 ml, 20 mM HEPES, pH 7.40, 150 mM NaCl) is assayed using S2251. The solution is then reacted with 2,6-dansyl-EGR-ck (30 nmol, as 6.0 mM solution in 0.010 M HCl). After 2 min, the enzymatic activity is again assayed with S2251 and has decreased to 45% of the original. Additional 30-nmol aliquots of inhibitor are added and the product is assayed. After 150 nmol total addition, no activity versus S2251 can be detected, even when the concentration of enzyme is increased fivefold. The reaction mixture is then chromatographed on Sephadex G-25 in HEPES/saline, giving a single protein band that is intensely fluorescent. This is pooled and the protein is precipitated by bringing the solution to 80% ammonium sulfate. The suspension is centrifuged and the resulting pellet is dissolved in 50% glycerol. Analysis, based on the extinction coeffi-

[11] S. Krishnaswami, E. B. Williams, and K. G. Mann, *J. Biol. Chem.* **261,** 9684 (1986).

cient of 2,6-dansyl (5700 M^{-1} cm^{-1}), indicates a ratio of 0.98 equivalents of fluorophore/protein.

For other protein/inhibitor combinations, complete elimination of protease activity required from 2 to 50 equivalents of chloromethyl ketone. Smaller equivalences could be obtained by longer incubations or by adding inhibitor in smaller aliquots. Proteins successfully labeled in our laboratory include thrombin, factor VIIa, factor IXa, factor Xa, activated protein C, TPA, and plasmin. In all cases, completion of the labeling was judged by complete loss of enzymatic activity, high levels of fluorescent yield, and purity of the original chloromethyl ketone derivatives.

Similar procedures can be used to label the proteases with biotinylated inhibitors.[12]

Preparation of Biotin-CAP-FPR-thrombin. A concentrated solution of thrombin in Tris (20 mM)/saline (0.15 M), pH 7.4, is treated with a five-fold excess of biotin-CAP-FPR-ck and reacted for 5 min at 37°. Loss of activity is verified using synthetic chromogenic thrombin substrate. (Note: In cases where activity remains, a second aliquot of inhibitor can be added to complete deactivation.) The derivatized protein is separated from excess inhibitor (and its hydrolysis products) by gel filtration on Sephadex G-25–150 (1.5 × 60 cm). The derivatized protein is analyzed by gel electrophoresis and blotting assay (see below) and collected by ammonium sulfate precipitation and centrifugation. Products can be stored in 50% glycerol solution at −20°. Similar preparations have been made with factor Xa, plasmin, rt-PA, and activated protein C.

Assay of Biotinylated Proteases

The presence of the biotin label on the protein can be determined either by fluorescence or by electroblotting assay. Both techniques employ the unique high-affinity binding of avidin to biotin.

Fluorescence Assay. Fluorescein-labeled avidin (F-avidin) increases in fluorescence by 75% when bound to biotin. A commercial preparation of F-avidin is assayed for fluorescent yield alone and with known concentrations of biotin and a standard curve is prepared. Biotinylated proteases prepared as above are then added to the F-avidin solution and the increase in fluorescence is compared to the standard curve to determine biotin concentration. When no caproyl spacer is employed, avidin–biotin interaction is very weak; however, addition of one or two caproyl units gave sufficient separation to allow full interaction.

Electroblotting Assay. Biotinylated proteases, pure or in mixtures with nonreactive zymogens and other proteins, are electrophoresed using a

[12] K. G. Mann, E. Williams, S. Krishnaswami, W. Church, A. Giles, and R. Tracy, *Blood* **76**(4), 755 (1990).

standard SDS-PAGE gel system. The developed polyacrylamide gels are electroblotted onto nitrocellulose as described by Towbin *et al.*[13] The nitrocellulose sheets are blocked for 1 hr with 1% BSA in HEPES (20 mM)/NaCl (0.15 M), pH 7.4, then reacted with a complex of avidin with biotinylated peroxidase. After two washes to remove unbound reagent, the blots are developed with 0.56 mg/ml 4-chloronaphthol, 0.02% (v/v) hydrogen peroxide in HEPES/saline. The visualized bands can be analyzed by densitometry.

Applications of Labeled Peptide Chloromethyl Ketones

Determining Activity of Protease/Zymogen Molecules. Either fluorescent or biotinylated peptide chloromethyl ketones can be used to distinguish active from inactive proteases. Reaction of a protein or protein mixture with the chloroketone inhibitor at a large excess (50–100×) ensures complete reaction of active protease molecules and labeling of these structures with the dye or other marker group. The inhibitor also rapidly blocks proteolysis of zymogens by traces of active protease. The resultant mixtures can be analyzed by SDS-PAGE electrophoresis followed by fluorescence scanning or electroblotting, depending on the label employed. We have used both methods to investigate the activity of zymogens of the coagulation proteases.[7] Labeling of the zymogens was not observed, even under extreme conditions of 50 M excess of chloromethyl ketone and prolonged reaction times.

Exhaustive Removal of Protease from Zymogen Preparation. Although a variety of methods exist for blocking the activity of active proteases in a zymogen preparation, the biotinylated chloromethyl ketones offer the best practical method for physical separation of the activated and zymogen molecules. Because only the active forms react with the chloromethyl ketone, biotin can be selectively attached to these structures, leaving the zymogens unlabeled. An affinity column containing avidin can then be used to bind the biotinylated proteases, whereas the zymogens pass through readily. In our hands, such procedures yielded zymogen preparations free of active enzyme at levels below detection limits, even for extremely sensitive blotting and immunoassay methods. Avidin-Sepharose affinity media are commercially available.

Fluorescence Polarization Techniques. An important application of the fluorescent labeled derivatives is the determination of kinetics of protein–protein or protein–lipid interaction using fluorescence polarization methods. When a fluorescent group is excited using polarized light, the degree of polarization of the emitted light depends on the fluorescent lifetime of the

---

[13] H. Towbin, T. Staelelin, and J. Gordon, *Proc. Natl. Acad. Sci. U.S.A.* **76**, 4350 (1979).

label and the rotation speed of the labeled structure. Because rotational speed is in part determined by mass and geometry, larger structures tend to show greater polarization than do smaller ones. Thus, while the labeled chloromethyl ketones generally rotate too rapidly to exhibit polarization, proteases labeled with dansyl or fluoresceinyl peptide chloromethyl ketones exhibit some polarization. As these proteins complex with their much larger binding proteins or with phospholipid vesicles, the level of polarization of the fluorescence increases markedly. Because "real time" measurement of polarization is possible, this property can be employed to follow directly the kinetics of association between proteins and between protein and phospholipid (PL) vesicle. In our laboratories, this method has been applied to studies of the assembly of a variety of complexes (Xa-Va-PL-Ca, IXa-VIIa-PL-Ca, and APC-Va-PL-Ca).[11,14,15]

Fluorescence Energy Transfer Techniques. When the emission frequency of one fluorescent group is close to the excitation frequency of a second, it is frequently possible to observe emission from the second group when the first is excited. This process of fluorescence energy transfer is strongly dependent on the distance between the two fluorophores as well as the respective frequencies. Energy transfer falls off rapidly as the distance increases. This allows the measurement of association between molecules or structures if the two fluorophores are attached to different molecules, and the measurement of structural variations within a molecule if the fluorophores are attached to different molecular subsections. By using fluorescently labeled phospholipid vesicles and proteases labeled at the active site with chloromethyl ketones, it is possible to assess the orientation of the protease on the phospholipid surface.[16] A number of good fluorophore pairs are available, including fluorescein–rhodamine, dansyl–fluorescein, and tryptophan (intrinsic fluorescence)–dansyl.

[14] M. E. Nesheim, E. Kettner, E. Shaw, and K. G. Mann, *J. Biol. Chem.* **256** 6537 (1981).
[15] M. E. Nesheim, P. B. Tracy, and K. G. Mann, *J. Biol. Chem.* **259**, 1447 (1984).
[16] S. Krishnaswamy, K. C. Jones, and K. G. Mann, *J. Biol. Chem.* **263**, 3823 (1988).

[29] Active Site-Specific Assays for Enzymes of Coagulation and Fibrinolytic Pathways

By RUSSELL P. TRACY, RICHARD JENNY, E. BRADY WILLIAMS, and KENNETH G. MANN

Introduction

Measuring the concentrations of coagulation and fibrinolysis enzymes in purified test systems is critical to our understanding of these processes. Because thrombosis and thrombolysis have been proved to be integrally involved with both the development and clinical manifestations of coronary artery disease,[1] it is likely that measurements in plasma may prove important as well.

In the past, measurement of plasma coagulation and fibrinolytic enzymes has been confounded by several factors. An active enzymatic species, e.g., thrombin, must be differentiated from inactive forms, such as prothrombin or thrombin–antithrombin III complex. If activity measurements are made directly, e.g., fibrinolytic activity, a particular enzyme such as tissue plasminogen activator (t-PA) must be differentiated from other components that might be present in a test system, which might manifest a similar activity toward a given substrate, such as urokinase.

With these issues in mind we have developed a system for assaying such enzymes that is characterized by specificity for active enzyme based on incorporation at the active site of a tripeptide chloromethyl ketone,[2,3] and specificity for a particular factor based on antibody recognition. This system has a format similar to the two-site immunoassay format, with the exception that either the capturing or the signaling process takes advantage of the interaction of avidin with biotin, the latter component having been attached to the tripeptide chloromethyl ketone.[4]

Briefly, the enzyme of interest is incubated in the presence of a tripeptide chloromethyl ketone, in our hands most frequently biotinyl-ε-aminocaproyl-D-phenylalanylprolylarginine chloromethyl ketone (BioCap-FPR-ck), which is reactive with this enzyme's active site. After active site labeling, there are two possible formats (Fig. 1). In the first format we have

---

[1] V. Fuster, B. Stein, J. Ambrose, L. Badimon, J. Badimon, and J. Chesebro, *Circulation* **82,** Suppl. II, 1147 (1990).
[2] C. Kettner and E. Shaw, this series, Vol. 80, p. 826.
[3] E. Williams, S. Krishnaswamy, and K. Mann, *J. Biol. Chem.* **264,** 7536 (1989).
[4] K. Mann, E. Williams, S. Krishnaswamy, W. Church, A. Giles, and R. Tracy, *Blood* **76,** 755 (1990).

METHODS IN ENZYMOLOGY, VOL. 222

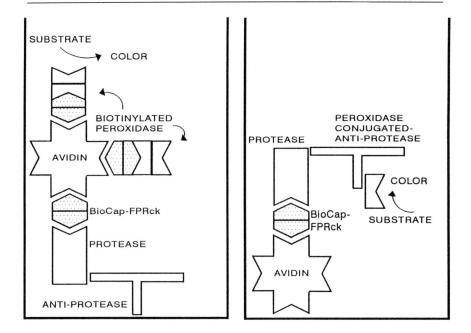

PROTEIN-SPECIFIC
CAPTURE

BIOTIN-SPECIFIC
CAPTURE

FIG. 1. Active site-specific assay formats. The schematic illustrates the two formats that are routinely used. In the protein-specific capture format, specific antibodies are adsorbed onto plastic microtiter wells and are used to capture the enzyme of interest, which has incorporated biotinylated tripeptide chloromethyl ketone, such as BioCap-FPR-ck. Avidin and biotinylated peroxidase are used to detect the biotinylated enzyme, with color generation resulting from the action of peroxidase on an appropriate substrate. The biotin-specific capture format utilizes solid-phase adsorbed avidin to capture the biotinylated enzyme. Specific antibodies, either conjugated directly to peroxidase or in conjunction with peroxidase-conjugated secondary antibodies, are used to detect the captured enzyme. Modified from Mann et al.[4]

described,[4] the biotinylated enzyme is captured by solid-phase avidin and detected by peroxidase-conjugated specific antibody, in a standard ELISA format. In the second format, the capturing is done by the antibody, while detection is accomplished by peroxidase-conjugated avidin. The first format has the advantage of capturing only active forms, but the presence of excess biotinylated chloromethyl ketone may interfere by competition for avidin binding sites; the second format protects against excess biotinylated reagent, but immunoreactive, inactive forms of the enzyme may interfere if present in large quantities, again by competition for the capturing reagent. This will be discussed in more detail below.

In either case, the antibody provides specificity for the particular factor, while the biotinylation by the chloromethyl ketone assures that only active forms of the enzyme will be detected. This assay system is a powerful tool to explore active enzyme species in complex mixtures, both *in vitro* and in plasma.

Preparing Biotinylated Chloromethyl Ketones

The synthesis and characterization of reagents such as BioCap-FPR-ck are described in detail by Williams *et al.*[3] and Williams and Mann.[5] The biotin moiety is attached to the active site–reactive tripeptide chloromethyl ketone via an ε-aminocaproyl spacer. We have found that both the absolute value and consistency of the specific activity of this labeling reagent are important factors, because they will affect the specific activity of the labeled enzyme, and therefore the signal that is generated in the assay. Williams and Mann[5] describe methods for determining the specific activity of labeled chloromethyl ketones, and particular care should be taken to ensure efficient labeling and consistent reagent.

Preparing Active Site-Labeled Enzymes

Solutions of 20 mM Tris-HCl, 0.15 M NaCl, pH 7.4, containing active enzyme [e.g., α-thrombin (IIa), tissue plasminogen activator, or plasmin] at a concentration of approximately 1 mg/ml or greater are incubated at 37° for at least 5 min with a 5- to 10-fold molar excess of BioCap-FPR-ck.

The optimal length of this incubation and the concentration of chloromethyl ketone are functions of the rate of inhibition of the particular enzyme by the particular biotinylated tripeptide chloromethyl ketone being used. For example, we have found that BioCap-EGR-ck (biotinyl-ε-aminocaproyl-D-glutamic acid glycylarginine chloromethyl ketone) is significantly more effective as an inhibitor of plasmin than is BioCap-FPR-ck.

We have also determined that, in some instances, the length of the spacer arm (i.e., caproyl moiety) is important in determining the rate of chloromethyl ketone incorporation. An example is the labeling of activated protein C. Incorporation of BioCap-FPR-ck proceeds relatively slowly, and does not go to completion; however, if BioCap-Cap-FPR-ck is synthesized (i.e., the same reagent with a double aminocaproyl spacer) and used in this reaction, incorporation is rapid and complete. Presumably the single spacer does not remove the biotin moiety far enough from the tripeptide

---

[5] E. Williams and K. Mann, this volume [28].

chloromethyl ketone to eliminate stearic hindrance during incorporation, but this has not been investigated in detail.

Following the labeling reaction, the incorporation of the active site-directed agent is analyzed using a chromogenic assay for the enzyme in question. For example, we have used an assay system containing 200 μM Spectrozyme TH(American Diagnostica) in 20 mM HEPES/saline, pH 7.4, to follow α-thrombin activity. Other substrates are available for other enzymes that might be of interest.[6] Enzyme assays may be done once at the end of the reaction, or at multiple times during the reaction, to follow tripeptide chloromethyl ketone incorporation over time. Enzyme inactivation is discussed in more detail in Williams *et al.*, this volume. The biotin-labeled, inactivated enzyme is separated from excess BioCap-FPR-ck by gel filtration (e.g., Sephadex G-25–150), collected and precipitated by ammonium sulfate [80% saturation using solid $(NH_4)_2SO_4$], resuspended in 50:50 (v/v) glycerol:water, and stored at $-20°$.

Further analysis has included sodium dodecyl sulfate electrophoresis (SDS-PAGE), followed by Western blotting and visualization with avidin–peroxidase conjugate to verify biotinylation at the appropriate apparent SDS molecular weight.[4] Briefly, SDS-PAGE is done by the method of Laemmli,[7] using either 10% acrylamide or 10–20% acrylamide gradient. Western blotting is accomplished essentially following the methods of Towbin.[8] Once blotted, the nitrocellulose sheets are blocked (1% bovine serum albumin, 0.05% Tween 20, 20 mM HEPES, 0.15 M NaCl, pH 7.4, 1 hr at room temperature) and then reacted with avidin-biotinylated peroxidase complex (Vector Laboratories, Burlingame, CA). After washing extensively to remove excess peroxidase reagent (washing buffer consists of blocking buffer without albumin), the blot is developed with 4-chloronaphthol (0.56 mg/ml) in 20 mM HEPES, 0.15 M NaCl, 0.02% H_2O_2, pH 7.4. This set of conditions yields consistent Western blots of biotinylated enzymes such as α-thrombin, and factor Xa. This method may be used to follow the activation of prothrombin and factor X over time. For example, prothrombin activation is accomplished with a mixture of factor Xa, factor Va, calcium, and phospholipid vesicles. At various times a sample is removed from the reaction mixture and analyzed for thrombin activity by chromogenic assay as well as incubation with BioCap-FPR-ck. The resulting samples are analyzed by SDS-PAGE and Western blotting to demonstrate incorporation of the tripeptide chloromethyl ketone in a manner consistent with the inactivation of the enzyme.[4]

---

[6] H. Hemker, "Handbook of Synthetic Substrates for the Coagulation and Fibrinolytic System." Martinus Nijhoff, Boston, 1983.

[7] U. Laemmli, *Nature (London)* **227,** 680 (1970).

[8] H. Towbin, T. Staehelin, and J. Gordon, *Proc. Natl. Acad. Sci U.S.A.* **76,** 4350 (1979).

Active Site-Specific Assay Method

As mentioned above, two different conformations are possible when constructing assays based upon the tripeptide chloromethyl ketones and specific antibodies (Fig. 1). We have called these "protein-specific capture" and "biotin-specific capture."

Protein-Specific Capture

Coating solution, 200 μl, containing approximately 10 μg/ml of the appropriate monoclonal or polyclonal antibody in 50 mM NaHCO$_3$, pH 9.5, is added to each well of a microtiter plate and incubated for either 3 hr at room temperature or overnight at 4°. The wells are washed with phosphate-buffered saline (PBS)–Tween (20 M sodium phosphate, 0.15 M NaCl, 0.1% (v/v) Tween 20, pH 7.4), and then blocked with PBS–Tween containing 5% bovine serum albumin by incubation at room temperature for 1 hr. If covered and sealed carefully, these plates can be stored at −20° for several weeks prior to use. After removing the blocking buffer, biotinylated enzyme is added, either in the form of a standard of known concentration, or an "unknown" from a reaction mixture. PBS–Tween containing 3% bovine serum albumin is used as the assay buffer for dilutions. After incubation at room temperature for 3 hr or at 4° overnight, the antigen solution is removed and the wells are washed with PBS–Tween. Biotinylated enzyme is detected by the addition of avidin-biotinylated peroxidase in PBS–0.05% Tween following the manufacturer's direction (Vector Laboratories), and, after washing, the addition of peroxidase substrate (0.2 mg/ml o-phenylenediamine, 0.003% H$_2$O$_2$, 0.1 M citric acid, 0.2 M sodium phosphate, pH 5.0). After approximately 10 min the reaction is stopped with sulfuric acid and the color is determined at 490 nm. For data analysis we use a microtiter plate reader connected to a computer, and software designed for immunoassay data (Titer-Calc, Hewlett-Packard, San Francisco, CA).

Biotin-Specific Capture

Avidin (Strepavidin, Calbiochem, La Jolla, CA), from 5 to 25 μg/ml in 50 M NaHCO$_3$, pH 9.5, is added to microtiter plate wells as a 200-μl sample and is allowed to incubate for 3 hr at room temperature or overnight at 4°. After blocking and washing as described above for protein-specific capture, samples containing the biotinylated enzyme of interest, either as a standard or an "unknown," are added and allowed to incubate, again as above. After aspirating excess sample and washing, primary antibody is added, depending on the antigen of interest. We have used both monoclonal and polyclonal antibodies, both conjugated directly with per-

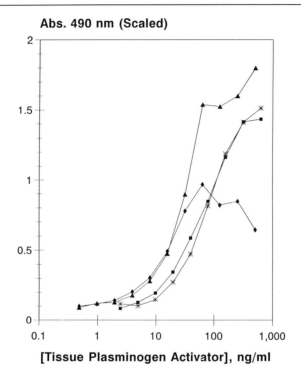

Abs. 490 nm (Scaled)

[Tissue Plasminogen Activator], ng/ml

FIG. 2. Active site-specific assays for tissue plasminogen activator. Standard curves for t-PA were prepared in both the protein-specific capture (*) and the biotin-specific capture (▲) formats, and compared to dilutions of plasma spiked with 0.5 μg/ml biotinylated t-PA, which was also measured in the protein-specific capture (■) and biotin-specific capture (♦) formats. The absorbance values on the y axis have been normalized by scaling for ease of inspection. Data taken from Mann et al.[4]

oxidase, or using a secondary peroxidase-conjugated anti-IgG, with success. For a directly conjugated antibody, prepared by the method of Nakane and Kawaoi,[9] usually a dilution of approximately 1000-fold will suffice; for unconjugated primary antibodies, we have used a concentration of 0.5–25 μg/ml, depending on the affinity/avidity of the immune reagent, and, if polyclonal, the actual concentration of specific antibodies. Incubation with primary antibody is usually done for 1 hr at room temperature, followed by extensive washing. If a secondary peroxidase-conjugated anti-IgG is used, this is also incubated for 1 hr at room temperature followed by washing. In either case, substrate addition, stopping, color quantitation, and data analysis proceed as described above.

Figure 2 illustrates two standard curves for t-PA, along with serial dilutions of spiked plasma, one prepared with the protein-specific capture

[9] P. Nakane and A. Kawaoi, *J. Histochem. Cytochem.* **22,** 1084 (1974).

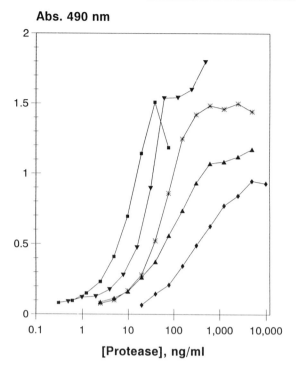

Abs. 490 nm

[Protease], ng/ml

Fig. 3. Active site-specific assays for factor Xa, tissue plasminogen activator, activated protein C, plasmin, and thrombin using the biotin-specific capture format. Standard curves were prepared for factor Xa (■), t-PA (▼), activated protein C (*), plasmin (▲), and thrombin (♦) as described in the text. Data taken from Mann et al.[4]

assay and the other prepared with the biotin-specific capture method. In both cases the signals respond to increasing concentrations of biotinylated t-PA in a consistent manner, when buffer and plasma are compared. As with many two-site ELISA assays, one should be careful to avoid the so-called "hook effect" that occurs at high concentrations of added antigen (note especially the curve representing plasma dilutions, in the biotin-specific capture assay method). Based on these initial experiments, recovery values of approximately 100% are routinely observed,[4] with spiked plasma levels of 0.5 μg/ml t-PA. It is apparent that the biotin-specific capture assay is considerably more sensitive (approximately two- to three-fold) than the protein-specific capture assay. We speculate that this may be due to the greater affinity of the biotin–avidin pair, compared to the protein–antibody pair, resulting in more complete capture.

Figure 3 depicts standard curves obtained for t-PA, IIa, plasmin, factor Xa, and activated protein C, plotted on the same scale, using the biotin-specific capture method. As can be seen, although biotinylated forms of all five enzymes may be used, there is a range of sensitivities. This may be due to differences in affinity/avidity of the recognition antibodies. Using these standard curves as examples, assay sensitivity (expressed as the concentration of analyte that results in a signal twofold over background) ranges from 1.5 to 30 ng/ml. Typical normal plasma concentrations for the zymogen forms of these analytes are factor X, 10 μg/ml; protein C, 3 μg/ml; plasminogen, 150 μg/ml; and prothrombin, 100 μg/ml. Therefore, these assays can detect active enzyme concentrations, expressed as a percentage of the circulating zymogen forms, of 0.01% for factor Xa and plasmin, 0.03% for thrombin, and 0.5% for activated protein C. In practice, these estimates are correct for buffer-based systems. However, in plasma-based systems, matrix effects from other plasma components serve to decrease overall sensitivity. The extent of this effect is partly a function of the antibody reagent used in the assay, as is true for other immunoassays.

By utilizing alternative antibody reagents we have been able to improve the sensitivity for a given assay, for example, the IIa assay, and this approach should prove useful for any assay. An antibody of high affinity that reacted only with the active form of a particular zymogen/enzyme pair would be optimal.

With each form of the assay, there are factors which limit the sensitivity. When the biotin-specific capture assay is used, excess biotinylated reagent, e.g., BioCap-FPR-ck, will complete with the biotinylated enzyme; because the reagent is usually present at a 5- to 10-fold (or greater) molar excess, this competition can prove significant. In those cases where the ratio of biotinylated enzyme:free BioCap-FPR-ck is high, and the assay sensitivity is high as well, simple dilution of the sample may prove useful, as the binding capacity of the solid-phase avidin may be sufficient to overcome excess reagent. However, in the case in which the ratio is low, such as in an attempt to measure an unknown but small amount of active enzyme in plasma using a relatively large amount of BioCap-FPR-ck to ensure quantitative incorporation, separation of the unincorporated BioCap-FPR-ck might be essential. We have used various methods for this separation, including gel filtration, dialysis, and ultrafiltration. In buffer-based systems, using purified reagents, all these methods work well, taking advantage of the difference in size between the labeled enzyme and the BioCap-FPR-ck. In plasma, however, we have been less successful in completely removing excess BioCap-FPR-ck, possibly because a small amount of this material may remain associated, nonspecifically, with the plasma protein fraction. We are continuing to study this issue.

The analogous problem arises when using the protein-specific capture format and capture antibodies that do not discriminate between zymogen and active forms of the enzyme of interest. If the labeled enzyme exists as a small fraction of the total zymogen concentration, competition for anti-

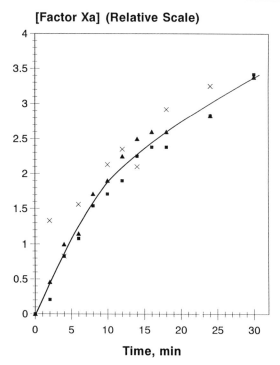

FIG. 4. Factor Xa generation. The generation of factor Xa by Russell's viper venom was followed over time by SDS-PAGE [as the change in the Coomassie blue-stained factor X band (×) and the increase in the biotinylated factor Xa band on Western blotting (▲)] as well as by active site-specific assay (■). The values on the y axis have been normalized to a common scale for the purpose of presentation. The line has been drawn by inspection. Data taken from Mann et al.[4]

body binding sites, between the labeled forms and the zymogens, might limit the sensitivity. In fact, using the factor Xa assay as an example, we have determined that at a ratio of 1:1, there is virtually no effect on the assay, given the avidity and binding capacity of the solid-phase rabbit polyclonal antifactor X/Xa we use. However, at ratios of zymogen:enzyme of 10:1 and 100:1, we have observed 25 and 80% reductions in signal, respectively. As mentioned above, antibodies which reacted with only the active form of the protein would be optimal, assuming that the avidity/affinity values were appropriate.

Measurements in Vitro

In in vitro experiments, we have successfully measured factor X activation and prothrombin activation, in complex assay mixtures, using the

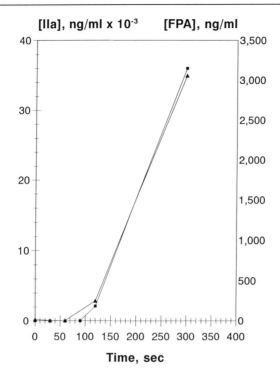

FIG. 5. Generation of thrombin and fibrinopeptide A in clotting plasma. Plasma was treated as described in the text and allowed to clot. At various time points samples were assayed for thrombin by active site-specific assay (■) and for fibrinopeptide A by ELISA (▲). The plasma began to clot visibly by approximately 2 min under these conditions. For comparison, a value of 40,000 ng/ml thrombin represents approximately 40% of the total plasma prothrombin concentration converted to thrombin, and a value of 3500 ng/ml fibrinopeptide A represents approximately 10–25% of the total fibrinogen converted to fibrin I.

methods described above.[4] Incorporation of the BioCap-FPR-ck was monitored electrophoretically. In the case of factor Xa, activation by the addition of purified Russell's viper venom (0.8 μg/ml) to a reaction mixture containing factor X (8.5 μM), HEPES buffer (20 mM, pH 7.4), saline, and CaCl$_2$ (5.0 mM) was monitored over time by Coomassie blue-stained SDS-PAGE, Western blotting for biotinylation, and factor Xa activity assays. At the time points noted in Fig. 4, samples were removed and stopped with either EDTA (10.0 mM), for activity assays, or with EDTA plus BioCap-FPR-ck. As can be seen in Fig. 4, all three methods yielded comparable results.

We have also monitored the formation of thrombin over time in clotting plasma using this method. Figure 5 illustrates one example of such an

TABLE I

MEASUREMENT OF CIRCULATING RECOMBINANT HUMAN t-PA IN DOG PLASMA BY ELISA
AND ACTIVE SITE-SPECIFIC ASSAYS[a]

Time point	ELISA[b] t-PA (μg/ml)	Active site[d] t-PA (μg/ml)
Preinfusion	ND[c]	ND
1 minute	39.2	38.6
2 minutes	35.2	38.2
5 minutes	23.4	23.6
15 minutes	5.7	5.5

[a] A dog was infused with 2 mg/kg recombinant human t-PA, and blood samples were collected at the times indicated. From Mann *et al.*[4]

[b] This t-PA ELISA measures t-PA antigen and is sensitive to free t-PA as well as to t-PA in complex with PAI-1.

[c] ND, Not detectable.

[d] The active site-specific assay was done in the biotin-specific capture format.

experiment. Citrated plasma was recalcified and allowed to clot in a glass test tube. At the indicated times, samples were removed, quenched with BioCap-FPR-ck, and incubated with 1.0 M barium chloride, 80 μl/ml plasma. After an incubation of 30 min on ice, the barium citrate pellet, with adsorbed prothrombin (as well as other vitamin K-dependent proteins), was collected by centrifugation and removed. Supernatant samples were subjected to active site-directed assays, using the protein-specific capture format. By removing the prothrombin with barium citrate precipitation, there was no potential for interference of thrombin reactivity by the zymogen in the protein-specific capture assay. We also measured fibrinopeptide A in these samples as an estimate of ongoing thrombin activity. The results indicate the progressive generation of thrombin over time in clotting plasma and illustrate the potential for assays of this type in complex mixtures.

Measurements *in Vivo*

We are particularly interested in circulating t-PA values in patients receiving recombinant t-PA as thrombolytic therapy, and have had success measuring t-PA in plasma samples from dogs and humans receiving the recombinant agent. The data in Table I describe results we have seen in dogs infused with recombinant human t-PA, where active site-directed assays and ELISAs have been performed at the same time. In these preliminary studies, results indicate that active site assays and t-PA-specific ELISA assays yield similar data. It should be pointed out that these t-PA values are quite high (a reference range for endogenous t-PA in normal human plasma is from 5 to 20 ng/ml), and therefore not confounded by circulating t-PA–inhibitor complexes.

Studies continue on improving the sensitivity of these active site-specific assays so that events which involve less dramatic changes in the circulating concentrations of these factors may be explored in detail.

[30] Kinetic Characterization of Heparin-Catalyzed and Uncatalyzed Inhibition of Blood Coagulation Proteinases by Antithrombin

By Steven T. Olson, Ingemar Björk, and Joseph D. Shore

Introduction

Antithrombin is a plasma glycoprotein belonging to the serpin (an acronym for serine proteinase inhibitor) superfamily of proteins.[1] The principal function of antithrombin is to inhibit the proteolytic enzymes of blood coagulation. Heparin or heparan sulfate glycosaminoglycans may contribute to this function by binding to antithrombin and greatly accelerating the rate at which the inhibitor inactivates these enzymes. This property is the basis for the widespread clinical use of heparin for anticoagulant therapy. Significant progress has been made in elucidating the mechanisms of antithrombin inhibition of proteinases and of the accelerating effect of heparin on these reactions since the last chapter on antithrombin appeared in Volume 45 of this series.[2] The current chapter focuses on the kinetic approaches that have been instrumental in defining these mechanisms. Such approaches should be applicable also to the study of other serpin–proteinase reactions either in the absence or presence of similar glycosaminoglycan effectors.

Purification and Properties of Antithrombin

Purification

Antithrombin is easily purified from plasma by a procedure based on affinity chromatography on matrix-linked heparin.[3] The procedure out-

[1] R. W. Carrell and D. R. Boswell, *in* "Proteinase Inhibitors" (A. J. Barrett and G. Salvesen, eds.), p. 403. Elsevier, Amsterdam, 1986.
[2] P. S. Damus and R. D. Rosenberg, this series, Vol. 45, p. 653.
[3] M. Miller-Andersson, H. Borg, and L. O. Andersson, *Thromb. Res.* **5**, 439 (1974).

lined below is optimal for human antithrombin, but should be largely applicable also to antithrombin from other species.[4] The human protein can be purified from either fresh-frozen or outdated plasma. For maximal purity, the affinity chromatography should be followed by ion-exchange and gel chromatography steps to remove possibly contaminating heparin released from the affinity matrix and aggregated protein, respectively.

Polyethylene Glycol Precipitation. An initial precipitation step is done by addition of 150 g polyethylene glycol (PEG) 4000 per liter of plasma. The precipitate is discarded, and antithrombin is precipitated by addition of a further 200 g polyethylene glycol per liter of plasma. The precipitates can be recovered by centrifugation at 10000 g for 30 min.[5] Omission of this step increases the final yield of antithrombin somewhat without decreasing the purity or affecting the properties of the protein. However, polyethylene glycol precipitation has the advantage that it removes fibrinogen, which otherwise can precipitate during the subsequent affinity chromatography step, and also reduces the volume of the sample that must be applied to the column in this step.

Affinity Chromatography. Heparin covalently linked to agarose can be purchased from Pharmacia LKB Biotechnology (Uppsala, Sweden) or Bio-Rad (Richmond, CA). Alternatively, the heparin–agarose can be made from commercial heparin and cyanogen bromide-activated agarose.[6] The latter can be obtained commercially or prepared by established methods.[7] A high heparin concentration, up to 40 mg/ml, should be used to ensure maximal antithrombin-binding properties of the gel.

The plasma or polyethylene glycol precipitate is dissolved in a minimum volume of 0.02 M Tris-HCl or sodium phosphate, 0.1 M NaCl, pH 7.4, and then applied to the heparin affinity column ($\sim$150 ml gel per liter of plasma) equilibrated in the same buffer. The column is extensively washed with the starting buffer until the absorbance is $\leq$0.05. Elution is performed with a linear gradient to 3 M NaCl, a total gradient volume of 2 liters, and flow rate of $\sim$3 ml/min, being appropriate for a 150-ml column. The major portion of antithrombin elutes at 0.8–1.0 M NaCl, although a minor variant of the protein with different glycosylation (see below) elutes at $\sim$1.4 M salt.[8,9]

Ion-Exchange and Gel Chromatography. The antithrombin pool is dialyzed against 0.02 M Tris-HCl or sodium phosphate, 0.02 M NaCl, pH 7.4 and applied to a column of DEAE-Sepharose (Pharmacia LKB Bio-

[4] R. E. Jordan, *Arch. Biochem. Biophys.* **227**, 587 (1983).
[5] E. Thaler and G. Schmer, *Br. J. Haematol.* **31**, 233 (1975).
[6] P. A. Craig, S. T. Olson, and J. D. Shore, *J. Biol. Chem.* **264**, 5452 (1989).
[7] S. C. March, I. Parikh, and P. Cuatrecasas, *Anal. Biochem.* **60**, 149 (1974).
[8] T. H. Carlson and A. C. Atencio, *Thromb. Res.* **27**, 23 (1982).
[9] C. B. Peterson and M. N. Blackburn, *J. Biol. Chem.* **260**, 610 (1985).

technology; ~200 ml per liter of starting plasma) in this buffer. Antithrombin is eluted at ~0.3 M NaCl by a linear gradient (total volume of 1 liter) to 0.6 M NaCl. After concentration by ultrafiltration with an M_r 30,000 cutoff membrane, the antithrombin pool is applied to a Sephacryl S-200 (Pharmacia LKB Biotechnology) column in a suitable buffer at near-neutral pH and physiological ionic strength, e.g., 0.02 M sodium phosphate, 0.1 M NaCl, 0.1 mM EDTA, pH 7.4. Usually, a 2.5 × 100-cm column is adequate for the pool from 1 liter of plasma. The monomeric protein elutes at a K_{av} of ~0.2–0.3 and is concentrated by ultrafiltration. The yield is usually ~60 mg from 1 liter of plasma, representing ~50% of the normal antithrombin concentration in plasma of ~0.12 mg/ml or 2.3 μM.[10] Antithrombin is stable for several months to years at −70°, although some aggregation may occur during prolonged storage.

Concentration Determination. Concentrations of purified antithrombin can be obtained from the absorbance at 280 nm. Absorption coefficients of 0.65 and 0.67 liters g^{-1} cm^{-1} have been measured for the human and bovine proteins, respectively.[11]

Properties

Antithrombin is stable between pH ~5.5 and ~9. At pH <5.5 the inhibitor undergoes an irreversible conformational change that abolishes both the proteinase- and heparin-binding activity.[12] The midpoint of the thermal transition leading to loss of activity is at ~59°, somewhat dependent on solvent conditions.[13]

Human antithrombin has 432 amino acid residues, as deduced from protein and cDNA sequencing.[14-17] The six cysteine residues of the protein form three disulfide bridges (Cys^8-Cys^{128}, Cys^{21}-Cys^{95}, and Cys^{247}-Cys^{430}).[14] Bovine antithrombin is one residue longer and has ~89% sequence identity and ~95% similarity with the human inhibitor.[18] The molecular weight of the polypeptide chain of human antithrombin is 49,045, but the

[10] J. Conard, F. Brosstad, M. L. Larsen, M. Samama, and U. Abildgaard, *Haemostasis* **13**, 363 (1983).
[11] B. Nordenman, C. Nyström, and I. Björk, *Eur. J. Biochem.* **78**, 195 (1977).
[12] B. Nordenman and I. Björk, *Biochim. Biophys. Acta* **672**, 227 (1981).
[13] T. F. Busby, D. H. Atha, and K. C. Ingham, *J. Biol. Chem.* **256**, 12140 (1981).
[14] T. E. Petersen, G. Dudek-Wojciechowska, L. Sottrup-Jensen, and S. Magnusson, *in* "The Physiological Inhibitors of Blood Coagulation and Fibrinolysis" (D. Collen, B. Wiman, and M. Verstraete, eds.), p. 43. Elsevier/North-Holland, Amsterdam, 1979.
[15] S. C. Bock, K. L. Wion, G. A. Vehar, and R. M. Lawn, *Nucleic Acids Res.* **10**, 8113 (1982).
[16] E. V. Prochownik, A. F. Markham, and S. H. Orkin, *J. Biol. Chem.* **258**, 8389 (1983).
[17] T. Chandra, R. Stackhouse, V. J. Kidd, and S. L. C. Woo, *Proc. Natl. Acad. Sci. U.S.A.* **80**, 1845 (1983).
[18] H. Mejdoub, M. LeRet, Y. Boulanger, M. Maman, J. Choay, and J. Reinholt, *J. Protein Chem.* **10**, 205 (1991).

molecular weight of the predominant, fully glycosylated form in plasma (see below) is 58,200.

The amino acid sequences of human and bovine antithrombin have four potential N-glycosylation sites, which are at Asn-96, Asn-135, Asn-155, and Asn-192 in human antithrombin and at homologous residues in the bovine protein. All these sites carry oligosaccharide side chains in the predominant form of antithrombin (antithrombin α) in plasma.[14,15] The four carbohydrate side chains of human antithrombin all have an identical, biantennary structure except in their terminal sialic acids.[19,20] This heterogeneity leads to a charge heterogeneity of the protein seen in electrophoresis or isoelectric focusing.[11,21] A minor fraction (antithrombin β) of the inhibitor lacks the carbohydrate side chain on Asn-135, resulting in a slightly higher heparin affinity than that of the predominant form.[8,9,22]

A preliminary report of the three-dimensional structure of bovine antithrombin, although in a modified form cleaved one residue carboxy terminal of the reactive bond (see below), has been published.[23] The structure is highly similar to that of an analogous modified form of human α_1-proteinase inhibitor, which has been reported in greater detail.[24] In these structures, the two residues of the cleavage site are at opposite ends of the molecule, separated by 60–70 Å,[23,24] indicating that a substantial conformational change accompanies this cleavage. A similar conformational change has been implicated in the trapping of enzymes in stable complexes, as discussed below.

Reaction with Blood Coagulation Proteinases

Antithrombin inactivates a target proteinase by forming an equimolar, tight complex, in which the active site of the enzyme is inaccessible to substrates.[25] Complex formation involves interaction between the enzyme and a specific, reactive bond of antithrombin, which in the human inhibitor is the Arg[393]-Ser[394] bond in the carboxy-terminal region and in the bovine protein is the homologous Arg[394]-Ser[395] bond.[26-28] The inactivation

[19] L. E. Franzén, S. Svensson, and O. Larm, J. Biol. Chem. 225, 5090 (1980).
[20] T. Mizuochi, T. J. Fujii, K. Kurachi, and A. Kobata, Arch. Biochem. Biophys. 203, 458 (1980).
[21] A. D. Borsodi and R. A. Bradshaw, Thromb. Haemostasis 38, 475 (1977).
[22] S. O. Brennan, P. M. George, and R. E. Jordan, FEBS Lett. 219, 431 (1987).
[23] L. Mourey, J. P. Samama, M. Delarue, J. Choay, J. C. Lormeau, M. Petitou, and D. Moras, Biochimie 72, 599 (1990).
[24] H. Loebermann, R. Tukuoka, J. Deisenhofer, and R. Huber, J. Mol. Biol. 177, 531 (1984).
[25] R. D. Rosenberg and P. S. Damus, J. Biol. Chem. 248, 6490 (1973).
[26] H. Jörnvall, W. W. Fish, and I. Björk, FEBS Lett. 106, 358 (1979).
[27] I. Björk, Å. Danielsson, J. W. Fenton, II, and H. Jörnvall, FEBS Lett. 126, 257 (1981).
[28] I. Björk, C. M. Jackson, H. Jörnvall, K. K. Lavine, K. Nordling, and W. J. Salsgiver, J. Biol. Chem. 257, 2406 (1982).

presumably is initiated by the enzyme attacking this reactive bond as in a regular substrate. However, antithrombin is induced to trap the enzyme by a conformational change, which greatly retards the cleavage, at some intermediate stage of the proteolytic reaction,[29–31] possibly at the tetrahedral intermediate.[32] Some target proteinase molecules apparently can escape trapping by completing the cleavage of the inhibitor reactive bond before the conformational change occurs.[33] As a consequence, a small amount of a noncomplexed, reactive-bond-cleaved form of antithrombin is formed concurrent with the antithrombin–proteinase complexes.[28,34–36] The amount of the cleaved inhibitor is markedly increased in the presence of heparin.[35,36]

Stoichiometry

The purity of the antithrombin preparation should be assessed by a determination of the apparent stoichiometry of inhibition of thrombin. This stoichiometry should be close to 1.0 in experiments done in the absence of heparin, as the inactivation of the inhibitor by cleavage of the reactive bond is small under these conditions.[35,36] The inhibition stoichiometry is best determined by a batchwise titration of a constant amount of thrombin with increasing concentrations of antithrombin, monitored by the loss of the activity of the enzyme against a synthetic substrate (Fig. 1).[35,36] A highly purified thrombin preparation[37,38] should be used in this assay, the enzyme concentration of this preparation having been determined by active site titration against p-nitrophenyl p'-guanidinobenzoate[39] or 4-methylumbelliferyl p-guanidinobenzoate.[40]

Thrombin, at a constant concentration (0.25–5 μM), is incubated at 25° with increasing concentrations of antithrombin (at molar ratios of inhibitor to enzyme of 0 to ~1.2) in a buffer of near-neutral pH and physiological ionic strength, containing 0.1% (w/v) polyethylene glycol

[29] G. B. Villanueva and I. Danishefsky, *Biochemistry* **18**, 810 (1979).
[30] P. Wallgren, K. Nordling, and I. Björk, *Eur. J. Biochem.* **116**, 493 (1981).
[31] S. T. Olson and J. D. Shore, *J. Biol. Chem.* **257**, 14891 (1982).
[32] N. R. Matheson, H. van Halbeek, and J. Travis, *J. Biol. Chem.* **266**, 13489 (1991).
[33] S. T. Olson and I. Björk, *in* "Heparin and Related Polysaccharides" (D. A. Lane, I. Björk, and U. Lindahl, eds.), p. 155. Plenum, New York, 1992.
[34] W. W. Fish, K. Orre, and I. Björk, *FEBS Lett.* **98**, 103 (1979).
[35] I. Björk and W. W. Fish, *J. Biol. Chem.* **257**, 9487 (1982).
[36] S. T. Olson, *J. Biol. Chem.* **260**, 10153 (1985).
[37] W. G. Owen, C. T. Esmon, and C. M. Jackson, *J. Biol. Chem.* **249**, 594 (1974).
[38] J. W. Fenton, II, M. J. Fasco, A. B. Stackrow, D. L. Aronson, A. M. Young, and J. S. Finlayson, *J. Biol. Chem.* **252**, 3587 (1977).
[39] T. Chase, Jr. and E. Shaw, *Biochemistry* **8**, 2212 (1969).
[40] G. W. Jameson, D. V. Robert, R. W. Adams, S. A. Kyle, and D. T. Elmore, *Biochem. J.* **131**, 107 (1973).

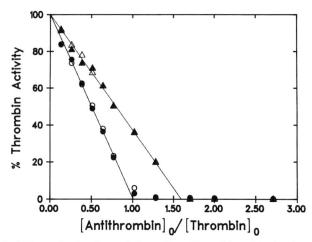

FIG. 1. Stoichiometric titrations of thrombin with antithrombin in the absence and presence of heparin. Thrombin (0.2 μM) was incubated with increasing molar ratios of antithrombin either in the absence (O,●) or presence (△,▲) of 0.2 μM high-affinity heparin (M_r ~8000) in 20 mM sodium phosphate, 0.1 M NaCl, 0.1 mM EDTA, 0.1% PEG 8000, pH 7.4, I 0.15, 25°. Residual enzymatic activity was assayed after 2 (open symbols) and 4 hr (closed symbols) by diluting samples into 100 μM S-2238 and monitoring the initial rate of substrate hydrolysis at 405 nm. Activities are expressed relative to control samples in which the inhibitor was absent.

4000–8000 to prevent adsorption of the proteins to surfaces. At 5 μM thrombin, the reaction should be >98% complete within 15 min, whereas the reaction time increases inversely proportional to the enzyme concentration at lower concentrations. The time required for completion of the reaction should be checked in suitable control experiments at near-equimolar concentrations of enzyme and inhibitor, where the reaction is slowest. After the appropriate time, an aliquot of the reaction mixture is transferred to a photometric cuvette, containing a thrombin substrate, such as D-phenylalanyl-L-pipecolyl-L-arginyl-p-nitroanilide (S-2238; Kabi, Franklin, OH) or N-p-tosyl-glycyl-L-prolyl-L-arginyl-p-nitroanilide (Sigma, St. Louis, MO), at a final concentration of 50–100 μM in buffer, to yield a thrombin concentration of ~2 nM. The initial rate of substrate hydrolysis (<10% substrate consumption) is then determined by recording the increase in absorption at 405 nm. Linear extrapolation of the initial rates at different molar ratios of antithrombin to thrombin to the abscissa gives the apparent stoichiometry of inhibition (Fig. 1). The sensitivity of the assay can be increased with the use of a fluorogenic thrombin substrate, such as N-p-tosyl-glycyl-L-prolyl-L-arginyl-7-amido-4-methylcoumarin (Sigma), at a concentration of 50 μM (λ_{ex} 380 nm, λ_{em} 440 nm), although a practical lower limit for the thrombin concentration during the incubation with antithrombin is 50 nM, which would require overnight incubations.

Kinetics

Both discontinuous and continuous assay methods can be used to measure the rates of inhibition of clotting proteinases by antithrombin. In the discontinuous method, the proteinase is incubated with antithrombin, and samples of the reaction mixture are taken at various times for measurement of residual enzyme activity. In the continuous method, a spectroscopic indicator that reports the concentration of active enzyme is included in the antithrombin–proteinase reaction mixture, and the spectroscopic changes resulting from enzyme inactivation are continuously monitored. The indicator may be a chromogenic or fluorogenic substrate or a spectroscopic probe which binds to the enzyme active site with altered spectroscopic properties. The continuous methods have the advantage that an entire progress curve for the reaction is generated by a single measurement, thereby resulting in greater precision of the data. However, such methods have the disadvantage that the measurement of enzyme inactivation is indirect and could be influenced by interactions of the indicator molecule with the inhibitor or heparin when present. Continuous methods should thus be validated by the more reliable discontinuous method.

Discontinuous Method. Antithrombin and the proteinase are incubated together at 25–37° in a buffer of near-neutral pH at physiological ionic strength. After different times, the reaction is quenched by diluting a sample of the reaction mixture into a cuvette containing a solution of a chromogenic or fluorogenic substrate in reaction buffer or other buffer optimal for the assay, and the residual enzyme activity is determined by monitoring the initial rate of substrate hydrolysis spectroscopically for several minutes (Figs. 2A and 2B). Concentrations of inhibitor and enzyme should be chosen to yield pseudo first-order conditions, which are most simply analyzed (see below). Such conditions are achieved by using an inhibitor concentration at least 5-fold and preferably 10-fold or more greater than the enzyme concentration. Reaction half-lives of > 10 min are obtained with inhibitor concentrations of ≤ 0.1 μM for the thrombin reaction and ≤ 0.5 μM for the factor Xa reaction, which allow sufficient samples to be taken to define a complete enzyme inactivation curve (~ 3 half-lives). Enzyme concentrations should satisfy the pseudo first-order condition, but be high enough to give activities that can be measured with good precision in the assay. For chromogenic substrates, 0.1–0.2 nM is a lower limit, whereas lower concentrations can be used with the fluorogenic substrates. Chromogenic peptidylarginine p-nitroanilide (Kabi or Ameri-

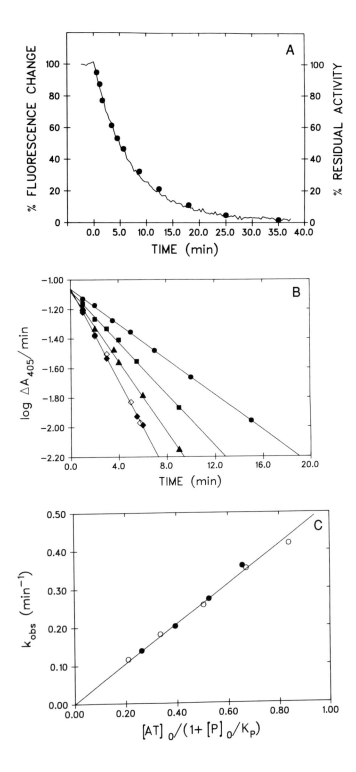

can Diagnostica, Greenwich, CT) or fluorogenic 7-amido-4-methylcoumarin substrates (Sigma or Bachem Biosciences, Philadelphia, PA) are available for assay of most coagulation enzymes by methods analogous to those described in the previous section. The linearity of substrate hydrolysis rates with enzyme concentration over the range of concentrations assayed should be confirmed to exclude artifacts related to adsorption of the enzyme, which can result in nonlinearity. Adsorption can usually be avoided by including 0.1% polyethylene glycol 4000–8000 in the reaction and substrate solutions and using polystyrene cuvettes and/or test tubes that have been coated with polyethylene glycol 20,000.[41] A further control against spontaneous losses in enzymatic activity not due to the inhibitor is to run a blank enzyme incubation without the inhibitor.

Reaction of antithrombin with proteinases over the range of concentrations that give reaction rates measurable with this assay can be described by the simple second-order reaction:

$$E + AT \xrightarrow{k} E\text{–}AT$$

where AT is antithrombin, E is the enzyme, and k is the second-order rate constant. Analysis of enzyme inactivation is based on the first-order rate equation,

$$[E]_t = [E]_0 \, e^{-k_{obs}t} \tag{1}$$

which in logarithmic form is

$$\log[E]_t = \log[E]_0 - k_{obs} t/2.303 \tag{2}$$

[41] R. Lottenberg, J. A. Hall, M. Blinder, E. P. Binder, and C. M. Jackson, *Biochim. Biophys. Acta* **742**, 539 (1983).

FIG. 2. Discontinuous or continuous measurements of the rate of thrombin inactivation by antithrombin. (A) Reaction of 0.2 μM thrombin with 1 μM antithrombin in the presence of 150 μM p-aminobenzamidine was monitored by discontinuous measurements of residual enzyme activity in 20-μl samples withdrawn at the indicated times and diluted into 1 ml of 76 μM S-2238 (circles) or by continuous changes in fluorescence (λ_{ex} 345 nm, λ_{em} 370 nm) calculated as $(F_t - F_\infty)/(F_0 - F_\infty)$ [Eq. (7), solid line]. Conditions: 20 mM sodium phosphate, 0.25 M NaCl, 0.1 mM EDTA, 0.1% PEG 6000, pH 7.4, I 0.3, 25°. (B) Semilog plots of residual enzyme activity measured from initial velocities of S-2238 hydrolysis (ΔA_{405}/min) as above vs. time for reactions of 0.04 μM thrombin with 0.27 (●), 0.41 (■), 0.54 (▲), and 0.68 (◆,◇) μM antithrombin. The reaction depicted by the open symbols contained 100 $\mu g/ml$ Polybrene to verify lack of contamination of the antithrombin with heparin. Conditions as in A. (C) Plot of k_{obs} as a function of the "effective" antithrombin concentration for reactions of 0.25–0.5 μM thrombin plus 700 μM p-aminobenzamidine with antithrombin monitored by continuous fluorescence changes as in A (○) or for the reactions in B (●).

TABLE I
INHIBITION RATE CONSTANTS FOR ANTITHROMBIN-PROTEINASE REACTIONS

| | Second-order rate constant | | | | |
| | −Heparin | | +Heparin | | |
Enzyme	$M^{-1}\ sec^{-1}$	Ref.	$M^{-1}\ sec^{-1}$	Ref.	Heparin rate enhancement
α-Thrombin	$0.7–1.1 \times 10^{4a}$	31, 124	3.7×10^{7a}	56	~4000
	1.4×10^{4b}	49	$1.5–4 \times 10^{7b}$	42, 99, 107	~2000
Factor Xa	$2.1–2.5 \times 10^{3a}$	6	1.3×10^{6a}	6	~600
	$3–4 \times 10^{3b}$	99,125	4×10^{6b}	99	~1000
Factor IXa	4.8×10^{2b}	99	5×10^{6b}	99	10,000
Factor XIa	$1.8–3 \times 10^{2a}$	126, 127	$0.6–1.5 \times 10^{4a}$	126, 127	~40
Factor XIIa	3.6×10^{1b}	128	4.2×10^{2b}	128	12
Plasma kallikrein	$1.6–3 \times 10^{2a}$	106, 129	2×10^{3a}	106	~10
Plasmin	6.7×10^{2b}	99	6.7×10^{4b}	99	100
β-Trypsin	1.5×10^{5a}	124	1.1×10^{6a}	124	7

[a] 23–25°.
[b] 37°.

where $[E]_t$ and $[E]_0$ are enzyme concentrations at time t and zero time, respectively, and k_{obs} is the observed pseudo first-order rate constant given by

$$k_{obs} = k[AT]_0 \tag{3}$$

where zero subscripts denote initial concentrations. The value for k_{obs} is determined from the slope of a semilogarithmic plot of the residual enzyme activity, given by the initial rate of substrate hydrolysis, versus time [Eq. (2); fig. 2B], or alternatively by nonlinear least-squares fitting of the enzyme inactivation curve to an exponential decay function [Eq. (1)]. Values for k_{obs} determined for reactions conducted over a range of inhibitor concentrations are then plotted as a function of the inhibitor concentration to confirm the predicted proportional relationship of Eq. (3) (Fig. 2C). The slope of such a plot provides the second-order inhibition rate constant. Alternatively, reaction of inhibitor and enzyme at comparable concentrations can be analyzed by the kinetic equation for a second-order reaction,[42,43] although such an analysis requires accurate quantitation of the inhibition stoichiometry, as detailed above. The validity of this latter type of analysis should be checked by verifying the invariance of the rate constant when reactant concentrations are changed. Table I shows the

[42] R. Jordan, D. Beeler, and R. Rosenberg, J. Biol. Chem. 254, 2902 (1979).
[43] D. A. Lawrence, L. Stranberg, J. Ericson, and T. Ny, J. Biol. Chem. 265, 20293 (1990).

second-order rate constants that have been measured, mostly by the discontinuous method, for the inhibition of various target enzymes by antithrombin.

Continuous Method 1. Use of a Chromogenic or Fluorogenic Substrate as an Indicator for Active Enzyme. Continuous monitoring of antithrombin–proteinase reactions can be done in the presence of a substrate for the enzyme that yields a product which can be detected spectrophotometrically.[44,45] The basis for this method derives from the following reaction scheme:

$$E + AT \xrightarrow{k} E\text{--}AT$$
$$+$$
$$S$$
$$K_m \left\updownarrow\right.$$
$$E \cdot S \xrightarrow{k_{cat}} E + P$$

where S is the substrate and P is the product of substrate hydrolysis. Under pseudo first-order conditions, and with negligible consumption of substrate (< 10%), the appearance of product is given by the exponential function:

$$[P]_t = [P]_\infty (1 - e^{-k_{obs}t}) \tag{4}$$

where $[P]_t$ and $[P]_\infty$ are the concentrations of product at time t and after completion of the reaction, respectively, and k_{obs} is the pseudo first-order rate constant. $[P]_\infty$ and k_{obs} are given by Eqs. (5) and (6):

$$[P]_\infty = (k_{cat}/K_m) \, [E]_0 [S]_0 / k[AT]_0 \tag{5}$$

$$k_{obs} = k[AT]_0 / (1 + [S]_0/K_m) \tag{6}$$

Thus, knowledge of the kinetic parameters, K_m and k_{cat}, for the enzyme–substrate reaction can allow determination of the second-order inhibition rate constant either from the total amount of product formed or from the pseudo first-order rate constant for the formation of this product.

Execution of the method involves mixing the inhibitor and the indicator substrate in buffer containing 0.1% polyethylene glycol in a polystyrene cuvette at constant temperature and recording the starting absorbance (405nm) or fluorescence (λ_{ex} 380 nm, λ_{em} 440 nm) of the solution. The reaction is then initiated with temperature-equilibrated enzyme with minimal dilution (< 10%), and the exponential appearance of absorbance or fluorescence is continuously monitored with time for 5 to 10 half-lives (Fig. 3). Substrate concentrations of 50–200 μM are usually high enough

[44] J. W. Williams and J. F. Morrison, this series, Vol. 63, p. 437.
[45] W.-X. Tian and C.-L. Tsou, *Biochemistry* **21**, 1028 (1982).

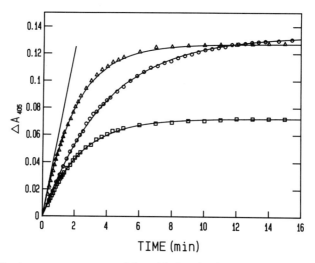

FIG. 3. Continuous measurement of thrombin inactivation by antithrombin in the presence of a chromogenic substrate. Reactions of 2.4 μM antithrombin with 3.3 ($\square$) or 6.6 ($\triangle$) nM thrombin and of 1.2 μM inhibitor with 3.3 nM enzyme (O) in the presence of 115 μM benzoyl-L-Phe-L-Val-L-Arg-p-nitroanilide were conducted in 20 mM sodium phosphate, 0.1 M NaCl, 0.1 mM EDTA, 0.1% PEG 6000, pH 7.4, I 0.15 at 25°. Solid curves represent nonlinear least-squares fits of the data by Eq. (4). The solid straight line shows the linear absorbance change resulting from a control incubation of 6.6 nM thrombin in the absence of inhibitor.

to give an appreciable absorbance or fluorescence change with negligible consumption of substrate. To ensure that significant substrate depletion has not occurred, the rate of substrate hydrolysis in the absence of inhibitor should be linear up to the maximum spectral change (Fig. 3). Concentrations of substrate below K_m are also preferable to avoid a large correction for the competitive effect of the substrate on the reaction rate constant [Eq. (6)]. Alternatively, such competitive effects can be exploited to allow measurement of fast inhibitor reactions on a slow time scale, e.g., in the presence of heparin, although with a reduced precision of the inferred inhibition rate constant. Inhibitor concentrations yielding reaction half-lives of $\geqslant 1$ min are readily measurable with conventional spectrophotometers, whereas faster reaction times require a stopped-flow instrument. With $[S]_0 \ll K_m$, antithrombin concentrations of $\leqslant 1$ and $\leqslant 5$ μM are suitable for monitoring reactions with thrombin and factor Xa, respectively. The appropriate enzyme concentration to use can be calculated based on Eq. (5). The frequent observation of a low level of substrate hydrolysis persisting at the end of the reaction has been attributed to the slow breakdown of the stable enzyme–inhibitor complex to release free enzyme and proteolytically modified inhibitor,[46,47] or to a modified or contaminating enzyme that reacts slower or not at all with the inhibitor.[48] The alternative suggestion that this residual rate reflects a reversible formation of the stable antithrombin–enzyme complex[49,50] is not supported by direct measurements of the products of complex dissociation.[46] Such behavior can be treated by adding a term to the equation for product formation which accounts for the limiting linear rate of substrate hydrolysis.[47,48,50] To obtain $[P]_\infty$ and k_{obs} from the product formation curve (Fig. 3), the end point or final linear rate of product formation is extrapolated to zero time, and the logarithm of the difference between the observed absorbance or fluorescence values from the progress curve and the extrapolated line at time t is plotted as a function of time. The slope of this plot yields $-k_{obs}/2.303$ and the intercept, $\log \varepsilon[P]_\infty$, where ε is the coefficient that relates the measured absorbance or fluorescence to the concentration of product. Assays conducted over a range of inhibitor concentrations should yield a linear dependence of k_{obs} on the ratio $[AT]_0/(1 + [S]_0/K_m)$. The ratio represents the "effective" inhibitor concentration or the inhibitor concentration that would yield the same k_{obs} in the absence of substrate.[6] The slope of this plot directly gives the second-order rate constant. A plot of $[P]_\infty$ vs. $(k_{cat}/K_m)[E]_0[S]_0/[I]_0$ should also be linear with a slope of $1/k$. Second-order rate constants determined from Eq. (5) or (6) should agree. Any disagreement may reflect errors in either the enzyme concentration or K_{cat} and K_m values for the substrate. Because increasing inhibitor concentrations produce inversely proportional changes in $[P]_\infty$ which can compromise the sensitivity of the assay [eq. (5)], the enzyme and inhibitor concentrations can be varied in a constant ratio to achieve an invariant level of product formation, since k_{obs} should be independent of enzyme concentration.

Continuous Method 2. Use of a Spectroscopic Probe to Follow Enzyme Inactivation. A classical method for following the combination of substrates or inhibitors with enzymes is from the spectroscopic changes accompanying the displacement of a probe reversibly bound to the enzyme active site. The arginine analog, *p*-aminobenzamidine, has been shown to be a sensitive and specific probe for the active site of trypsinlike serine proteinases.[51] Binding of this probe to trypsin, thrombin, and factor Xa is accompanied by substantial red shifts and large enhancements in the

[46] Å. Danielsson and I. Björk, *Biochem. J.* **213**, 345 (1983).
[47] M. C. Naski, J. W. Fenton, II, J. M. Maraganore, S. T. Olson, and J. A. Shafer, *J. Biol. Chem.* **265**, 13484 (1990).
[48] P. J. Hogg and C. M. Jackson, *Proc. Natl. Acad. Sci. U.S.A.* **53**, 1238 (1989).
[49] J. Jesty, *J. Biol. Chem.* **254**, 10044 (1979).
[50] C. Longstaff and P. J. Gaffney, *Biochemistry* **30**, 979 (1991).
[51] S. A. Evans, S. T. Olson, and J. D. Shore, *J. Biol. Chem.* **257**, 3014 (1982).

fluorescence emission spectra of 50-, 230-, and 200-fold, respectively, which resemble the shifts and enhancements of the fluorescence of the probe in apolar solvents.[6,32,52] Substantial fluorescence enhancements have also been demonstrated for p-aminobenzamidine binding to factor IXa,[53] plasmin,[54] plasma kallikrein,[54] and tissue plasminogen activator.[54] The magnitude of these fluorescence enhancements together with the affinity of enzyme–probe interactions ($K_D \sim 10$–$100~\mu M$) allow substantial saturation of the enzyme without introducing a large background fluorescence due to free probe. Moreover, the affinities are in a range compatible with a rapid equilibration of the enzyme–probe interaction even on a millisecond time scale, so that the rate of probe dissociation from the enzyme is not expected to limit the rate of inhibitor binding. The conditions for employing the method are based on the following reaction scheme:

$$E + AT \xrightarrow{k} E\text{--}AT$$
$$+$$
$$P$$
$$\Big\updownarrow K_P$$
$$E\cdot P$$

where P is p-aminobenzamidine and K_P is the dissociation constant for the enzyme–p-aminobenzamidine interaction. This scheme predicts that the decrease in p-aminobenzamidine fluorescence resulting from displacement of the probe from the enzyme by antithrombin should directly report enzyme inactivation, as reflected by Eq. (7):

$$\frac{[E]_t}{[E]_0} = \frac{(F_t - F_\infty)}{(F_0 - F_\infty)} \tag{7}$$

where F_0, F_t, and F_∞ represent the fluorescence at time zero, at time t, and at the reaction end point, respectively (Fig. 2A).[6,55] Under pseudo first-order conditions ($[AT]_0 \gg [E]_0 \ll [P]_0$) an exponential inactivation of enzyme and loss of probe fluorescence is predicted, with k_{obs} a function of total inhibitor and probe concentrations, as given by Eq. (8)[31,51]:

$$k_{obs} = k[AT]_0/(1 + [P]_0/K_P) \tag{8}$$

[52] D. H. Atha, J.-C. Lormeau, M. Petitou, R. D. Rosenberg, and J. Choay, *Biochemistry* **26**, 6454 (1987).
[53] D. M. Monroe, G. B. Sherrill, and H. R. Roberts, *Anal. Biochem.* **172**, 427 (1988).
[54] S. T. Olson and J. D. Shore, unpublished observations.
[55] S. T. Olson, *J. Biol. Chem.* **263**, 1698 (1988).

The method involves preincubating enzyme ($\geq 10^{-7} M$) and p-amino-benzamidine ($\sim 100 \mu M$ or around K_P) in reaction buffer at constant temperature and recording the initial fluorescence of the solution at wavelengths optimal for the bound probe. Excitation and emission wavelengths of 325 and 370 nm, respectively, appear suitable for a number of enzymes (pH 7.4, 25°), although inner filter effects due to the absorbance of the probe may require excitation at longer wavelengths when higher probe concentrations ($> 200 \mu M$) are used.[51] For submicromolar enzyme concentrations, it is desirable to offset the free probe fluorescence with a solution of the probe without enzyme and amplify the difference between bound and free probe fluorescence for optimal sensitivity. The reaction is then initiated with temperature-equilibrated antithrombin with minimal dilution($< 10\%$), and the exponential decrease in probe fluorescence is continuously recorded until a stable end point fluorescence is obtained (5–10 half-lives, Fig. 2A). The progress curve is analyzed, as with the other continuous method, either graphically as a semilogarithmic plot or by nonlinear least-squares fitting by an exponential function to obtain k_{obs}. Values of k_{obs} measured over a range of inhibitor concentrations can then be plotted as a function of the "effective" antithrombin concentration, given by the ratio $[AT]_0/(1 + [P]_0/K_P)$ (Fig. 2C). The slope of this linear plot yields the second-order rate constant. The value of K_P needed to calculate the "effective" inhibitor concentration has been determined for several enzymes at pH 7.4, 25°.[6,51] Alternatively, this value can be determined by direct binding measurements, by kinetic measurements of k_{obs} as a function of the probe concentration, or by classical steady-state kinetic analysis of the competitive effect of the probe on the reaction of the enzyme with a synthetic substrate.[6,51,56]

Resolution of Encounter Complex and Stable Complex Formation Steps. The inactivation of proteinases by antithrombin appears to involve a simple one-step bimolecular association over the range of reaction rates that can be measured by conventional spectrophotometric methods. Extension of the accessible range of reaction rates with the use of rapid kinetic methods has shown, however, that the inactivation of thrombin by antithrombin is actually a two-step process,[31] similar to other proteinase–proteinase inhibitor reactions.[57] Thus, an encounter complex is initially formed, in which antithrombin is bound at the active site of the enzyme, prior to the conversion to a stable complex. The experimental basis for distinguishing a one-step from a two-step process follows from a consider-

[56] S. T. Olson and I. Björk, *J. Biol. Chem.* **266**, 6353 (1991).
[57] M. Laskowski, Jr. and I. Kato, *Annu. Rev. Biochem.* **49**, 593 (1980).

ation of the reaction scheme for the latter in the presence of the probe, p-aminobenzamidine:

$$E + AT \xrightleftharpoons{K_{E,AT}} E \cdot AT \xrightarrow{k_I} E-AT$$

$$+ $$

$$P$$

$$\Big\Updownarrow K_P$$

$$E \cdot P$$

where $K_{E,AT}$ is the dissociation constant for the encounter complex intermediate and k_I is the first-order rate constant for conversion of the encounter complex to the stable complex. In this scheme, it is assumed that the probe and encounter complex equilibria are rapid compared with the rate of stable complex formation. According to this scheme, the reaction will appear to be a one-step bimolecular process when inhibitor concentrations are much less than $K_{E,AT}$, i.e., when negligible intermediate is formed. Under such conditions, the apparent second-order rate constant is equal to the ratio, $k_I/K_{E,AT}$ [Eq. (9)]. At inhibitor concentrations approaching $K_{E,AT}$ and which yield pseudo first-order conditions, a two-step process should be evident from a hyperbolic rather than linear dependence of k_{obs} on the inhibitor concentration (Fig. 4):[31,58]

$$k_{obs} = \frac{k_I[AT]_0}{K_{E,AT}(1 + [P]_0/K_P) + [AT]_0} \tag{9}$$

Moreover, for such a two-step reaction, displacement of the probe should occur in two phases, consisting of an initial "instantaneous" phase due to formation of the encounter complex, followed by a slower observable phase reflecting stable complex formation.[31,58] This behavior would be evident from a hyperbolic decrease in the fluorescence amplitude of the observable second phase that parallels the hyperbolic increase in k_{obs}. Such behavior would be indicated from a linear dependence of the reciprocal of the fluorescence amplitude on the inhibitor concentration (Fig. 4), as given by Eq. (10):

$$\frac{\Delta F_0}{\Delta F} = 1 + \frac{[AT]_0}{K_{E,AT}(1 + [P]_0/K_P)} \tag{10}$$

where ΔF_0 is the fluorescence due to active site-bound p-aminobenzamidine in the absence of inhibitor and ΔF is the bound probe fluorescence which remains after the initial rapid probe displacement by the inhibitor and which decays in the observable second reaction phase.[56]

[58] U. Quast, J. Engel, H. Heumann, G. Krause, and E. Steffen, *Biochemistry* 13, 2512 (1974).

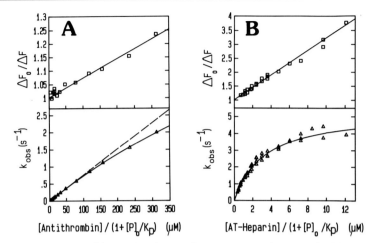

FIG. 4. Dependence of k_{obs} and reciprocal fluorescence amplitudes on inhibitor concentration for the two-step reactions of antithrombin (A) and antithrombin–heparin complex (B) with thrombin. At least a 10-fold molar excess of antithrombin (A) or antithrombin–heparin ($M_r \sim 8000$) complex (twofold molar excess of inhibitor to heparin; B) was mixed with $0.25–2\ \mu M$ thrombin and 30 (A) or 150 (B) μM probe in the stopped-flow fluorimeter, and the fluorescence changes accompanying the displacement of p-aminobenzamidine from the enzyme were continuously monitored (λ_{ex} 330, $\lambda_{em} > 350$ nm). Rate constants and amplitudes were measured by fitting fluorescence progress curves to a single exponential decay function. Amplitudes were normalized to the extrapolated fluorescence change at zero inhibitor concentration. Solid lines are nonlinear least-squares fits to Eq. (9) or to the untransformed, hyperbolic form of Eq. (10), although the latter fits are plotted according to the linear relation of Eq. (10). The dashed line in A represents the initial linear portion of the hyperbolic curve. Conditions were as in Fig. 2. Adapted from Olson and Shore[31] with permission.

For the antithrombin–thrombin reaction, a dissociation constant of $\sim 1.4 \times 10^{-3}\ M$ for the initial encounter complex interaction and a rate constant of $\sim 10\ \mathrm{sec}^{-1}$ for conversion of the encounter complex to a stable complex have been determined using the above approaches, as shown by the data in Fig. 4.[31] The slow inhibition of thrombin by antithrombin is thus due to the low affinity between the two proteins in the initial complex. An analogous two-step process could not be shown for the antithrombin–factor Xa reaction, presumably because of a much weaker affinity of the initial complex and a higher rate constant for complex stabilization.[6] In addition to the p-aminobenzamidine displacement method for characterizing two-step inhibitor–proteinase reactions, methods based on chromogenic or fluorogenic substrate indicators of active enzyme have been described.[45] In this case, saturation of an intermediate encounter complex is again evidenced by a saturable dependence of k_{obs} on inhibitor concentration that is described by an equation analogous to Eq. (9), i.e., with the term $[P]_0/K_P$ replaced by $[S]_0/K_m$. Product amplitudes can provide a further diagnostic for an intermediate complex.

Dissociation of Antithrombin–Proteinase Complexes. The rate of dissociation of stable antithrombin–proteinase complexes can be measured by continuously monitoring the release of active enzyme from the complex with a synthetic substrate under initial velocity conditions, i.e., less than 1% complex dissociation.[46,49] Whereas early studies attributed the observed slow dissociation of the stable antithrombin–thrombin complex ($t_{1/2} \sim 3$ days) to a reversible equilibrium with free enzyme and inhibitor,[49] subsequent studies which examined the products of this dissociation found that only cleaved antithrombin was detectable.[46,59] Antithrombin–enzyme complexes thus appear to represent only kinetically and not thermodynamically stable intermediates of a regular substrate reaction between the inhibitor and enzyme.

Antithrombin–thrombin complex, prepared with an approximately fivefold molar excess of inhibitor over enzyme (to minimize complex degradation by free enzyme during the reaction), is diluted to $\leq 0.1~\mu M$ into a solution of $\sim 400~\mu M$ S-2238 at $25-37°$ in near-neutral pH buffer containing 0.1% polyethylene glycol in a polystyrene cuvette, and the absorbance increase at 405 nm is continuously monitored for 10–20 min. A parabolic increase in the absorbance, reflecting a linear rate of appearance of active enzyme, is observed which is governed by Eq. (11)

$$\Delta A_{405} = \varepsilon_{405}[v_0 t + k_d(v_f - v_0)t^2/2] \tag{11}$$

where ΔA_{405} is the difference between the absorbance at time t and at zero time, ε_{405} is the p-nitroaniline absorption coefficient (9920 M^{-1} cm^{-1} in $I \sim 0.15$ buffer),[60] and v_0 and v_f are the initial velocities of substrate hydrolysis corresponding to the initial free enzyme level ($[E]_0$) and final enzyme level ($[E]_0 + [E-AT]$) after complete dissociation of the complex, respectively. Equation (11) can be transformed to linear form by dividing both sides of the equation by t. A plot of $\Delta A_{405}/t$ vs. t is thus linear, and from the slope and intercept of this plot, together with the calculated value of v_f,

$$v_f = k_{cat}([E-AT] + [E]_0)[S]_0/(K_m + [S]_0) \tag{12}$$

k_d can be determined. The slopes of such plots should increase linearly with complex concentration, and k_d values calculated from these slopes should be shown to be independent of substrate concentration. The k_d values should also be measured at different times following complex formation to correct for any time dependence which may result from complex aggrega-

[59] Å. Danielsson and I. Björk. *FEBS Lett.* **119**, 241 (1980).
[60] R. Lottenberg and C. M. Jackson, *Biochim. Biophys. Acta* **742**, 558 (1983).

tion.[46,49] This method can be extended to other antithrombin–enzyme complexes provided that a substrate sensitive enough to detect ~1% complex dissociation is used. The observation that some serpin–proteinase complexes do appear to dissociate to regenerate native inhibitor[61] suggests that the observed values of k_d measured by this method could include contributions from both the forward dissociation of complex to cleaved inhibitor as well as the reverse dissociation of complex to native inhibitor. Analysis by SDS gel electrophoresis of the reaction products resulting from dissociation of stable antithrombin–enzyme complexes under conditions where more extensive dissociation occurs[46] can allow quantitation of the fraction of the apparent k_d resulting from each of these two dissociation pathways.

Effect of Heparin on Antithrombin–Proteinase Reactions

Heparin is a highly sulfated glycosaminoglycan of repeating disaccharide units consisting of a hexuronic acid, either glucuronic or iduronic acid, linked to glucosamine. Marked heterogeneity in the polysaccharide results from variability in the distribution of iduronic and glucuronic acid residues as well as in the pattern of sulfation.[62-64] Commercial heparin preparations are also polydisperse, with molecular weights ranging from ~5000 to ~30,000, i.e., ~15 to 100 monosaccharide residues per chain. The well-established anticoagulant activity of heparin results from the polysaccharide accelerating the reactions of antithrombin with its target enzymes. Such accelerations, which range up to ~10,000-fold (Table I), depend largely on the presence of a unique pentasaccharide sequence which constitutes a specific site for antithrombin binding,[65-70] and, to a lesser extent, also on

[61] B. H. Shieh, J. Potempa, and J. Travis, *J. Biol. Chem.* **264**, 13420 (1989).

[62] I. Björk and U. Lindahl, *Mol. Cell. Biochem.* **48**, 161 (1982).

[63] D. A. Lane and U. Lindahl, eds., "Heparin: Chemical and Biological Properties; Clinical Applications." Edward Arnold, London, 1989.

[64] D. A. Lane, I. Björk, and U. Lindahl, eds., "Heparin and Related Polysaccharides." Plenum, New York, 1992.

[65] U. Lindahl, G. Bäckström, L. Thunberg, and I. G. Leder, *Proc. Natl. Acad. Sci. U.S.A.* **77**, 6551 (1980).

[66] B. Casu, P. Oreste, G. Torri, G. Zopetti, J. Choay, J. C. Lormeau, M. Petitou, and P. Sinaÿ, *Biochem. J.* **197**, 599 (1981).

[67] L. Thunberg, G. Bäckström, and U. Lindahl, *Carbohydr. Res.* **100**, 393 (1982).

[68] J. Choay, M. Petitou, J. C. Lormeau, P. Sinaÿ, B. Casu, and G. Gatti, *Biochem. Biophys. Res. Commun.* **116**, 492 (1983).

[69] D. H. Atha, A. W. Stevens, and R. D. Rosenberg, *Proc. Natl. Acad. Sci. U.S.A.* **81**, 1030 (1984).

[70] D. H. Atha, J. C. Lormeau, M. Petitou, R. D. Rosenberg, and J. Choay, *Biochemistry* **24**, 6723 (1985).

the length[71-74] and charge density[75] of the polysaccharide chain. Heparin chains bearing the sequence-specific pentasaccharide bind the inhibitor with ~ 1000-fold higher affinity than those lacking the sequence and are responsible for the bulk of the anticoagulant activity of the preparation.[42,76,77] Commercial heparin preparations are readily fractionated according to size and affinity for antithrombin, and the resulting more homogeneous fractions should be used for kinetic characterization of the accelerating effect of the polysaccharide on antithrombin–proteinase reactions.

Molecular Weight and Affinity Fractionation of Heparin

Heparin purified from porcine intestinal mucosa or bovine lung is available from a number of commercial sources. Heparin chains of defined molecular weights which contain the pentasaccharide binding site for antithrombin can be isolated from such preparations most simply by gel filtration followed by affinity chromatography on matrix-linked antithrombin.[71,78,79]

Molecular Weight Fractionation. The commercial heparin is chromatographed on a Sephadex G-100 column eluted with a buffer containing 1 M NaCl.[71,80] About 1–3 g of polysaccharide dissolved in ~ 10 ml equilibrating solvent can be fractionated on a column with dimensions of 5 × 100 cm. Heparin is eluted from the column as a broad band, which can be most simply monitored by an Azure A dye binding assay[81] or by analysis for uronic acid.[82] The broad elution band is divided into 5 to 10 equal volume fractions, depending on the narrowness of the molecular weight distribution desired. Accumulated fractions from several runs of starting

[71] T. C. Laurent, A. Tengbald, L. Thunberg, M. Höök, and U. Lindahl, *Biochem. J.* 175, 691 (1978).
[72] G. M. Oosta, W. T. Gardner, D. L. Beeler, and R. D. Rosenberg, *Proc. Natl. Acad. Sci. U.S.A.* 78, 829 (1981).
[73] R. E. Jordan, L. V. Favreau, E. H. Braswell, and R. D. Rosenberg, *J. Biol. Chem.* 257, 400 (1982).
[74] M. E. Nesheim, M. N. Blackburn, C. M. Lawler, and K. G. Mann, *J. Biol. Chem.* 261, 3214 (1986).
[75] R. E. Hurst, M. C. Poon, and M. J. Griffith, *J. Clin. Invest.* 72, 1042 (1983).
[76] B. Nordenman and I. Björk, *Biochemistry* 17, 3339 (1978).
[77] Å. Danielsson and I. Björk, *Eur. J. Biochem.* 90, 7 (1978).
[78] M. Höök, I. Björk, J. Hopwood, and U. Lindahl, *FEBS Lett.* 66, 90 (1976).
[79] L. O. Andersson, T. W. Barrowcliffe, E. Holmer, E. A. Johnson, and G. E. C. Sims, *Thromb. Res.* 9, 575 (1976).
[80] Å. Danielsson and I. Björk, *Biochem. J.* 193, 427 (1981).
[81] L. B. Jacques, *Methods Biochem. Anal.* 24, 203 (1977).
[82] T. Bitter and H. M. Muir, *Anal. Biochem.* 4, 330 (1962).

material are each concentrated by lyophilization or by ultrafiltration with a membrane that retains the heparin, desalted by chromatography on Sephadex G-25 equilibrated in water or 10% ethanol (bed volume at least five times the sample volume), and again lyophilized. The G-100 chromatography step is then repeated for each fraction, and the central portion of the heparin peak is collected. Salt-free polysaccharide is again obtained by lyophilization followed by G-25 chromatography. The resulting heparin fractions should be resolvable by analytical gel chromatography and show a barely discernible polydispersity when characterized by sedimentation equilibrium in the ultracentrifuge.[80]

Affinity Fractionation. Molecular-weight-fractionated heparin chains with low and high affinity for antithrombin are separated by chromatography on an antithrombin – agarose column.[78,79] Immobilization of the inhibitor with a 70% coupling efficiency may be done with commercially activated agarose resins such as Affi-Gel (Bio-Rad) or cyanogen bromide-activated Sepharose (Pharmacia LKB Biotechnology), or else the agarose may be chemically activated using established procedures.[7] Maximal binding activity of the immobilized antithrombin is achieved by coupling the inhibitor in the presence of acetylated unfractionated heparin[83] at concentrations sufficient to saturate the heparin binding site.[78] The molecular-weight-fractionated heparin is applied to the affinity column equilibrated in $0.02\ M$ Tris-HCl or sodium phosphate, $0.25\ M$ NaCl, pH 7.4, in an amount containing high-affinity heparin below the column capacity. The column is subsequently washed with the equilibrating buffer to elute the low-affinity binding fraction, which lacks the antithrombin-binding pentasaccharide sequence. Alternatively, application of an amount of heparin containing sufficient high-affinity material to exceed the binding capacity of the affinity matrix can ensure isolation of the highest affinity form of the polysaccharide and thereby minimize variant high-affinity forms, possibly resulting from structural differences in a secondary interaction site outside the pentasaccharide region.[42,84,85] The high-affinity binding fraction containing the pentasaccharide is eluted at $\sim 1\ M$ NaCl, preferably with a salt gradient to $2\ M$ NaCl, although, with the latter method, step elution with buffer containing 1 to $3\ M$ NaCl is also possible.

The molecular weights of affinity-fractionated heparin are best determined by sedimentation equilibrium in the analytical ultracentrifuge[71,80,86]

[83] I. Danishefsky and H. Steiner, *Biochim. Biophys. Acta* **101,** 37 (1965).

[84] P. Gettins and J. Choay, *Carbohydr. Res.* **185,** 69 (1989).

[85] S. T. Olson, I. Björk, R. Sheffer, P. A. Craig, J. D. Shore, and J. Choay, *J. Biol. Chem.* **267,** 12528 (1992).

[86] E. Braswell, *Biochim. Biophys. Acta* **158,** 103 (1968).

or in a Beckman Airfuge.[87] Heparins with molecular weights characterized in this manner can provide a series of standards which may be used to calibrate an analytical gel chromatography column for routine determinations of heparin molecular weights. Molar concentrations of the polysaccharide can then be calculated from dry-weight measurements of the salt-free polysaccharide.[78,86]

Binding of Heparin to Antithrombin

Heparin chains containing the unique pentasaccharide bind tightly to antithrombin with a dissociation constant of $\sim 10-20$ nM at pH 7.4, I 0.14–0.15, 25°.[42,85,88,89] This binding is readily quantified spectroscopically by monitoring the changes in the ultraviolet absorbance, circular dichroism, or fluorescence of aromatic amino acids of the inhibitor which accompany the binding of the polysaccharide.[76,88,90,91] Alternatively, analysis of the competition between heparin in solution and immobilized heparin for binding to antithrombin can provide a rigorous quantitative measure of the stoichiometry and affinity of the interaction.[92] Because of the greater sensitivity of fluorescence detection and the strong affinity of antithrombin for the pentasaccharide binding site, binding constants and stoichiometries are best quantified using the $\sim 40\%$ protein fluorescence enhancement as the detection method.

Fluorescence Titration of Antithrombin with Heparin. A solution of antithrombin ($\sim 0.05-0.1$ μM) in 0.05 M Tris-HCl or 0.02 M sodium phosphate, 0.1 M NaCl, pH 7.4, 25°, is titrated with small aliquots of a concentrated high-affinity heparin solution up to a four- to eightfold molar ratio of polysaccharide to protein with minimal ($<10\%$) dilution. The protein fluorescence (λ_{ex} 280 nm, λ_{em} 340 nm) is measured before heparin addition and then after each addition of heparin, following careful mixing of the solution (Fig. 5). Suitable corrections for buffer blanks and dilution are applied to these measurements. Possible mixing artifacts due to protein adsorption to the cuvette or to light scattering by denatured protein can be minimized by limiting the number of titrant additions in a single titration and pooling results from multiple titrations in which the titrant additions have been staggered. The dependence of the relative fluorescence change, i.e., $\Delta F/F_0 = (F_{obs} - F_0)/F_0$, where F_{obs} and F_0 represent the corrected observed and starting fluorescence values, respectively, on the total heparin

[87] S. T. Olson, H. R. Halvorson, and I. Björk, *J. Biol. Chem.* **266**, 6342 (1991).
[88] B. Nordenman, Å. Danielsson, and I. Björk, *Biochim. Biophys. Acta* **672**, 227 (1978).
[89] S. T. Olson, K. R. Srinivasan, I. Björk, and J. D. Shore, *J. Biol. Chem.* **256**, 11073 (1981).
[90] R. Einarsson and L. O. Andersson, *Biochim. Biophys. Acta* **490**, 104 (1977).
[91] G. B. Villanueva and I. Danishefsky, *Biochem. Biophys. Res. Commun.* **74**, 803 (1977).
[92] S. T. Olson, P. E. Bock, and R. Sheffer, *Arch. Biochem. Biophys.* **286**, 533 (1991).

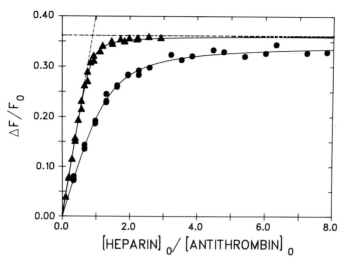

FIG. 5. Equilibrium binding of heparin to antithrombin monitored by fluorescence titrations. High-affinity heparin ($M_r \sim 8000$) was titrated into solutions of antithrombin at concentrations much greater than K_D (1 μM, ▲) or comparable to K_D (42 nM, ●), and the relative increase in antithrombin protein fluorescence (λ_{ex} 280 nm, λ_{em} 340 nm), corrected for dilution, was measured after each addition. Titrations were done in 20 mM sodium phosphate, 0.1 M NaCl, 0.1 mM EDTA, 0.1% PEG 8000, pH 7.4, I 0.15 at 25°. The apparent stoichiometry determined by fitting the former titration to Eq. (13) (0.85 mol heparin/mol AT, equal to the intersection of the lines defining the initial and final regions of the titration) was fixed when fitting the latter titration by this equation. This gave a K_D of 8.8 ± 1.3 nM for the interaction.

concentration, can then be analyzed by nonlinear least-squares computer fitting of the data by the binding equation:

$$\frac{\Delta F}{F_0} = \frac{\Delta F_{max}}{F_0}$$
$$\times \frac{[AT]_0 + n[H]_0 + K_D - \{([AT]_0 + n[H]_0 + K_D)^2 - 4[AT]_0 \, n[H]_0\}^{1/2}}{2[AT]_0} \quad (13)$$

where $[AT]_0$ and $[H]_0$ represent total concentrations of antithrombin and heparin, respectively, K_D is the dissociation constant and n is the apparent stoichiometry for the interaction, and $\Delta F_{max}/F_0$ is the maximum relative fluorescence change. Because the K_D, but not the stoichiometry, is well determined when the antithrombin concentration is around K_D, it is best to determine the stoichiometry independently by performing titrations at inhibitor concentrations where essentially complete binding of the added heparin occurs, i.e., at ~ 100 times K_D. An inhibitor solution of ~ 1 μM is sufficient for this purpose. In this case, the stoichiometry is obtained as the ratio of heparin to antithrombin concentrations where the initial linear increase in fluorescence intersects the maximum fluorescence change corresponding to saturation of the inhibitor (Fig. 5). The resulting stoichiometry can then be fixed in the binding equation above to allow determination of K_D and $\Delta F_{max}/F_0$ by computer fitting.

Properties. Most high-affinity heparin chains contain a single pentasaccharide binding site for antithrombin, distributed randomly in these chains, although some higher molecular weight chains may contain two to three such sites.[73,74,80,93] The binding of antithrombin to oligosaccharide and full-length heparin chains containing the pentasaccharide site is independent of the length of these chains or the presence of other such high-affinity sites on the chain, indicating that these sites function independently and are noninteracting.[80,94] The synthetic pentasaccharide also binds the inhibitor with nearly the same affinity, indicating that the bulk of the free energy of the binding ($\sim 95\%$) is provided by the pentasaccharide.[70,85] However, an additional nonspecific ionic interaction outside the pentasaccharide binding region may possibly make a further minor contribution to the interaction.[84,85,94,95] Binding of antithrombin to heparin is strongly dependent on the ionic strength and pH and less so on the temperature.[12,42,56,88,89] The affinity is decreased about 10-fold from that at physiological ionic strength at an ionic strength of 0.3 and about 100-fold at an ionic strength of 0.5. The ionic strength dependence suggests that four to five ionic interactions contribute to the binding,[12,56,85] consistent with the number of charged groups on the pentasaccharide shown to be essential for the binding. However, a large nonionic contribution to the binding energy is also evident from this dependence. The binding affinity is maximal at pH 5.5, decreases about 25-fold as the pH is increased to 8.5, and rapidly decreases outside this range.[12]

The spectroscopic changes accompanying the binding of heparin to antithrombin are compatible with the polysaccharide inducing a conformational change in the inhibitor rather than simply perturbing the environment of surface residues of the protein. This suggestion is supported by the observations that the fluorescence enhancement arises from buried and not surface tryptophan residues[96] and that the susceptibility of certain inhibitor residues to chemical modifying agents[97] or to proteolysis[98] is

[93] L.-G. Oscarsson, G. Pejler, and U. Lindahl, *J. Biol. Chem.* **264**, 296 (1989).
[94] U. Lindahl, L. Thunberg, G. Bäckström, J. Riesenfeld, K. Nordling, and I. Björk, *J. Biol. Chem.* **259**, 12368 (1984).
[95] A. L. Stone, D. Beeler, G. Oosta, and R. D. Rosenberg, *Proc. Natl. Acad. Sci. U.S.A.* **79**, 7190 (1982).
[96] S. T. Olson and J. D. Shore, *J. Biol. Chem.* **256**, 11065 (1981).
[97] J. Y. Chang, *J. Biol. Chem.* **264**, 3111 (1989).
[98] L. F. Kress and J. J. Catanese, *Biochemistry* **20**, 7432 (1981).

altered by heparin binding. The kinetics of heparin binding to antithrombin further indicate a two-step binding process, consistent with an initial weak binding of the polysaccharide with a dissociation constant of $2-4 \times 10^{-5}$ M inducing a subsequent conformational change at a rate constant of ~ 500 sec^{-1}.[85,89] The conformational change step enhances the affinity of heparin for antithrombin ~ 300- to 1000-fold and is thus responsible for the high-affinity interaction.

Reaction of Antithrombin–Heparin Complex with Proteinases

Heparin acceleration of the reaction between antithrombin and proteinases results from heparin binding to antithrombin promoting the association of the inhibitor with the proteinase.[42,55,99,100] This promotion appears to involve two distinct mechanisms, whose relative contribution depends on the proteinase. For enzymes such as factor Xa, plasma kallikrein, and possibly also factor XIIa, the antithrombin conformational change induced by heparin binding appears to be mostly responsible for the promotion, as judged by the ability of the heparin pentasaccharide to produce a rate-enhancing effect comparable to that of full-length heparin chains containing this sequence.[68,72,101–106] For other enzymes, such as thrombin, factor IXa, and probably factor XIa, however, the promotion appears to be mainly due to the binding of the proteinase to heparin, which acts to bridge or approximate the enzyme with antithrombin bound to the same heparin chain.[55,56,71,98–104,107–109] Thus, the pentasaccharide is ineffective in accelerating antithrombin inhibition of these enzymes, despite the indistinguishable conformational changes induced by pentasaccharide and full-length heparins containing this unique sequence.[85] Instead, heparin chains containing the pentasaccharide plus an additional 13 saccharide residues are required for a substantial heparin rate enhancement of these reactions.[71,72,102–104] Consistent with the involvement of proteinase binding

[99] R. E. Jordan, G. M. Oosta, W. T. Gardner, and R. D. Rosenberg, J. Biol. Chem. 255, 10081 (1980).

[100] L. C. Petersen and M. Jørgensen, Biochem. J. 211, 91 (1983).

[101] E. Holmer, U. Lindahl, G. Bäckström, L. Thunberg, H. Sandberg, G. Söderström, and L. O. Andersson, Thromb. Res. 18, 861 (1980).

[102] E. Holmer, K. Kurachi, and G. Söderström, Biochem. J. 193, 395 (1981).

[103] D. A. Lane, J. Denton, A. M. Flynn, L. Thunberg, and U. Lindahl, Biochem. J. 218, 725 (1984).

[104] Å. Danielsson, R. Raub, U. Lindahl, and I. Björk, J. Biol. Chem. 261, 15467 (1986).

[105] V. Ellis, M. F. Scully, and V. V. Kakkar, Biochem. J. 238, 329 (1986).

[106] S. T. Olson and J. Choay, Thromb. Haemostasis 62, 326 (1989).

[107] M. J. Griffith, J. Biol. Chem. 257, 7360 (1982).

[108] M. E. Nesheim, J. Biol. Chem. 258, 14708 (1983).

[109] M. Hoylaerts, W. G. Owen, and D. Collen, J. Biol. Chem. 259, 5670 (1984).

to heparin in the latter but not the former class of enzyme reactions, heparin rate enhancements, measured under conditions where heparin chains are fully saturated with antithrombin, strongly depend on salt for the latter class of enzymes, but are independent or only weakly dependent on salt for the former class of enzymes.[33,85] For those proteinases which must bind to heparin for rate enhancement to occur, antithrombin binding to heparin prior to the proteinase is the preferred kinetic pathway for the reaction because of the greater affinity of antithrombin than of the proteinase for heparin.[55,99,100]

Rapid kinetic studies have shown that the reaction between antithrombin–heparin complex and proteinases is a two-step process, like that of the reaction involving the free inhibitor.[31] Thus, an encounter complex between the heparin-bound inhibitor and the enzyme is initially formed, followed by conversion to a stable complex (Fig. 4). Heparin acceleration results mainly from the polysaccharide promoting the association of the inhibitor with the proteinase in the ternary encounter complex,[6,31] although an additional effect of heparin on the rate of conversion of the ternary complex to the stable complex is also possible for some proteinases.[6] Formation of the stable complex is accompanied by a large reduction in heparin affinity, which promotes the concomitant release and catalytic recycling of the polysaccharide.[42,99,110,111]

Because heparin acts as a catalyst in accelerating antithrombin–proteinase reactions, i.e., one heparin molecule is sufficient to accelerate the reaction of a large number of antithrombin and proteinase molecules, a useful paradigm for kinetic studies of heparin catalysis is to consider the reactions as equivalent to a two-substrate enzyme reaction, where heparin is the enzyme, i.e., catalyst, and antithrombin and the proteinase are the substrates.[111-114] The kinetic theory developed for two-substrate enzyme reactions can thus be used to analyze heparin-catalyzed reactions. Within the context of this paradigm, a kinetic description of these reactions can be obtained either by rapid kinetic methods, in which "substrate" levels of the catalyst resulting in only a single turnover are studied, or by steady-state methods, in which the levels of the heparin catalyst are well below the levels of inhibitor and proteinase so that multiple turnovers of the catalyst occur. Each of these approaches for obtaining kinetic information on heparin-accelerated antithrombin–proteinase reactions will be described.

[110] A. S. Carlström, K. Lieden, and I. Björk, *Thromb. Res.* **11**, 785 (1977).
[111] S. T. Olson and J. D. Shore, *J. Biol. Chem.* **261**, 13151 (1986).
[112] J. R. N. Evington, P. A. Feldman, M. Luscombe, and J. J. Holbrook, *Biochim. Biophys. Acta* **870**, 92 (1986).
[113] M. J. Griffith, *J. Biol. Chem.* **257**, 13899 (1982).
[114] C. H. Pletcher and G. L. Nelsestuen, *J. Biol. Chem.* **258**, 1086 (1983).

Kinetics of Single Heparin Turnover Reactions. As discussed above, the heparin-accelerated inhibition of proteinases by antithrombin in a single catalytic cycle proceeds according to the following reaction scheme:[31,111]

$$AT \cdot H + Pr \underset{}{\overset{K_{ATH,Pr}}{\rightleftharpoons}} H \cdot AT \cdot Pr \xrightarrow{k_{I,H}} AT - Pr + H$$

where Pr is the proteinase, H is heparin, $K_{ATH,Pr}$ is the dissociation constant for the ternary encounter complex interaction of the proteinase with antithrombin – heparin complex, and $k_{I,H}$ is the first-order rate constant for conversion of the encounter complex to the stable complex and release of heparin. The kinetic parameters characterizing the single-turnover reaction are best determined under pseudo first-order conditions where the observed rate constant, k_{obs}, is a hyperbolic function of the antithrombin – heparin complex concentration given by Eq. (14)[31,56]:

$$k_{obs} = \frac{k_{I,H} [AT \cdot H]}{K_{ATH,Pr} + [AT \cdot H]} \tag{14}$$

A single catalytic cycle of the heparin-accelerated antithrombin – proteinase reaction is thus quantitatively described by the same kinetic equation as the uncatalyzed reaction, although with the antithrombin – heparin complex rather than antithrombin acting as the inhibitor of the proteinase [Eq. (9); Figs 4 and 6]. Free heparin not bound to antithrombin in these analyses can compete with the inhibitor – heparin complex for the proteinase and thereby reduce the reaction rate for those proteinases whose binding to heparin is essential for rate enhancement. Equation (14) therefore applies only when the free heparin concentration is either well below the dissociation constant for the binary proteinase – heparin interaction ($K_{Pr,H}$) or minimized by saturating the polysaccharide with a molar excess of antithrombin.[55] Otherwise, $K_{ATH,Pr}$ in Eq. (14) must be multiplied by a factor, $1 + [H]_{free}/K_{Pr,H}$, to account for the competitive effect of free heparin chains on ternary complex formation. Similarly, proteinase binding to nonspecific sites on the polysaccharide component of the antithrombin – heparin complex can compete with binding of the enzyme to the single "specific" site located adjacent to the inhibitor. To account for such nonspecific binding, $K_{ATH,Pr}$ in Eq. (14) should be multiplied by the term $1 + [AT \cdot H]/K_{HAT,Pr}$, where $K_{HAT,Pr}$ represents the dissociation constant for proteinase binding to these nonspecific sites. However, because $K_{HAT,Pr} \sim K_{Pr,H} \gg K_{ATH,Pr}$, this competitive term can usually be ignored.[56] Under conditions where heparin is saturated with inhibitor, the antithrombin – heparin complex concentration in Eq. (14) can be approximated by the total heparin concentration. It is best, however, to calculate the actual concentration of the antithrombin – heparin complex from the dissociation constant for the binary complex interaction. Although free

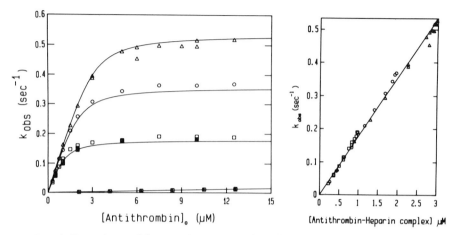

FIG. 6. Dependence of k_{obs} on the concentration of antithrombin or antithrombin–heparin complex for the heparin-accelerated antithrombin–thrombin reaction. *Left:* Varying concentrations of antithrombin in the absence (■) or presence of 1 (□,■), 2 (○), or 3 (△) μM heparin ($M_r \sim 8000$) were reacted with 0.05–0.3 μM thrombin and 500 μM *p*-aminobenzamidine in the stopped-flow fluorimeter under pseudo first-order conditions and k_{obs} determined, as in Fig. 4. Open and closed squares represent experiments in which heparin was premixed with the inhibitor or enzyme, respectively. Solid lines are computer fits to Eq. (15), with substitution of the quadratic expression for the concentration of the antithrombin–heparin complex.[55] Fitting of these data to a more general equation that accounts for inhibition by free heparin as well as for the uncatalyzed reaction (lowest curve) yields a slight improvement in the fit, but with indistinguishable kinetic parameters, indicating that such corrections are minimal.[55] Adapted from Olson[55] with permission. *Right:* The data on the left are plotted as a function of the calculated concentration of the antithrombin–heparin complex, based on a measured K_D of 0.23 μM at the ionic strength (I 0.3) of these experiments.

inhibitor present in this experimental design contributes to the observed inhibition of the proteinase, most heparin accelerations are large enough that corrections for the free inhibitor reaction usually can be ignored.

For those proteinases studied, the limiting first-order rates of conversion of the ternary encounter complex to the stable complex occur in the stopped-flow time range, i.e., milliseconds to seconds.[6,31,56] Rapid kinetic studies are thus required to resolve the kinetic parameters for saturation of the intermediate ternary complex and for conversion of the ternary complex to products in a single heparin turnover reaction. However, when the concentration of antithrombin–heparin complex is well below that required to saturate the ternary complex, i.e., $[AT \cdot H] \ll K_{ATH,Pr}$, the half-time for proteinase inhibition can be substantially increased.[55] Under such conditions, Eq. (14) reduces to

$$k_{obs} = \frac{k_{I,H}[AT \cdot H]}{K_{ATH,Pr}} \tag{15}$$

indicating that the reaction will appear to be a simple one-step bimolecular association between antithrombin–heparin complex and the free proteinase, due to the small amounts of ternary complex formed. Only second-order inhibition rate constants, equal to $k_{I,H}/K_{ATH,Pr}$, are thus measurable over this range of inhibitor–heparin complex concentrations and are obtained from the slope of the linear dependence of k_{obs} on the concentration of the binary complex (Fig. 6). Because the antithrombin–heparin complex concentration is a saturable function of the antithrombin and heparin concentrations and the dissociation constant for the interaction [Eq. (13)], the dependence of k_{obs} on the inhibitor or polysaccharide concentrations, under conditions where Eq. (15) applies, parallels a saturation curve for the formation of the antithrombin–heparin complex (Fig. 6).[55] Values of k_{obs} are determined as for the uncatalyzed reactions by monitoring the rate of proteinase inhibition by the discontinuous or continuous methods described earlier. In the case of the discontinuous method, the polycation, hexadimethrine bromide (Polybrene), should be included in the substrate solution for assaying residual enzyme activity at a concentration of ~100 μg/ml to quench the heparin-dependent reaction. When continuous methods are used, appropriate terms reflecting the competition by substrate or probe must be included in Eqs. (14) and (15) [cf. Eqs. (6), (9), and (11)].[31,55]

Kinetics of Multiple Heparin Turnover Reactions. Kinetic characterization of heparin-catalyzed antithrombin–proteinase reactions may also be performed by the classical steady-state approaches developed for two-substrate enzyme reactions, i.e., at concentrations of polysaccharide (the "enzyme") much lower than those of antithrombin and the proteinase (the "substrates"). In the case of those proteinases for which binding to heparin is necessary for the accelerating effect of the polysaccharide, it appears that a rapid equilibrium binding of the substrates to the catalyst in a random order accounts for the observed kinetic behavior, except possibly when multiple proteinase molecules bind to the catalyst.[112,113,115] The rapid equilibrium model thus appears to be obeyed when thrombin concentrations remain below $K_{Pr,H}$. Deviations from this model are apparent when thrombin concentrations approach $K_{Pr,H}$, as evidenced by anomolously high K_m values for antithrombin.[112] Such deviations can be accounted for by the binding of multiple proteinase molecules to a single heparin chain under the latter conditions,[87] which would be expected to reduce the affinity of antithrombin for the specific pentasaccharide binding site. Due to the much higher affinity of antithrombin than of the proteinase for heparin, a preferred binding of first the inhibitor and then the proteinase to

[115] I. Björk, S. T. Olson, and J. D. Shore, *in* "Heparin: Chemical and Biological Properties; Clinical Applications" (D. L. Lane and U. Lindahl, eds.), p. 229. Edward Arnold, London, 1989.

the polysaccharide is typically observed under most reaction conditions, whether the proteinase binds directly to the catalyst or indirectly by just binding to bound antithrombin,[6,111,114,115] as discussed above. Casting the steady-state rate equation for such a mechanism in a form appropriate for the heparin-catalyzed antithrombin–proteinase reaction and including the contribution of the uncatalyzed reaction yields Eq. (16)[116]:

$$v = \frac{k_{cat}[H]_0}{1 + \dfrac{K_{m,AT}}{[AT]} + \dfrac{K_{m,Pr}}{[Pr]} + \dfrac{K_{AT,H}K_{m,Pr}}{[AT][Pr]}} + k_{uncat}[AT][Pr] \qquad (16)$$

where v is the velocity of proteinase inhibition, k_{cat} is the apparent first-order rate constant for heparin-catalyzed proteinase inhibition when the catalyst is saturated with both substrates, $K_{m,AT}$ and $K_{m,Pr}$ are Michaelis constants of heparin for antithrombin and proteinase, respectively, $K_{AT,H}$ is the dissociation constant for the antithrombin–heparin binary complex, and k_{uncat} is the second-order rate constant for the uncatalyzed proteinase inhibition. $K_{m,AT}$ and $K_{m,Pr}$ correspond to dissociation constants for ternary complex formation from antithrombin and proteinase–heparin binary complex and from proteinase and antithrombin–heparin binary complex, respectively. For those proteinases which need not bind to heparin for acceleration of their inhibition reactions, the $K_{m,AT}/[AT]$ term drops out of the equation.[116] The three binding constants in Eq. (16) are related to the fourth possible binding constant in this system by the following relation:

$$K_{AT,H}K_{m,Pr} = K_{Pr,H}K_{m,AT} \qquad (17)$$

As with enzyme-catalyzed reactions, k_{cat} and K_m values for the protein substrates can be determined by analyzing the initial velocities of proteinase inhibition as a function of the concentrations of both substrates (Fig. 7). Discontinuous or continuous methods described above can be used for this purpose. However, accurate values of initial velocities can be difficult to measure by direct monitoring of the loss of proteinase activity, because only small losses in activity ($<10\%$) must be measured. Polynomial fitting of more extensive proteinase inactivation to yield the initial reaction velocity can somewhat obviate this problem.[6] Alternatively, full reaction progress curves can be analyzed by the integrated Michaelis–Menten equation.[111] Such an approach is possible because the product of the reaction, the stable antithrombin–proteinase complex, binds the catalyst considerably (at least 1000-fold) more weakly than antithrombin, so that the competitive effect of the product on substrate binding to the catalyst can typically be ignored.[42,99,110,111] A further possibility for applying this

[116] M. Dixon and E. C. Webb, "Enzymes," 3rd ed. Academic Press, New York, 1979.

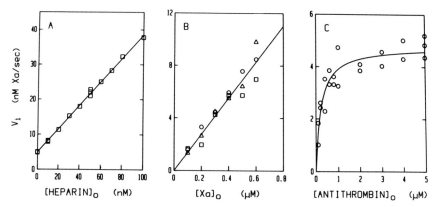

FIG. 7. Dependence of initial velocities (v_i) for the heparin-catalyzed reaction of factor Xa with antithrombin on the heparin (A), factor Xa (B), and antithrombin (C) concentrations. Fixed reactant concentrations were (A) 5 μM antithrombin, 0.5 μM factor Xa; (B) 20 nM heparin ($M_r \sim 8000$) and 1 (□), 2.5 (△), or 3.5 (○) μM inhibitor; and (C) 20 nM heparin and 0.3 μM factor Xa, with 40–80 μM p-aminobenzamidine included in all reactions. Reactions were conducted in I 0.3, pH 7.4, sodium phosphate buffer (Fig. 2) at 25°. Initial velocities were measured by fitting the first 40% of p-aminobenzamidine fluorescence decay curves to a second-order polynomial function and using Eq. (7) to convert the fluorescence change to the change in factor Xa concentration. For those reactions where the inhibitor concentration was at least five times that of the enzyme, the initial rate was determined from the product, $[Xa]_0 \times k_{obs}$, with k_{obs} obtained by fitting progress curves to a single exponential decay function. Reaction rates were corrected in B and C for the uncatalyzed reaction and in all cases for probe competition by multiplying by the factor $1 + [P]_0/K_P$, using a value of 80 μM for K_P.[6] Solid lines represent linear or nonlinear regression fits by Eq. (18). Reprinted from Craig *et al.*[6] with permission.

approach is to analyze complete progress curves under conditions where the proteinase concentration can be maintained well below the K_m of heparin for the proteinase, i.e., $[Pr] \ll K_{m,Pr}$, and where the antithrombin concentration greatly exceeds that of the proteinase.[6,117,118] Under such conditions, the $K_{m,Pr}/[Pr]$ terms in the denominator of Eq. (16) greatly exceed $1 + K_{m,AT}/[AT]$, thereby reducing the equation after algebraic rearrangement to the following form:

$$v = \frac{k_{cat}[H]_0}{K_{m,Pr}} \left(\frac{[AT]_0}{K_{AT,H} + [AT]_0} \right) [Pr] + k_{uncat}[AT]_0[Pr] \qquad (18)$$

where the total antithrombin concentration, $[AT]_0$, appears, because negligible inhibitor is consumed in the reaction. This equation indicates a

[117] I. Björk, S. T. Olson, R. G. Sheffer, and J. D. Shore, *Biochemistry* **28**, 1213 (1989).
[118] I. Björk, K. Ylinenjärvi, S. T. Olson, P. Hermentin, H. S. Conradt, and G. Zettlmeissl, *Biochem. J.* **286**, 793 (1992).

first-order dependence of the proteinase inactivation rate on the proteinase concentration, and is readily integrated to yield:

$$[Pr]_t = [Pr]_0 \, e^{-k_{obs}t} \tag{19}$$

where

$$k_{obs} = \frac{k_{cat}[H]_0}{K_{m,Pr}} \left(\frac{[AT]_0}{K_{AT,H} + [AT]_0} \right) + k_{uncat}[AT]_0 \tag{20}$$

Progress curves for the reaction are thus exponential, with k_{obs} being a linear function of the heparin concentration and a hyperbolic plus linear function of the antithrombin concentration. The catalytic efficiency, $k_{cat}/K_{m,Pr}$, and the dissociation constant for the antithrombin–heparin binary complex interaction, $K_{AT,H}$, can thus be determined under these conditions by analyzing the inhibitor concentration dependence of k_{obs}.[118] Alternatively, because k_{uncat} and $K_{AT,H}$ are typically known from independent measurements, $k_{cat}/K_{m,Pr}$ can be directly determined from the slope of the dependence of k_{obs} on heparin concentration when the antithrombin concentration is saturating. Attainment of such conditions can be confirmed from the independence of this slope on inhibitor concentration. Figure 7 shows initial velocities for the heparin-catalyzed antithrombin–factor Xa reaction measured as a function of heparin, factor Xa, and antithrombin concentrations under conditions where Eq. (18) is valid; i.e., $[Xa] \ll K_{m,Pr}$. Thus, initial velocities are linearly dependent on the heparin (Fig. 7A) and factor Xa concentrations (Fig. 7B), but depend hyperbolically on the antithrombin concentration (Fig. 7C), in accordance with this equation. Appropriate fitting of these data yields values for $k_{cat}/K_{m,Pr}$ that are in agreement with values for $k_{1,H}/K_{ATH,Pr}$ determined in single-turnover reactions, and a value for $K_{AT,H}$ indistinguishable from that measured by equilibrium binding.[6] The agreement between $k_{cat}/K_{m,Pr}$ and $k_{1,H}/K_{ATH,Pr}$ arises from k_{cat} and $K_{m,Pr}$ values measured under multiple-turnover conditions corresponding to the first-order rate constant for stable complex formation, $k_{1,H}$, and the dissociation constant for ternary complex formation from antithrombin–heparin complex and free proteinase, $K_{ATH,Pr}$, respectively, measured under single-turnover conditions. This correspondence is due to the release of heparin from the stable complex occurring concomitant with the rate-limiting formation of this complex.[6,111]

Contribution of the Substrate Reaction Pathway

The formation of a noncomplexed, proteolytically modified form of antithrombin, cleaved at the reactive bond, concurrent with the formation of stable inhibitor–enzyme complexes in antithrombin–proteinase reactions is increased in the presence of heparin.[34-36,104,119] The contribution of this competing substrate pathway can be determined by measuring the apparent inhibition stoichiometry, as described earlier for the reaction in the absence of heparin. Apparent stoichiometries of ~1.5 at physiological ionic strength,[35] and considerably higher at lower ionic strengths,[36,85] are usually found in the presence of heparin (Fig. 1). The extent to which such apparent inhibition stoichiometries exceed a value of 1 mol inhibitor per mole of enzyme thus defines the amount of reactive-bond-cleaved antithrombin formed in the reaction. Apparent inhibiton stoichiometries in the presence of heparin are measured by titrations of the proteinase with antithrombin in a manner analogous to that done in the absence of heparin. Less than stoichiometric amounts of heparin are sufficient to promote fully formation of the reactive-bond-cleaved inhibitor at ionic strengths ≥0.15, although stoichiometric amounts are required at ionic strengths <0.1.[36] The incubation time can be appreciably shortened in these experiments, as heparin greatly accelerates the reaction. As discussed earlier, Polybrene should be included in the substrate solution for assaying residual enzyme activity to neutralize the heparin.

The amount of reactive-bond-cleaved antithrombin formed during the reaction with enzymes can also be estimated by electrophoresis. The cleaved inhibitor has a somewhat faster mobility than intact antithrombin under reducing conditions on 7.5% gels in the phosphate buffer system,[120] due to the loss of the carboxy-terminal peptide (M_r ~5000), which is disulfide-linked to the amino-terminal part.[35] This mobility difference is less apparent in the Tris buffer system.[121] However, for unknown reasons, the cleaved inhibitor migrates somewhat more slowly than intact antithrombin on 7.5–10% gels in the Tris buffer system, but not in the phosphate buffer system, under nonreducing conditions (Fig. 8).[36] Usually, the best separation is obtained with the Tris buffer system under nonreducing conditions. The amount of cleaved antithrombin formed can be obtained by scanning the gels. The amount of protein applied to the gels for such quantitation should be kept in a range where the integrated intensities of protein bands are proportional to the protein load.

The methods of kinetic analysis of antithrombin–proteinase reactions in the absence and presence of heparin discussed above do not account for the influence of the substrate pathway on the measured rate of proteinase inhibition. However, the contribution of the substrate reaction is at most ~30% at physiological or higher ionic strength. Inhibition rate constants

[119] E. Marciniak, *Br. J. Haematol.* **48**, 325 (1981).
[120] K. Weber and M. Osborn, *J. Biol. Chem.* **244**, 4406 (1969).
[121] U. K. Laemmli, *Nature (London)* **227**, 680 (1970).

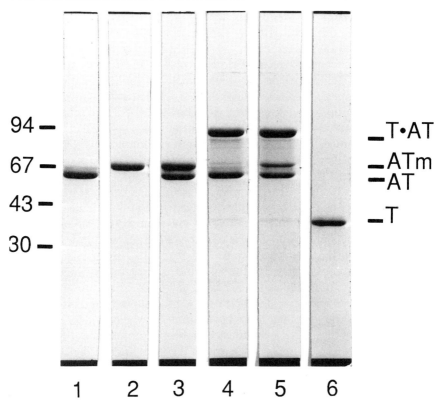

FIG. 8. SDS gel electrophoresis of the products of the antithrombin–thrombin reaction in the absence and presence of heparin. Antithrombin (10 μM) and thrombin (5 μM) were reacted in the absence or the presence of 1 μM high-affinity heparin for ~90 min in the I 0.15, pH 7.4, buffer of Fig. 1 at 25°, after which <0.2% thrombin activity remained. Reactions were then quenched by boiling in 1% (w/v) SDS for 2 min. Nonreducing SDS gel electrophoresis (10% gel) of samples of these reaction mixtures or equivalent volumes of 5 μM proteins similarly treated, containing ~3–7 μg protein, was then performed according to Laemmli.[121] Lane 1, antithrombin; lane 2, reactive site cleaved antithrombin (AT_m); lane 3, mixture of lanes 1 and 2; lane 4, antithrombin and thrombin reacted in the absence of heparin; lane 5, antithrombin and thrombin reacted in the presence of heparin; lane 6, thrombin.

may therefore not be significantly affected in most cases, except where the increased consumption of the inhibitor by the substrate reaction is not taken into account; e.g., in assuming that the inhibitor concentration does not significantly change under pseudo first-order conditions or in assuming a stoichiometric reaction under second-order reaction conditions where inhibitor concentrations are comparable to the proteinase. The reactions of antithrombin and other serpins as both inhibitors and substrates of their target proteinases indicate that serpins behave as suicide substrates of these enzymes, i.e., they are initially recognized as regular substrates but become inhibitors as a result of activation by the enzyme.[56,115,122] The kinetic theory developed for enzyme reactions with such substrates is thus in theory applicable to serpin–enzyme reactions. This theory affirms that in the case of suicide substrates, which are efficient inhibitors of their target enzymes and undergo little substrate reaction, like the serpins, the substrate reaction will only modestly affect the kinetics of inhibition.[123]

Acknowledgments

We wish to thank Roberta Sheffer and Ann Marie Francis for their expert technical assistance in obtaining some of the data presented in the figures. We also thank Frances DeVos for assistance in preparing the manuscript.

[122] P. A. Patson, P. Gettins, J. Beechem, and M. Schapira, *Biochemistry* **30**, 8876 (1991).
[123] S. G. Waley, *Biochem. J.* **227**, 843 (1985).
[124] Å. Danielsson and I. Björk, *Biochem. J.* **207**, 21 (1982).
[125] J. Jesty, *Arch. Biochem. Biophys.* **185**, 165 (1978).
[126] C. F. Scott and R. W. Colman, *Blood* **73**, 1873 (1989).
[127] S. T. Olson and J. D. Shore, *Thromb. Haemostasis* **62**, 381 (1989).
[128] R. A. Pixley, M. Schapira, and R. W. Colman, *Blood* **66**, 198 (1985).
[129] M. Schapira, C. F. Scott, A. James, L. D. Silver, F. Kueppers, H. L. James, and R. W. Colman, *Biochemistry* **21**, 567 (1982).

Author Index

Numbers in parentheses are footnote reference numbers and indicate that an author's work is referred to although the name is not cited in the text.

Le Bonniec, B. F., 127, 317, 326(28), 327, 359, 376, 377(64)
Lecocq, J.-P., 243
Lecocq, J. P., 67, 451, 454(15)
Lecocq, P., 476
Leder, I. G., 543
Leder, P., 240, 245(50)
Ledley, F. D., 399
Lee, L. S., 457
Lee, P., 66
Leeksma, O. C., 358
Leffert, C. C., 12
Legaz, M. E., 67, 129, 187, 261
Lehrman, S. R., 438
Leibman, H. A., 469, 477
Leinbach, R. C., 339
Leis, J., 89
Lekutis, C., 237
Lentz, B. R., 326(30), 327
Lentz, S. R., 326(29), 327, 362, 371(35), 372(35), 377
LeRet, M., 527
Lerman, L. S., 168, 169
Levinson, A. D., 457
Lewis, B. A., 31
Lewis, J. H., 67, 69(21), 238, 397
Lewis, M. L., 306
Lewis, M. S., 37
Lewis, S. D., 23, 24, 341, 342(5), 342(14, 15), 343, 344(5, 14), 345, 345(14), 346, 346(14, 15), 347(14), 348(35), 349(35), 350(14, 35), 351(14), 352(14), 353(5, 14, 29), 354(14), 356(15, 35), 357, 357(14, 29), 358, 358(41)
Leyte, A., 238, 239
Leytus, S. P., 13
Li, L. C., 281, 282(5)
Liebman, H. A., 167
Liedén, K., 550, 554(110)
Lijnen, H. R., 65, 398, 436
Likert, K. M., 199, 204(8), 206, 206(8), 207, 207(8), 208, 208(24)
Lilja, H., 387
Lillicrap, D. P., 144(3), 145, 147(3), 149(3), 152(3), 154(3), 155(3), 158(3), 159(3), 160(3), 162(3), 163(3), 164(3), 166(3), 417
Limentani, S. A., 469
Lin, P.-H., 410
Lin, S.-W., 107, 111, 149, 168, 417

Lindahl, U., 359, 362(5), 371(5), 543, 544, 545(71, 78), 546(78), 548, 549, 549(71), 557(104)
Lindhout, T., 306
Linse, S., 366, 418, 434(29), 435(24, 29)
Lipkowitz, K. B., 88
Lipscomb, M. S., 69
Littmann, B. J., 286
Litwiller, R., 408, 409(7)
Liu, C. Y., 66, 344
Liu, L.-W., 362, 376, 376(36), 382(36)
Ljung, R., 158, 162(30, 31), 164(31), 165(30), 168(30, 31)
Llinas, M., 18, 19(24–30, 32, 35), 20
Loebermann, H., 528
Logan, J., 452
Lollar, P., 128, 131, 132, 132(11), 134(11, 23), 137(23), 138, 140, 141, 141(11, 12), 238
Long, G. L., 397, 398, 399
Longstaff, C., 537
Lorand, L., 1, 2, 22, 23, 24, 24(1, 21), 25, 25(20, 26), 31, 33, 33(1, 26, 27, 34, 35), 34(19, 20, 45), 35, 36, 37, 50, 341, 342(15), 343, 346, 346(15), 348(35), 349(35), 350(35), 356, 356(15, 35), 357, 358(41)
Lord, S. T., 164
Lormeau, J.-C., 538
Lormeau, J. C., 528, 543, 549(68)
Loskutoff, D. J., 398
Losowsky, M. S., 1, 22, 24(1), 33(1), 36, 341
Lottenberg, R., 198, 329, 356, 375, 484, 533, 542
Lottspeich, F., 344, 358, 358(26)
Louie, G. V., 53, 127
Lovrien, E. W., 149, 150(12), 151(12), 158(12), 159(12), 166(12)
Lowry, O. H., 25
Lozier, J. N., 417
Lu, R., 381, 382(81), 384(81)
Lucas, J., 358
Ludwig, M., 107, 108(28), 119(28), 127(28), 128(28), 144(3), 145, 147(3), 149(3), 152(3), 154(3), 155(3), 158(3), 159(3), 160(3), 162(3), 163(3), 164(3), 166(3), 417
Lukacs, K. D., 83
Lund, B., 12, 15
Lund, E., 259

Subject Index

isolation, 363–368
module organization, 12
monoclonal antibodies to
immunization and screening, 363–364
in protein C isolation, 364–365
partially carboxylated, separation from
functional protein C, 367–368
plasma concentration, 521
protease domain, 360–361
purification, 102–103
recombinant, 367–368
expression in mammalian cells, 452
cell lines for
characterization, 459–468
generation, 454
in vivo labeling and gel analysis, 462–
466
structure, 360–361
structure–function analysis, 450–451
vitamin K-dependent γ-carboxyglutamic
acid domain, 360–361
Protein engineering, 14–15
Protein-glutamine γ-glutamyltransferase, see
Factor XIII, activated
Protein S
anticoagulant function, 9–10
purification, 102–103
Proteins
immunoprecipitation, 250–251
modular, 14
mosaic
distant homologies of modules in, detection, 20–21
homologies of
functional implications, 20
structural implications, 19–20
intronic recombination in, 16
modular assembly, 15–17
modular organization, and gene structure, 17
multidomain, modules in
folding autonomy, 17–18
structure–function correlations, 18–19
plasma, immunochemistry, 400–416
vitamin K-dependent
amino acid modification, 435–436
γ-carboxyglutamic acid analysis
direct, 442–445
immunologic methods, 445–447
carboxylation, 450–452

carboxylation assay in vitro, 438, 447–
449
expression in mammalian cells, cell
lines for, characterization, 459–
468
expression vectors, 452–453
immunoaffinity chromatography, 408–
409
posttranslational modification, 450–
451
propeptides
as carboxylation recognition site,
437–438
mutational analysis, 450–451
sequence homology, 437
structure, 436–437
in vitamin K-dependent carboxylation, 437–449
recombinant
barium citrate precipitation, 466–468
expression, 439–441
expression in mammalian cells, 450–
477
mutational analysis, 450–451
production by cell lines, generation,
452–459
production, 451
purification, 441–442
reversed-phase HPLC, 443
sequence homology, 436
structure–function analysis, 450–451
synthesis, 435
Protein-tyrosine sulfotransferase, 252
Protein Z
gene structure, 12
module organization, 12
reactive site, structure, 398
Prothrombin, 6
activation, 224, 281–282, 299, 328
by factor Xa, intermediate species
formed in, 299–302
monitoring, 301–302
pathways, 299–302
activation products, 299–300
activators, 315–316
bovine
factor Xa cleavage sites, 300
thrombin cleavage site, 300
γ-carboxyglutamic acid analysis
direct, 444–445

immunologic methods, 445–447
defects, 312–313
phenotypic expression, 313
functional assay, 315–317
gene
  chromosomal localization, 313
  DNA sequencing, in characterization of
    thrombin mutants, 325–326
  mutant, characterization, 317
gene structure, 12
human, thrombin cleavage site, 300
isolation, 327
kringle 2, synthetic peptide representing,
  solution-phase immunoassay, with
  antikringle monoclonal antibody,
  415–416
kringle domains, 54
module organization, 12
mutant, expression in mammalian cells,
  327
plasma, quantitative immunoelectropho-
  resis, 315
plasma concentration, 521
propeptide
  as carboxylation recognition site, 437–
    438
  secondary structure, 439
purification, 105
site-specific mutagenesis, 327
substrates, 315–316
Prothrombinase
activation products, 299–300
binary macromolecular interactions
  within, 270–280
complex, 299
  assembly, 260–281
    in adherent monocytes, 287–290
    in cells in suspension, 286–288
    fluorescence anisotropy measure-
      ments, 266–267
    rapid kinetic measurements, 269–
      271
    steady-state fluorescence intensity
      measurements, 266–269
  function
    in adherent monocytes, 287–291
    in cells in suspension, 286–288,
      291
ternary interactions within, 266–270,
  280

R

Rabbit, femoral arteries, contractile activity
  of meizothrombin in, assay, 307–308,
  310
Radioimmunoassay, competition, γ-carbox-
  yglutamic acid in vitamin K-dependent
  proteins, 446–447
Recombination, intronic, in modular assembly of proteins by exon shuffling,
  15–17

S

Sepharose CL-4B
  activation with cyanogen bromide, 405–
    406
  antibody immobilization on, 405
Serine protease, see also Protease
  in coagulation cascade, 3–4
  complexed, catalytic efficiencies, 7
  domains, amino acid sequence alignment,
    120–123
  in extrinsic pathway of coagulation, 177
  inhibitors, see Serpin
  trypsinlike, 2, 11
    gene structure, 16
  in vitamin K-dependent enzyme com-
    plexes, 5–7
Serine proteinase, see also Proteinase
  active site-selective labeling, 478–503
    with thioester peptide chloromethyl ke-
      tones
      principle, 479–480
      procedure, 488–500
  active site-specific, labeled with fluores-
    cence probes, as reporters of interac-
    tions, screening for, 500–501
  trypsinlike, active site, spectroscopic
    probe for, 537
Serpin
  interaction with proteinases
    contribution of substrate reaction path-
      way, 558–559
    kinetics, 525
  as suicide substrate of proteinase, 558–
    559
  superfamily proteins
    reactive sites, amino acid sequences
      around, 397–398